VISUAL CULTURES IN SCIENCE AND TECHNOLOGY

Visual Cultures in Science and Technology
A Comparative History

KLAUS HENTSCHEL

OXFORD
UNIVERSITY PRESS

OXFORD
UNIVERSITY PRESS

Great Clarendon Street, Oxford, OX2 6DP,
United Kingdom

Oxford University Press is a department of the University of Oxford.
It furthers the University's objective of excellence in research, scholarship,
and education by publishing worldwide. Oxford is a registered trade mark of
Oxford University Press in the UK and in certain other countries

First Edition published in 2014
Impression: 1

Published in the United States of America by Oxford University Press
198 Madison Avenue, New York, NY 10016, United States of America

British Library Cataloguing in Publication Data
Data available

Library of Congress Control Number: 2014938233

ISBN 978–0–19–871787–4

Printed and bound by
CPI Group (UK) Ltd, Croydon, CR0 4YY

CONTENTS

LIST OF ILLUSTRATIONS

Color plates   I-XVI

0

Preface and acknowledgments

This book attempts a synthesis. I decided to make this study differ from most of my other books in the history of science. Rather than working on primary material in dusty archives, I delve deeply into the rich reservoir of case studies that have been amassed over the past couple of decades by historians, sociologists and philosophers of science on visual representations in scientific and technological practice. This book's goals are thus located on a meta-level. First and foremost, I aim at an integrative view on recurrently noted general features of visual cultures in science and technology, something that has hitherto been unachieved and has been believed by many to be a mission impossible. Furthermore, I have broadened the view from myopic microanalysis to a search for overriding patterns extracted from a comparison of many such case studies, again something that many of my colleagues have given up on doing. Readers already somewhat familiar with the broad field of visual studies will undoubtedly recognize some of these cases, but I am quite certain that not all will be known to any one of them, since I touch upon many different disciplines and research areas ranging from mathematics to technology, from natural history to medicine, and from the geosciences to astronomy, chemistry and physics. At the same time, this book should also be perfectly readable to beginners still looking for basic orientation in the maze of pertinent analyses published during the last two or three decades. I hope to have produced a text of potential interest to both audiences. Given this agenda, the bibliography necessarily grew to rather lengthy proportions. It cannot claim completeness, though, but is – I hope – a fair and broad selection. References to specially pertinent primary literature are included along with indications of where further sources on each of these many and very diverse topics can be found. I would like to acknowledge a special debt to the following colleagues and scholars whose work on specific case studies was most helpful to me. In alphabetical order, they are: Svetlana Alpers, Kirsti Andersen, Rudolf Arnheim†, Charlotte Bigg, Horst Bredekamp, Olaf Breidbach, Jimena Canales, Jordi Cat, Lorraine Daston, Margaret Dikovitskaya, Meghan Doherty, Monika Dommann, Samuel Edgerton, James Elkins, Eugene Ferguson†, Peter Galison, David Gooding†, Jochen Hennig, Ludmilla Jordanova, Bruno Latour, Michael Lynch, Omar W. Nasim, Kärin Nickelsen, Alex Soojung-Kim Pang, Nicolas Rasmussen, Frances Robertson, Tim Otto Roth, Martin Rudwick, Simon Schaffer and Aaron Wright.

The notes to the main narrative always make clear from whom I took which example and where (and how) I agree with or perhaps differ from their interpretation. Most of them have not seen a draft of my text, so I alone am responsible for any errors. In my comparative approach, I have tried to be as comprehensive and all-inclusive as possible, but of course there will be gaps and missed opportunities for further references. I apologize to all scholars whose pertinent work was not referred to – given the enormous size to which literature on visual culture(s) has grown in recent years, I could not go beyond the explicit inclusion of c. 2000 entries listed at the end of this book. In the brief section preceding my detailed bibliography, I give some personal recommendations on where to start excursions into the thicket of existing literature on visual cultures in science and technology.

I would like to point out, though, that this selection of examples was not based purely on personal taste but also on a more hidden agenda of establishing fruitful interconnections. There are surprisingly many cross-linkages throughout this book, which evidence a high degree of interconnectivity between science cultures otherwise not brought into association. In my opinion, this is exactly one of the strengths of a comparative approach. I hope the reader will appreciate this effort of weaving together many strands, so far only discussed in isolation. My aim is to provide the reader with just the right number of primary and secondary references so as not to overtax him or her with material, yet without omitting essential hints for further study. Each of the hundreds of historical case-studies touched upon in this book is so complex and interesting that they urge further work, especially where they venture beyond the necessarily limited pool of examples here. You will notice that these examples, picked from different periods, typically even from different centuries within each chapter, also make a fair attempt at representing all the natural sciences, medicine and technology. My main emphasis and expertise is in the physical sciences, though.

It is indicative that representatives of many fields have claimed their discipline to be "by nature" most particularly visual. Thus, for instance, the lead-in to a 10-line announcement about a recent workshop on the history of astronomical imaging: "Of all the sciences it is arguable that images have played the greatest role in astronomy, both for the professional and for the interested public."[1] Similar claims could be made of anatomy, botany or zoology, microscopy, mineralogy and crystallography, perspectival theory, technical drawing or the designing of bridges. All these fields, and many more besides, feature prominently in this book.

---

[1]Quoted from the public announcement by the RAS about a special discussion meeting in London, January 13, 2012, distributed via MERSENNE.

Given the strong interest by art and science historians, sociologists, and philosophers in issues of visual representation for several decades now, it is surprising how few studies have attempted a comparative approach. By this I mean more than a mere compilation of impressive and/or famous images as found in anthologies such as Baynes & Pugh's *The Art of the Engineer* (1981), Brian Ford's *Images of Science* (1992) or Harry Robin's *The Scientific Image* (1992) including some 150 selections "from cave to computer." In my opinion, the most convincing attempts at a truly comparative approach have been made in relatively specific fields. An *intra*disciplinary comparison is drawn within a given field encompassing various actors and possibly larger spans of time. Examples would be Robert Brain's doctoral dissertation, which documents and analyzes the emergence and diffusion of the graphic method within ballistics and physiology, or Eda Kranakis's comparative exploration of French and American cultures of civil engineering in the 19th century. The latter was partly prompted by the German historian of technology Ulrich Wengenroth's call for comparative studies,[2] as "a panacea against technological determinism" and – more positively speaking – as a convenient and methodologically satisfactory pathway toward recognizing "general patterns in the interlocking of [science,] technology and society" or "social and material components in the construction of a technology" – or science, I would add.[3] Further calls for a comparative history of science and technology mostly sought to transcend the fixation on national contexts by explicitly addressing international comparisons, for example, of education, research or innovation systems.[4] Even broader, intercultural comparisons – e.g., between Western and far-Eastern cultures – are even rarer.[5] Within the context of this book, I have refrained from such far-comparisons. I stick to the Western hemisphere, but choose my examples from very many different national, regional and disciplinary contexts from the early-modern era until the 20th century – a field already extremely broad.[6]

[2] See Wengenroth (1993); cf. Kranakis (1997) and Nikolow & Bluma (2002/2008b) p. 44 for similar calls.

[3] See Brain (1996) and Kranakis (1997), with all previous quotes from the latter, p. 6. My own books on spectroscopy as a visual culture (Hentschel, 2000*a*) and on astronomy (Hentschel & Wittmann, eds. 2000) follow a similar agenda.

[4] See, e.g., Pyenson (2002a) on comparative history of science, (2000b) about "an end to national science," Emmerson (1973), König (1997) & Olesko (2009) on technical education, or Simon (ed. 2012, 2013) on cross-cultural and comparative history of science education.

[5] See, e.g., Lloyd & Sivin (2003) for one such daring and impressive attempt.

[6] For synchronic studies and comparisons within the early-modern period, see, e.g., Baldasso (2006), P. H. Smith (2006), Kusukawa & MacLean (2006), Kusukawa (2012); on the 18th c.: Stafford (1994); on the 19th c.: Schwartz & Przyblyski (eds. 2004), Lightman (2000, ed. 2007), Morus (2006);

A similarly broadly ranging, but – in my opinion – only partly successful attempt at an *inter*disciplinary comparison was made in a paper that appeared in *Representations*, one of the key journals for the new field of 'visual studies.' Lorraine Daston and Peter Galison's work on the history of objectivity, first appearing in 1992, was later expanded into a book-length study (Daston & Galison, 2007). These widely noted and discussed publications focus on visual sources, such as atlases; they study the historical development in the making, intentions and usage of scientific atlases. Their claim is that three distinct phases of development exist:

1. The search for the typical or representative exemplar that is chosen for depiction in an atlas, for example, of natural history. This homotype is carefully crafted to bring out the features deemed essential or characteristic, and to suppress other features deemed unimportant, irrelevant or idiosyncratic. Such atlases were prevalent in the 18th and early 19th centuries.

2. The goal of so-called 'mechanical objectivity,' i.e., self-registration of scientific objects. New technologies such as photography, allegedly obviating intervention by the observer, dominated the second half of the 19th century.

3. The constructive use of and work with the observer's subjectivity during the 20th century.

The courage and energy motivating these two top-notch historians of science to search for patterns in the broadly scattered material of atlases in anatomy, astronomy, botany and radiology is impressive. Yet their resulting claims are not quite so convincing.[7] These differences of opinion nevertheless motivated me to likewise step beyond the myriad of details in a micro-historical case study and search for deeper historical patterns holding true and telling more than in a single instance. The following are the results of this decade-long search in material ranging from the late Middle Ages throughout the early-modern period into the 20th century and touching on all kinds of disciplines and research fields in science, technology and medicine.

The following features characterize my comparative approach to the history of science and technology:[8]

- avoidance of the pitfalls of a local microstudy;
- parallel analysis of many comparable cases;

on the 20th c., e.g., Galison (1997), etc. and a myriad of further references in each of these.

[7]See, e.g., Hentschel (2008*b*) for a detailed and critical review of Daston & Galison (2007).

[8]For a deeper methodological reflection on historical comparisons, see Hentschel (2003*a*) and further methodological literature on comparative analysis cited there.

- no forced analysis, but category choices motivated by the sources (actor's categories wherever available);
- a bottom-up approach, starting out from the material;
- intentional generalizability beyond the pool of selected cases; and
- context-sensitive, cautious structural description of historical processes.

Following this methodology, history of science can produce results of relevance to neighboring fields and not exhaust itself in internal dialogue within a narrow niche of specialists. I fully agree with H. Floris Cohen, professor of comparative history of science at the University of Utrecht, who described the proper way to make historical comparisons:

> Rather than importing a conceptual apparatus from the outside, a better approach is to develop it from the inside. Such a procedure allows us to avoid current categories as much as an a priori limitation to actors' categories. [...] In the writing of history patterns must be discerned, not imposed.[9]

This is what Clifford Geertz (1926–2006) meant when he urged us "not to generalize across cases but to generalize within them." Generalizable patterns are most likely to be found on the medium scale of time and level of abstraction. Alluding to the same phenomenon, the famous *Annales* historian Fernand Braudel (1902–85) remarked that "history does not repeat itself, but it has its habits."[10]

Acknowledgments

The considerations set forth in this volume had time to mature over the course of a decade. The plan for such a work was originally conceived since completion of my studies on spectroscopy in 2002, but teaching duties and other vagaries of academic life hindered its execution to now. I thank all the students participating in my Stuttgart lecture cycle on visual cultures of science and technology in the summer term of 2009 and the participants of the Sixth European Spring Academy on Visual Science Cultures in Menorca in May 2011 for intense and fruitful discussions of these ideas. This feedback helped very much to fine-tune my claims and at the same time to broaden my empirical basis.

---

[9] Cohen (2007) p. 496.
[10] Cf. Geertz (1973) §6, Hentschel (2003a) p. 257, and Jürgen Kocka (1989) p. 27: "Comparing means arguing, with sharply defined concepts and a willingness to abstract at the medium level."

Writing a book of this scope and breadth requires access to a very broad spectrum of publications, many of them quite inaccessible. My cordial thanks to the interlibrary office of our Stuttgart University library for the excellent and prompt service concerning roughly 300 interlibrary loans. I also thank the staff of the *Württembergische Landesbibliothek Stuttgart*, especially the staff in the manuscripts and rare books reading room. Reproduction permissions have been granted by courtesy of the following copyright holders:

Museums:

George Eastman House, International Museum of Photography and Film, Rochester, New York; MIT Museum, Cambridge, Massachusetts; Musée d'histoire des sciences de Genève, Switzerland; Museum for the History of Sciences (Ghent University); National Maritime Museum, Greenwich, London; Naturhistorisches Museum, Vienna; Newark Museum, Newark, New Jersey; Science Museum/Science & Society Picture Library, London; Teylers Museum, Haarlem.

Institutions:

Albert-Ludwigs-Universität Freiburg im Breisgau; American Physical Society, College Park, Maryland; American Society of Clinical Oncology, Alexandria, Virginia; Austrian National Library. Liechtenstein: The Princely Collections, Vaduz-Vienna; b p k – Bildagentur für Kunst, Kultur und Geschichte, Berlin; Bundesanstalt für Materialprüfung, Berlin; Deutsche Akademie der Naturforscher Leopoldina, Halle; FEI Company, Hillsboro, Oregon; Fonar Corporation, Melville, New York; Historical Metallurgy Society, Gateshead, England; Institution of Mechanical Engineers, London; Koninklijke Nederlandse Akademie van Wetenschappen, Amsterdam; Library of Congress Prints & Photographs, Washington, DC; Los Angeles Public Library, Photograph Collection, Los Angeles, California; Massachusetts Institute of Technology (MIT), Cambridge; Oregon State University Libraries, Special Collections & Archives Research Center (Pauling Papers), Corvallis, Oregon; Pepys Library, Magdalene College, Cambridge, England; Princeton University Archives, Rare Books & Special Collections, Princeton, New Jersey; Real Jardín Botánico, CSIC, Archivo, Madrid; Royal Society, London; Wellcome Institute, London.

Publishers:

Cambridge University Press, Cambridge, England; Elsevier Limited, Oxford, England; John Wiley & Sons, Inc.; Maney Publishing; MIT Press, Cambridge, Massachusetts; Oxford University Press, Oxford, England; Pacific Press, New York; SAGE Publications, Ltd, London; Sinauer Associates, Inc., Sunderland Massachusetts, Springer Science + Business Media, BV.

The following individuals have also generously granted permissions:

Prof. Dr Reto Bale, MD, University of Innsbruck, Austria; Ms Dee Breger, Micrographic Arts, Saratoga Springs, New York; Prof. Derek E. G. Briggs, New Haven, Connecticut; Dr Raymond V. Damadian, president, FONAR Corp., Melville, New York; Prof. Christoph Gerber, Basel; Prof. Jürgen Klaus Hennig, Freiburg; David Jasmin, Institut National de Recherche Pédagogique, Paris; Prof. Hans-Holger Jend, MD, Hamburg; Prof. Dr Willi A. Kalender, Institut für Medizinische Physik, University of Erlangen; Prof. Thierry Lefebvre, Université Paris-Diderot; Prof. Charles W. Misner, University of Maryland; Prof. Kärin Nickelsen, University of Munich; Mr Joe Orman, Chandler, Arizona; Mr Mike Peyton, Essex; Prof. Martin J. S. Rudwick, Cambridge University; Prof. Bradley E. Schaefer, Louisiana State University, Baton Rouge; Prof. Günter Schnitzler, University of Freiburg; Prof. Roger N. Shepard, Stanford University; Prof. Klaus Staeck, Heidelberg; Prof. Kip S. Thorne, Caltech, Pasadena, California; Mr Simon Weber-Unger, Vienna, Austria; Mr John Wetzel, WikiPremed; Prof. John White; Dr Martijn Zegel, Haarlem.

Furthermore, I am very grateful for the kind assistance of Prof. Kenneth W. Ford, Philadelphia; Johanna Norborg, Volvo Car Corp., Gothenburg, Sweden; Mrs Kathleen M. Peyton, Essex; Prof. James E. Wheeler, MD, executor of the John Wheeler Estate, Philadelphia; and Ms Lauren Amundson, Lowell Observatory, Flagstaff, Arizona. It was a great pleasure to cooperate with the team at Oxford University Press. I would like to thank in particular Sonke Adlung and Jessica White as well as Saravanan Omakesaven of Integra Co. My wife Ann kindly checked the style, translated some quotes from German, French and Spanish originals and helped to obtain copyrights of images. Hearty thanks to all of the above and – needless to say – all remaining errors are mine.

A quick guide through this book

In the following introductory chap. 1, I give a concise survey of the vast literature on visual representations in science and on visual cultures more generally. I define fundamental terms (in sec. 1.1), discuss basic polarities such as visual versus textual, iconophile vs. iconophobe (in sec. 1.2), and thematize their interplay (sec. 1.3). Visual rhetorics and visual argumentation are analyzed (in sec. 1.4) with respect to 2D images and 3D models alike. In sec. 1.7, which is a must even for those readers already familiar with the literature, I summarize roughly two dozen deep insights from early visual studies. The work of Svetlana Alpers on the Dutch connection and its mapping impulse is singled out and identified as a useful point of departure (in sec. 1.5), whereas many later works are criticized by me as "wrong turns of the visual turn" (in sec. 1.8).

In chap. 2, I develop my own notion of 10 historiographic layers of visual science cultures. My claim is that their historical analysis should delve into all 10 of these layers, not just focus on a few of them as is normally done. Only then can we hope to get a 'thick description' of visual cultures in science, medicine and technology. Chapter 3 discusses how visual science cultures are formed. It starts with a summary of Martin Rudwick's paradigmatic analysis of geology, then switches to stereochemistry (in sec. 3.2), metallography (sec. 3.3) and geometrodynamics (sec. 3.4). A few carefully selected pioneers of visual cultures are portrayed in chap. 4. After four biographical case-studies in sec. 4.1, I offer a more systematic prosopography of 30 spectroscopists from the 19th and early 20th centuries, which already aims at the identification of common features. Section 4.4 continues to search for a generalizability of these claims and is another must for all readers.

The transfer of visual techniques is exemplified with the complex shifts back and forth between arts, architecture, mathematics, optics in the history of perspectival drawing (in sec. 5.1). Indicator diagrams mark a point of transition between secret industrial practice of steam-engine development and the new science of thermodynamics established in the middle of the 19th century (in sec. 5.2). The trajectory from NMR to MRI, i.e., from physics and chemistry to medicine, is studied in sec. 5.3, with CT and PET scanners as a late 20th-century endpoint (in sec. 5.4). Chap. 6 traces the role of images and their makers in the business of science and chap. 7 explores their evolutive histories.

Among the 10 historiographic layers expounded in chap. 2, quite a few are treated in chapters of their own: the methods of practical training in visual skills (in chap. 8); the embuing mastery of pattern recognition (in chap. 9), of visual thinking in scientific and technological practice (in chap. 10), the aesthetic fascination of practitioners (in chap. 12) and issues of visual perception (in chap. 13). Recurrent color taxonomies are also given their own chapter (11). A concise summary of my main results is attempted in the final chap. 14 about visuality through and through. This should also be compulsory reading for all who might otherwise suit their reading to their own fields of interest or follow the many cross-references in the text that also trace the complex interconnections. Before the c. 2000 item bibliography and the name index I give personal recommendations of where to start further reading.

1

INTRODUCTION

1.1 Cultures, scopic regimes and visual domains

Whole libraries have been written on the meaning and connotation of the term 'culture' or 'cultures,' without any agreement being reached on its precise definition. I should therefore begin this systematic introduction by saying a few words about it, as it figures so prominently not only in this book's title but also in the field called 'visual culture,' to which the present work aims to contribute. This is by no means an easy proposition.[1] The term 'culture' derives from the Latin *cultura*, initially denoting husbandry, i.e., the practice of "cultivating" land and soil, and tending plants or crops.[2] Since the 16th century, and increasingly since the 18th century, this term was also in use metaphorically in the sense of cultivating or developing the human mind, faculties or manners. 'Culture' was then understood as the result of such development, training or refinement. During the Enlightenment, the concept of culture and especially its French companion *civilisation* became loaded with the notion of progress and unidirectional development away from supposed barbarism or "savagery," in which process all tribes and nations undergo many steps of mental cultivation. Against this Eurocentric and ethnocentric ideology, Johann Gottfried Herder (1744–1803) defended the idea of a plurality of cultures, each with its own way of life, its own traditions and its own artistic and intellectual achievements. In the 19th century, culture increasingly came to be seen in material and social terms, and thus in reference to the products of communal activity. Twentieth-century 'cultural studies' began as a revolt against the self-imposed limitation by academic disciplines like art history or the philologies to analyze only select products of "high" culture such as the fine arts, poetry or tragedy. The material culture of the populace was sought, the vernacular imagery in children's books and popular media, along with all kinds of

---

[1]On the historiography of 'culture' see the early but still useful literature survey by Kroeber & Kluckhohn (1952). Exemplary for current approaches, but by no means exhaustive, are Geertz (1973), Hall (1980, 1990), Storey (ed. 1996), Hardtwig & Wehler (ed. 1996) and Fried & Stolleis (eds. 2009), each with further references. Finally, see the "What's culture" webpage: www.wsu.edu/gened/learn-modules/top_culture/culture-index.html (last accessed April 29, 2011) with links to key texts in this debate.

[2]See *Oxford English Dictionary*, 2nd ed., vol. 4 (1989) pp. 121ff.; cf. also Rampley (2005) chap. 1.

practices found in everyday life. Nowadays, culture is understood as encompassing "anything that is meaningful to more than one person" or "everyday objects
and practices of a group of people, or of an entire way of life."[3] The singular, with
which the concept of culture had once been mobilized as a banner against barbarism (or what was rhetorically demarcated as such), has given way to a plurality
of cultures, sometimes explicitly defined as a "patchwork of situated, disparate,
locally organized practices, in which knowledge is constituted through a variety of
social and political processes."[4]

The concept of culture historically comprehends both a process and material
artifacts. Throughout this book, both these meaning variants are applied. Our
understanding of culture will not be limited to denoting skills or products of
"high" culture but responds to the impulse from 20th-century cultural studies
and social history to broaden the concept.[5] Not only will the elitist products of
top-notch science and technology be looked at, but an attempt will also be made
to find a level of description that fits typical examples of "normal" science (in
Thomas S. Kuhn's sense) from the everyday lives of scientists and engineers.

From the ethnologist Clifford Geertz, in particular, we learned that human
cultures are complex systems of knowledge in which the members converse about
and among themselves in a partly self-referential mode. Visual representations
of all kinds play a vital, but not isolated, part in these 'autopoietic' systems of
communication. Since many layers of meaning are superimposed in all cultural
objects and processes, ethnologists and historians alike need a 'thick description'
to unpack the various layers of signification. Cultural objects carry a high degree
of symbolic content. Consequently, Geertz has defined 'culture' as the "self-spun
webs of significance" that humans inhabit, as "a system of inherited conceptions
expressed in symbolic forms by means of which people communicate, perpetuate,
and develop their knowledge about and attitudes toward life."[6] Seen from this
perspective, visual elements in science and technology are "efflorescences of informational images in general."[7] Barbara Stafford's portrayal of the Enlightenment

---

[3]On these definitions by Raymond Williams 1958 and Malcolm Barnard 1998, see, e.g., Storey
(ed. 1996), Schwartz & Przyblyski (eds. 2004) pp. 6ff., as well as W. J. T. Mitchell and Brian Goldfarb
in Dikovitskaya (2005) pp. 1, 33, 57, 79f., 164, 245, 288.

[4]Thus, for instance, the anthropologist Charles Goodwin (1994) p. 608, in a study about
professional vision and encoding schemes.

[5]For a good survey of cultural studies methodology, see Storey (ed. 1996). Cf. also Dikovitskaya
(2005) pp. 16f., and Rampley (2005) for further references.

[6]See Geertz (1973) pp. 5 and 89, respectively.

[7]Elkins (1999) p. 5. The distinction between informational and artistic (or expressive) images is
by no means a dichotomic one for art historian James Elkins: see here p. 16.

as a period of "artful science" and as the "eclipse of a visual education" can be understood as an example of this kind of multifaceted approach. Associatively spun cross-connections are aimed at between spectators at a market fair or diorama, between instrument-makers of gadgetry for visual entertainment and children's books and between museums, collections and what remained of the *Wunderkabinett*, soon to disappear forever in the much more rigid textual classification and institutionalization of the new sciences in the 19th century.[8] Methodologically, these heroic efforts at a thick description of a whole cultural period are riddled with problems, since they presume a unity of experience and visuality throughout this period despite many (often conflicting) cultural strata, domains of knowledge and practices. Impressionistic case studies and mostly associative linkages between these domains and practices[9] don't suffice to prove the relevance, say, of an artist's visuality to that of a tradesman, a naturalist or an engineer. Therefore I choose not to pursue my own analyses in this broad Staffordian approach of looking at *all* strata of a population at once, rather I limit my reference groups to people somehow contributing to a systematic study of nature or technological artifacts. This is by no means restricted to the elite, but also includes mechanics and assistants, etc., and hence it still constitutes a very large group. Following the lessons of a 'materialized epistemology,' we will rather base our comparisons on "in-depth studies of the people and practices involved in making particular sorts of images and of the ways in which those images form both what and how we know."[10] As will become evident throughout this book, there are already plenty of cultural studies in this vein, which I take to be richer in insight than would be a broad cleavage of whole strata of epistemes in a Foucaultian manner.

The close interplay between 'culture' and 'practice' makes 'cultural studies' of all aspects of human societies go hand in hand with 'practice studies' – both these trends having intensified since the 1970s. For the purposes of history of science and technology, a special branch of these cultural studies has proven to be of special relevance, namely those dealing with 'knowledge cultures,' i.e., cultures more broadly conceived than cultures of science, but still directed toward an increase

---

[8]See Stafford (1994) and the review symposium in *Metascience* 1994, issue no. 6, pp. 46–60. On the radical restructuring of natural history since the Sattelzeit of 1770–1830, see, e.g., Lepenies (1994), Stichweh (1984) and Rudwick (2007).

[9]Such as are practiced by Stafford. In all her publications, she likes to play the "game of back and forth." This is a quote from Stafford (1999) p. 1 in a (to me unbearable) book on visual analogy.

[10]Norton Wise (2006) p. 82 in a focus section on science and visual culture. Cf. also Dupré (2010) p. 621 and Elkins (2007) for many examples of "visual practices across the university."

in knowledge of some sort, whether theoretical or practical.[11] The sociologist Nico Stehr (*1942) noted the growing importance of canonized knowledge in modern societies, as contained in the term 'knowledge society' (*Wissensgesellschaft*) to denote this concentration of highly skilled and well-informed labor in the modern and postmodern world.[12] Specialists and experts gain central importance in modern societies; and generating, stabilizing and transmitting such knowledge from one generation to the next becomes of vital importance to their survival in an increasingly global market of expanding systems of expertise. Science and technology are, of course, special cases in point. The American historian of science Peter Galison (*1955) has shown, for instance, how 20th-century high-energy physics can be fruitfully studied as a "material culture of microphysics." His British colleague Andrew Warwick has demonstrated how to transfer this approach to "cultures of theory." The American historians Kathryn Olesko and David Kaiser as well as the Spanish historian Josep Simon have broadened this further to integrate science pedagogy into this science–culture package.[13] From the technological angle, there are portrayals of material cultures as well as technological training.[14] The German historian of technology Wolfgang König (*1949) has published a nice documentation comparing different traditions of education in machine building and construction between 1850 and 1930. France and Germany are both characterized as "school cultures" and contrasted against a "production culture" typical of the USA; the British "practice culture" trains more engineers on the job, in industry.[15] Wolfgang König and the American engineer and historian of technology Eugene S. Ferguson (1916–2004), as well as several historians of education, have pointed out how very important practical exercises in free and technical drawing as well as in 3D model-making were in the curricula of (poly)technical schools of the time.[16] At the general, prosopographic level,

---

[11] In German: *Wissenskultur(en)* as opposed to cultures of science; see, e.g., Knorr-Cetina (1999*a*), who prefers to speak of 'epistemic cultures.'

[12] See Stehr (1994) and further sources cited therein.

[13] See Galison (1997), Warwick (2003) and the special issue of *Studies in the History and Philosophy of Modern Physics* 29B,3 (1998), Olesko (1991, 2006), Kaiser (1998, 2005*a*, *b*), Kaiser (ed. 2005) and Simon (ed. 2012).

[14] See, e.g., Booker (1963), Hindle (1983) on the US context, both examples for the latter, comparatively organized, and comprising examples from the USA, Great Britain, Germany, France and Switzerland. Cf. Dalby (1903), Emmerson (1973), Ferguson (1992), König (1997), Olesko (2009) and A. J. Angulo in Simon (ed. 2012).

[15] See again König (1997) p. 11 for the quotes, and part II for his international comparison.

[16] See König (1997) pp. 81f. and Ferguson (1992); also Ulrich (1958), Feldhaus (1959), Booker (1963) and Lipsmeier (1971) for general histories of technical drawing; cf. here secs. 5.1 and 8.1.

technical education is already densely documented. It becomes more impressive still by zooming in on individual biographies including these kinds of visual training. For instance, it is still not very widely known that the young Ludwig Wittgenstein (1889–1951), later to become one of the foremost philosophers of the 20th century, underwent thorough training in descriptive geometry at the Berlin Polytechnic in Charlottenburg, where he was receiving his undergraduate training in architecture in 1906/07 after having passed his school-leaving exams at the Linz *Oberrealschule*, a high school placing heavy emphasis on applied geometry and drawing instruction.[17]

Like language, mentality and habitus, visual representations are shaped by cultural factors: globally by the distinctive culture of scientific disciplines, and more locally by subcultures and local clusters.[18] In chapter 6 we will see in detail how fruitful it is to look beyond the higher echelons of leading scientists or engineers and view them embedded within a broader context of other carriers of culture. These include illustrators, photographers and image technicians. Many of them might not have received any 'scientific' or 'academic' training, having rather acquired their skills at an artisanal workshop or polytechnic institution, or by private tuition from a practicing master in their field.[19] Since the Renaissance, the hitherto strictly separated social milieus of craftsmen, scholars and naturalists became increasingly intercalated: a) Ambitious craftsmen and artist-engineers such as Leonardo da Vinci, Francesco di Giorgio Martini and Lorenzo Ghiberti became interested in acquiring higher knowledge such as in anatomy, alchemy, perspectival theory or optical theory, either for their own purposes or to raise the social status of their work.[20] b) Systematic observers of nature, such as Galileo or Hooke, Agricola or Huygens, concentrated on the fine arts in order to improve their own drawing and observing skills.[21] Cultures of science and technology, the center of attention in this book, will always be understood here as embedded within these broader cultures of knowledge.

---

[17] On Wittgenstein's technical training, see Hamilton (2001) pp. 56ff., who claims that this training also left its mark on his *Bild* conception in the philosophy of science; cf. Consentius (1872), Müller (1872), Hindle (1983) pp. 13f. and Vincenti (1993) for other examples.

[18] See, e.g., Knorr-Cetina (1991, 1999), Pickering (ed. 1992) or Galison & Stump (1996) for exemplary studies on these local cultures of science and technology.

[19] See Hentschel (ed. 2008) and further references listed there on 'invisible hands,' furthermore Gotz & Gotz (1979) and Stenstrom (1991) for a statistical analysis of their typical family background.

[20] On artisanal skills and traditions in the early-modern period, see, e.g., Findlen (1994) and Pamela H. Smith (2004).

[21] On the plurality of motifs and functions of drawings in the late Middle Ages and early-modern period, see, e.g., Lefèvre (ed. 2004) and Dupré (2008).

Attentive readers will notice that I tend to speak of cultures and sciences in the plural, avoiding the singular that philosophy of science had started out with at the beginning of the 20th century. Logical empiricists, who dominated philosophy of science from the 1930s to the 1970s, had always insisted on the notion of a unity of science, guaranteed by a purported unity of method and by collective submission to the dictate of truth as the only legitimate authority. All components of this belief have come under heavy fire; but I will refrain from going into these debates, which would lead us far astray.[22] What is necessary for the broadly comparative approach practiced here is an acceptance of the great variety of scientific and technological practices rather than the presupposition of any common denominator under which all can be subsumed. I sympathize with art theorist James Elkins (*1954), who demands that "we need to let individual image-making practices exist in all their splendid particulate detail" and "simply listening to the exact and often technical ways in which images are discussed."[23] This plurality of image practices does not imply boundless heterogeneity, though. We will watch out for patterns and spell them out wherever we spy them, but we won't hypothesize them before they have been traced down in a sufficiently broad array of analyzed cases. Thus a log must also be kept of the differences and specificities of each case. Our starting point will be the plural of both sciences and cultures – emphatically so, for all its messiness and exasperating variety.

Logical empiricists also tried to demarcate their singular science, which they contrasted to its obverse: despised 'metaphysical,' i.e., unproven and unprovable pseudo-science. By limiting our object of study in the title of this book to "cultures of science and technology," we implicitly also seem to fall prey to this charge of drawing a simplistic and narrow-minded distinction that cannot be upheld if one looks carefully enough at scientific and technological practice. I would retort to this charge that this title only implies a set focus, not a sharp demarcation line. This point will be corroborated in the further discussion of many cases trespassing this imaginary borderline between science and technology. In particular, we will notice that their fringes and skills in visual cultures clearly extend beyond both, into artisanal workshops and deep into the everyday life-worlds of our actors.

A similar demarcation problem has haunted the history of art, incidentally, with the traditional distinction drawn between 'high art' and 'low art.' Even within the

---

[22]Book titles are indicative, such as *A Social History of Truth* by Shapin (1994), *The Disunity of Science* by Galison & Stump (1996), and the three editions of the *Handbook of Science and Technology Studies*, Cambridge, Mass.: MIT Press 1977, 1995, 2007.

[23]Elkins (2007) p. 8; he then presents more than 30 different examples across academe; cf. also Hentschel (2011*b*) for a survey of other recent studies of image practices.

first category, lines are drawn between first-rate artists, second-raters and no-names at the very fringes of what was considered worthy of study by a well-educated art historian. These strong qualitative judgments of connoisseurship have been under fire for a long while now. Aby Warburg's and Erwin Panofsky's pioneering efforts to broaden the subject of art history date back to the first half of the 20th century. Renewed attacks came with Ernst Gombrich's, Samuel Edgerton's and Martin Kemp's persistence within the history of perspectival seeing. To this day, controversies continue to rage over the issue of perspectival representation. On the one hand, there are those who argue (like Panofsky and the neo-Kantian philosopher Ernst Cassirer) that cultural preconditions determine what and how objects are seen and represented, or (like the cultural relativist Nelson Goodman) that all forms of representation are heavily overloaded with conventionality. On the other hand, there are those who see linear perspective as the only "true" form of representation, "corresponding to a high degree of approximation to the way we actually see the world around us."[24] Maurice Henri Pirenne (1912–78) had a point when he declared that "the strange fascination which perspective had for the Renaissance mind was the fascination of truth" or perfection of perspectival representations.[25] It is important to keep in mind that this fascination was not limited to a few "great" minds but was fairly broadly disseminated down to the level of architects and surveyors, painters and draftsmen, goldsmiths, carpenters and stone masons.

The demarcation between art and non-art was also openly combatted since the 1990s by James Elkins at the Department of Art History, Theory and Criticism of the *School of the Art Institute of Chicago*. He points out that the overwhelming number of images around primarily convey information and thus are not images of art, created with the aim to please (or displease) aesthetically. These 'utility graphics' include musical scores, floor plans, money, bond certificates, seals and stamps, pictographic signs indicating one-way streets or escalators, etc., as well as geometric diagrams, astronomical charts and engineering drafts. Earlier generations of historians of art tended to dismiss the latter as "intrinsically less interesting than paintings," engravings or sculptures. Consequently such profane images were ignored as mere "half-pictures or hobbled versions of full pictures, bound by the necessity of performing some utilitarian function and therefore unable to mean more freely, [...] as incapable of the expressive eloquence that

---

[24]See, e.g., Panofsky (1927), Cassirer (1955), Goodman (1976), Veltman (1980*b*) or Moxey (2001) vs. Derksen (2005), Gombrich (1960, 1982), Krieger (1984), Turner (1992) or Pirenne (1952), with the last quote from p. 170. On Kemp's publications, see also Dupré (2010).

[25]Pirenne (1952) p. 185. See here sec. 5.1 and further references there.

is associated with painting and drawing."[26]   Illustrators were only included in biographical dictionaries compiled by historians of art if they had succeeded in transcending the busy world of commerce and everyday life, having thus striven for acceptance as "real" artists, not as lowly artisans. Consequently, only a few scientific illustrators are found in handbooks and dictionaries of the fine arts (for more on this prosopographic point, see chap. 6). Precisely in order to fill this historiographic gap, I instituted the compilation of a *Database of Scientific Illustrators* (DSI) at the University of Stuttgart. DSI covers five centuries from 1450 to 1950, thus excluding illuminators of medieval books and professionally still active illustrators. Its 20 search fields, ranging from name to region of activity, education and patronage, allow the user to search for illustrators (many of them not listed anywhere else), their clients and relatives, their techniques and their regions of activity.[27]

A small minority of historians of art countered these prejudices against the applied arts. They argued in favor of many apparently dry, mundane "informational" and "utility" representations: "far from being inexpressive, they are fully expressive, and capable of as great and nuanced a range of meaning as any work of fine art."[28]   A first move proving this point would be to show how often they are actually inspired by the fine arts. Andreas Vesalius's (1514–64) famous skeleton figures and muscle-men in *De humani corporis fabrica* (1543) become allusions to older Italian compositions; medical illustrations more generally become the "shadow of fine-art depictions of the body." Likewise, "computer software developers recapitulate the history of art in various particulars" such as the adoption of perspectival space construction and quasi-mylar layering.[29]   A next step could be to show how scientific motifs enter the fine arts, such as cartographic maps adorning the interiors of some of Jan Vermeer's (1632–75) famous paintings. They were depicted in such detail that the original prints could be traced down by

---

[26] All preceding quotes are from James Elkins (1995) p. 553 (also republished as chap. 1 of Elkins 1999), who then proceeds to refute each of these quotes. A good analysis of the full breadth of images with an effort towards their taxonomy is provided by Gottfried Boehm (1994).

[27] Currently (as of December 2013) the database contains over 6,200 entries, with more to come in the next months and years. See www.uni-stuttgart.de/hi/gnt/dsi and its subpages, online since April 2011, and Hentschel (2012) for a brief introduction on how to work with this database.

[28] Elkins (1995) p. 554. Cf. also Elkins (1999) pp. 6, 10f. and Schwartz & Przyblyski (eds. 2004) pp. 4f. on the social history of art. Boehm (1994) tries to retain some distinctions and emphasizes the limits of utility graphics.

[29] All preceding quotes are from Elkins (1995) p. 556. On Vesalius and his illustrators, see, e.g., Kemp (1970), O'Malley (1964), Harcourt (1987), P. H. Smith (2006) pp. 86f. and the interestingly annotated online version of Vesalius's main oeuvre of 1543 at http://vesalius.northwestern.edu.

specialists. Whether and how scientific devices like the *camera obscura* might have been used by the same artist would be another avenue.[30] The final move along these lines would be to concede to the very occasional scientific image a strong enough impact and symbolic weight to leave its mark in works of art. Well-known examples of this feedback loop between the sciences and the arts include James Watson (*1928) and Francis Crick's (1916–2004) double-helix DNA, photographs of an atomic bomb mushroom cloud or some portraits of Einstein.[31]

None of these three argumentative strategies does away with the hierarchic distinction between art and non-art though. They all work with it and just proudly present some point of contact, overlap or a limited transfer from one realm into another. Edgerton's, Elkins's and Kemp's critique aims at this very dualism, however. They see it as categorically misplaced. It simply does not make sense to keep apart Leonardo da Vinci (1452–1519) the acute observer and proto-scientist from Leonardo the artist. Any effort at such a distinction is artificial and useless.[32]

How do I position myself in these debates? On one hand, I sympathize with those who warn against arbitrary distinctions in continuous fields. I do not want to throw out the baby with the bathwater, though, and dispense with all disciplinary borders, especially not for periods from the 19th century onward where these borders were very much in place, both cognitively and socially.[33]

In focusing on what is loosely termed "visual cultures of science and technology" in the title of this book, I do not want to fail to define the latter differently from how they were considered during the period involved. The boundaries become sharper the more modern our examples become. But even during the 17th century there is already a difference between speaking about a workshop for the painter Vermeer or one for the microscopist Leeuwenhoek, although both were embedded in the broader context of early-modern Dutch culture.[34] To continue within this example – our focus will clearly remain on the Leeuwenhoek side of

---

[30]See, e.g., Steadman (2001). Cf. here sec. 1.6, note 184 suggesting Vermeer used a *camera obscura*.

[31]On the preconditions for this to happen, see Bredekamp in Ullrich (2003*a*) pp. 18f. On Watson and Crick's DNA model as an icon, see Nelkin & Lindee (1995). Van Dijck (1998) coined the term 'imagenation' with respect to popular images of genetics. See Kemp (2000) pp. 254ff. and Soraya de Chadarevian's paper on Watson & Crick's model in Chadarevian & Hopwood (eds. 2004). On atomic bomb images, see part III of Bigg & Hennig (eds. 2009), as well as here p. 61 on the Einstein iconography and processes of idolization.

[32]See, e.g., Kemp (1990, 2004, 2007) and further Leonardo literature listed there in making this point. Cf. also Fehrenbach (1997, 2002), Dupré (2010) and here p. 242 on Leonardo's ingenious efforts to capture motion.

[33]On this process of disciplinary differentiation, see Stichweh (1984).

[34]For more on the latter, see Alpers (1983), commented upon in greater detail here in sec. 1.5.

the argument. Nevertheless, obvious links to his artist contemporaries will not be ignored, such as his role as the official executor of Vermeer's last will and testament. Nor will the broader cultural strata in which both these masters were embedded. The plural in 'visual cultures' is very important to me. There is always more than just one 'culture' even during one and the same period, and one and the same place; here Delft around, say, 1670.[35]

In order to distinguish specific spheres of activities performed by small groups of specialists in particular ways of seeing, representing and recognizing patterns, I have introduced the concept of visual or scopic domains.[36] In my book on spectroscopy, for instance, I distinguished at least 14 different 'spectro-scopic domains,' all having to do with spectra, their recording and interpretation, but differing in minor details about the instrumentation, spectral resolution and range, sensitivity, goals and tacit knowledge not easily transferable from one domain to the next.[37] Viewed from this angle, the visual culture of spectroscopy becomes "contested terrain" over which instrumentation linked to specific observing and measuring practices should compete. Within each visual domain, there exist stages of familiarity and immersion, from novices in a discipline (i.e., beginner students), or apprentices and journeymen (roughly corresponding to the academic undergraduate or postdoc stages), to the final stage of expert in a visual culture.

A similar idea lies behind Jonathan Crary's study on *Techniques of the Observer*. Taking examples mainly from the 19th century, Crary presents a plethora of modes of vision, created by means of various optical devices ranging from the *camera obscura* (known since the 17th century) to the stereoscope, kaleidoscope and phenakistoscope.[38] Taken as a historical progression, each of these gadgets increasingly distances the observer from his object of observation, yielding increasingly artificial and synthetic views embedded in increasingly complex systems of conventions of representation and limitations of seeing. Each of these gadgets creates another view of the world and its parts. Each of them plays on human perception with a sequence of deconstruction (or analysis) and reconstruction (or synthesis).[39] Each of them is bound to a different user group, fascinated with very

---

[35] See, e.g., Mitchell (1995) p. 543 on this issue of the singular vs. plural of culture.

[36] See Hentschel (2002a) pp. 434ff. This concept incorporates both Martin Jay's concept of 'scopic regimes' (see below on p. 27 and James Elkins' (1999) 'domain of images' (see here p. 16).

[37] See Hentschel (2002a) pp. 434–6 for a survey summary; and Elkins (2007) pp. 50f. on "families of visual technologies" within which such transfers are relatively easy to perform.

[38] See Crary (1988, 1995); cf. Liesegang (1920) pp. 53ff., Füsslin (1993); cf. Lenoir (1997) p. 164 for von Helmholtz's 'spectrascope' for determining complementary colors, and Brücke's variant of it, the 'schistoscope.'

[39] On this feature, see esp. Timby (2005), who parallels 19th-century stereoscopy and color

special features and goals of the various instruments. The thaumatrope is a simple disk, with two printed images, one on each of its faces. They appear to merge when the disk is rapidly spun around itself. The phenakistoscope ('spindle viewer,' also metaphorically called a 'fantascope' or 'wheel of life,' *Lebensrad*) gives the illusion of seeing objects in motion by watching a rapid succession of images printed or mounted on the circumference of a disk (see fig. 1 middle). The observer looks at these images through a second disk with as many thin slits as there are images on the other disk. Because both disks are mechanically coupled and thus rotate at the same speed, the observer's eye held fixed sees only the image sequence and is not disturbed by the white spaces in between the images, which are blocked by the black intervals between the slits of the second disk.[40] The zoetrope (also called *daedaleum* or in German *Wundertrommel*; fig. 1 right) has the image sequence inside a rotating drum; the observer looks through thin slits along its circumference. Both gadgets are early examples of stroboscopic viewing instruments that – when rotated fast enough – generate the impression of continuous motion.[41]

The stereograph gave the closest 3D impression obtainable by 2D representations throughout the 19th century. Two images (usually photographs) taken from slightly different angles (for an example, see fig. 3) are viewed with a special binocular device at the appropriate distance such that each eye looks only at one of the two photographs. In the viewer's mind, these two pictures are superimposed and create a nearly perfect 3D impression of the depicted objects. Stereo-viewing under controlled conditions creates a sense of depth that simulates normal human perception of objects in space. This technique was invented in the 1830s by the English scientist Charles Wheatstone (1802–75) in the context of his studies on binocular vision.[42] Since Wheatstone published the first part of his findings in June 1838, which was roughly half a year before Daguerreotypes and Talbotypes became publicly known, his illustrations were simple line drawings. The technique would have remained an obscure, odd invention and soon been forgotten if photographic processes had not allowed relatively simple produc-

---

photography. A variant of stereoscopy, anaglyphs, use two images taken with different filters from slightly differing perspectives in conjunction with color-tinted viewing binoculars (cf. here p. 23).

[40] For photographs of historical phenakistoscopes, see Füsslin (1993); cf. also Rocke (2010) p. 213 on Kekulé's use of this device to visualize the motions of atoms in molecules.

[41] See Plateau (1831, 1832, 1833) and Horner (1834). The terms 'stroboscope' and 'stroboscopic' were introduced by Stampfer (1833). On these two and various other related devices, cf. Liesegang (1920) pp. 54ff., Crary (1995) chap. 4, Füsslin (1993) and the website courses.ncssm.edu/gallery/collections/toys/opticaltoys.htm . Cf. here sec. 7.4 on the long history of efforts to depict motion. On Plateau's biographical background and strong visuality, see here p. 156.

[42] See Wheatstone (1838/52), Brewster (1856), Gill (1969) and Wade (1983).

Fig. 1: Left: A thaumatrope. Middle: A simple phenakistoscope of Joseph Plateau. Right: William G. Horner's zoetrope. All from the mid-1830s. From Liesegang (1920) pp. 55, 57.

tion of such stereo-images. In the late 1840s, substantially improved versions of stereoscopes were developed by the Scottish physicist and spectroscopist David Brewster (1781–1868). He replaced Wheatstone's double mirror with a pair of adjacent half-lenses that allowed direct viewing of the two stereophotographs, side by side, which considerably facilitated their mounting. Around 1850, the Parisian opticians and instrument-makers Duboscq and Soleil started to market these gadgets, which became quite popular in the second half of the 19th century.[43]

The first stereophotographs were advertised at the 1851 Crystal Palace Exhibition in London. They were perceived as providing a "truthful, yet wondrous experience for the at-home viewer."[44] The professor of anatomy and art critic Oliver Wendell Holmes (1809–94) was one of the most outspoken and enthusiastic advocates of the new viewing device. He developed a particularly light and cheap variant of it that became known as the 'American stereoscope.' Holmes was fascinated by how much stereoscopes made flat surfaces look "solid," or, as we would rather put it, three-dimensional:

> The first effect of looking at a good photograph through the stereoscope is a surprise such as no painting ever produced. The mind feels its way into the very depths of the picture. The scraggy branches of a tree in the foreground run out at us as if they would scratch our eyes out. The elbow of a figure stands forth so as to make us almost uncomfortable. Then there is such a frightful amount of detail, that we have the same sense of

---

[43]On the history of stereoscopes, see Reynaud et al. (eds. 2000), esp. Pellerin (2000) and Timby (2000, 2005) and Halsband (2008). On Brewster's visual culture, see here p. 144.

[44]See Halsband (2008), quoting early enthusiastic commentaries, most notably by Queen Victoria.

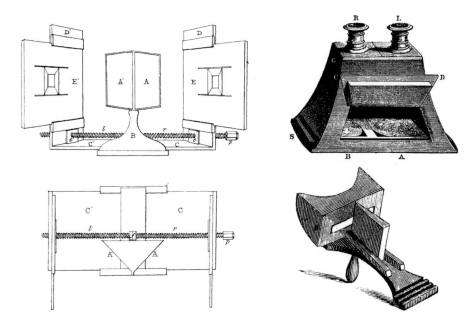

Fig. 2: Left: Wheatstone's reflecting stereoscope (1838). Right top and bottom: Brewster's lenticular (1849) and Holmes's handheld American stereoscope (1861), respectively. From Wheatstone (1838/52) p. 10, figs. 8 and 9; Brewster (1856) p. 67 and Holmes (1869) p. 24.

> infinite complexity which Nature gives us. A painter shows us masses; the stereoscopic figure spares us nothing – all must be there, every stick, straw, scratch, as faithfully as the dome of St. Peter's, or the summit of Mont Blanc, or the ever-moving stillness of Niagara. The sun is no respecter of persons or of things.[45]

Stereoscopes were used intensively. They were certainly not limited to the realm of popular images. Being an expert in prosthetics, Holmes made use of stereographs to depict people performing various motions, in order to improve the design of his artificial limbs, which were in high demand as a consequence of the American Civil War (1861–65).[46] During the era of professional stereoscopy (1852–69), when this art was practiced by just a few specialists, Charles Piazzi Smyth's account of *An Astronomer's Experiment* was the very first book worldwide with reproductions of select stereographs, taken on the Canary Islands in 1858 (see fig. 3). They were simply glued into the volume as adjacent prints on albumen paper. The book documents the intricate fauna and flora as well as Smyth's expedition as Astronomer Royal of Scotland to the mountain tops of Tenerife, where he made geological, topographic, spectroscopic and astronomical observations.

[45] Holmes (1859) p. 744.
[46] On this application of stereography, see Albrecht Hoffmann (1990) p. 32.

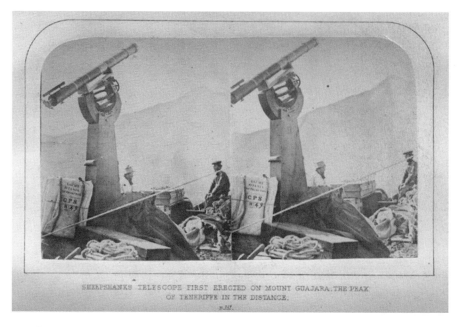

SHEEPSHANKS TELESCOPE FIRST ERECTED ON MOUNT GUAJARA, THE PEAK
OF TENERIFFE IN THE DISTANCE.

Fig. 3: Stereograph taken on Tenerife by C. P. Smyth, albumen print, 6 cm × 7 cm 1858. Courtesy of George Eastman House, International Museum of Photography and Film, no. 1995:0152:0005. See pp. 142ff. below on Smyth's other visually oriented activities.

In the late 1850s and early 1860s, the British astronomer, chemist and print expert Warren de la Rue (1815–89) and the American amateur astronomer and expert photographer Lewis Morris Rutherford (1816–92) obtained the first successful stereographs of the Moon. The large distance of this object precludes the usual procedure of choosing two different perspectives for the left and right photographs. Instead, they took photographs at different times, making use of the Moon's libration. Improved versions of such lunar stereographs were made by Henry Draper (1837–82) in New York and by John Adams Whipple (1822–91) at the *Harvard College Observatory* in Cambridge, Massachusetts. Around the turn of the century, lunar stereographs were commercially mass-marketed just as were topographic and other motifs by companies such as *E. & H.T. Anthony* or *Underwood & Underwood* in New York.[47]

In this case, 3D perception in the human mind went beyond mere mimicry. It in fact amplified the image. Holmes wrote enthusiastically: "the [Moon's] sphere rounds itself out so perfectly to the eye that it seems as if we could grasp

[47]See A. Hoffmann (1990) p. 35; for examples see www.londonstereo.com/modern_stereos _moons.html and www.geh.org/ne/mismi2/moon_sum00003.html

Fig. 4: Lunar stereograph by Warren de la Rue. Left: on February 27, 1858 at 13:50 at a lunar age of 14.2 days; right: on September 11, 1859, 11:20 at 14.8 days. The lunar libration led to a difference of c. 6° latitude and 2.5° longitude between two exposures, creating the stereoscopic effect.

it like an orange."[48]   By 1875, more than 100 American photographers were trading in stereograph sets totaling over 1,000 different motifs, mostly landscapes and other sightseeing topics allowing virtual tours throughout the continent.[49] The London-based *Stereoscopic Society* (founded in 1893) advertised with the telling slogan: "No home without a stereoscope." The invention of anaglyphic prints in which the stereo images were superimposed on the same surface, but in two complementary colors (such as red and cyan or green), spread stereoscopic 3D effects to new user groups at very low cost (see Lorenz (1985) for examples). The only accessory needed to obtain the 3D effect is a pair of cheap color-coded anaglyph glasses with a filter in those two colors, one for each eye.[50] But travel impressions, vivid portraits or erotic scenes were only some of what stereographs could depict. One pioneer, Charles Wheatstone, envisioned as early as 1852, "works on crystallography, solid geometry, spherical trigonometry, architecture,

[48]Holmes (1861) p. 27; on astronomical photography and stereography see, e.g., de Vaucouleurs (1961), Pang (1997b), Hentschel (1999a) and Hentschel & Wittmann (eds. 2000).

[49]See Jenkins (1975) and Halsband (2008) pp. 20f. on the USA and Albrecht Hoffmann (1990) pp. 16f. on the UK and Germany.

[50]Anaglyph images were developed in 1852 by Wilhelm Rollmann in Leipzig, Germany, but became widespread only in the 20th century: see http://en.wikipedia.org/wiki/Anaglyph_3D ; see Reynaud et al. (eds. 2000) pp. 121–31 for references, examples and links to modern applications.

machinery, &c, might be thus rendered more instructive."[51] In the period of mass-produced stereographs (c.1880–1920), their omnipresence in popular contexts led to a veritable craze in various areas of science to improve visualization of complex objects. Complex molecules, for instance, or crystallographic grids were more clearly understandable and almost graspable through a stereoscope. Paul Groth (1843–1927) was the first crystallographer to embellish later editions of his textbook with a set of stereoscopic plates of the crystal structures then known.[52] These plates allowed practitioners and researchers alike rapid localization and spatial contextualization of individual atoms in a complex crystallographic lattice. Thus they could dispense with the imaginary "internal view" [*inneres Sehen*] to visualize such structures.[53] Even as this "stereomania" was subsequently declining, due to the diffusion of new technologies such as cinematography, stereography continued to be used in certain niche visual science cultures. In crystallography, William and Lawrence Bragg issued sets of stereoscopic photographs of crystals as late as 1928 and 1930,[54] and in Germany, Max von Laue and Richard von Mises also published such stereoscopic plates in two bilingual sets in 1926 and 1936.[55]

Within the somewhat related field of photogrammetry, in which the exact dimensions of an object are determined by accurate measurements of two perspectival views of it, truly stereoscopic applications had an even later start. Stereoscopic aerial photographs were taken as a convenient base from which to set out, for surveying large objects such as trenches, mountains or other difficult terrain.[56] In 1899, the Jena physicist Carl Pulfrich (1858–1927) introduced his first device for determining spatial distances based on stereoscopic photographs. In 1901, the fully fledged stereo-comparator followed, marketed by the optical company *Zeiss*,

---

[51] See Wheatstone (1838/1852) p. 6; Brewster (1856) likewise foresaw technical applications.

[52] See Paul Groth's *Physikalische Krystallographie und Einleitung in die krystallographische Kenntniss der wichtigsten Substanzen*, Leipzig: Engelmann, 1st edn 1876 with stereoscopic plates in its 3rd and 4th edns of 1895 and 1905.

[53] See Herlinger (1928) p. 165 on this scopic domain within crystallography: "Thereby the problem mentioned at the beginning falls away of having to put oneself into a position to get an internal view, so to speak, of the precise manner in which the individual atoms arrange themselves around each other in the lattice." ("Damit fällt die eingangs erwähnte Schwierigkeit fort, daß man sich gewissermaßen durch ein inneres Sehen hineinversetzen muß in die Art und Weise, wie die einzelnen Atome sich im Gitter gegenseitig umgeben.")

[54] See Bragg & Bragg (1928/30). For the periodization of the invention, innovation and diffusion of this visual technology see Halsband (2008) p. 41.

[55] See von Laue & von Mises (eds. 1926/36), with the assistance of Clara von Simson, drawn by Elisabeth Rehbock-Verständig (1897–1944), and translated into English by Gabriel Greenwood.

[56] See, e.g., Rudolf Burkhardt in Kemner (ed. 1989) pp. 33–42, Reynaud et al. (eds. 2000), pp. 200ff. for examples from Paris (the earliest dating from 1923) and further literature.

Pulfrich's employer. Later variants, such as the Zeiss 'stereoplanigraph,' made it possible to transform stereographic images of landscapes, taken by airplane, into topographic maps with height profiles.[57] Nowadays, computer-aided programs for digital photogrammetry are even easier.[58] Other aerial stereophotographs allowed meteorologists to reconstruct cloud shapes in 3D, whereas stereoscopic x-ray photographs yielded a 3D image of the interior of the human body long before the advent of computed tomography.[59] The fusion of stereoscopy with electron microscopic techniques allows a 3D impression of microorganisms or of microstructures of materials.[60] In all these cases of objects either at very close range or far away, the usual recipe of shifting the camera for the second exposure by an amount equaling the mean distance between the two eyes on a face (c. 2.5 inches) would not work. As a rule of thumb, stereophotographers chose an offset of roughly 1/30 of the distance to the objects in the foreground of their image, but sometimes even larger offsets were taken in order to heighten the effect. This practice became very popular among landscape photographers, but also highly controversial, since it led to a kind of enhanced perspective or 'hyperspace' that some observers regarded as "distorted," if not "monstrous."[61]

Jonathan Crary has described how each of these technical gadgets actually created a very specific *visuality* among their users at the height of popularity. Crary takes this term from Hal Foster's anthology on *Vision and Visuality* (1998), where "visuality" is defined roughly as the variegated bundle of social factors involved in the process of seeing, whereas "vision" is supposed to denote all of its anatomical, physical and geometric aspects. Foster's central idea was thus to historicize and to "socialize vision" by pointing out how the allegedly objective physical act of visual perception is heavily loaded with personal and social layers that infuse vision subjectivity and mold – if not warp – our sensorial impressions. Although this concept of visuality is most frequently taken to be a postmodern notion that has taken center-stage in the debate on visual cultures, it actually has much older roots: As Nicolas Mirzoeff has pointed out, the word had been coined by the Scottish essayist and historian Thomas Carlyle (1795–1881) in his lectures *On Heroes* (1841) simply to denote the increasing importance of visual representations

---

[57] See, e.g., Pulfrich (1902, 1911), Lorenz (1985) pp. 25ff., and A. Hoffmann (1990) pp. 33–7.

[58] See, e.g., Jörg Albertz in Kemner (ed. 1989).

[59] On meteorological applications, see Lorenz (1985) pp. 47–57 and in Kemner (ed. 1998) pp. 61–70; on CT scanning see below, pp. 200.

[60] On these applications, see, e.g., Jo-Gerhard Helmcke in Kemner (ed. 1989) pp. 71–8.

[61] On this controversy rooted in different norms of visuality, see Silverman (1993) pp. 748ff.

in the construction of heroic figures in the Victorian era.[62]

More recent discussants tend to agree that people from different eras as well as different subcultures, even at roughly the same times, often differ drastically in their 'visuality,' i.e., in the complex cultural baggage carried along in the process of visual perception. Members of different cultures select and conceptualize what they see differently. The same applies to their anticipations and associations, and even the intensity with which they observe, discriminate and recognize what they see. Our examples later will provide ample evidence for this claim. But here are a few findings by others also corroborating this claim of competing "visual subcultures," first advanced by members of the Birmingham School of Cultural Studies. Xiang Chen recently contrasted a visual tradition of optical measurements against a geometric tradition in his analysis of instrumental conventions and theories of light during the 19th century. The visual tradition regarded man-made optical instruments as "aids to the eye, and evaluated [them] according to how well they produced images suitable for the perception of the eye," as the ultimate "goal of the optical system." The geometric tradition strove to "reduce and eventually eliminate the role of the eye in optical experiments."[63] Following a similar vein, the Berlin historian of medicine Thomas Schlich distinguished between the subcultures of microscopic and photographic vision in the early history of bacteriology. To belong to the latter meant "to know how to 'read a photograph' and [...] to share certain presuppositions as to which interventions by the photographer were to be tolerated and which ones constituted forgery."[64] Likewise, we may regard the various types of spectroscopes, spectrographs and spectrometers encountered as competing 'visual technologies' in the sense spelled out by Mirzoeff,[65] which lead to different manifestations of visuality.

Interestingly, most of the above examples happen to come from the same era, the 19th century. This demonstrates how manifold and polyphonic these various visual domains can be, even in a limited comparative analysis of nearly synchronous case studies. If one starts to compare diachronically across longer time spans, the variations in the actors' visuality become even stronger. We

---

[62]For a closer analysis of these roots and possible definitions, see Mirzoeff (2006).

[63]See Chen (2000) pp. 121–8, esp. pp. 124f. for the quotes.

[64]See Schlich (2000) p. 50; cf. also Schlich in Rheinberger et al. (eds. 1997) pp. 165ff. and here p. 53 and color pl. XIII on Robert Koch's bacteriological drawings and photographs, Jordanova (1990) on medical practitioners' ways of looking, or Evelyn Fox Keller (1996) on the "biological gaze."

[65]"any form of apparatus designed either to be looked at or to enhance natural vision, from oil painting to television and the Internet." From the introduction to Mirzoeff (ed. 1998).

immediately recognize whether some visual representation shown to us comes from pharaonic Egypt or from Roman Pompeii, from a scriptorium of a medieval monastery or from a Renaissance painters' workshop. Why? Because each of these four settings is different, in fact, mutually exclusive 'scopic regimes.' This very helpful term has been introduced by the Californian historian Martin Jay (*1944) to distinguish drastically different 'visual domains' of seeing, if you will, as the visual analoga of Foucault's epistemes, characterizing whole centuries in their fundamental approach to nature and knowledge. Jay's pathbreaking essay from 1988 on the "scopic regimes of modernity" argued for a reconceptualization of visuality "as a contested terrain, rather than as a harmoniously integrated complex of visual theories and practices."[66]   Jay focused on the distinguishing features of three highly influential visual cultures in the modern era, namely Renaissance perspectivalism (dealt with here in sec. 5.1), 17th-century Dutch art with its narrational quality (cf. sec. 1.5) and baroque art with its strongly haptic qualities. This tripartite division already debunked the often-heard opinion that Cartesian perspectivalism was the sole reigning visual model of modernity.

Jay's 'scopic regimes' aim at visuality *tout court*, at broad cultural strata extending beyond national and local contexts. It is a macro-concept on a very coarse-grained scale to describe visual practices persisting over many decades, indeed, centuries. In comparison with this, my coinage of 'visual domains' is a related concept on a finer scale, capturing the more subtle distinctions between, say, a gauging photograph for quantitative spectrochemistry and a plate with various types of stellar spectra. Rather than 'regimes' *tout court* – with its imperial connotations[67] – what we will be exploring throughout this study are less 'regimes' than 'domains,' i.e., practices on a much closer scale of differentiation, in the social as much as in the cognitive sense. These visual domains define a certain mode of visuality, practiced by more than one person, often by larger groups of practitioners within a scientific, medical or technological field.   In a way, these domains are thus the visual equivalent of 'styles of thought' defining scientific communities and subcommunities (in the sense of Ludwik Fleck's 'thought collectives'), with the difference that here it is not the concepts or theories used that demarcate the group but their joint predilections for certain types of images, patterns or recording techniques. Sometimes these domains are stabilized and transmitted to the next generation of practitioners within this field through training classes and other

---

[66]See Jay (1988*a*) p. 4 or (1988*b*) p. 66; cf. also Lynch (1985), Lynch & Woolgar (eds. 1990) and Brain (1996) pp. 11ff. on 'regimes of visuality.'

[67]See the *Oxford English Dictionary*, 2nd edn, vol. 13 (1989) p. 508.

pedagogic means (cf. here chapter 8 for exemplary studies on this process). If a whole paradigm agglomerates around these visual domains, they may evolve into 'research schools'[68] or longer-term 'research traditions.'[69] Often, these visual practices are too rapidly changing for this accretion process to happen, though, and they will be abandoned for the next such visual domain.

1.2 Visual versus textual

After this careful scrutiny of the nouns in the main title of my book, it is high time to turn to the equally difficult adjective: "visual." In the literature, we find a surprising manifold of efforts to bifurcate human artifact representations. Indeed even the actors often sunder them into two mutually exclusive classes, such as:

- visual vs. textual (or verbal)[70]
- graphic vs. legible
- iconic vs. symbolic
- diagrammatic vs. sentential[71]
- imagistic vs. verbal code[72]
- logocentric vs. iconocentric[73] and finally
- iconophile vs. iconophobe[74]

Human history has gone through various phases of iconoclasm in which fanatic groups actively destroyed religious or political icons that they perceived as blasphemous, politically provocative, authoritarian, uncouth or simply misleading. Consequently, certain types of images were banned in certain societies. We might immediately think of various waves of clashes since AD 730 up to the Reformation between iconoclasts and iconophiles (or "iconodules" as they were called in Byzantium). Other cases in point are the Islamic interdiction on depicting

---

[68]in the sense of Gerald L. Geison (1981); Geison & Holmes (eds. 1993).

[69]in the sense of Larry Laudan (1977) chap. 3, (1989).

[70]See, e.g., Hindle (1983) pp. 34, 60ff. and Root-Bernstein (2001) p. 303.

[71]See, e.g., Larkin & Simon (1987) and Bauer & Johnson-Laird (1993).

[72]See, e.g., Gardner (1985) p. 177, and cf. Gardner (1985) on the opposition of the right and left brain hemispheres, functionally linked to these two different representational systems.

[73]See Hofmann (1999) pp. 7–11: "logozentrisch vs. ikonozentrisch"; see also Bredekamp in Ullrich (2003a) p. 12.

[74]See, e.g., Hagner (2003). On the stark contrasts between both, see also Latour's "iconoclash."

God – which, incidentally, enormously boosted the intricacy of ornamental images in Islamic cultures – or more recent complaints today about media-hypes,
"videocracies" and "tele-revolutions."[75]

Several researchers have warned against overstressing these distinctions and
construing too rigorous a dichotomic divide, since quite often an intense text–
image interplay exists; so "the dual structure is necessarily a fiction."[76] James
Elkins argued for a gradual progression from "pure scripture" to "pure picture."[77]
Others have tried to rework the unhappy bifurcation into a trident: "Thus, scientists who practice the visual arts and musical arts, tend to be visual thinkers; those
who practice literary arts tend to be verbal thinkers; those who practice sculpture
tend to be kinaesthetic and imageless thinkers."[78] Artisans, instrument-makers
and experimentalists learn the tricks of the trade by emulation rather than from
a textbook. The historian of technology Brooke Hindle (1918–2001) studied the
sources and procedures of innovation in steam-engine and steam-boat design,
for instance, or electrical telegraphy. He pointed out the importance of mental
representation and mental manipulation of objects and processes as a skill not
acquired by bookish learning but by close observation and physical interaction.
The artisan and mechanic David Rittenhouse (1732–96) once asserted: "much
writing ill suits a Mechanic." Brooke Hindle adds: "Most mechanics lacked an
extended verbal education; their craft heritage encouraged secrecy; and their work
was primarily a series of exercises in spatial thinking, not talking."[79]

We will return to these issues of text–image interplay and ekphrasis in sec. 1.3,
but let us first collect a few facts about the frequency of visual representations of all
kinds (other than text) in science and technology publications. In the early-modern
period, printing from copper plates was a very expensive business,[80] so images
tended to be scarce and very select. But even then, a heavily illustrated classic
like Robert Hooke's *Micrographia* (1665) had altogether 38 plates, most of them in
full-page size and a few (like Hooke's most famous portrayal of a flea) even in fold-

[75]On iconoclasm, see Latour (2002), www.iconoclash.de/ and http://en.wikipedia.org/
wiki/Iconoclasm (both last accessed June 3, 2011); on the latter, see Bredekamp (1997) pp. 225–31.

[76]See, e.g., Alpers (1978) pp. 152ff., esp. p. 154 for the quote.

[77]See Elkins (1999) p. 257; cf. also Lüthy & Smets (2009) pp. 401–4 and passim for their own
taxonomy of medieval and early-modern text, line, diagram and image types.

[78]Root-Bernstein (2001) p. 301. N. Goodman (1969) divided into writing, notation and picture.

[79]Hindle (1983) p. 34.

[80]On the economy of early-modern prints and on publisher's calculations that sometimes led to
bankruptcy, see Kusukawa (2012) chap. 2.

outs, 18 inches long.[81] At a total of 246 pages (excluding index and errata), this amounts to c. 15% of this paradigmatic text. This might be close to the maximum of what was possible in the early-modern period. But beware: many other books, even by wealthy authors like Robert Boyle, only had one or two such plates, quite often none at all. Furthermore, since these copper plates could only be printed separately and had to be bound into the volume after the text and plates had been printed separately on different presses, the interplay of text and images was also fairly limited, with features of the objects depicted often labeled for purposes of reference in the accompanying text. Because of this lack of integration between text and graphics, it was not easy to use plates. In the few parsimoniously illustrated journals, like the *Philosophical Transactions of the Royal Society* (appearing since 1660) or the *Acta Leopoldina* (appearing since 1670), plates were bound at the end of a work or even issued as separate plate volumes. Many other important journals, like *Acta Eruditorum* (appearing in Leipzig since 1682) or the *Göttingische Gelehrte Anzeigen* (appearing since 1739), were completely devoid of illustrations apart from a title vignette. We have some statistics for survey journals in biology and physics from the modern period. On average, there are 14.8 or 12 visual representations for every 10 pages of scientific text in biology or physics, respectively, of which 4.2 and 10, respectively, are histograms, scatter plots and line graphs; the remainder are more iconic forms such as photomechanical reproductions of photographs. For medicine and general science journals, the corresponding averages for numbers of graphics per 10 pages are 11 and 12, respectively.[82] With each of these illustrations taking up to one-third of a printed page, we approach a 50% quota of illustrations in certain biology survey journals. Thus, extraordinarily high weight is placed on visual representations in modern science. Top journals like *Nature* and *Science* (founded in 1869 and 1880, respectively) were pioneers in the use of state-of-the-art printing techniques; and spectacular glossy color images adorn their modern-day covers.[83]

[81] On Hooke's *Micrographia* (first published in 1665 as one of the first publications by the newly founded Royal Society and as the first illustrated book on microscopy altogether, see Dennis (1989), Harwood (1989) and Doherty (2012*a*).

[82] See Roth, Bowen & McGinn (1999) for biology, based on a survey of more than 2,500 pages from ecology research journals around 1995, and Lemke (1998) for medicine, physics and general science journals from 1992; cf. also Roth (2004) p. 596 and Roth & Bowen (2003) for further analysis of graphing practices in scientific research.

[83] On the history of *Nature* and *Science*, see, e.g., Meadows (1972) chap. II and www.nature.com/nature/history/ ; for an analysis of select images from *Nature*, see Kemp (2000).

Of course, these percentages are gross averages, and each scientific or technological subdiscipline will have its own characteristic dose of necessary and admitted visual representations, some fields more, others less. According to the statistician William S. Cleveland (*1943), the fractional graph area in chemistry journals is 0.18, whereas in humanities and social sciences, the proportion of the total page area in an article that is devoted to graphics is much lower, only 0.01 in sociology or for the *Journal of Counseling Psychology*, for instance. The *Journal of Geophysical Research* devotes 31% of its total page area to graphics, whereas *Behavioral Neuroscience* yielded 13% just like the average of all biology journals. Sociology, economics and psychology taken together led to an average of only 0.03, i.e., only 3%. To some extent, the use of graphs in journal articles thus seems to correlate highly with the "hardness of scientific fields."[84] But even within one closely defined field (such as, in fig. 5, ecology), specialized scientific journals and high-school textbooks will differ substantially in the specific kinds of inscription devices used. As fig. 5 shows, typically "scientific" forms of representation, such as mathematical equations, histograms or scatter plots to visualize statistics, occur frequently in scientific journal articles but significantly less frequently in textbooks, where photomechanically reproduced photographs, drawings or other iconic or diagrammatic content dominate. In the survey on ecology textbooks in fig. 5, there is, on average, a photograph on every second page or more!

Psychological and ethnomethodological studies have revealed strong differences between experts and nonexperts in the speed, efficiency and precision with which, for instance, nontrivial graphs, histograms or maps can be "read," i.e., understood unambiguously. Each person has an interpretative horizon of visual competency acquired by training as much as by routine use and general familiarity with the scientific domain. But it often does not suffice, even for experts, just to look at a graph in order to immediately understand what it is supposed to display: a dialectic interplay ensues between an inscription and its verbal and nonverbal context in which it is embedded in the publication involved: "Readers must engage in a reflexive elaboration in which the main text and caption provide iterable instructions for where to look and how to read the graphical display."[85] We will return to this issue when we address text–image interrelations in sec. 1.3.

---

[84]See Cleveland (1984) pp. 263ff. and L. D. Smith et al. (2002) p. 749ff. for data and the two quotes.

[85]Quote from Roth, Bowen & McGinn (1999) p. 1009. Cf. also Lemke (1998) and Roth & Bowen (2003), who examined the use and interpretation of graphs in scenarios where users were confronted pairwise with unfamiliar or unusual graphs and asked to state out loud their interpretational steps.

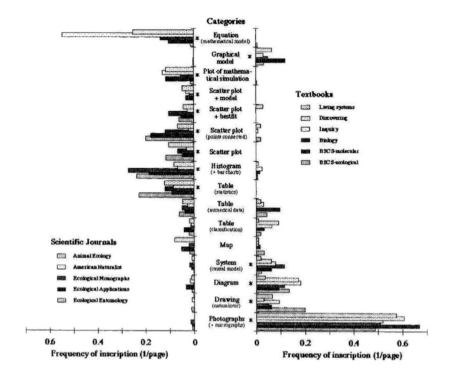

Fig. 5: The frequency of various types of inscriptions in ecology-related scientific journals (1995) and high-school textbooks. From Roth, Bowen & McGinn (1999) p. 987. © Wiley & Sons

Approximately 60% of the input into the human brain comes from vision. Humans can only digest this massive influx of visual data because of a highly non-linear form of information processing concerning colors, contours and brightness contrasts that allow quick recognition of shapes, distances and speeds in the world around us, especially for potential hazards. What actually does happen in visual recognition? How is a 3D impression obtained from a 2D trace of light rays focused by the human eye onto its retina? The following four-staged model of vision, developed in the early 1980s by David Marr (1945–80), is still the basis of much more refined modeling:[86]

(i) Human vision is based on the retinal image, generated by a kind of *camera obscura* projection of the outer world onto the spherically concave back wall of the eyeball. Information on light intensity and color is recorded by light-sensitive

---

[86]See, e.g., Marr (1982) and cf. de.wikipedia.org/wiki/David_Marr (accessed on May 29, 2011). On the neurophysiological aspects of vision, cf., e.g., Zeki (1999), esp. pp. 59ff. on the various visual areas of the brain and their functional specialization.

rods and cones making up the retina.

(ii) A first draft or "sketch" (somewhat similar to a line drawing) of the viewed object is construed by our nervous system, by a multi-stage combination of connected borderlines between areas of equal illumination (edges), topologically connected surfaces and surface textures.

(iii) A $2\frac{1}{2}$D mental model is formed by combining the retinal information from both eyes. This mental model already contains features like spatial orientation and approximate depth estimates, so that an initial, coarse "image" of the outer world is generated.

(iv) A 3D model is formed on the basis of a more systematic comparison of the two retinal images, yielding more precise information about the perceived depths, with the constraint that the perceived object remains invariant despite rapid eye and body movement.

This minute, irregular and subconscious movement of the eyeball, referred to as 'physiological nystagmus,' is actually very important for human depth perception. If it is artificially suppressed with drugs, human sight can even break down. "the eye, it seems, is less a passive screen upon which things are projected than an active scanner taking things in."[87]

You will notice that stages (i) and (ii) are actually identical for letters and nontextual signs of all sorts, since letters as well as drawings consist of broken lines and points in specific spatial orientation relative to each other. There is thus a common denominator in our visual perception of these signs, with the difference between both types of sign only coming into play in the last two stages of the four-stage process. This ontogenetic observation has its correspondence in the phylogenetic finding that images, numbers and words have common roots in the earliest cultural relics of *homo sapiens*: The earliest signs for words like man or woman, hill or fruit were simple images of them, with more and more symbolic shorthand signs forming over time in slow processes of cultural development through many generations.[88] The anthropologist and philosopher Hans Jonas has actually declared humans to be *homo pictor*, thus making the ability and desire to draw a *differentia specifica* of mankind.[89] Certain intimate relations between textual and visual elements of communication remain to this day. Diagrams or

---

[87]Turner (1992) p. 142.

[88]See, e.g., Haarmann (1991), cf. also Elkins (1999) part II.

[89]See Jonas (1961) p. 162 on his claim that no animal would voluntarily start to draw – not disputing that some higher animals can be taught to draw somewhat.

graphs, for instance, are strange hybrid forms.[90] Our rigid organization of texts in visual schemes of lines, paragraphs and pages preserve a spatial, and hence visual, order in addition to the syntactic logic of a concatenation of letters, words and sentences.[91]

1.2.1   Differences in visual information retrieval and inferences

Philosophers of mind and representation as well as cognitive scientists have pondered much about the epistemic (dis)advantages of various types of representation.

Jill Larkin and Herbert A. Simon (1916–2001), both pioneers of cognitive science and artificial intelligence from *Carnegie Mellon University*, provide an interesting discussion of different types of representation. In particular, they distinguish between diagrammatic and sentential representations, as well as between paper-and-pencil representations and mental representations. According to them, the major advantage of visual representation over nonvisual representation is not qualitative or categorial but merely quantitative, simply rooted in quicker information retrieval from images. Users can infer topological and geometrical relations from diagrams – at least from well-drawn diagrams – at a glance, whereas a lot more work has to be done to infer the same information from sentential representations in the form of long strings of words – "a diagram is (sometimes) worth ten thousand words."[92] Just how much can be extracted from an image is determined by what is called 'informational equivalence.' The comparative speed at which this extraction is possible is rather determined by 'computational equivalence.' This latter, in turn, is determined by the particular encoding of information in the graph of the image.[93] Both might be interchangeable from the point of view of 'informational equivalence' the times needed to find a specific datum or to recognize a certain feature strongly differ. It is by no means the case that images are always superior with respect to computational accessibility, depending on whether one focuses on a quick survey or on numerical detail.

---

[90]On diagrams as mediating hybrids of text and image, see Bonhoff (1993) and Bogen & Thürmann (2003).

[91]On this point, see, e.g., Lemke (1998) p. 95 and Hessler & Mersch (eds. 2009) p. 10.

[92]See Larkin & Simon (1987) – this is a quote from the title of their paper on p. 65, allegedly a Chinese proverb. On the claim in the quote, cf. also Kitcher & Varzi (2000) and Hentschel & Wittmann (eds. 2000) p. 6.

[93]See Simon (1978) pp. 4f. and Larkin & Simon (1987) p. 67 for the quote. See Tufte (1983), Roth & Bowen (2003) and Kulvicki (2010) pp. 302ff. for examples. A. Müller et al. in Liebsch & Mößner (eds. 2012) pp. 209ff. demonstrate the different responses of laymen and experts to "pretty" and "informative" images.

Fig. 6: Abacus vs. written calculation as competing representations of numbers, surveyed by an allegory of arithmetica. Pythagoras and Boëtius are listed as authorities on the two practices. From Gregor Reisch: *Margaritha Philosophica*, 1508.

For instance, tomorrow's weather can be inferred at a glance from a weather chart, whereas a tabular list of average temperatures will be quantitatively more precise but less easily screenable. In general,

> immediate availability of a great many pieces of abstract information accounts for some of the epistemological weight given to images and graphs, not to mention photographs. Descriptions, by contrast, are very selective in the pieces of abstract information that they provide. [...] Images are tools for figuring out what is important whereas descriptions and lists are used when we already know what matters.[94]

---

[94]Kulvicki (2010) p. 298; this text offers an abundance of other interesting examples. Cf. also Lima (2014) on the multifarious use of tree diagrams since the middle ages.

What counts in the end is the accessibility of the information to be extracted from the representation, the speed at which this extraction is achievable, and the ease with which reorganization or manipulation of this information is possible.[95] All of these features depend on syntactic, semantic and pragmatic factors that are not always easy to separate in specific examples. In general, scientific representations have a dual aim: to convey information about real features of the world – i.e., objective representation – *and* conveying it in a conveniently manipulable form – i.e., subjective functionality.[96] Inferential relations are generally expressed in strings of sentences. In the early 19th century, though, Venn diagrams were invented to represent inclusion and exclusion relations between sets of objects; and they do so much more clearly than sentences can. Quantitative experiments confirming this have been provided by Malcolm I. Bauer and Philip N. Johnson-Laird (*1936) at Princeton University. These cognitive psychologists confronted two test groups with either verbal or diagrammatic representations of double disjunctions: "The subjects responded faster (about 35 s) and drew more valid conclusions (nearly 30%) from the diagrams than from the verbal premises."[97]

Digital images are nothing but visualizations of matrices of numbers representing shades of gray or colors. Especially in the early days of computers, everyone knew how awfully long it can take to "make" an image on the screen on the basis of numbers fed into the computer, even though the actual number-crunching by the machine was incredibly fast. Despite the homomorphism existing between image and numbers, there certainly was no 'calculational equivalence' in Simon's sense. It is hopeless to transmit complex two- or three-dimensional spatial relations as strings of words, but they are easily depicted in an iconic representation preserving at least some of the relations in two-dimensional projection. To infer such 3D relations from 2D representations is something one has to learn, though, and we will get back to this when we speak about the history of perspectival representations (in secs. 5.1 and 8.1) and about x-ray atlases of the human body (in sec. 8.3).

---

[95] See, e.g., Pylyshyn (1973) p. 3, who later concedes that in a psychological laboratory experiment a visual display of a letter of the alphabet permitted extraction of information within a period of 10 milliseconds, whereas 100 ms were needed to name the letter – in my opinion, clear evidence of their categorical difference.

[96] On the formal distinction between informational vs. functional theories of scientific representation, as well as on the claim that both are complementary, both contributing to a general understanding of scientific representation, see Chakravartty (2010).

[97] See Bauer & Johnson-Laird (1993) p. 372.

1.2.2   Unavoidable interpretation

In science, both iconic images and textual descriptions of objects or processes are supposed to be observer-neutral, 'objective' and 'realistic,' i.e., they are supposed to depict the objects' or processes' properties as closely and correctly as possible.[98] But the criteria for what counts as most 'realistic' or 'naturalistic' have changed over time, and, even in the same era, the intuitions of different persons might drastically differ on this issue.[99] In the era of woodcuts, the approximate capturing of a plant's outline might pass as the then-optimal kind of representation in print. But once the much finer options of copper engraving or later of lithography and chromolithography became available, these standards and the actors' intuition about what is 'realistic' shifted – repeatedly so with each new technological invention in the printing sector. Furthermore, deviations from this norm of objectivity abound, to which most scientists and technologists would unhesitatingly subscribe today. There are several reasons for this. One is the need to concentrate on important features in representations, thus forcing researchers to skip other properties and observations considered either unimportant, irrelevant or even potentially misleading. If a map were a perfect duplicate of the landscape represented, it would ultimately have to be as large and messy as that landscape and thereby lose its advantageous quality as a condensed, transportable and manipulable re-presentation of reality.[100]

Moreover, even if this possibility of a full mapping were given, no human observer could possibly grasp it all at once. These limitations in representation become most obvious with objects or processes never before seen. Such representational firsts are inextricably bound within visual or textual conventions of the medium used and the pictorial or textual styles in which they are recorded. Witness the artful painting of a kangaroo by the London painter and illustrator George Stubbs (1724–1806) in fig. 7.

Stubbs had earned a reputation as a skilled painter of horses, dogs and other Central European animals. He had also traveled widely himself, making it as far away as Morocco in Northern Africa. So he was a straightforward choice as an illustrator for the botanist Joseph Banks (1742–1820), who had accompanied

---

[98]On the history of this scientific norm of 'objectivity,' see Daston & Galison (1992, 2007), here p. 4 and Hentschel (2008b) for a critical review.

[99]See below p. 211 on Dürer and Weiditz, who were among the first scientific illustrators to draw plants directly from life, whereas Fuchs and his illustrators chose to depict idealized types.

[100]This weird fantasy first appeared in a novel by José Luis Borges, translated into English as *A Universal History of Infamy* (1972). It was later parodied by Umberto Eco (1990).

Fig. 7: A kangaroo sighted along Endeavour River on June 23, 1770 during Cook's first expedition to Australia. Painted by George Stubbs in 1773 on the basis of a reconstruction from the preserved pelt; here reproduced from an engraving based on the painting in John Hawkesworth's *Account of the Voyages of James Cook*, London, 1773, vol. III, pl. 20.

James Cook (1728–79) on his expedition to Australia and had brought back, among other specimens, a stuffed kangaroo. Yet Stubbs did not fully succeed in grasping the to him totally unfamiliar *Gestalt* of this animal. He inadvertently lapsed back into elemental shapes from his animal studies in Europe – and the product on his oil canvas was a strange mouse-headed creature with rabbit ears and legs.[101] Similar episodes could be told about the surprisingly static depictions of apes. Edward Tyson's (1650–1708) pretty helpless rendition of what he called an orang-

---

[101] On the historical background of this episode, and for a comparison against the pencil sketch of a kangaroo by Sydney Parkinson, who had accompanied Cook on his voyage, see Lysaght (1957), www.mezzo-mondo.com/arts/mm/stubbs/stubbs.html and www.mezzo-mondo.com/arts/mm/stubbs/stubbs.html on Stubbs. Cf. the engraving by John Hawkesworth for the published account of Cook's travels in 1773, also www.sciencephoto.com/images/imagePopUpDetails.html?pop=1& id=670069005.

outang (in fact a chimpanzee!) from 1699 was still the model for illustrators one and a half centuries later when Cuvier's *Animal Kingdom* was published in English in 1827–35 and when Thomas Henry Huxley later defended the Darwinian theory of evolution.[102] When unnamed Chinese illustrators working for John Reeves (1774–1856), tea inspector of the *East India Company* in Canton in the 1820s, drew an Asian leopard cat, a ruffled lemur or a lion-haired macaque,[103] they involuntarily but unavoidably mixed their traditional stylized Chinese painting with the techniques of Western natural history illustration to produce hybrid images as strange and fascinating as Stubbs's kangaroo of 1773.

Verbal descriptions of a finding never yet seen will be likewise strongly tainted by the interpretative frame of the observer. It will depend on the comparisons or analogies drawn, and these in turn will depend on what had been seen before, what is familiar, what is expected, etc. Witness early Chinese stargazers, who were documentably the first humans to consciously register and record observations of what we today call 'sunspots.' This term itself bears a kind of interpretation. It implicitly equates the dark blotch on the solar surface with an ink spot on paper, although we normally are not aware that this term is a dead metaphor. The first Chinese record of a sunspot observation, and possibly the earliest recorded human sunspot observation in the world, was made in the year 43 B.C. Initially these surprising irregularities on the Sun's surface, observed through clouds or mist, were called *wu*, meaning crow or black. By 28 B.C., the observers were beginning to call them *hei qi*, roughly meaning black essence or black subtle matter.[104] For instance, the imperial scholar Wang Chong (A.D. 27–97), who was working and recording for the Han imperial dynasty, wrote about observations by the royal astronomer/astrologists: "The scholars [*ru zhe*] say that the Sun has a three-legged crow inside." Searching for an interpretation of this finding, he then asked himself: "But the Sun is just fire in the sky, and is no different from fire on Earth. If terrestrial fire does not generate anything within it, how can the celestial fire produce a crow?" He then consoled himself by assuming that these appearances must indicate some kind of *qi* [subtle matter or pneuma] on the Sun.[105] Later, during the Jin dynasty (A.D. 265–420), the term *hei zi* became standard usage, denoting a black birthmark or mole. Other expressions found

[102] I owe this reference to David Knight (1985) p. 111; cf. also Knight (1977) for further examples; on Huxley, see below p. 155.

[103] For these examples see Magee (2009) pp. 150–3 as well as Whitehead & Edwards (1974).

[104] All preceding quotes from the summary in Ancient Sunspot Records Research Group (1977) p. 347.

[105] Ibid., pp. 347f., quoting from Lun Heng: Chapter Shuo Ri [Discourse on the Sun].

in the Chinese literature include the following: "flying magpie," "black patch," described as being as big as a plum or as big as a jujube, a "black mass like a hen's egg" or "black mores on the Sun rolling about like goose eggs," or, finally, "Light of the Sun churned about and turned into a black cake."[106] Each of these expressions carry along interpretative ballast, illustrating how important it was for these early observers to link surprising features to objects more common to them.

Let me emphasize that this finding is not restricted to the early history of human civilization or even prehistory.[107] Such bold interpretations also occur during later periods of scientific observation. Soon after the telescope was invented, Saturn continued to be recognized as "a planet with a ring" for nearly 50 years. It was alternately interpreted as:

- three distinct globes that almost touch (Galileo 1610)
- sometimes solitary or like an egg, now triple-bodied (Locher 1614)
- like a silk-egg (Gassendi 1633)
- two overlapping rings and a central body (Fontana 1638)
- a handled shape (Hirzgarter 1643, Gassendi 1649, Hevelius)
- one elliptical body with two black spots (Odierna 1657)
- a planet surrounded by a thin flat ring (Huygens anagram 1656 and publication 1659)
- a planet with six satellites (Fabris 1660)
- a kind of ring, but a bit flattened around the center (Ball 1666)

This ambiguity can partly be explained by the insufficient optical resolution of early telescopes. But other factors also played a part in this episode, such as an obsession with specific *Gestalten* (like circular whirls for Cartesians, or Satellite-like companions for Copernicans).[108] To give yet another example from the age of exploration, sea-faring nations like Portugal, Spain and Great Britain sent expeditions to faraway regions such as South America, India and China and trained

---

[106] Ibid., excerpts from the listings on pp. 348–52; cf. here p. 379 on later ambiguities in the representation of sunspots.

[107] For anthropological and evolutionary perspectives on the function of images for humans in the stone-age see, e.g., Jonas (1961), who relabeled human beings as *homo pictor*, and Franz M. Wuketits in Sachs-Hombach (ed. 2009).

[108] See van Helden (1974*a, b*) for references and a detailed analysis, esp. pp. 110f., 119 on the indubitable influence of optical resolution, p. 108 for a quote from a Scheiner student from 1614, who mocked astronomers about this strange apparent metamorphosis of Saturn, and p. 120 on "Galileo and Gassendi [and others] looking for different things."

local artists in the skills of western-style visual documentation. Many of these plants, animals or other subjects were portrayed beautifully and with the greatest care for detail, yet the overall flair remains somewhat foreign to us. Somehow the native traditions of these artists shines through on these plates, either in the coloring, grouping or overall style.[109]

In secs. 12.2 and 13.1 we will meet two other superb examples from 19th-century astronomy. They attempt to resolve an ambiguity in the visual interpretation of fundamentally new findings by likening them to *Gestalt* features in the already-known world. These visual analogies can occasionally work extremely well, but they often led scientists astray.[110]

1.3   Text–image interplay and ekphrasis

Many of the above bifurcations implicitly assume that a sharp opposition exists between visual and nonvisual communication. Semiotically, this is very problematic, since (written or printed) words and (drawn or printed or projected) images as well as three-dimensional objects or models are perceived alike by the same bodily sense, i.e., optical vision by a pair of eyes.

Beyond these physiological points in common, studies on the usage of images for scientific purposes have shown that images of all kinds are very often intimately coupled with textual elements. It starts with figure captions below an illustration and ends with labels or commentary carefully superimposed on these images by the scientist or his graphics expert assistants.[111] W. J. T. Mitchell has tried to come to grips with this finding by claiming that in science, nearly "all media are mixed media" and coining two new terms that still deserve broader recognition. He speaks of 'imagetext' for sources, frequently found in science, in which "pictures and text interact symbiotically," whereas he speaks of 'image/text' whenever this symbiosis is broken or disturbed, i.e., when gaps or clashes between both of these components appear.[112]   Philipp Prodger and Jonathan Smith have applied this

---

[109] Good examples for these cultural hybrids are found in Archer (1962), Whitehead & Edwards (1974) and Mutis (1954ff.).

[110] One fascinating example exhibiting both facets is Paul Ehrlich's use of visual analogies in cell physiology and immunology: see Cambrosio et al. (1993) and Müller & Fangerau (2010); in technology, interesting, sometimes playful examples of visual analogies by Fritz Kahn (1888–1968) are discussed by Rolf F. Nohr in Liebsch & Mößner (eds. 2012) pp. 157–68.

[111] For case studies of this interplay in biology, see Maienschein (1991), Cambrosio et al. (1993) or Schlich (2000); cf. Knight (1993) or Roche (1993) in Mazzolini (ed. 1993) on chemistry and physics.

[112] See Mitchell (1994) p. 5; cf. also Smith (2006) pp. 39f.

approach to Darwinian visual culture.[113]

From a more cognitively oriented point of view, David Gooding has also claimed that "word–image–object hybrids," as they occur so often in science and technology, "have a cognitive function and a socio-cognitive one: they integrate different types of knowledge and expertise gained in different sensory modalities, and they support movement between the personal domain of "mental" representations and the social domain of public tokens of meaning."[114]

1.3.1 'Pictures without words' and ekphrasis

Yet another form of interplay has only one of the two partners present. The other is invoked. In a 'picture without words,' one consciously abstains from any text and lets the image "speak for itself." Usually, a very "telling" picture is chosen. (For a critique of this metaphor, see below.) In the other extreme, words alone are used to evoke an image powerfully or to describe a process so vividly that the listener or reader gets a concrete image of what is described. As a literary strategy of visualization, this is called ekphrasis (from the Greek term $\epsilon\kappa - \phi\rho\alpha\sigma\iota\varsigma$ for "out" and "speak," i.e., literally, to proclaim or call an inanimate object by name; more freely translated as vivid "description," in Latin *descriptio*). The 4th-century Byzantine scholar and bishop Nicholas of Myra said that a good ekphrasis turns the listener into an eye-witness. Art historian Gottfried Boehm traces its various forms from the old art of describing (*Beschreibungskunst*) up to today's 'art description' (*Kunstbeschreibung*).[115] When the Saxon humanist Janus Cornarius (1500–58) published his own commentary on Pedanius Dioscorides's antique *De materia medica*, he still objected to the use of images, even though several transcriptions of the early *pharmacopeia*, such as the one made in Constantinople in A.D. 512/513 (now preserved in Vienna), had been lavishly illustrated. The plant "emblemata" Cornarius gave in his published commentary were devoid of woodcuts or engravings. They were purely verbal descriptions of the forms, scents and effects of plants.[116] Medieval astronomical schemata and tree diagrams, Renaissance treatises on architecture and even technological literature in the modern period have also been analyzed with regard to their relations between text and

---

[113]See Prodger (1998) and Smith (2006).

[114]Gooding (2004*b*) p. 581; cf. Dupré (2008) on images in theory-dominated contexts.

[115]See, e.g., Boehm (1995). Cf., e.g., the very different German and English Wikipedia entries on ekphrasis for further literature.

[116]On this example, see Kusukawa (2012) pp. 126f. and further sources quoted there.

image(s).[117] Qualitatively, printing technologies such as woodcut and engraving make it much harder to achieve this integration, since engraved plates were usually printed separately and bound as a frontispiece or in a gallery at the end of the volume, if not into separate plate volumes. Woodcuts could at least be inserted into the printing block just like movable type. But, practically, they often blocked half or whole pages because setting floating text was complicated and increased the necessity for hyphenation, etc. Such plates were expensive anyway, so pictorial material was usually very limited. According to the Dresden historian of technology Karl-Eugen Kurrer, it was only with the invention of lithography that the "leaden armor of algebraic scholasticism" was dismantled, leading to a "disarmament process favoring the image." It would ultimately lead to an inversion of their relations, with a text now dominated by its images.[118]

Later in that same century, the introduction of cheaper and more enduring forms of wood and steel engravings made a tighter interplay of text and image possible – first in pioneering science journals with a higher density of illustrations such as *Nature*, edited for half a century by the spectroscopist Joseph Norman Lockyer (1836–1920). Perhaps also because of this well-balanced interplay of words and images, *Nature* has remained a leading science journal and is still very heavily illustrated, and since the 1990s even mostly in color.[119] According to recent estimates, today about one-third of the available space on the pages of leading scientific journals is devoted to nontextual material, be it glossy photographs, plots of functional dependencies of variables or schematic diagrams.[120]

1.3.2 Categorical differences between text and image

Taxonomically, Mitchell has suggested a differentiation between images into graphic, optical, perceptual, mental and verbal images. He thus explicitly includes purely verbal descriptions in the "family tree of images"[121] – a problematic move that I find counter-intuitive and not helpful toward our goal of understanding the image-text interplay. Of course it is possible to evoke a "mental" image of something merely by speaking about it, but that justifies neither the nonsensical

---

[117]See, e.g., Kurrer (1996), Müller (2008) and Lima (2014); cf. also Remmert in Kusukawa & Maclean (eds. 2006), esp. on frontispieces.

[118]All quotes from Kurrer (1996) p. 74, who, on pp. 90–8, also offers representative examples from the technical literature 1833–1925.

[119]On Lockyer and the early history of the journal see, e.g., Meadows (1972) pp. 26ff.; for an analysis of more recent images in *Nature*, see Kemp (1999).

[120]For the estimate above see, e.g., Meadows (1991) p. 23, Kemp (2000) and Bredekamp (2005c).

[121]Mitchell (1984) p. 505.

classification of the "verbal image" as a kind of image nor the derogatory dismissal of such an image as a "mere derivation" of a graphic or perceptual image.[122] Misleading metaphors such as "telling" or "eloquent image" also don't help us further – it is not the image *per se* that tells us something, but rather those who interpret it by reacting to visual input with some form of verbal output. In practice, however, both visual and verbal components are typically used side by side, or rather, they are intertwined. In scientific articles and books, photographs or diagrams are usually accompanied by a caption that is read as the image above it is studied. Only by taking in both inputs together does the user get sufficient information for a scientific evaluation of the meaning, importance and epistemological impact of an image or diagram with regard to the author's argument.[123] Subtle changes in the caption can modify, indeed invert, any "message" that an image supposedly "sends," and they can also redirect the attention to details that would otherwise go unnoticed.

These two forms of mutual rapprochement between words and images notwithstanding, semiotic and information-theoretical analyses have also revealed categorical differences that are important to bear in mind. Assuming the widely held dual-code model, according to which mental representations are either verbal or pictorial,[124] at least normally or typically:[125]

- Graphs or images provide fine-grained, detailed, often also spatial information, whereas verbal descriptions are selective, often minimalist.
- Images only represent singular entities, not whole classes. If images or parts of images are supposed to represent whole classes, then this has to be clarified, implicitly by setting them in stark contrast to another class within the image, or explicitly by an explanatory legend.
- (Single) images are static and thus tend to represent synchronous states, not processes. If the latter are the aim, either longer sequences of images have to be presented (further along the same line) or a cinematographic film.

---

[122]Bal (2003) strongly criticizes visual essentialism as a prejudice often in-built in visual studies; see also the intense debate about this paper in the same issue of *Journal of Visual Culture* 2.2 (2003).

[123]On the systematic evaluation of pictorial illustrations, cf. Goldsmith (1984) and Dragga (1992). They suggest a heuristic of checking a 12-cell matrix based on syntactic, semantic and pragmatic criteria, ranging from unity, location and emphasis to image–text relations and parallels; Luc Pauwels in Margolis & Pauwels (eds. 2011) p. 5 focuses on the origin and nature of visuals, research focus and design, format, and purpose.

[124]For a clear exposition but unsuccessful critique of this dualism, see Pylyshyn (1973) pp. 3ff., 22, where an attempt is made to reinterpret (mental) images as quasi-propositional data structures.

[125]On the following, see, e.g., Harms (ed. 1990), Boehm (1994), Frank (2006) p. 81 and Kulvicki (2010) and further references there.

- (Wordless) images do not contain any propositions as such. Any propositional content that is linked to an image first has to be construed by relating striking image-content markers. If an artist or other creative illustrator wants to make a "statement" with an image, this is done indirectly, tentatively, and often supported by further explanation or commentary, to get the message across more clearly.

This last point motivates many political caricaturists and poster artists, such as Klaus Staeck (*1938),[126] to use image–text combinations to express their statement less ambiguously. In fig. 8, for instance, an unaltered visual "quotation" of Dürer's well-known portrait of his elderly mother is taken out of its context and presented to the viewer with the question: "Would you rent out a room to this lady?" The "message" of this image only unfolds by several interpretative stages. First, one looks at the rather haggard lady and feels inclined to reply in the negative. Next, one realizes that this is not just any old lady, but happens to be Dürer's mother. This leads to the insight that the spontaneous answer initially given may have been too rash and hasty. The end of this thought process might be deep insight into the mechanisms of prejudicial rejection and – possibly – a change in future behavior with respect to needy people. This example also shows that the interpretative sequence is long, tenuous and indirect, presupposing sufficient knowledge about the artist Albrecht Dürer (cf. here pp. 174ff.) and his oeuvre to be able to recognize his anagram, none of which can normally be assumed. It is no accident that this poster was designed in 1971, the quincentenary of Dürer's birth in 1471. Staeck's poster was displayed in Nuremberg, Dürer's native town, where such background knowledge could be assumed to be fairly widespread, at least during that anniversary. In another context, this poster would not "work."

1.3.3 Epistemic differences between images and texts

The differences between texts and images listed above also help us understand their quite different typical usage scenarios. In setting up a thorough classificatory system for all plants on Earth, Carl Linnaeus (or Linné 1707–78) mostly worked with long lists of systematic descriptions and a carefully chosen terminology based on a numeric classifier – here the number of sexual organs in a plant. Images are surprisingly sparse in his extensive publications in natural history. Any such Linnean image would only depict a representative of a particular class of plants; and one of the rare color plates directly linked to his oeuvre (here reproduced

---

[126] On Staeck's career and oeuvre see www.klaus-staeck.de/biografie/index.html.

Fig. 8: Poster by Klaus Staeck, 1971, asking the viewer: "Would you rent out a room to this woman?" Courtesy of Prof. Klaus Staeck, Heidelberg.

as color pl. IV) summarizes the system of classification for which he became world-famous.

In other fields such as geography, maps had been in use since antiquity. They rely on an intricate interplay between a coordinate grid (be it latitude and longitude or some other kind of 2D coordination of surfaces in space) and contour lines (denoting regions, countries, coastlines or sociopolitical borders), written-out labels and encoded symbols (denoting lakes, streets, towns, special buildings or whatever).[127] Early-modern times saw a major upswing in different cartographic mapping systems such as Mercator's projection of 1569/70. They are essentially different mathematical solutions to the intricate problem of projecting a curved

---

[127] On the history of cartography and for good examples of early maps, see, e.g., Talbert (ed. 2008, 2012); Woodward (1975), and Woodward (ed. 2007) on early-modern map making techniques; on the evolution of cartographic symbols, cf. Blakemore & Harley (1980).

3D world onto a flat 2D map.[128] The astronomer Edmund Halley (1656–1742) was probably the first to transfer the Cartesian coordinate approach into natural history by including measurement data on wind direction and intensity of the trade winds and monsoons (see fig. 9). In another plate, he incorporated local magnetic declinations into a world map, thus turning geographical maps into meteorological and geophysical charts.[129]

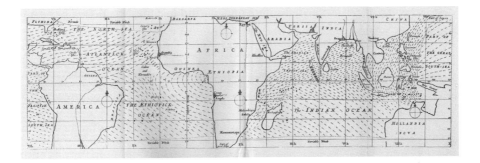

Fig. 9: Halley's first meteorological map, charting the directions of the trade winds and monsoons. This fold-out plate from Halley (1686) exhibits an intricate interplay between iconic, symbolic and indexical signs (for wind direction) with textual elements such as country and region labels.

Halley carried the symbolic decoding of numerical data even further by plotting barometric readings against elevation above sea-level. This visualization of his measurements in a graph led him to realize that a hyperbola could be fitted to his data – graphic display thus helped find a formula.[130] In the late 17th and early 18th centuries, such graphic analysis of data remained the exception, though.[131] It started come into common use only after Lambert applied it, initially for barometric data, but soon also for the visualization of other strings of data taken over long time series.[132] By plotting these discrete data as points or columns, with the column's length proportional to the measurement value, a quasi-continuous series of points or column-ends results that can be "read" as a curve illustrating

---

[128]See, e.g., Monmonier (1996), Pickles (2004), and Michael Rottmann in I. Reichle (ed. 2007).

[129]Cf. http://libweb5.princeton.edu/visual_materials/maps/websites/thematic-maps/quantitative/meteorology/meteorology.html (accessed Sept. 6, 2013) for further commentary.

[130]On this episode, finalized by Halley's publication of his findings in 1686, see Beninger & Robyn (1978) p. 2.

[131]Tilling (1975) p. 194 concludes an exhaustive search of 18th-century texts by stating that "the use of experimental graphs, never became commonplace in that age."

[132]See below p. 121 on Lambert as an early protagonist of this technique of graphical analysis.

the variation of this quantity as a function of time.

It takes longer to infer the water level of the river Oder in a given week out of Johann Albert Eytelwein's (1764–1848) early graph, plotting its levels between 1782 and 1791 (see fig. 10), than it does from a list.[133] The exact location along the x-axis (divided up into ten-day intervals) is much harder to pinpoint than it is to infer it from a tabular listing, on the basis of which this graph had been prepared as a kind of visual summary. This graph is, after all, supposed to give a good overview of relative and absolute minima and maxima, the periodicity, and irregularities, which would be hard to infer from the table.

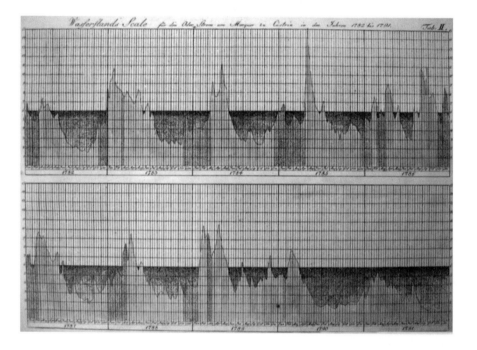

Fig. 10: Graph tracing the water levels of the river Oder 1782–91, from Eytelwein (1798) plate II. The y-axis indicates water levels measured in Rhineland feet, and the x-axis defines years, months and ten-day intervals, the smallest time unit.

Such graphs of quantitative data were still quite new and were used much more hesitantly than nowadays, where making a plot is the most natural thing to do. Around 1800, William F. Herschel (1738–1822) plotted his measurements

---

[133]This example comes from Kathryn Olesko's marvelous analysis of Prussian technical education in the late 18th century. See Olesko (2009) p. 31 and Olesko (in prep.).

of thermometric heat taken along the visible spectrum and beyond its red end. By comparing this curve with one illustrating the intensity of sunlight gauged photometrically, he was able to infer the presence of a different type of radiation in the space apparently beyond the range of visible light, a species that he then reluctantly called "invisible light" or "dark heat."[134]

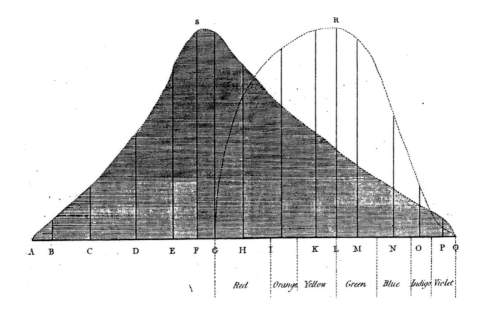

Fig. 11: William Herschel comparing visual (curve R) and thermometric (curve S) intensity in the spectrum. From W. Herschel (1800*d*) pl. xx, facing p. 538.

At roughly the same time, in 1786, the Scottish engineer and economist William Playfair (1759–1823) also started to use innovative diagrams, line graphs and bar charts in order to illustrate basic facts, mostly from the realm of economics and 'political arithmetic' (what would nowadays be called the social sciences). By 1801, he had enriched this inventory of graphic representations by pie charts to represent percentages and proportional relations. Several of these graphic displays have become so standard today in commerce and trade that it is hard to realize how brilliantly innovative those graphical means were at the time, to represent complex issues in the simplest terms.

Playfair's graphic talent has been often and very justly celebrated by experts

---

[134]See W. Herschel (1800*a–d*). See Hentschel (2002*a*) pp. 61–3, (2007) for further analysis of this.

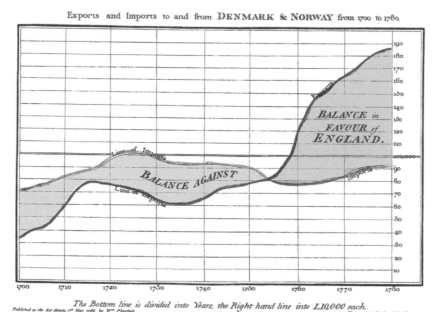

Fig. 12: William Playfair's temporal graph of English exports and imports. From Playfair (1786*a*) unnumbered plate, re-appearing as a more compressed pl. 12a in the 3rd ed. of 1801.

as exemplary.[135] In his *Commercial and Political Atlas* (1786), He embellished such a dry topic as the balance of trade from 1700 to 1780 between England, on the one hand, and Denmark and Norway, on the other, by a color-coded graph. The trade imbalance for Denmark and Norway is rendered in red (in the left half of fig. 12), and for England in yellow (in the right half of the image). The time scale at the bottom immediately portrays the temporal development of the surplus, with the intersection point between the export and import curves in this trade (in 1754) clearly shown. Playfair's diagram also shows the rising absolute rates along the vertical axis in units of £ 10,000.[136]

Such clear and innovative graphs became more frequent in scientific papers only in the 19th century. The dropping cost of plates with the advent of lithography and, towards the end of that century, photomechanical printing techniques,

---

[135]Most importantly in Tufte's various manuals on good graphic design, e.g., Tufte (1983) pp. 32ff., 45ff. passim; cf. also Beniger & Robyn (1978) pp. 3f.

[136]As a preview to chaps. 3 and 4, I would like to point out here that William Playfair's own background was in engineering. 1777–81 he was assistant and draftsman to James Watt in the Birmingham steam engine factory; William's brother John was a geologist, another field that was just on the verge of becoming a visual culture, so William was an ideal transmitter.

made this development possible: "the number of printed pictures between 1801 and 1900 was probably considerably larger than the total number of printed pictures before 1801."[137] According to a condensed survey of the use of graphics in science, the earliest examples in the 17th and 18th centuries (such as Halley's world map showing lines of magnetic declination, 1701) present data in spatial distributions. They are followed by visualizations of discrete quantitative comparisons (such as Playfair's charts of trade balances in 1786) in the late 18th and early 19th centuries. These, in turn, are followed by representations of continuous distributions of measured variables (such as Herschel's temperature or Eytelwein's water-level readings) throughout the 19th century. Finally come multivariational distributions and complex correlations in the 20th century.[138] These last examples provide a good transition to another aspect that has to be covered in this introductory chapter: visual rhetoric.

1.4 Visual rhetoric – arguments with images and models

An image may include decipherable words as a direct message to the spectator; yet many images may appeal to the viewer or repel him or her, excite curiosity, happiness or anger. Images built into texts about science or technology are more than "mere" illustration. They often function as part of the argument. Thus, it is fair to speak of 'visual rhetoric' in analogy to the much older discipline of 'rhetoric,' one of the basic trivium in the seven *artes liberales*. These were exclusively bound to textual forms of argumentation in scholastic disputes at universities of the Middle Ages, or as more formalized ways to compose a speech, a letter or an appeal. Visual rhetoric, however, deals with ways to argue by means of diagrams, icons or models. Aside from these contexts of persuasion and reinforcement of beliefs, "visual rhetoric is concerned with understanding how images communicate, how they function in a social and cultural environment, and how they embody meaning."[139]

---

[137] Ivins (1973) p. 94; for more on this interplay between forms of representation and printing processes, see, e.g., Jussim (1974), Gascoigne (1986), Meadows (1991) and Hentschel (2002a) chap. 4.

[138] See Beniger & Robyn (1978); cf. also Shields (1937), Tilling (1975), Meadows (1991), Roche (1993) and Hankins (1999) more generally on the early history of graphs.

[139] Terence Wright in Margolis & Pauwels (eds. 2011) p. 316, substantiated with examples from journalistic photography; cf. also Cambrosio et al. (2005) on Linus Pauling's "arguing with images" in his theory of antibody formation, here sec. 6.2 on his use of 3D models in chemistry, and many other examples from science, technology and medicine given in this book. On the rhetorics of 3D models: Chadarevian & Hopwood (eds. 2004), Fischer (1986) and Sattelmacher (2013).

William Herschel's graph (fig. 11) is a case in point. The lighter of the two curves indicates the dependence of visual intensity on various color ranges of the visible spectrum labeled (below the *x*-axis) as the seven Newtonian colors, ranging from violet at the far right end to red (around the middle of the figure). The intersection of this light curve with the *x*-axis at the very end of the red spectrum range "means" that there is nothing visible beyond this point. However, when heat measurements are taken with a fine thermometer at carefully positioned points in the various color regions of the spectrum, the maximum thermometer reading – here indicated at S, as opposed to the maximum reading, at R, of the visual intensity as registered by the human eye – is clearly outside the optical part of the spectrum in the region henceforth labeled infrared, i.e., beyond the red end of the spectrum. As one retreats from this end into the infrared, the thermometer continues to indicate the presence of some ominous heat rays obviously not coupled with anything visible. Herschel's graph shows this finding in the clearest possible terms, in fact, much better than his accompanying text, with which he supported the tentative conclusion reached.

> Now when these [two curves] are compared, it appears that those who would have the rays of heat do also the office of light, must be obliged to maintain the following arbitrary and revolting propositions; *viz*, that a set of rays conveying heat, should all at once, in a certain part of the spectrum, begin to give a small degree of light; that this newly acquired power of illumination should increase, while the power of heating is on the decline; that when the illuminating principle is come to a maximum, it should in its turn also, decline very rapidly, and vanish at the same time with the power of heating. How can effects that are so opposite be ascribed to the same cause? First of all, heat without light; next to this, decreasing heat but increasing light, then again, decreasing heat and decreasing light.[140]

The contrasting behaviors of visible radiation and heat radiation are "translated" graphically in this famous image as a contrast between the light region to the right and the heavily darkened region leftwards of it. Overlapping and nonoverlapping regions are easily discernible, and the claim that there is something out there in the solar radiation beyond visible light is brought out most unmistakably. The multiple reproductions of this image in other articles and textbooks ever since is a good measure of the success of this rhetoric.[141]

---

[140] Herschel (1800*d*) p. 508. In Hentschel (2007) pp. 362f., this interesting verbal argument is classified as a 'counter-tendency argument' – *Gegenläufigkeitsargument*. It is, in fact, a verbalization of something more easily apprehended in the graph than in tables or lists.

[141] For some examples and a detailed discussion of the controversial discussion following Herschel's paper, see Hentschel (2002*a*) pp. 63f., (2007) sec. 5.1.

Herschel's diagram is a good example of an image performing the feat of making the invisible visible. Robert Koch (1843–1910) had a similar task ahead of him, when he tried to convince himself and his contemporaries of the existence of bacteria.[142] By 1867, he had discovered "spore" formation in the corpses of humans and animals, victims of anthrax. He hypothesized the existence of a causative agent, termed *bacillus anthracis*. Experimentally, he could show that anthrax bacteria normally remained dormant, but under specific conditions of heat, nutrition, etc., he could activate them and they then invariably caused the lethal disease. To examine and isolate this causative agent, Koch dry-fixed bacterial cultures onto glass slides, dyed them to increase the contrast between the bacteria and their context, and then observed them through a high-quality Zeiss microscope. In 1876, Koch published colored lithographs of his colored drawings of these microorganisms (cf. color pl. XIII). Although a skilled draftsman, Koch realized the inherent limitation of these images as subjective drawings. So he started to experiment with photographing his microscope images. Up to this point, only he himself had looked through his microscope ocular. From the company *Seibert & Krafft*, he obtained a microphotographic camera that allowed him to photograph the anthrax bacillae. In a subsequent paper of 1877, Koch published 24 of these photographs on three plates, photomechanically printed as Albertype *Photogramme* (in German: *Lichtdruck*).[143] These photographs became Koch's crown witness in his argument for the existence of such organisms: The images documented all stages of the bacteria's development, which aided other microbiologists enormously in their microscopic identification. In order to increase the persuasiveness of these photographs, Koch did not remove traces of the photographic processes of chemical development and optical enlargement, such as scratches on the paper by the clamps during the photo-bath, dust on the lens, etc. All temptations to retouch the images to improve their sharpness, contrast and to get rid of disturbing features were stoically withstood. In fact, as Koch argued in a follow-up article five years later containing no less than 84 photographic images: "photography is not merely an illustration, but a piece of evidence – a kind of document [...] that does not permit the least doubt as to its authenticity."[144] Koch's rhetoric of mechanical

[142]On the following: Koch (1876, 1877, 1881); cf., e.g., Münch & Biel (1998), Hentze (2000), Schlich (2000), Rheinberger et al. (eds. 1997) pp. 165–90, Gradmann (2009) and refs. there.

[143]See the small inset on the lower right corner of color pl. XIII; on Koch's interactions with the chromolithographer Albert Schuetze (1827–1908) and with the engraver and draughtsman Wilhelm Grohmann (1835–1918), see Hentschel (2002*a*) pp. 153, 220; cf. also their entries in the Stuttgart University Database of Scientific Illustrators (DSI).

[144]Koch (1881) p. 14: "nicht allein eine Illustration, sondern in erster Linie ein Beweisstück, [...] ein Document [...], an dessen Glaubwürdigkeit auch nicht der geringste Zweifel haften darf.."

objectivity went so far as to exclude photographs as legitimate evidence in cases where any form of human intervention is traceable. When it turned out, in 1906, that some kind of dye coloring was indispensable to bring out traces of piroplasmae in a photograph, Koch reverted back to colored microscopic drawings by himself and a prominent Berlin artist, the painter and sculptor Max Landsberg (1850–1906).[145] Koch's visual rhetoric paid off: The existence of bacteria – which so far had been a matter of some controversy – became accepted and Koch was widely hailed as a pioneer of bacteriology.[146]

Other types of images perform rhetorical services of a different kind to such use by researchers of supposedly "similar" iconic images as epistemic "evidence" for the alleged existence of hitherto unknown and unseen objects. Otto Neurath's simplified graphic language *Isotype* in posters visualizes complicated aspects of population development and other statistical insights in the social sciences in terms immediately understandable to the layman. Marey's chronophotography decomposes the moving human body into a time series of stills freezing otherwise imperceptibly rapid motions.[147] If the additional temporal dimension does not suffice, a third spatial dimension will turn flat displays into highly compelling demonstration objects: Remember the impact that touchable and freely turnable terrestrial globes had on you when you were learning that the Earth is actually spherical and not flat. The cultural history of terrestrial and stellar globes is another superb example of a scopic domain and a marked component of the visual cultures of cartography also comprising geography, astronomy and planetology.[148]

1.4.1 Visual analogies

A visual argument of a quite different sort is the visual analogy. I define it as an effort to systematically map the relations between image elements from one image, the familiar base, onto another image, the target.[149] The point of a visual analogy

---

[145]On this episode: Bredekamp (2011) p. 42 and Bredekamp & Bruns in Maar & Burda (eds. 2004) p. 373; on Landsberg, cf. again the Stuttgart database DSI and the *Allgemeines Künstler-Lexikon* (AKL). On 'mechanical objectivity' as a typical late 19th-century argumentative stance, see Daston & Galison (2007) and Hentschel (2008*b*).

[146]For a parallel to the work of Edgar Crookshank (1858–1928), whose textbook *Photography of Bacteria* (1887) included 68 photographs reproduced as autotype, see Keller et al. (eds. 2009) pp. 42f.

[147]See Oestermeier (2000) p. 67 or Nikolow in Hessler (ed. 2006) on Neurath, or Douard (1995) on Marey (cf. here pp. 249ff.).

[148]On terrestrial and stellar globes, whose production took off in early-modern times, see Mokre (2008); on lunar and Martian globes, cf. Blunck (1995/96, 1999) and here pp. 304f.

[149]On analogies in general, and on visual analogies in particular, see K. Hentschel's introduction in Hentschel (ed. 2010). On Maxwell's and von Helmholtz's use of hydrodynamic analogies, see Cat

is thus *not* any single transfer of features or properties, but a mapping of a whole web of relations between such single nodes. When James Clerk Maxwell used the hydrodynamic visual analogy of vortices to illustrate the interactions between electric and magnetic fields (see fig. 13), the point was to model this complex interaction, not to claim that magnetic fields "really" are vortices.[150]

The basic idea behind Maxwell's analogy between electromagnetic fields and hydrodynamics was to find structural similarities in the relations between electric charges, currents, fields and magnetic fields on the one hand, and hydrodynamic vortices in an idealized friction-free, inviscid and incompressible fluid on the other. Magnetic field lines were thus reinterpreted as the axes of more or less circular vortices in such an idealized fluid. This mapping of magnetic fields into vortices was highly intuitive. Magnetic fields such as those around a needle or a compass are traceable as closed, more or less circular field lines, for instance by Michael Faraday's technique of scattering fine iron filings on a sheet of paper wrapped around a magnet. The effect of a magnetic field on a magnetized body was in a certain sense likewise circular. In the Ørsted effect, for instance, a magnetized needle attempts to turn in response to a magnetic field induced by a current that is switched on or off. It was a pretty obvious and relatively uncontroversial step in this physical modeling to represent magnetic fields as quasi-circular vortices. Depending on the direction of the current or the direction of the north and south poles of the magnet generating the field, we can imagine clockwise or anticlockwise orientations, symbolized as − and + in the right half of fig. 13.

As a next step in the development of this intuitive model we want to understand the propagation of magnetic fields, i.e., we want to make the model demonstrate other facets of the phenomena. There comes the rub. Just look at one horizontal row of such magnetic vortices in the left half of fig. 13 without the small balls in between the bigger magnetic vortices. Imagine that we start to rotate the uppermost vortex at the top right; then the one situated left of it would start to rotate, too (friction is idealized away), but – alas – the adjacent one would have to start rotating in the counter sense. Globally, the individual effects of two adjacent vortices of the magnetic field would thus cancel each other out, contrary to what we observe in the propagation of magnetic fields.

(2001), furthermore M. Heidelberger in Hentschel (ed. 2010).

[150]On the following, see Maxwell (1861/62). Cf. also Kargon (1969) on the context, Chalmers (1986), Siegel (1991) on the physics, Nersessian (2008) pp. 21–60, 135–50, Davies, Nersessian & Goel (2005) and Cat (2014) for plausible reconstructions of the genesis of Maxwell's famous model.

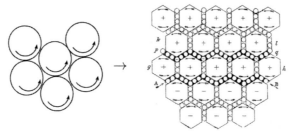

Fig. 13: Left: A – too simple – model of the magnetic field as an ensemble of circular vortices. Right: Maxwell's hydrodynamic model of the aether in 1861. From Maxwell (1861/62*b*) pl. VIII.

In order to get adjacent vortices to turn in the same direction, i.e., in order to model the propagation of magnetic fields properly, Maxwell was forced to introduce a new element in his hydrodynamic model: an additional layer of friction balls or "idle wheels," i.e., small spheres situated between all the magnetic vortices. In order to make the space-filling optimal, Maxwell represented these vortices not as circles but as hexagons, representing the 2D cross-sections of 3D dodecahedrons as a good approximation to spatial spheres. Maxwell's inspiration to insert the little "idle wheels" had come from the realm of mechanical engineering, as becomes clear from Maxwell's explicit reference to epicyclic trains or to Siemens's governor for steam engines.

With this simple mechanical analogy, Maxwell could now assume that all hexagonal magnetic vortices could spin in the same direction without friction, in agreement with what we observe in the propagation of fields. This amplified model also helped explain several other features already noted by Faraday but never properly understood.

First, the combined motion of these little "idle wheels" could be physically interpreted as equivalent to an electric current in conducting media, thus explaining an odd finding of Faraday that had given rise to one of Maxwell's famous equations for electromagnetism: Every current gives rise to a magnetic field, or, symbolically, in one of Maxwell's equations, $\nabla \times \vec{B} \sim \vec{j} + \partial\vec{E}/\partial t$.

Second, the two sides above and below the current (let's say, the central third row *gh* in fig. 13) rotate in opposite directions. The magnetic field changes orientation. Above the current, all vortices are oriented counterclockwise, symbolized as positive; below the current, all the vortices run clockwise, symbolized as negative. (Note that the engraver made mistakes with some of the arrows on the second-to-last vortex row at the bottom, corrected only in later editions!)

Third, all changes in the magnetic field (such as our thought experiment in which we started to rotate one of the marginal vortices and observed what happens in due course) cause changes in the electric flow, symbolized by the motion of the "idle wheels." Thus, another of Maxwell's equations could be intuitively understood by means of this model: $\nabla \times \vec{E} \sim -\partial \vec{B}/\partial t$.

Finally, in nonconducting media, where such a chain-like transport of charges was impossible, one had to assume that the idle wheels just moved from one end of the hexagon to the other, thus amassing at one side of the hexagons and deforming them slightly. Maxwell reinterpreted this as analogous to the build-up of tension in the medium, i.e., as the 'electromotive force,' in his parlance. In fact, without the background of Maxwell's idiosyncratic modeling of electromagnetic phenomena, concepts like the so-called Maxwellian 'displacement current' must have remained utterly mysterious. It is consequently wrong to try to confine Maxwellian electrodynamics to his set of equations, as the German physicist Heinrich Hertz (1857–94) later suggested.

These four conclusions nicely illustrate the crucial point that all such models are only useful if they yield a surplus of valuable inferences *beyond* what had been put into them at the outset. In order to extract this surplus value, one has to play with the model, to see what kind of implications come with it, often unintentionally. Sometimes it proves exceedingly helpful in gaining a new perspective on objects or processes apparently already known inside out. That is why Maxwell and other creative physicists from the 19th and early 20th centuries worked with such models. Often they used several, sometimes even incompatible, models side by side, alternating between them, in order to compare their usefulness or "likeness" to the phenomenon at hand.[151] For Maxwell and Maxwellians, an analogy was not an unidirectional mapping, but a mutual mapping of source and target domain, each illustrating and illuminating the other in this switching back and forth:

> By a physical analogy I mean that partial similarity between the laws of one science and those of another which makes each of them illustrate the other.[152]

On a more general level, Maxwell recommended tolerance for different modes of expressing physical insights and flexibility in converting between them. In his

---

[151]One of the deepest discussions on instrumental use versus ontological belief in such models is provided by George Francis FitzGerald (1851–1901), published posthumously in his *Collected Writings*: see FitzGerald (1902).

[152]Maxwell (1855/56*b*) p. 156.

address to the Mathematics and Physics Sections at the annual meeting of the *British Association for the Advancement of Science* in 1870, he made the appeal:

> For the sake of persons of different types, scientific truth should be presented in different forms, and should be regarded as equally scientific, whether it appears in the robust form and the vivid coloring of a physical illustration, or in the tenuity and paleness of a symbolic expression.[153]

When Galileo observed the surface of the Moon with a new telescope of his own making, which had a magnification of about 16×, later even about 30×, he immediately drew a visual analogy between what he was able to observe with what he knew from observations of terrestrial landscapes. Volcanic craters seen from a neighboring hill in strongly skewed evening light exhibit one brightly illuminated rim, whereas the other flank is enveloped in deep shade on the side pointing away from the light source. What Galileo saw on the Moon's surface reminded him of this pattern, so he suspected the Moon's surface, too, was riddled with craters. Continuing to develop this visual analogy, he even tried to infer the height of the largest of these craters from the length of the shadows it cast.[154] That Galileo was the first to draw this visual analogy is rooted not only in the superiority of his telescope. He had been trained in perspectival drawing at the *Accademia delle Arti del Disegno*, where his early mentor Ostilio Ricci (1540–1603) gave lessons in applied geometry.[155] This prior training in perspectival seeing, including tricky effects of foreshortening and shading, preconditioned Galileo to drawing this analogy quasi-automatically. Other observers of the lunar surface with instruments of similar quality at roughly the same time, such as Thomas Harriot (1560–1621) in England, drew these same surface features in some flat fashion, not yet able to "see" their three-dimensional depth. Once the *Gestalt* of a lunar crater is known, recognizing these craters in telescopic images of the Moon becomes trivial – indeed unavoidable. In Galileo's day, however, it was a far from trivial achievement to draw this analogy and to see this "strange spottednesse in the moone" (Harriot's own terms) as lunar craters. Conversely, presenting such intuitive visual analogies in the form of woodcuts, together with very detailed descriptions of his findings, certainly helped Galileo to convince his contemporaries of the plausibility of his bold claims. This is beautifully illustrated by a comparison of Harriot's drawing from July 26, 1609 with the one from July 17, 1610, which shows the lunar craters very much in the fashion of Galileo's published woodcuts. Harriot's later drawing

---

[153] Maxwell (1870*b*) p. 220, commented upon in Root-Bernstein (1999) p. 289.
[154] See Galileo (1610) fig. 6. Cf. Winkler & van Helden (1992) and Spranzi (2004).
[155] On this Ricci–Galileo link: Settle (1971), Edgerton (1984) and Kemp (1990) pp. 93ff.

was done after he had access to Galileo's publication.[156] Galileo's *Sidereal Messenger* ultimately revolutionized the world-view.[157]

Visual analogies and visual rhetorics are by no means confined to the early-modern period with Galileo and his Jesuit or Aristotelian opponents. Woodcuts and copper engravings were expensive and thus remained a relatively rare feature of rather expensive publications. The development of cheaper printing processes led to a dramatic jump in the number of visual representations appearing in scientific publications. For instance, wood and steel engraving, allowing higher print runs than copper plate engravings, or more efficient recording processes, such as photography in the mid-19th century or CCD cameras in the 20th century.[158] Today most science magazines have glossy color illustrations, whereas a hundred years ago their typeset galleys were still veritable lead deserts with much fewer illustration blocks or plates.[159]

In the heyday of photography, this new medium was especially praised as the "veritable retina of the scientist."[160] At least the public at large was not aware of the extent of manipulability of photographs by means of superpositioning and retouching.[161] Another promise of early photography, namely the vague hope that soon "every man [could become] his own printer and publisher" and that this new art of 'photogenic drawing' would "enable poor authors to make facsimiles of their works in their own handwriting,"[162] could not be fulfilled at the time, either. Photography long remained a relatively expensive affair unsuitable for higher runs of prints and still fairly limited in the quality achievable. The traditional ways of printing woodcuts, engravings or lithographs – all ultimately based on hand-drawn templates – still remained important and often preferred alternatives in many branches of science.

A new wave of distrust in photographs as reliable evidence set in after the

---

[156] On Galileo as eye-opener and Harriot's "borrowed perceptions" thereafter, see Bloom (1978).

[157] See, e.g., Bredekamp (2007) and Heilbron (2009). Cf. also Mann (1987), Byard (1988), Ostrow (1996), Reeves (1997) and further sources cited there on the role of the visual arts in this *Gestalt* switch in the perception of the Moon around the world.

[158] Interesting examples from the Victorian period are found in Lightman (ed. 2007), esp. pp. 174, 188 for visual analogies. In the 1920s and 1930s, Fritz Kahn's playful visual analogies, e.g., between human digestion and industrial production processes, gained popularity.

[159] On this process from an art historian's point of view, see, e.g., Bredekamp (1997), Bredekamp (2005c) pp. 155f. and Kemp (2000).

[160] On this famous quote by the astronomer Jules Janssen, see note 11 on p. 379 below.

[161] On the technical possibilities in retouching and on its procedures see, e.g., Stein (1888).

[162] These are quotes from a letter by William Henry Fox Talbot, the inventor of photography on paper, to his friend John Herschel, March 21, 1839, preserved at the Royal Society, Herschel collection, and quoted in Schaaf (ed. 1985), p. 8.

digitization of this medium. Retouching of negatives or photographic prints usually leaves clear, albeit often inconspicuous, traces, which at least the experts could identify. The digitalized photograph radically widened the scope of potential alteration and manipulation. In 1991, Fred Ritchin claimed that a new era had set in, sounding the death knell of the "photograph as we know it" and (over)emphasizing the range of new options for undetectable digital manipulation techniques: changes in color, focus or resolution, and insertion or removal of depicted figures. "Now the photograph is as malleable as a paragraph, able to illustrate whatever one wants it to."[163] Blowing into the same horn, William Mitchell argued in 1992 that digitization caused a cultural reconfiguration in the "post-photographic" era. They were right.

In recent science and technology since the 1990s, complex and glossy visualizations have become relatively easy to produce, and thus form part and parcel of conference presentations and websites of individual researchers or their institutions. In the research strand of computer simulations, practically no result is presented without such animated images and film-like illustrations of the underlying dynamic process. Stills, snapshots and low-resolution movies have likewise become a powerful medium in the popularization of these research results.[164] Annemaria Carusi has shown that enhanced images and computed simulations act as a kind of "common currency" or as a "trading zone" between various research communities. These images or animations seem to be universally understood, whereas details of a scientific practice are otherwise by no means so easily grasped by their disciplinary neighbors. Newly founded centers and/or computer programs for visualization thus "have an important role in building up a common way of seeing among the disparate groups that are needed as collaborators."[165] Animated simulations or glossy images of nano-machines or other objects construed by modern science and technology, and visualized in a quasi-realistic mode of representation that makes one forget that these objects do not (yet) exist and will never be observable quite so directly as these images seem to suggest, are thus "misleading in that they are seductive, and may persuade you to buy into a

---

[163]Ritchin (1991) p. 12; he also gives a couple of spectacular and notorious examples of faked images. Cf. Mitchell (1992) on the electronic tools and technical possibilities.

[164]For examples from computational biology and further analysis see Carusi (2008, 2011). On the in-built tendencies of modern visualization to objectify and naturalize artificial objects and processes, cf. Araya (2003).

[165]Carusi (2008) p. 244. Cf. also Friedhoff & Benzon (1989) on the techniques of modern computer-aided, interactive visualization and Carusi (2012) on simulations.

particular way of seeing – and therefore understanding – the phenomena."[166] In fact, science journals such as *Nature* have even started to publish warnings and admonitions to scientists not to overdo the pseudo-realism technically possible in digital media.[167] Likewise, neuroscientists have become beware of the the danger of the public's inherent fascination with brain images, which have been proven to have a strong impact on judgments of scientific reasoning. The reductionistic explanation of cognitive phenomena that the inclusion of such images in scientific articles rhetorically suggests has in fact not yet been achieved.[168]

Of course, visual rhetorics occur elsewhere besides these still fairly academic contexts of textbooks, research journals and academic talks. They can be found in many other visual media as well. The enormous historical impact of early-modern printed flyers comes from their gripping woodcut illustrations. Postcards, posters and the illustrated press since the advent of photomechanical printing are likewise replete with examples. A few scientific images have made it far beyond the confines of conference halls and textbooks. They are used as backgrounds for posters, office artwork or advertisements, as logos and cultural identifiers of social groups. A still from an animation of the millennium simulation of the Big Bang by a Potsdam group of astrophysicists, for instance, has gained particular fame. Watson and Crick's helical DNA model, some of Einstein's portrait photos or Darwin's finches are other examples of cultural icons in science.[169]

In 1917, in the midst of World War I, General Erich Ludendorff (1865–1937), head of the German army, ordered 700 cinemas be built along the lines of battle, as "the war has demonstrated the overwhelming power of images and the film as a form of reconnaissance and influence."[170] For some cultural theorists, cinematography even became the hallmark of modernity, whereas for others it rather appeared to be an outgrowth of decadence and superficiality. Increased interest in complex and fleeting phenomena such as cell division, metabolism, growth or Brownian motion did indeed lead to a greater awareness of the temporality of natural processes, whose time-scales could be stretched out or shortened at

[166]Carusi (2011) p. 327; cf. also ibid. on the limits of shared vision and on resistance to crossing over subdomains.

[167]Ottino (2003); cf. Bredekamp (2005c) pp. 155ff. on the ongoing issue of hand-drawing vs. photography and digital visualizations.

[168]See, e.g., Beaulieu (2002), Racine et al. (2005), McCabe & Castel (2008) and Peter Hucklenbroich in Liebsch & Mößner (eds. 2012) pp. 265ff.

[169]On this process of idolization see, e.g., Nelkin & Lindee (1995), Voss (2007/10) and Carusi (2011) p. 316.

[170]See Bredekamp (2003) p. 422, quoting Kittler (1986) pp. 129, 197 in Bredekamp's English translation.

will by such film features as slow or fast motion. In our context, celluloid films and television play an important part in some areas of 20th-century science, not only in research, but even more so in the dissemination of popular images of nature and science.[171] We will briefly return to these topics when we discuss the urge by engineers and scientists to generate a dynamic – thus, strictly speaking, a four-dimensional – mental image of complex processes. This had, in fact, been one of the driving motivations for the Lumière brothers, Marey and others in their pioneering experiments eventually leading to motion pictures and the cinema in the late 19th and early 20th centuries. This ability of the cinematographic method to capture rapid and complex processes has led to a "natural affinity between film and scientific research" and to an abundance of scientific films,[172] even though the spread was – and is – by no means equal in all disciplines (see sec. 7.4).

1.5   Alpers on the "Dutch connection"

A main source of inspiration for this study was the masterful study by the art historian Svetlana Alpers on Dutch culture in the 16th and 17th centuries.[173] Although trained as an art historian and a practicing professor of art history at one of the world's leading universities, Alpers by no means limited herself to a traditional account of Dutch painting, famous and impressive though it is for its fine naturalism. She was after something deeper and much more original – she was probing the visuality of the artists and the ways in which this visuality was generated and trained. She was trying to understand the sense in which "seeing" had become plainly synonymous with "knowing" during the Dutch Republic, and what repercussions this focus on describing, mapping or mirroring had on Dutch culture in general during the early-modern period.[174] Consequently, she did not limit herself to the study of the "high" visual arts, but also incorporated maps – after all, the Netherlands was one of the world's centers of early-modern map-making. Telescopes and microscopes also figure there along with illustrated

---

[171]See sec. 7.4. Cf. also Lefebvre (1993), Martinet (ed. 1994), Tsivian (1996), Gunning (1999) and in Schwartz et al. (eds. 2004), Curtis (2005), Boon (2008) and Wellmann (2011); more specifically on American wildlife films, cf. Mitman (1999). More generally, on film and mass-media studies as the starting point of the visual culture movement, see also Balázs (1924) and Frank (2005); furthermore Mitchell (1995) p. 543.

[172]See Curtis (2005) p. 25, Landecker (2005) p. 122 and Boon (2008).

[173]On the following, see mainly Alpers (1983), but also her earlier papers Alpers (1976/77, 1978).

[174]On "seeing as knowing," see Alpers (1978), esp. pp. 154ff. on the Dutch separation of description from what that description evokes.

children's books and optical gadgets such as eye-glasses or the *camera obscura*.[175] It is this integrated analysis of a whole culture – actually a prime example of an outstanding visual culture in its full breadth – that I found so impressive when I first came across this work in the mid-1990s. Only retrospectively, when looking back at this classic example of the new art history, did I realize that this text also contains one of the first occurrences of the term 'visual culture.' It is to be found in the following quote from her introduction:

> What I propose to study then is not the history of Dutch art, but the Dutch visual culture – to use a term that I owe to Michael Baxandall.
>
> In Holland the visual culture was central to the life [...; if] we look beyond what is normally considered to be art, we find that images proliferate everywhere. They are printed in books, woven into the cloth of tapestries or table linen, painted onto tiles, and of course framed on walls. And everything is pictured – from insects and flowers to Brazilian natives in full life-size to the domestic arrangements of the Amsterdamers. The maps printed in Holland describe the world and Europe to itself.[176]

In Baxandall's earlier work on *Painting and Experience in Fifteenth-Century Italy*, which Alpers cites as her source in the above quote, this term 'visual culture' is not yet explicitly used; it is only implied. Baxandall's aim was a "social history of pictorial style." In his notion of the "period eye," he bundled culturally variable factors of human seeing and visual interpretation:

> Some of the mental equipment a man [or a woman – this text predates the age of political correctness] orders his visual experience with is variable, and much of this variable equipment is culturally relative. Among these variables are categories with which he classifies his visual stimuli, the knowledge he will use to supplement what his immediate vision gives him, and the attitude he will adopt to the kind of artificial object seen.[177]

Thus, for instance, the use of royal blue in Renaissance paintings immediately signaled a particularly important, often sacred, figure to viewers, because everyone at that time knew how extremely valuable the deep-blue paint, made from pulverized lapis lazuli, was. This knowledge is generally lacking among historically unenlightened present-day viewers. This color no longer bears a special significance for us. This focus on deep, culturally embedded habits of vision and interpretation

---

[175] Cf. here the following section on the interplay of instruments with visuality.

[176] Alpers (1983) p. XXV.

[177] Baxandall (1972*a*) p. 40. On the following, ibid., sec. II.8; on Baxandall's pioneering role in visual culture studies, cf. also Dikovitskaya (2005) pp. 9f.

led Baxandall to a broad perusal of sources well beyond the traditional confines
of art history. Thus, for instance, he tried to link the growing mathematization
of perspectival painting to the contemporaneous spread of applied mathematical
practices in double-entry bookkeeping, and the representation of the human figure
in Renaissance paintings to dance manuals of the time.[178] Baxandall and Alpers
both exemplify an emerging research practice aimed at an integrated portrayal of
a whole society's visuality embedded in everyday life. 'Visual culture' happened to
be a very fitting term to encompass this set of practices and objects. This precise
term had cropped up earlier, in the late 1960s, in discussions about new modes of
education by television,[179] but Alpers was the first to use it in a context outside the
postmodern realm of new media. She showed how productively it could operate
for an art historian's endeavor.

    When over a decade later Alpers was asked to contribute to the " Visual
Culture Questionnaire" of the culture journal *October*, she reflected back on her
former usage of the term and on her reasons for using it:

> When, some years back, I put it that I was not studying the history of Dutch
> painting, but painting as part of Dutch visual culture, I intended something
> specific. It was to focus on notions about vision (the mechanism of the
> eye), on image-making devices (the microscope, the camera obscura), and
> on visual skills (map making, but also experimenting) as cultural resources:
> related to the practice of painting.[180]

Alpers understood 'visual culture' as "a culture in which images, as distinguished
from texts, were central to the representation (in the sense of the formulation of
knowledge) of the world."[181] It is this simple, straightforward, and fine definition,
to which I would like to return here, and elaborate it further in a direction of
eminent use to the history of science.

1.6   Instruments for creating and recording images

Svetlana Alpers happened to pick the Dutch culture during the early-modern
period for her study of visual culture. She thus inevitably had to cover various

---

[178]See Baxandall (1972); cf. also Matthew Rampley in Rampley (ed. 2005) pp. 11f.

[179]See Gattegno (1969); I owe this reference to the excellent literature surveys by Dikovitskaya
(2005) pp. 6f. and Frank (2006). Cf. also Mitchell (1995) on studies of films, movies, television and
other mass media as the starting point for visual studies, and sec. 7.4. below.

[180]Alpers (1996) p. 26.

[181]Alpers (1996) p. 26. For reflections on true or realist (better yet: naturalist) representation, see
Alpers (1976/77) and, of course, Gombrich (1960).

instruments that made Dutch instrument-makers of the period world-famous. Most prominently among these are telescopes and microscopes, both invented shortly after 1600 by spectacle-makers in the Dutch border town of Middleburgh, where an Italian glass-maker had settled, thereby transferring to this region the skill of making clear crystallo-glass. Once the secret of how to combine two lenses in order to produce magnification effects was revealed in patent submissions, these new instruments soon spread throughout the continent. Initially, they were still considered as curious gadgetry, but after substantial improvements were made in the workshop of Galileo Galilei and their theoretical codification was completed by Johannes Kepler, René Descartes and others, telescopes and microscopes became highly praised scientific instruments.[182]

Apart from these two famous instruments, Alpers also mentions another optical gadget commonly used in early-modern Holland: the *camera obscura* (literally, dark room). This camera-like device generates images by projecting the light off well-illuminated objects through a fine pinhole onto a flat surface inside a darkened room. Alpers sees this device as specially important not only as a handy aid in quickly depicting shapes and forms, which was very much appreciated at the time. It is, moreover, the incarnation of the contemporary ideal of what a "truly natural painting" was supposed to achieve:

> While the perspective system of the Italian Renaissance conceived of the picture as a window and began by positing a viewer of a particular size located in space looking through it to the reconstruction of the world beyond, the northern artist felt more at home with the world seen and then laid out onto a flat surface. The extent, the edges of the picture, were nowhere accounted for, and neither was the artist-viewer.[183]

The painter, engraver and art critic Samuel van Hoogstraten (1627–78) applauded as particularly "natural" paintings designed with the aid of this device, and even the superb master Vermeer seems to have used it at least occasionally.[184] A long list of other scientific instruments for the creation and recording of images can be provided for later periods. The following proposes just a few of the most

[182] On the context of discovery and early conceptualization of the telescope, see Biagioli (2006) and van Helden & Dupré et al. (eds. 2010).

[183] Alpers (1978) p. 162. On the history of the *camera obscura* as a technical gadget: Hammond (1981), Lefèvre (ed. 2007) and Breidbach et al. (2013); on its role as an "epistemology engine," i.e., as a paradigm for the process of (re)cognition: Ihde (2000).

[184] On this still controversial issue, see, e.g., Mayor (1946), Seymour (1964), Schwarz (1966), Fink (1971), Wheelock (1977*a*) chap. 7, Wheelock (1977*b*), Steadman (2001) and www.essentialvermeer.com/camera_obscura/co_one.html (last accessed April 29, 2011).

important ones, and this list could be extended at will. I include invented and continuously improved instruments that were not only important in their own day but were also an emblem of their time and are considered somehow characteristic of their cultural knowledge and contemporary productivity:

- telescope and microscope
- camera obscura
- laterna magica
- camera lucida
- kymograph
- ballistic galvanometer
- photographic camera
- stereoscope
- x-ray tube and screen
- oscilloscope
- computer in conjunction with computer screen or display
- computed tomography
- NMR and MRI
- PET scanner, etc.

The first items on this list are just instrumental *aids* for human observation, allowing access to orders of magnitude not discernible by the naked eye, from the very small in microscopic detail to the very large in the solar system and beyond.[185] Various fascinating studies on the telescope and the microscope have shown how the complex history of these two early-modern instruments is multiply intertwined with the history of experimental examinations at the limits of seeing and theoretical reflections on distinguishing real effects from artifacts,[186] – most famously, the early-modern Dutch physician Nicolaas Hartsoeker's (1656–1725) observations of sperm around 1694. Using the simple screw-barrel microscope of his own invention, Hartsoeker sketched what he saw: a homunculus, a miniaturized fully prefigurated human being waiting inside to grow into the larger foetus and human baby. His speculative (and, as it turned out, false) preformation theory of

---

[185] For a systematic exploration of these orders of magnitude ("powers of ten") crossed by modern scientific instrumentation, see Morrison (1982).

[186] See, e.g., van Helden (1977), van Helden & Dupré et al. (eds. 2010), Keller (1996) 110ff., Wilson (1995) and Schickore (2007).

embryological development had led him astray and influenced his vision to the extent that he was convinced he saw what he believed to be true.[187]

The kymograph is listed here as a representative of a whole class of instruments, in which a measurable quantity, such as pressure or motion, is continuously recorded, often by means of a pen or needle leaving a trace on blackened paper spooled on a roller system. The first such instrument was developed by Thomas Young (see sec. 4.1). It later came into much more common use as a 'kymograph' (literally, wave writer) when the physiologist Carl Ludwig (1816–95) started to build instruments for recording blood pressure, muscle contraction and other physiological processes around 1846. Other physiologists with whom Ludwig was closely associated, most notably Hermann von Helmholtz (1821–94), Ernst Wilhelm von Brücke (1819–92) and Emil du Bois-Reymond (1818–96), also began to use similar kinds of registering instruments to visualize and record such dynamic processes as signal propagation along nerves.[188] It is quite striking that around the middle of the 19th century, self-registering instruments, designed to automatically record changes or parameters of motion by means of an indicator leaving continuous traces, appeared in many different disciplines, from geophysics, where seismometers recorded minute terrestrial vibrations, to sensory physiology. The historians of science Lorraine Daston and Peter Galison have argued that this craze for automatic self-registration instruments is a clear indicator of a globally reigning norm during this period.[189]

The literature is copious on how this new recording system transformed visuality in photography and camera technology among active photographers as well as viewers, users and customers, especially when such companies as *Kodak* turned photography from a specialized branch requiring optical and chemical skills into a mass market after 1900.[190] With the widespread distribution of cheap cameras and the easy-to-load film roll, the tourist's experience changed as drastically as did scientific field work. In the mid-20th century, photography became "the fullest way to comprehend the world [...]. Armed with a camera [as

---

[187] According to Hill (1985), Hartsoeker had only postulated their existence as part of his spermist theory of conception and never claimed to have seen them. Many others fell into this trap, however.

[188] On precursors of the kymograph, see Hoff & Geddes (1959, 1960). On the work of these four physiologists, see, e.g., Borell (1986), Brain & Wise (1994), Holmes & Olesko (1994), Schmidgen (2009, 2011) and http://vlp.mpiwg-berlin.mpg.de/experiments. On Marey's later work along the lines of this "méthode graphique," cf. Brain (1996), Braun (1992) and here pp. 249ff. on his subsequent chronophotographic studies.

[189] See Daston & Galison (1992, 2007) on mechanical objectivity. Cf. Hentschel (2008*b*) for a critical review.

[190] See, e.g., Jenkins (1975) on the Kodak slogan: "You press the button, we do the rest."

the Polaroid SX-70 ad has it], you see a picture everywhere you look."[191] Later, the
cinematograph and the associated media technologies of filming led to another
marked change in overall visuality.[192]

The last examples on this list represent a wide array of modern instruments
designed not to strengthen or amplify, but substitute human vision. NMR and
PET scanners are complex machines designed to create artificial images from an
input of data-streams not recorded by the human eye but by CCDs, sensors or
other gadgets able to convert incoming radiation into electronic signals. When
the sociologist Bruno Latour (*1947) conducted his ethnomethodological visit in
Roger Guillemin's biomedical laboratory at the *Salk Institute* in La Jolla, California
in the mid 1970s, the microbiologists working in these laboratories appeared to
him a "strange tribe of [...] compulsive and manic writers [...] who spend the greatest part of their day coding, marking, altering, correcting, reading, and writing."[193]
To denote this extreme importance placed on recording devices of all kinds, from
pens to graphic plotters and computer printers, Latour used the ethnological
term 'inscription device.' All scientific instruments were to him ultimately such
'inscription devices,' not just the first on this list, for which this attribute is relatively obvious. Even large and expensive laboratory equipment (e.g., bioassays,
spectroscopes or mass spectrometers) "have the sole purpose of transform[ing] a
material substance into a figure or diagram." In fact, the whole molecular biomedical laboratory Latour was visiting "began to take on the appearance of a system of
literary inscription."[194] These last two quotes are characteristic of the exaggera-
tion and misinterpretation in Latour's approach. Obviously, complex instruments
like mass spectrometers do not have as their "sole" purpose the creation of a
figure or diagram. However, Latour and Woolgar were never really interested in
the many other functions of this complex technology, only in its effects on the
interrelations among the people in the laboratory. So all the material culture of
their functions and their handling disappeared behind a strange curtain of semiotics. Kathryn Henderson, another sociologist of technology who conducted
quasi-ethnographic studies in various design centers of US industry, to study the
visual culture of engineers, consequently coined the term 'conscription devices'
for that Latourian subclass of inscription devices that draw people together and

---

[191] Quote from Alpers (1978) p. 148.

[192] On this rupture, to some people only comparable with the invention of the printing press in
the 15th century, see, e.g., Balázs (1924), Frank (2005), Wellmann (2011) and here sec. 7.4.

[193] Latour & Woolgar (1979) pp. 48–9.

[194] Quotes from Latour's and Woolgar's *Laboratory Life* (1979), pp. 51f.; for a summary of their
overall approach, cf. also http://en.wikipedia.org/wiki/Laboratory_Life .

"enlist group interaction" with a common goal, often centered around images or other physical inscriptions.[195] For sociologists, this might be a useful and indeed fruitful way to go about extricating all the intricacies and problems of instrumentation while just focusing on laboratory shop-talk or ethnomethodological studies of graph usage, i.e., on the social impact of images, models, etc. But this one-sided approach to scientific practice won't suffice for historians of science.[196]

Long dominated by an emphasis on theory and conceptual development, the history of science turned to a closer study of experimental practice since the 1980s. The resonance was remarkable: On the one hand, there was the 'pictorial,' 'iconic' or "visual turn" of W. J. T. Mitchell, Martin Jay, Gottfried Boehm and other art historians,[197] paralleled by Martin Rudwick, Peter Galison and a few other historians of science;[198] on the other hand, there was the 'experimental' or 'practical turn' of such historians of science as Peter Galison, Simon Schaffer and Steven Shapin, to name just a few.[199]

1.7   A few deeper insights from early 'visual studies'

In the preface to the recommendable *19th Century Visual Culture Reader*, its two editors suggest that one distinguish between historically oriented studies of visual cultures and more broadly culturally oriented 'visual studies.'[200] Insofar as the latter mostly take their examples from everyday life in realms such as advertisement, cinema or fashion, all far away from the key interests pursued here, I willingly adopt this terminological distinction here. Pioneering studies from the 1980s and 1990s on visual science cultures – that is, before the "contagion" of visual studies spread epidemically – contained contributions from several disciplinary fields, including the history of science,[201] the sociology of science (which since

---

[195] Henderson (1999) pp. 53, 74ff., 204.

[196] See my essay review on new historical studies of image practice, Hentschel (2011*b*); cf. furthermore Heintz & Huber (eds. 2001), Hessler (ed. 2006), etc.

[197] For a good introduction to common motifs in these papers from the 1990s, see Bredekamp and Boehm in Maar & Burda (eds. 2004).

[198] On Rudwick (1976) see here sec. 2.1. On the insights of other historical contributions from the 1980s and 1990s, see here sec. 1.8.

[199] For detailed surveys of the literature on studies about scientific experiments and their interplay with instrument making and theory formation, see Hentschel (2000*b*) and (1998*a*) chap. 1.

[200] See Schwartz & Przyblyski (eds. 2004) p. xxii and pp. 26ff.; on this distinction cf. here, chap. 2.

[201] See Rudwick (1976) on the history of geology, Mazzolini (ed. 1993), entitled *Non-Verbal Communication in Science Prior to 1900*, Ann S. Blum (1993) on American zoology in the 19th century, Alex Pang (in the 1990s) on astronomy and astrophysics, and Peter Galison (1997) on image vs. logic traditions in high-energy physics.

the 1980s applied the ethnomethodological approach also to the usage of images in scientific practice),[202] the philosophy of science[203] and, of course, the history of art.[204] A few important and lasting insights from these pioneering studies are the following:

1. Rather than referring to pictures as (mere) **illustrations**, we have to study (full-blown) **images** in their own contexts of production and reception (Alpers 1983, Pang 1996).[205]

2. They must be studied not just as to their impact on their users, but also as to the procedures that led to their production (cf. Svetlana Alpers (1983) on the move from **pictures** to the process of **picturing**).

3. Images, even photographs, are **not simply "taken" but elaborately constructed** using special skills and after much labor.[206]

4. Visual representations require skills not only in their rendering but also in their adequate interpretation (Gombrich 1960, Perini 2013).

5. Visual perception and recognition of patterns and *Gestalten* are **active processes**, in which past experiences, future expectations and current styles of representation all play their part (Gombrich 1982).

6. New shapes or **patterns will be assimilated to familiar schemata** and patterns, or "the familiar will always remain the likely starting point for rendering the unfamiliar" (Gombrich 1960, pp. 77–82).

7. Evidence and transparency of visual representations have to be generated and established – they don't just come naturally but have to be **acquired and inculcated in cultures of seeing** (Lynch 1985).[207]

---

[202] Aside from the afore-cited Latour & Woolgar (1979), one should also note Lynch (1985), Lynch & Edgerton (1988), Lynch & Woolgar (1990) and the work by Klaus Amann and Karin Knorr-Cetina (1988) in Bielefeld, Germany.

[203] Pioneered by Ian Hacking: *Representing and Intervening* (1983); cf. also Brian Baigrie (1996) on the epistemology of images and Laura Perini (2004ff.) on the epistemic truth value of photographs.

[204] Erwin Panofsky on iconology aside from iconography, Samuel Edgerton on the heritage of Giotto's geometry (1991), Martin Kemp (1972, 1977, 2004, 2007) on Leonardo and on *Nature* images (Kemp 2010), J. V. Field on linear perspective (1987–1997), H. Bredekamp and co-workers on Galileo's drawings (2007), and, on technical images, Bredekamp et al. (eds. 2008).

[205] See also Bryson, Holly & Moxey (eds. 1994) p. xvi, Lynch & Woolgar (eds. 1990) or Staubermann (2007), and the essay review of Hentschel (2011*b*) on image practice.

[206] See, e.g., Lynch & Edgerton (1988), Yoxen (1987) and here p. 238.

[207] Exemplarily on x-rays: Henderson (1988), Eisenberg (1992), Dommann (2001), Dommann (2003) pp. 14f. and Vera Dünkel in Bredekamp, Schneider & Dünkel (eds. 2008), pp. 136–47.

8. **"There is no innocent eye"**: Pictorial conventions and perceptual habits are important – in fact, inescapable (Gombrich 1960).[208]

9. Scientific illustrations, in particular, are more or less **'theory-laden'** in that they not merely depict objects or processes but rather contain various **levels of symbolic encodings** that one has to learn in order to be able to "read" such images appropriately.[209]

10. The **non-uniqueness of representation**: many different representations are possible and legitimate (Goodman 1969); thus the choice of images in this "thicket of representation" (Wimsatt 1990) is often difficult and depends on syntactical, semantic, pragmatic and aesthetic criteria.

11. "Representation of" and "representation as" always come together, i.e., any **re-presentation is an inter-pretation**, mediating between object and subject (Morgan & Morrison 1999).

12. Images embody information and encode conventions, but sometimes they also support and signify detection; thus, certain types of images, such as photographs, electron micrographs or x-ray diffraction patterns, are often **presented and epistemically accepted as evidence** for the reality of object or process features (Chaloner 1997, pp. 367–71; Perini 2012, 2013).

13. Likewise, medical imaging can also be interpreted "as instrumentally aided perception"; imaging technology then becomes **"visual prosthesis,"** i.e., an aid to see deep inside the human body, granting noninvasive visual access to hidden organs such as the brain or the intestines (Semczyszyn 2010, pp. 66ff., 210ff.).

14. Many phenomena of nature are in fact "never seen but through the 'clothed' eye of **inscription devices**" such as bubble chamber photographs, spectrograms or chromatographs (Latour in Lynch & Woolgar (eds. 1990) p. 42, Golinski (1998) chap. 5).

15. One image rarely comes alone: We have to look at the **full chain (sometimes veritable cascades) of representations** in various forms/media.[210]

---

[208] Aside from Gombrich (1960), cf. also Krieger (1984) pp. 183ff. on Nelson Goodman's strong reading of Gombrich, and on his denial of any distinction between natural and conventional signs.

[209] See, e.g., Knight (1985) on "visual language," and Topper in Baigrie (ed. 1996) on visual epistemology; cf. Lüthy & Smets (2009) on epistemic and metaphysical meanings and allusions in images.

[210] For good examples, see, e.g., Latour (2002) pp. 37ff., Carusi (2011), Nasim (2012/14) chap. 2, Perini (2012) and here chap. 7.

16. Images are **"immutable mobiles"** (Latour) or **"effective proxies"** (Hineline 1993, Rudwick 2005): they easily circulate, objects don't. Images travel so that people or objects don't have to.

17. **"Paper tools"** (Klein) can act as visible and maneuverable substitutes for scientific objects.

18. Images are embedded into <u>v</u>iscourses (Karin Knorr-Cetina) (analogous to embedding of utterances in <u>d</u>iscourses)

19. **Iconology:** all images have a prehistory in other images (Panofsky).[211] Thus, certain motifs have a history, their own aesthetic lineages (Elkins 2007), or – as others put it – images have a "biography" or "life-cycle" (Hagner, Mazzolini & Pogliano (2009), Nasim (2012/14) pp. 14, 53, 125ff.)

20. Images exhibit **contextuality, intertextuality and text–image interplay** (cf. sec. 1.3 for further detail and literature).

21. **Transitions in changes of media** (e.g., "translations" from photography to lithography to line-drawing) are difficult; correspondingly, frictions between scientists and their illustrators are frequent (see sec. 6.3).

22. There is an interplay of images and seeing as (*Gestalt* recognition), embedded in visual domains (good examples from microscopy or medical imaging are provided in Schickore (2007) and Semczyszyn (2010); cf. here chap. 9).

23. Vision and pattern recognition are embedded in "endogenous communities of practice" (Goodwin 1994, p. 606) in which consistent "ways of seeing" are obtained through stabilized techniques, joint activities, laboratory practices, routines and scientio-technical cultures (Burri (2000, 2008) examined this exemplarily for MRI in Switzerland; cf. Hentschel (2011*b*) for a survey).

24. Visual representations play an important role in cognition (Gooding, Twchey).

I do not want to go into every one of the above points right away, as several will be discussed in detail below. But a few of them are worth commenting upon further here. The first three of these points are marvellously summarized in a famous quote from Svetlana Alpers:

> I employ the word "picturing" instead of the usual "picture" to refer to my object of study. I have elected to use the verbal form of the noun for essentially three reasons: it calls attention to the *making* of images rather than to the finished product; it emphasizes the inseparability of maker, picture and what is pictured; and it allows us to broaden the scope of

---

[211] See also H. Bredekamp in Ullrich (2003*a*) p. 16: "There are no images without any prehistory in imagery." ("Es gibt kein Bild ohne Vorgeschichte der Bilder").

> what we study since mirrors, maps, and [...] eyes also can take their place
> alongside of art as forms of picturing so understood.[212]

Not only did Alpers plead programmatically for broadening the scope of history of art. She showed how a cultural history could also effectively include maps, lenses, and other objects of the world of early-modern Netherlands, all usually not classified as art. Only thus was it possible to cope with what she had identified as the "mapping impulse" of this early-modern society so very dependent on exact triangulation of land and its mapping. Her claims and methodology were later taken up by historians of cartography, who now understand cartographic devices as part of a "history of spaces" and of "geo-coding the world."[213] No. 1 could provocatively be reformulated thus: We need a **history of images** rather than a history of art. Consequently, the latter field is currently engaged in deep internal debates on its future as a broadened *Bildwissenschaft* (only roughly translatable as "image science").[214]

Several of the succeeding points on the complexity of images could be summarized by the statement "images have a life of their own," thus closely paralleling a famous statement of Ian Hacking (1983) about experiments, for which this is equally true.[215] We have to look at all stages of image production, optimization and diffusion of images to get hold of this intricate dynamics of image generation and reception.

Another very fundamental insight was provided by the sociologist Bruno Latour when he coined his concept of **representation chains** or **cascades of images**. He put it most provocatively in his introduction to the "Iconoclash exhibition" in Karlsruhe in 2002:

> The cascade of images is even more striking when one looks at the series
> assembled under the label of science. An isolated scientific image is mean-
> ingless, it proves nothing, says nothing, shows nothing, has no referent.
> Why? Because a scientific image [...] is a set of instructions to reach another
> one down the line. A table of figures will lead to a grid that will lead to a
> photograph that will lead to a diagram that will lead to a paragraph that will

---

[212] Alpers (1983) p. 26.

[213] See, e.g., Pickles (2004) and Monmonier (1996); on the mapping metaphor, cf. Godlewska (1999), Rupke (2001) and Hentschel (2002a).

[214] On these debates, see below pp. 79f. and, e.g., Bredekamp (2003), Sachs-Hombach (ed. 2005), Dikovitskaya (2006) and further pertinent sources listed there.

[215] See Hacking (1983) p. 150; cf. Lefèvre et al. (eds. 2003) and Boehm in Maar & Burda (eds. 2004) on "power"/*Macht* and *Eigenleben* of images.

lead to a statement. The whole series has meaning, but none of its elements
has any sense [by itself].[216]

At first sight, this statement looks obviously wrong, as we are accustomed to
looking at individual images, to analyzing and interpreting them without referring
to other images as we do so. But actually, a lot of other images are in our heads
while we are viewing them. We cannot suppress this cultural baggage; and if
we could, we would indeed be helpless in interpreting the image at hand, since
any interpretation utilizes many unwritten conventions of image construction and
stylistic features which we are not conscious of as we look. Furthermore, in texts
in science and technology, there is constant alternation between very different
types of representation, also between visual and textual elements.[217] Aside from
the chain of published visual representations, there are also complex chains of
what Omar Nasim called "working images," i.e., all kinds of provisional sketches or
drawings, diagrams, outlines, schematics and 3D models created during the often
long process of "familiarization" with frequently strange and unknown scientific
objects or processes. These working images are intensely discussed, processed
and reworked or transformed, compared, multiplied and selected before this chain
of representations finally settles on an image deemed ripe for publication.[218]

Since the cave paintings of Chauvet (c. 30000 B.C.), Lascaux (c. 15000 B.C.)
and Altamira (10000 B.C.), the very act of drawing by hand – one of mankind's
oldest forms of culture – has been interpreted as a form of cognition. Drawing an
object is not only a passive rendering and re-presentation, but an active grasping
(in German: "*be-greifen*") of its main features. Thus, for Galileo's painter-friend
Ludovico Cigoli (1559–1613), a mathematician without the trained skill in drawing
by hand is not only "merely a half-mathematician, but also a man without eyes."[219]
A good draftsman not only has a skilled hand, but also a mental reservoir of
patterns and *Gestalten* needed in order to understand what he or she sees – any
drawing is thus also already a kind of proto-conceptualization prefiguring and
molding later stages of cognitive processing. Even excellent draftsmen such as
Leonardo da Vinci or the astronomers working for Lord Rosse in the 19th century

---

[216]Latour (2002) p. 34. Agreement with this statement does not imply acceptance of Latour's
radical anti-realist claim that there is "no world beyond the image wars."

[217]We will get back to the latter type of interplay in sec. 1.3 and postpone the discussion of the
former chains of different images to chap. 7.

[218]Nasim (2008–2012/14) and Nasim in Hoffmann (ed. 2001) has wonderful examples of these
processes of familiarization and ripening.

[219]Cigoli, according to Bredekamp (2005*c*) p. 156, who also quotes Leonardo da Vinci to similar
effect.

often needed to produce long series of drafts before they felt they had grasped their object of study. Early drawings of unfamiliar objects or processes, occasionally preserved in notebooks, letters or on scrap paper, are an excellent documentation of these early stages of cognition and understanding.

The historian of chemistry Ursula Klein (*1952) enriched the discussion by introducing the useful concept of 'paper tools' for chemical formulas and structure notation, as visible and maneuverable devices that represent the material objects under investigation. That they do not actually interact with these objects is an important distinction from physical tools, scientific instruments and apparatus. Nevertheless, paper tools are exterior to mental processes and allow some kind of manipulation on the page. A chemist can work with his formulas to check for completeness (chemical budgeting), a full accounting of the valences, bonding, and perhaps even structural features of stability, and their combinability or sym-metry infractions (in the case of structural formulas).[220] It is interesting that in the early debates on the introduction of structural formulas during the late 1870s and early 1880s, this feature of shifting chemical work to some extent from the labo-ratory bench to the writing desk had been noticed and heavily criticized as a step downwards to flimsy "paper chemistry."[221] Similar discussions could be observed when the theoretical physicist Richard Feynman (1918–88) introduced symbolic diagrams into quantum electrodynamics as graphic aids in the calculation and summation of higher-order corrections to scattering processes. These discussions only ended when it could be proven that these "Feynman diagrams" were mathe-matically equivalent to – but pragmatically far easier to handle than – the longish sets of algebraic formulas in the more traditional form of QED as practiced by Feynman's colleagues Julian Schwinger (1918–94) and Shin'ichiro-Tomonaga (1906–79).[222]

---

[220] For a clear exposition of this concept and rich examples from the history of chemistry, see Klein in Heidelberger & Steinle (eds. 1998), Morgan & Morrison (eds. 1999), Klein (2001) and Klein in Klein (ed. 2001) pp. 16ff., 28ff. Her coinage of 'iconic symbols' for certain chemical formulae is a misnomer, though – cf. Hentschel (2003b) p. 112.

[221] See, e.g., Kolbe (1877) or (1881a) p. 353 on his charge of a "papierne Chemie" aimed against Dumas, Kekulé and other propounders of a structural chemistry. For more on Kekulé and Kolbe, see, e.g., Rocke (1993, 2010) and here secs.3.2 and 4.2. On Kekulé's debates with Alexander M. Butlerov (1828–86) and on the latter's introduction of the term 'chemical structure,' see Benfey (ed. 1966) pp. 4–8, 16, Rocke (1981) and Rocke (2010) pp. 135–42.

[222] See Kaiser (2005b) and primary sources quoted there.

1.8   Later wrong turns of "the visual turn"

Unfortunately, the further trajectory of 'visual studies,' as this branch of the "new art history" founded by Alpers, Baxandall and a few others was soon called, soon left these fertile beginnings behind.[223] A few words about the wrong turns taken by visual cultures studies in the 1990s cannot be omitted from this discussion, but I can be brief because various readers and anthologies give this terrain quite thorough coverage already, unfruitful and heavily mined though it is.[224]

What 'visual culture' certainly is *not* (or ought not to be):

- "An 'undisciplined' conjunction of art and theory, critique and practice, new technologies and cultural everyday life"[225]
- An "interdiscipline" or "indiscipline [...] of turbulence or incoherence of the inner and outer boundaries of disciplines"[226]
- "A panicky, hastily considered substitution of image history for art history"[227]
- "The study of the social construction of visual experience"[228]
- "A tactic with which to study the genealogy, definition and functions of postmodern everyday life"[229]
- A nearly arbitrary projection surface of misguided 'cultural studies' on spectacles, lifestyle, design, advertisement, pornography, and what have you.
- An exclusive study of photography, film and other media, video, computer animation, Internet, etc.[230]

James Elkins – professor of art history, theory, and criticism at the *School of the Art Institute in Chicago* – declared in his *Skeptical Introduction to Visual Culture* that

---

[223] For good surveys of the further trajectory with a special focus on art history, see Cherry (2005) and Dikovitskaya (2005) with interviews of 17 important protagonists in the USA.

[224] See, in particular Mirzoeff (ed. 1998), Evans & Hall (eds. 1999), Rampley (ed. 2005), and Sturken & Cartwright (eds. 2001).

[225] Mooshammer & Mörtenböck (2003), cover.

[226] Mitchell (1995) p. 541.

[227] Art historian Thomas Crow as quoted in Dikovitskaya (2005) p. 14.

[228] Mitchell (1995) p. 540; cf. also Mitchell as quoted in Dikovitskaya (2005) pp. 15–17, 238–57.

[229] Mirzoeff (ed. 1998) p. 5; analogously in the *Visual Culture Questionnaire* (1996): "an updated way of talking about postmodernism," and in Dikovitskaya (2005) pp. 22–6, 224–37.

[230] As evidenced, for instance, by the selection of topics in Mirzoeff (ed. 1998), Evans & Halls (eds. 1999) or Sturken & Cartwright (eds. 2001). Cf., e.g., Jay (2002) p. 87 on "peephole metaphysics" and "totalizing gaze."

visual studies is predominantly about film, photography, advertising, video and the internet. It is primarily not about painting, sculpture or architecture, and it is rarely about any media before 1950.[231]

It doesn't have to be this way, though! It is not an unwritten rule that 'visual studies' have to limit themselves to "a particular slice of the sum total of visual productions,"[232] nor is it an axiom that 'visual culture' only comprise products of the new media of the last three decades, as some people seem to assume. We do not have to follow "visual studyists" in their strange predilection for the most vulgar and popular strata of cultural productions such as advertisements or pornography. Elkins makes the explicit criticism that his colleagues mostly tend to exclude visual representations stemming from science and technology:

> Visual culture can include documents (the visual appearance of passports, bureaucratic forms and tickets) but in general it sticks to art and design – it does not encompass engineering drawing, scientific illustration or mathematical graphics.[233]

Why should we categorically exclude the latter and abstain from studying some of the most versatile and impressive forms of culture practiced on Earth? I applaud Martin Jay's democratic impulse to call for the study of "all manifestations of optical experience, all variants of visual practice" and anything that can "imprint itself on the retina."[234] This widened terrain should also include visual representations and the skills of seeing and recognizing developed in science and technology. Fortunately, Elkins himself (and a few other influential propagandists of visual studies) has taken up images in astronomy, electron microscopy, quantum mechanics and particle physics in his latest books, with the explicit aim of bridging the widening gulf between the two cultures and "speaking at once to both the sciences and the humanities."[235] The deep divide between traditional art history and visual studies causes several protagonists of the latter, although trained and raised in art history, such as James Elkins, Barbara Stafford and W. J. T. Mitchell, no longer to be perceived as regular art historians by their own scholarly community, but

---

[231] Elkins (2002) p. 94, quoted from the prepublication of this foreword in the new *Journal of Visual Culture*; his book subsequently appeared in 2003.

[232] Ibid. This programmatics is then pursued in Elkins (2007), with 30 examples from all fields.

[233] Ibid.; cf. also Elkins (2007) pp. 7ff.

[234] Jay (1996) p. 42 in his response to the famous *Visual Culture Questionnaire* (1996).

[235] Elkins (2008) p. 1. I share this aim and his concerns about the limits of this undertaking (ibid., p. 18), but differ in my (comparative) methodology. For a more detailed critique of Elkins (2008), see my essay review Hentschel (2011*b*).

rather as "outsiders," as "heretical 'visual studyists'."[236] Seen from art history, the new field of visual studies can be construed either as an (in)appropriate effort (or chance) to expand its territory and/or methodology, as a rivaling and potentially threatening challenge, or as a neighboring field specializing in digital and other 'postmodern' sources and practices.[237]

When the Amsterdam cultural critic and theorist Mieke Bal (*1946) hypothesized that the lack of a clear object of research explains why "visual culture and its study remain diffuse" and has to remain an "interdisciplinary movement,"[238] a storm of protest raged in the subsequent issue of the newly founded *Journal of Visual Culture*; but not much of substance was raised against Bal's "verbal pyrotechnics" with which she had attacked 'visual essentialism.' While some seem to believe that the new field lacks any specific object, others want to make an all-encompassing move to include virtually "everything related to the cultural and the visual," against which Matthew Rampley and others protested that then visual studies run the great risk of becoming totally inconsistent and unmanageable. Should visual studies really look at *all* images "without making qualitative distinctions between them," or just at those "for which distinguished cultural value has been or is being proposed"?[239] My own position in these debates is that, yes, we should be nonelitist and admit visual sources from all levels of scientific practice and teaching; and I have taken pains to follow up on this program here. That does not mean giving up all qualitative distinctions or boundary-markers, though, which are, after all, often set by the scientific and technological disciplines and actors themselves. For the purposes of my comparative analysis, those visual representations are most valuable and important that can be shown to be representative of typical scientific practice. I thus rather focus on unspectacular examples than on famous, singular cases for which no parallel in other corners of the science-universe can be found. The comparative perspective will help us get a feel for what is typical and what isn't.

---

[236] On this unfortunate tendency to encapsulate art history onto precious little islands, and on the amusing constriction of the term within quotation marks, see Bredekamp (2003) p. 428 and Dikovitskaya (2005) p. 27.

[237] All three positions are actually represented in Dikovitskaya (2005); see, e.g., p. 3 and the full range of her interviews with "visual studyists" in her 'appendix,' pp. 123–284.

[238] Bal (2003) pp. 9 and 6, respectively. On the following; cf. the contributions by Norman Bryson, James Elkins, Michael Ann Holly, Nicolas Mirzoeff and others in *Journal of Visual Culture* vol. 2 (2003) no. 2, quotes from p. 247.

[239] On this debate, which mirrors older debates within the history of art on how to delimit its subject area, see Rampley (ed. 2005) and Dikovitskaya (2005) pp. 2, 14, 16ff., etc.

Aside from these predominantly Anglo-Saxon developments in 'visual studies,' the German context has generated a competing approach, dubbed *Bildwissenschaft*[240] by one of its main protagonists, Klaus Sachs-Hombach (*1957), professor of cognitive philosophy at the *Technische Universität Chemnitz* and founder of the visual institute for 'image science' (www.bildwissenschaft.org). *Bildwissenschaft* tries to encompass a revised philosophical aesthetics with a visual semiotics and a highly complex art of creating images; this is indeed quite an odd mix of components that are very hard to integrate.[241] Most unfortunately, several protagonists of *Bildwissenschaft* have tried to establish their new field in a manner that completely excludes art history, if not history altogether. Critics of *Bildwissenschaft*, such as the Berlin art historian Horst Bredekamp (*1947), somewhat irritatedly insisted that history of arts in the tradition of Aby Warburg (1866–1929), who considered himself to be a "picture-historian, but not an art historian,"[242] has in fact always covered subjets outside the trodden terrain of the fine arts. According to Bredekamp, art history is not an enclave of elitist connoisseurs, but rather "a laboratory of cultural-scientific picture-history."[243] Consequently, Bredekamp himself has published extensively on topics ranging from the iconography of power and visual strategies in Thomas Hobbes's *Leviathan*, Galileo Galilei's drawing of lunar craters and the manifold problems with their "translation" into print media of the time, or about sketches and images in the writings of Gottfried Wilhelm Leibniz, to take only three examples of many fascinating topics, all three clearly not part of canonical art history.[244]

As the discussion of the history of the terms 'visual culture' and 'visual studies' has shown, there is unfortunately still no general agreement on their scope and limits. In frustration, some theorists and practitioners in the field are beginning to sound desperate, or at least "sceptical of the tenets, methodology, assumptions and promise of much of the new scholarship, [...] of some of the directions in which the new scholarship grows, especially those that rip visuality free of its

---

[240]Translated literally: "picture science" or "image science," but left untranslated here for reasons of clarity. On the differences between German, French and English-language scholarship, see Elkins (2007) pp. 287–90.

[241]See, e.g., Sachs-Hombach (ed. 2005), Frank (2006) pp. 30–4 on its first steps towards institutionalization, and Probst & Keuner (eds. 2009) on its intellectual roots.

[242]See Bredekamp (2003) p. 423 for this quote from Warburg's correspondence, and Ullrich (2003*a*) pp. 9f. on earlier efforts to establish a science of images.

[243]See Bredekamp (2003) for his own translation of "Laboratorium kulturwissenschaftlicher Bildgeschichte," a formulation from a German catalogue of Renaissance images.

[244]See Bredekamp (1999, 2004, 2005*a–c*, 2007), respectively, or Ullrich (2003*a*) for the transcript of an informative conversation with Bredekamp.

moorings in history and theory."[245]   Quite a few are already starting to mourn
its virulent crisis or even to prognosticate its imminent death.[246]   A host of
critics from text-bound disciplines, which the more radical advocates of visual
culture studies were trying to marginalize or even replace, has joined the chorus
becrying "an unjustifiable and ungraspable stupidity" or "a new decadence of
illiteracy."[247]   Others, coming more from art history, often miss the old skills of
careful observation and other *fortes* of their discipline in this "new lunapark of visual
culture"[248] with its overemphasis on postmodern media, which can indeed easily
be denounced as "oneiric, anamorphic, junk-tech aesthetics of cyber-visuality."[249]
It might be time now to try something else and to seek out the fruitful roots of
the movement, which somehow got buried under the rubble.

[245] Elkins (2002) p. 93; cf. Dikovitskaya (2005) p. 2: "no consensus among its adepts with regard
to the scope and objectives, definitions and methods."

[246] For instance, Renate Brosch, professor of *Anglistik* at Stuttgart University, in a personal
communication, who until 2007 headed a visual studies program at Potsdam University.

[247] Garcia-Düttmann (2002) pp. 101 and 103; this charge deriving from the Spanish poet José
Bergamin's poised praise of the cinematograph as an "admirably illiterate invention" (ibid., p. 102).

[248] Willibald Sauerländer 1995 as quoted in Dikovitsakaya (2005) p. 15.

[249] See the discussions in the *Visual Culture Questionnaire* (1996).

2

Historiographic layers of visual science cultures

I do share the frustration over lost opportunities to come to grips with the conceptual foundations of this field, but not the doomsday feelings about its future. Rather I suggest we return to a solid, sharp definition of visual cultures as our object, and correspondingly of visual studies as the methodological arena. In this book, I work toward a definition that does not exclude our intended area of application in history *per constructionem*, but allows a systematic comparison with all the many existing studies of visual cultures, be they popular or esoteric. In our careful circumscription, we do not want to throw out the baby with the bathwater. We need a fitting delineation of visual science cultures, capturing their specificities without exaggerating them and without building up new artificial fences. When the 'visual studies' movement started out as a movement in art history, it rebelled against the traditional distinction between high and low art. This might have been necessary at the time, in order to create the freedom and intellectual space for the study of other cultural strands, from cinema to cookbooks, but it blinded some of the protagonists to the fact that there is undoubtedly a "high–low distinction in visual culture." I thus fully agree with W. J. T. Mitchell that this is indeed a "crucial topic" in visual culture studies.[1] That there are mixtures of high and low, as well as borderline cases, and occasional transgressions of these borders does not refute but rather proves this point, as each transgression presumes existent distinctions. The important thing is, though, that it is not some external experts in aesthetics or visual cultures who make these distinctions, but the actors themselves. They are highly discriminatory in what they accept as valid visual representation worth discussing and what not to heed. Analogously to these obvious distinctions between high and low culture, there are intrinsic differences between high and low science and between high and low tech. Both of these can either be visual, or nonvisual. So, what we are after is indeed orthogonal to these old cultural divisions. And Mitchell is right, too, when he points out that this high–low distinction is by no means coincident with qualitative differences of good and bad, or between well done and badly done: In this sense, there are good and bad diagrams in science and technology, just as there are good and bad paintings in art museums.

---

[1]Both preceding quotes are from Mitchell in Dikovitskaya (2005) p. 252.

We will scan the full spectrum of visual representations, from superb exem-
plars of top-notch scientific research to fairly down-to-earth examples of technical
drawings, since they all can contribute to our structural inquiries into the mechanisms of formation and stabilization of visual cultures in science and technology.
Some of these examples will be outstanding in their clarity and informativeness,
others rather meager or even obscure and intransparent, although I must admit
that I did try to concentrate on examples that are somehow digestible for the
nonspecialist.

2.1   'Visual culture' vs. 'visual studies'

One place to start is a clear distinction between 'visual culture' and 'visual studies.'
The canonic reference on this is Mitchell's paper from 2002:

> I think it's useful at the outset to distinguish between visual studies and
> visual culture as, respectively, the field of study and the object or target of
> study. Visual studies is the study of visual culture.[2]

As explained above, 'visual studies' clearly originate from 'cultural studies.' This
led some commentators to demand that their field should actually be called "visual
culture studies."[3] I do not concur with this suggestion. It has remained a lone call
in the dark. Thus, I stick to the briefer and more common term 'visual studies'
whenever I refer to the methodological aspects of a study of 'visual cultures,' be
they cultures of science or technology or whatever else. This understanding of
'visual studies' as a decidedly cross-disciplinary endeavor is quite in line with the
pronouncement of virtually all proponents of this new field, regardless of their
specific affiliation. Thus historian of art James Elkins understands visual studies in
the introduction to his textbook explicitly as "the study of visual practices across all
boundaries."[4] His introduction is – by the way – called "skeptical" exactly because
of this missing disciplinary core, leading to gaping openness, if not to say total
arbitrariness, in its methodology and a kind of "anything goes" mentality of its
practitioners. Actually, it was openly defended by another influential protagonist
of this field as an "indiscipline," as a "moment of breakage or rupture" where the

---

[2]Mitchell (2002) p. 166; analogously, James D. Herbert 2003 as quoted in Dikovitskaya (2005)
pp. 53, 181f.: "visual studies is a name for the academic discipline that takes visual culture as its
object of study"; cf. also the introduction to Schwartz & Przyblyski (eds. 2004) quoted on p. 69.
[3]Most pronouncedly, Frank (2006) p. 43, who suggests the acronym VCS for this.
[4]Elkins (2003) p. 7.

continuity is [...] broken and the practice comes into question."[5] "Indiscipline" can be construed as a shorthand for the hybrid "in[ter]disciplinary" as much as an allusion to the "undisciplined." We have seen in sec. 1.8 that the latter reading has unfortunately become the favorite in recent years.

Now, 'visual culture' might be defined more closely as "a way of seeing that simultaneously both reflects and shapes how members render the world." This, at least, is the suggestion by sociologist Kathryn Henderson, developed with designers and engineers in mind. The "members" in this first vague circumscription (rather than definition) are meant to be members of a 'thought collective' in the sense of the microbiologist and sociologist Ludwik Fleck (1896–1961), or of a group of practioners of science or technology sharing a paradigm in the sense of Thomas S. Kuhn (1922–96).[6] Folded into this understanding of visual cultures is Bruno Latour's earlier insight that "a new visual culture redefines both what it is to see and what there is to see."[7] This is to be understood both verbatim, i.e., referring to *Gestalt* recognition, in general, differing from culture to culture, and metaphorically with respect to the ontology of the world, formed on the basis of such heavily interpreted insights. Strictly speaking, adherents to such a strong cultural reading of seeing and visuality would thus insist on members of different cultures living in different worlds. Indeed, seeing of and seeing as diverge strongly if one compares a layman and a radiologist both looking at, say, an x-ray image of a hypothalamus or a PET image of the human brain. Various disciplines and subdisciplines of science, medicine and engineering strongly differ in these interpretational skills and in the forms of training offered to students and beginners to acquire these visual skills.[8] As explained in sec. 1.1, in the case of radically different 'scopic regimes' they constitute different 'visual cultures' or, in the case of a considerable overlap, merely different 'visual domains' within one and the same visual culture as a reference frame. These visual cultures can be found in all historical epochs, although not all cultures are visual. They encompass specific modes of seeing and surveying, pattern recognition and visual interpretation, favorite aids and instruments for improving or supporting this viewing (e.g., telescopes) and for temporary or permanent recording of these observations (e.g., cameras), adopted media or the distribution and multiplication of these visual records (inscriptions,

---

[5] All quotes from Mitchell (1995) p. 541. Cf. Frank (2006) pp. 63ff. on the consequences of this on the lack of institutionalization, especially in German-speaking universities with their rigid disciplinary core structure.

[6] See Henderson (1995, 1999) p. 25.

[7] Latour (1986) pp. 9f.

[8] For 32 examples across the whole range of 20th century visuality see Elkins (2000).

print, digital storage, etc.), and preferred visual objects, as well as modes of viewing and communicating. How to structure this big convoluted cultural package?
In chap. 3 I will present my suggestion to unpack it as a series of superimposed
layers of culture that can be analytically separated even though in scientific and
technological practice they form a multiply interconnected tiered grid.

2.2   My account of visual cultures as superimposed layers

I have been in search of an integrative definition of visual cultures in science
and technology as well as a sharpened methodology of visual studies since first
entering this field in the mid-1990s. While writing microhistorical case studies,
mostly on spectroscopy and related fields, I already felt the urge to reach the level
of characteristic features of my stories of validity not only to the case at hand, but
representative or typical of a broader class. Here I will make an attempt to spell
out features on this level of non-uniqueness. How far this is actually generalizable
remains to be seen.

The following facets of visual cultures seem to me to be present in very many
case studies in science and technology, although hardly any single out all at once:

1. Highly developed skills of pattern recognition
2. Mastery of visual thinking – *anschauliches Denken* (Rudolf Arnheim)
3. Practical training in such skills (Eugene Ferguson)
4. High prestige attributed to atlases and plates
5. Obsessive quality improvement of representations
6. broader context of specialized print establishments or media experts
7. Interest in the physiology of visual perception and in related research
   instruments
8. Aesthetic pleasure in the scientific procedure
9. A fusion of profession and pastime, of labor and leisure.

These nine layers are, on the one hand, simply different aspects of visual cultures.
All need to be looked at to get a holistic view of these cultures. On the other
hand, each of these layers also defines a different historiographic angle of inquiry
from which we approach these cultures as historians.

Skills in pattern recognition, for instance, lead straight to considerations well
known among *Gestalt* psychologists and cognitive scientists but relatively uncom-

mon among historians of science.[9] I would argue, though, that we need to delve much deeper into these aspects to understand one of the essential strengths of visual science cultures, i.e., the ability for rapid discernment of basic features and their discrimination from diffuse background noise. Such is only possible with outright mastery in the skills of pattern recognition and visual thinking.

These visual skills are a precious resource, and one that isn't just there at the outset but has to be created by carefully conceived and rigorous training. Consequently, historian of science, technology and medicine have to study much more intensely the contexts of practical education in laboratory courses and hands-on sessions in which these *Gestalt*-recognition skills are learnt by novices as well as the visual strategies of experts in their presentations. A whole chapter (8) will deal with the practical training of technical drawing, spectroscopy and radiology as examples from all three base areas.

The high prestige placed on atlases and plates can already be inferred from their costliness. They are often exceedingly pricy, exquisitely produced masterpieces at the time of their production, not even mentioning how expensive they can be for the present-day collector. More important than these economic indicators are the high praise bestowed on such atlases and plate volumes by their users, some-times even centuries later, notwithstanding continued efforts to improve upon the visual representations.[10] When new printing technologies such as lithography (invented in 1796) and cost-reducing variants of long-existing techniques (such as the replacement of copper engraving by steel engraving, with which much higher print-runs could be obtained) became available, the number of printed atlases rose significantly: for geography from less than 3 per year before 1820 to more than 5, sometimes as many as 25, in the years thereafter.[11] Cognitively, these atlases fulfilled an important function as guides for the beginner toward recognizing what is important and dismissing what is irrelevant. To quote from a study on *Medical Knowledge in Britain in the Twentieth Century*:

> The anatomical atlas directs attention to certain structures, certain simi-larities, and not others, and in so doing forms a set of rules for reading

[9] A pioneering effort in this direction was made in Arthur I. Miller's study of *Imagery and Creativity in Science and Art* (1996). More recent examples are the epistemological inquiries by Baigrie (1996), but these initiatives have not been taken up by many colleagues in history departments so far.

[10] See, e.g., Choulant (1920), Herrlinger & Putscher (1972) or Thornton & Reeves (1983) on anatomical atlases from six centuries, and Daston & Galison (1992) on efforts to periodize atlases.

[11] For statistics of cartographic atlases based on the holdings of the Perthes Library in Gotha, see Rupke (2001) pp. 94f. For other highly visual fields such as anatomy, botany or zoology, the corresponding figures might even be larger.

the body and making it intelligible. The atlas enables the anatomy student,
when faced with the undifferentiated amorphous mass of the body, to see
certain things and ignore others. In effect what the student sees is not the
atlas as a representation of the body but the body as a representation of the
atlas.[12]

As is often the case in science, each of these masterworks stimulated obsessive
efforts to raise the level of perfection. Thus, a long and impressive string of sequels
resulted out of the contemporary research, each new atlas striving to improve on
the quality or bring out some still lacking detail, to broaden the scope, scale or
resolution, or to add some subtlety or whatever else for successive generations in
other spheres of the culture.

That this was technically feasible is due to the rich context of the duplicative
media. Specialized draftsmen, engravers and lithographers, working for top-notch
printing establishments, were in high demand to reproduce certain scientific works
according to the highest standards of the day. During the early-modern period,
first Italy, then England and France were leaders in the field. Later, the Netherlands
followed, where a more liberal environment freed printers and their authors from
the political constraints of Catholic Italy and France, for instance.

The remainder of this book (from chap. 4 onwards) will systematically explore
these nine historiographic layers of visual cultures. Each will be individually
examined by means of examples of practices in science and technology across the
board. The similarities among them encountered along the way will – I propose
– underpin this comparative approach to historical analysis.

---

[12]David Armstrong 1983, drawing on Foucault's 1977 analysis of 'political anatomy' as quoted
in Waldby (2000) p. 98.

3

Formation of visual science cultures

The canonical study on the formation of a visual culture in science, and consequently also my first example in this chapter, is, of course, Martin Rudwick's pioneering paper: "The Emergence of a Visual Language for Geological Science 1760–1840."[1] After briefly summarizing his findings for the geological sciences, we will look out for these same elements in two other areas of science and technology.

3.1 Rudwick on geology

The starting point of Rudwick's paper was a broad survey of published materials in the earth sciences of the 18th and 19th centuries. This was combined with an inspection of unpublished materials among the private papers of at least some of the key actors plus a scattering of materials from other actors, either in personal estates, or in institutional files such as the Parisian *Musée d'Histoire Naturelle*, the *École des Mines* or the *Geological Society of London*. Rudwick observed that around 1830, geologists were increasingly making use of graphic images, i.e., maps and diagrams of various kinds, including traverse sections. It was not just the increased number of such visual representations that caught his attention. The changed forms of usage struck him even more. Some publications of the 18th century, such as the *Observations on the Volcanos of the Two Sicilies* (1776) and the eruption of the volcano Vesuvius in 1779 by the traveler–naturalist William Hamilton (1730–1803), were illustrated quite impressively with superb hand-colored engravings.[2] Not serving any particular scientific purpose, these older illustrations rather followed the tradition of landscape painting and gentlemanly travel accounts – one of the most popular types of publication in the early-modern period. Other texts that were greatly admired during the Enlightenment, such as Horace Bénédict de

---

[1] On the following, see Rudwick (1976), later followed *inter alia* by equally masterful analyses of the *Great Devonian Controversy* (1985) and the *Early Pictorial Representations of the Prehistoric World* (1992) and a study on the comparative anatomist Cuvier (1997). Rudwick's pioneering role in the history of the earth sciences has been taken up by various other historians with respect to the study of controversies, but not so intensely with respect to the visual culture of geology.

[2] These two oversize books had more than 50 such colored plates, done by the reputed engraver Pietro Fabris (1740–92) and privately financed by Hamilton, who spend more than £1300 on this luxury edition.

Saussure's *Voyage dans les Alpes* (appearing in four fat volumes of highly verbose text between 1779 and 1796) contained surprisingly few illustrations and relatively mediocre ones at that. By contrast, from c. 1830 onwards, images printed in scientific papers and textbooks and hand-drawn sketches in the private notebooks or correspondence of geologists began no longer merely to function as pretty add-ons, as supplements to the text. They became "an essential part of an integrated visual-and-verbal mode of communication."[3]

The major rise in numbers of images might be explained away by an increasing availability of cheaper printing technologies. Lithography, for instance, was invented in the late 18th century but only came into more common use after the 1830s.[4] It is this altered usage in combination with the emergence of new types of hitherto unknown visual representation that marks a deep change in scientific practice, a change that Rudwick was the first to notice and analyze so pointedly. This transition process is what this chapter will focus on: the formation of a 'visual culture' (as defined in chap. 2), combined with the emergence of a specific "visual language" for that same field – in Rudwick's case the geological sciences (which also encompasses mineralogy, crystallography and paleography).

There is no need to repeat Rudwick's detailed analysis of this transition, which is very recommendable reading in the widely accessible journal *History of Science* (1976). What I aim to bring out here is the structural aspects of his story, i.e., only the parts that might crop up again in other historical episodes of some other scientific or technological field.

First of all, Rudwick discussed the "emergence of a visual language," thus indicating that not only the number and frequency of visual material increased substantially over time. New types of visual representation appeared:

(i) traverse sections through terrain, revealing not only what one sees on the surface, but also the kind of stratification underneath;

(ii) more schematic block diagrams, executed in the style employed in engineering and designed not so much for the sake of representing a succession of layers as to bring out "the possible classes of interactions between various faulted and dipping strata";[5]

(iii) carefully color-coded geological maps; and

(iv) similarly produced mineralogical and petrographic maps.

---

[3]Rudwick (1976) p. 152. On de Saussure and his cyanometer, see here sec. 11.1.

[4]On the history of lithography and its use for purposes of scientific illustration, see, e.g., Jackson (1975) and Blum (1993).

[5]See Rudwick (1976) fig. 22. Today we would speak of geophysical dynamics.

None of these new types of visual representation appear instantaneously on the scene. All of them have precursors, in Rudwick's case often in mining, surveying, assaying and other contexts dominated by practitioners. They did not tend to publish at all and their papers are rarely preserved in archives, so we know far too little about these roots. According to Rudwick, the most decisive step – away from mere illustration toward a structural grasp of the layered features beneath the Earth's surface – was most easily taken by such practitioners:

> a fully structural approach to the interpretation of the complex phenomena of geology was most readily attained within a social context of practical mining and mineral surveying, by individuals whose familiarity with engineering practice 'pre-adapted' them, as it were, to the three-dimensional visualizing that structural geology required.[6]

So in a sense, Rudwick's story of the emergence of a new visual language is by no means one of 'catastrophism,' but "something more uniformitarian."[7] More specifically, it is a gradual transfer of visualization techniques already prevalent among subgroups (here miners or surveyors) adopted by others (here gentleman scientists) who modify and then canonize the new types of representation. Around 1830–40, they were integrated into a new code on how to represent such findings that was accepted by a broad array of people working with those features. This canonization, this consensus on how to and how not to represent, i.e., to draw and to print, produces a new 'visual language.' This "language" is by definition not only used privately by one individual or two, but by a larger group that has agreed on certain standards and rules about translating the findings of others into their own "language."[8] This mutual agreement was coupled with an "increasingly esoteric" character of the codes by which the experts in this culture communicated with each other, experts who came to call themselves "geologists" only from this point forward.[9]

---

[6] Rudwick (1976) p. 169. Cf. here chap. 4 for my broader claims on the generalizability of this finding.

[7] See Rudwick (1976) p. 159 for this nice historiographic parallel to the great conflict in late 18th-century theories on the genesis of the Earth, and ibid., pp. 159ff. for Rudwick's exemplary tracing of the prehistory of geological maps, which "emerged in a more gradual way."

[8] See Wittgenstein (1953) on the problem with "private" languages.

[9] See Rudwick (1976) pp. 178ff. Rudwick (1985) takes up this theme of inclusion of in-members and rejection of outsiders.

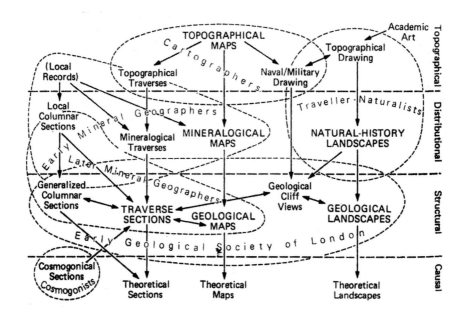

Fig. 14: Rudwick's graphic summary of the emergence of a visual language for geology. His scheme shows a stepwise integration of various types of visual representations (here in the columns) and different social groups (indicated by dashed fields), in four temporally succeeding goals (here indicated as horizontal regimes): first traditionally topographical, then distributional – i.e., the surveyal of mineral deposits, then structural and finally causal. From Rudwick (1976) p. 178.

As indicated in fig. 14, Rudwick was fully aware that this transition of geology to a highly visual field was superimposed on several other transitions that happened to occur at the same time. This immediately raises the chicken-and-the-egg question: What came first? And which was the cause of which? Simultaneously with the emergence of a visual language, the loose agglomerate of various practices, such as surveying, mining, assaying and classifying rocks and minerals, congregated into a new disciplinary unit called "geology." After 1800, several new institutions were founded, such as the *Geological Society of London* (in 1807, the first volume of its splendidly illustrated *Transactions* appearing in 1811), exclusively dedicated to this specialty. Prior to that, it had been studied within a much broader context (at the Parisian *Muséum National d'Histoire Naturelle*, founded in 1793, or at the *Royal Society of London for Improving Natural Knowledge*, founded in 1660). One could thus suspect that the formation of these new institutional contexts transformed the field, and that the increased use of visual material was an incidental side effect of these institutional and social developments. Rudwick raises this possibility, but justifiably rejects it as implausible. The drive to endow the *Transactions of the Geological Society of London* with rich illustrations had been there from the start. Thus, it was not the institution that created this wish (and perhaps contributed the means for it), but the other way round. These institutions only responded to a demand that was already forming. Rudwick's masterful map shows at a glance how representational practices in the emerging field of geology changed over time, in resonance with the social strata of practitioners and their skills. Initially, they were borrowed, then transferred from the academic or gentlemanly art of topographic drawing or of military surveying – eventually new types of visual representations were developed – such as the traverse section and the geological map that then became central for a professionalized and institutionalized study of geology, which thus progressed from a descriptive via a structural to a causal mode.

The lesson we can take away with us from this episode is to pay careful attention not only to the usage and types of visual representation, but also to the social and institutional contexts in which they are embedded. Each emergence of visual cultures is accompanied by a transfer of visual techniques, with an import of images and mental models.[10] Once a new 'visual language' has been established and paradigmatic examples have been given and accepted by the actors, new institutions, scientific societies and perhaps whole disciplines will be formed around this stabilized cluster of visual practices.Acceptance (or rejection) of new types of images or visualizations can depend strongly on cultural contexts, as is shown by the "striking cultural divide in attitudes to visual thinking and visual communication" among the reviewers of Rudwick's book *Great Devonian Controversy* (1985). They were "sharply divided between those who found the diagrams illuminating and valuable and those who hated them and found them unintelligible. Many years later, the most controversial of my Devonian diagrams was included in an exhibition of drawings by scientists, at Cambridge's modern art gallery Kettle's Yard!"[11] It is this fascinating process of forming new 'visual cultures' that we will now study in three other examples from the 18th and 19th centuries, respectively.

---

[10]For nice examples from 18th- and 19th-century technology in the USA, see Hindle (1983).

[11]Quotes from an email by Martin Rudwick dated Jan. 22, 2014; cf. here p. 113 on iconophile and iconophobe environments.

3.2   The architects of stereochemistry

Our next example takes us to the transformation of chemistry from a practice-oriented discipline excelling in instrumental techniques for the purification, analysis and synthesis of substances to a theoretically based field of knowledge in which three-dimensional modeling of the spatial structure of molecules and their chemical bonding gained increasing importance. In this transformation, in which links to architecture will appear repeatedly, a crucial role was played by the organic chemist August Kekulé (1829–96).[12] Born in Darmstadt as the youngest child of a senior councillor of war for the local grand duke and educated at the *Gross-herzogliche Gymnasium*, Kekulé manifested since childhood a vivid imagination and remarkable graphic and mathematical talent.

Fig. 15: A pencil drawing by August Kekulé at 13 years of age. From Hafner (1979) p. 644.

As the sample (in fig. 15) from his early graphic oeuvre demonstrates, Kekulé excelled in the visual arts and in drawing from a young age. Considering that he had already designed houses for the old part of town in Darmstadt while still a high-school pupil and that his father had close connections with famous architects, it was a straightforward decision for Kekulé to study architecture. From 1847 to 1848, he attended classes at the nearby University of Giessen (with the architect Hugo von Ritgen 1811–89) and then one term at the Darmstadt Polytechnic.

---

[12]On Kekulé's vita see, e.g., Kekulé (1890*a*) p. 1307, Schultz (1890), Japp (1898), Anschütz (1929), Hafner (1979), Gillis (1966), Wizinger-Aust (1966), Rocke (2010) chap. 2 and further sources cited there.

By then, the high-quality lectures and practical exercises offered by the chemist Justus Liebig in Giessen had already coaxed Kekulé into chemistry. His preference for visual thinking and three-dimensional spatial intuition remained, however, and proved to be important in his future work. As one biographer put it: "His thoughts turned into pictures and at times he could visually observe his thoughts."[13]

When he later turned to the theory of valency as private lecturer at the University of Heidelberg and then during his ten years as associate professor of chemistry at the State University of Ghent in Belgium, this general disposition bore fruit, again. Kekulé took into account systematic differences in the valencies of chemical elements such as hydrogen, nitrogen, carbon and oxygen. Since these are the four most frequently occurring elements in organic compounds, he searched for a concise notation to depict such valency. Such a "rational formula" would allow one to see at a glance how many atoms of different kinds could combine with each other without leaving uncoupled loose ends, so to say. The result of this search was the so-called "sausage diagrams," jokingly called either *Wurst-Formeln* by Otto Nikolaus Witt (1853–1915), one of its early adherents, or "bread-rolls" by Hermann Kolbe (1818–84), one of its most resolute enemies.[14] Methane (then still called marsh gas), chloromethane (chloride of methyl), phosgene, carbonic acid, and hydrogen cyanide (prussic acid) were denoted as shown in Fig. 16.

Marsh Gas.   Chloride of Methyl.   Phosgene.   Carbonic Acid.   Prussic Acid.

Fig. 16: A few simple sausage diagrams by August Kekulé, 1861. The black unit of four denotes carbon, the gray-shaded single, double or triple units oxygen, and the white circles hydrogen. From Kekulé (1861) p. 162; as redrawn by Alexander Crum Brown in his 1861 MD dissertation, for clarity's sake here taken from the reset online version, p. 17. Cf. also Ritter in Klein (ed. 2001) p. 38 for Crum Brown's original drawings by hand.

These sausage diagrams were initially designed to be maximally compressed so as not to take up much print space, and originally only oriented along one axis. However, some molecules of organic chemistry eventually forced Kekulé to move away from this horizontal axis into the second, the vertical dimension.

---

[13]Wizinger-Aust (1966) p. 8 (my translation). Hein (1966) p. 2 also insisted that Kekulé "retained an architectural sense all his life" and fittingly called Kekulé "the architect of chemistry."

[14]On the coinage of these terms, see Kekulé (1865), Hafner (1979) p. 647 and further sources related to Witt and Kolbe, respectively, listed there in notes 39–40 as well as in Rocke (1993, 2010).

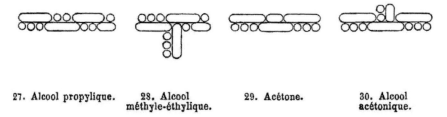

27. Alcool propylique.   28. Alcool   29. Acétone.   30. Alcool
                            méthyle-éthylique.   acétonique.

Fig. 17: More complex sausage diagrams of alcohols. The two on the left and on the far right are isomers, with three carbon atoms (already depicted more symbolically), eight hydrogen atoms and one oxygen atom, differing only in spatial orientation. From Kekulé (1865a) p. 110.

This shows a convenient side effect of the new notation: Kekulé began to distinguish molecules as different kinds of exactly the same number and sorts of constituents, differing only in their spatial orientation: so-called 'isomers' (literally translated as equal weights). Spatial thinking soon became imperative to keep orientated within the jungle of compounds of organic chemistry, and Kekulé's notation was a first step in the right direction. Another use for these sausage diagrams was a new classification of organic molecules according to structurally identical core parts, differing only in what was then termed 'radicals,' i.e., easily interchangeable appendages.[15] One molecule, in particular, forced Kekulé to step beyond the move from linear to planar molecules: the aromatic resin known since the late 15th century by apothecaries and tradesmen under the names benzoine, benzin, benzol or benzene. Michael Faraday had first successfully isolated it chemically and given it the new name, "bicarburet of hydrogen," already noting that the substance was only composed of carbon and hydrogen. By 1865, Liebig's methods of quantitative analysis of carbon compounds had revealed that the sum formula for this molecule was C_6H_6. That presented the problem. Since 1858, Kekulé had already inferred from many other chemical carbon combinations that carbon had four free valences – hydrogen only one.[16] Thus, there was no way to combine six carbon and six hydrogen atoms in his sausage diagram without leaving a few valences open. That went against the syntactic rules of the notational system. Playing with the options left open within his notational system – in other words, using creative visual thinking (cf. chap. 10) – he came up with the solution of recoupling the terminal end of the open chain with its likewise-open head (see

[15]On this point, see, e.g., Hafner (1979) p. 648, Goodwin (2010) pp. 623f., and critically Kolbe (1877), Kolbe (1881a) pp. 311, 353. Cf. also Rocke (1993, 2010).

[16]See Kekulé (1858a) and (1858b) p. 379 on the *nature quadriatomique* of carbon. Cf. Couper (1858a) p. 1158: "la limite de combination du carbone est égale à 4."

fig. 18). Thus, he was the first to suggest a cyclic ring structure for benzene in 1865. The courage to tinker with his notation beyond what it was built for, as well as virtuosic spatial thinking, had induced him to this step. He broke an unwritten rule of chemistry, which until then had been devoid of any such topologically closed ring structures.

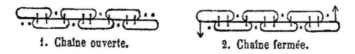

1. Chaîne ouverte.   **2. Chaîne fermée.**

Fig. 18: The step from an open chain of six carbon atoms (the oblongs, each with four valences) and six hydrogen atoms (the single dots, denoting a valence of one) to a closed chain (right). Topological closure is to be imagined between the two open slots at the far right and left, marked by arrows in the right diagram. From Kekulé (1865a) p. 108.

How Kekulé had come up with this ingenious solution is a matter of endless discussion, particularly since the so-called *Benzolfest* of 1890, the celebration of the 25th anniversary of his discovery – therefore a safe distance away from the actual event. At this occasion, Kekulé told his surprised chemist colleagues that he had dreamt it up. The image of a serpent or dragon devouring its own tail had suddenly flashed before his eyes in a daydream.[17] When Kekulé gave this famous speech on the importance of imagination (the German term he used was *Phantasie*), he encouraged his colleagues, many of them still educated in the old paradigm of Berzelius's symbolic formulas, to use their creativity and imagination, but not totally without restraint:

> let us learn to dream, gentlemen, then perhaps we shall discover the truth; but let us beware of publishing our dreams before they have been scrutinized by our vigilant intellect. [...] Let us always allow the fruit to hang until it is ripe. Unripe fruit brings even the grower but little profit; it damages the health of those who consume it; it endangers particularly the youth which cannot yet distinguish between ripe and unripe.[18]

Kekulé had waited long with this publication of his own private thoughts on structural issues and often only reluctantly exposed his ideas in public. By contrast,

[17]The ouroboros (or uroborus) symbol has a long alchemical tradition. The often not reliable literature on this episode is endless, including chemists like Scherf (1965), Wizinger-Aust (1966) pp. 21ff. and Wotiz & Rudofsky (1984) as well as the famous psychoanalyst Alexander Mitscherlich. For a survey with convenient excerpts, see www.sgipt.org/th_schul/pa/kek/pak_kek0.htm (last accessed May 14, 2011).

[18]Kekulé (1890) p. 1307, Engl. transl. by Klaus Hafner (1979) p. 641. Cf. also Rothenberg (1995).

Archibald Scott Couper (1831–92) had rushed his ideas on structural formulas into print without first seeking empirical validation; that was something Kekulé did not want to endorse.[19] The historian of chemistry Alan Rocke (*1948) has devised a hyphenated spelling of imagination, "image-ination," to denote Kekulé's conscious use of the heuristic power of models and structure formulas as a graphical language of chemistry.[20] The 1860s, when Kekulé made his ingenious leaps of imagination, were a time of epistemological shifts from chemical approaches to increasingly physical approaches, that is, from issues of chemical bonding and valency to questions of the actual physical structure of matter. Kekulé's sausage formulas were only a half-step in this direction away from Berzelius's and Daumas's type formulas toward full structural formulas.[21]

Fig. 19: The benzene ring in three dimensions. Left: A perspectival woodcut of the benzene ring structure from Kekulé's (1861ff.) *Lehrbuch*, vol. 2 (1866) p. 515; Right: Kekulé's 3D model of benzene still in the mode of his sausage model (c. 15 cm diameter and 2 cm height). Courtesy of the Museum for the History of Sciences (Ghent University, inv. no. MW 95/118).

Kekulé soon ceased to use his own sausage diagrams in favor of improved structural formulas, first suggested by Alexander Crum Brown (1838–1922) in his

---

[19]See Kekulé (1858a, b) for a commentary on Couper (1858a). Cf. Rocke (1985) p. 378. Kekulé's student Richard Anschütz (1909) later acknowledged Couper's priority: "without any doubt, Couper deserves the credit of having introduced into constitutional chemistry the lines indicating union of atoms, and of having thus produced what are now called structural formulae." Cf. also Dobbin (1934) and Larder (1967) pp. 114f.

[20]Rocke (1985) and (2010) p. 324. Kekulé himself used arguments of homology and symmetry.

[21]On this transition see, e.g., Meyer (1890), Crosland (1962), Ramsey (1974, 1975) and Ursula Klein, Christopher Ritter and Pierre Laszlo in Klein (ed. 2001).

Edinburgh MD thesis of 1861. Edward Frankland (1825–99) then simplified it in his *Lecture Notes for Chemical Students* (2nd ed. 1872) and it was soon canonized in a host of other textbooks. In his thesis, Crum Brown criticized Kekulé's sausage diagrams as "a most artificial [representation which] certainly does not represent the actual arrangement of the atoms."[22] As an alternative way to represent what he dubbed the "chemical position of the atom" in a molecule, Crum Brown suggested representing each atom in the molecule by its chemical symbol (i.e., H for hydrogen, C for carbon or O for oxygen), surrounding each such chemical symbol by a circle to denote the chemical atom and attaching as many lines to each circle as that atom has free valences or "equivalents," as Crum Brown then called it. "When equivalents mutually saturate one another, the two lines representing the equivalents are made continuations of one another, thus water is Ⓗ–Ⓞ–Ⓗ."[23] This might at first look like a fairly roundabout way to rewrite the old Berzelian chemical sum formulas, yet his splitting up what in the old sum formula for water (H_2O) would have been H_2 (for two hydrogen atoms) into two separate H's already forced him to think about how to place the oxygen atom. Of course, the O must be set centrally between the two H's. While this might sound fairly trivial, for more complex molecules, this spatial distribution of all the atoms of a larger molecule onto a plane soon led to insights into the fact that various spatial relations are possible. These inherent ambiguities could then be linked to cases of isomerism, where more than one molecule was known to have the same chemical sum formula, i.e., where the same number and types of constituents had led to chemically different substances just because of their various spatial structures. In his dissertation, Crum Brown was very careful not to overinterpret his own notation:

> I do not intend it to be supposed that this represents correctly, or even more correctly than Kekulé's method does, the actual arrangement of the atoms, but it is at least as probable; and all that I wish to show is, that his is not the only possible arrangement.[24]

After all, Crum Brown's new structural notation was still limited to a planar two-dimensional distribution of atoms, whereas in reality they were, of course,

---

[22]Crum Brown (1861*b*) p. 17. Cf. ibid. pp. 18ff. for his visual thinking with these structural formulas, here color pl. XIV, for one of his 3D mathematical models, and Gill (1969) p. 641 for his involvement in stereoscopy.

[23]Crum Brown (1864) p. 708. Cf. also Walker (1923) and Larder (1967) pp. 126ff. for Crum Brown's alternative resembling our modern way of drawing structural chemical formulas.

[24]Crum Brown (1861) p. 24. Cf. also Larder (1967) p. 126 and Rocke (2010) pp. 91, 143ff. on these interpretational subtleties.

positioned in three-dimensional space. As Kekulé's papers and lectures from the 1860s show, he himself was very interested in the 3D structure of molecules and matter. He built 3D models of his sausage diagrams with a metal clamp for each valence that fit into prepared holes in the sausages (cf. fig. 19). In his lectures, he apparently also used simple versions of the ball-and-stick model (see fig. 20). In his textbook on organic chemistry, Kekulé wrote already in 1861:

> It makes intrinsic sense that one cannot depict the arrangement of atoms in space – even if it had been figured out – on the plane of lettering set one after another on paper; rather that, for that, one would at least need a perspectival drawing or a model. But it is equally clear that one cannot find out the positions of atoms in an existing bond by studying the metamorphoses.[25]

By 1865, Kekulé had adopted Crum Brown's structural notation. Using his own version of the ball-and-stick model, he came up with a plausible structure of mesitylene in 1867 (fig. 20)

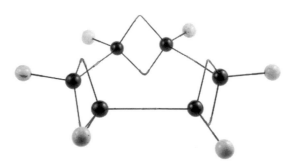

Fig. 20: Kekulé's ball-and-stick model (c. 60 cm diameter and 15 cm height) of mesitylene (1,3,5-trimethylbenzene), cyclically combining three acetone molecules, dated c. 1867. Courtesy of the Museum for the History of Sciences (Ghent University, inv. no. MW 95/116).

At roughly the same time, Kekulé's former pupil and chemist colleague August Wilhelm Hofmann (1818–92) also used ball-and-stick models. He made them out of colored croquet balls to demonstrate, very effectively, the iterative construction

---

[25] Kekulé (1861ff.), quote from vol. 1 (1861) pp. 157f. (transl. by A. M. Hentschel).

[26] See Kekulé (1867, 1869). On Kekulé's steps in the direction of stereochemistry, see, e.g., Meyer (1890) pp. 8, 12, 91, Anschütz (1929), Wizinger-Aust (1966) pp. 23ff., Gillis (1966) pp. 37, 44f., 48ff., Benfey (ed. 1966) pp. 7ff., 72ff., Ramsey (1974) pp. 8–11, Hafner (1979) pp. 647f., Paolini (1992) and Rocke (2010).

of organic molecules for his lectures held at the London *Royal Institution*.[27] By the way, Hofmann – like Kekulé – had also studied architecture until Liebig's lectures had lured him away to chemistry. Hofmann's architect father had, in fact, built Liebig's laboratory extensions! The full breakthrough to stereochemistry was achieved by Kekulé's student Jacobus Henricus van 't Hoff (1852–1911) with his manifesto on the arrangement of atoms in space.[28]

I would like to mention as an aside that it has been conjectured that the contemporaneous introduction of pea models into the German *Kindergarten* by the German pedagogue Friedrich Fröbel (1782–1852) had influenced this move to ball-and-stick models.[29] If this interesting conjecture is true, it would be a further example of feedback from cultural contexts into the scientific imagination. Much more certain is the strong influence of architectural inclination and training on Kekulé's visuality, a thing he cherished very much.

3.3 Sorby: Microscopic petrography and metallography

If in the year 1800 someone had predicted that it would soon be a routine matter to examine rocks by microscope, that person would have been considered crazy. Why look at mountains through a microscope?[30] Nevertheless, by the mid-19th century, the discovery of polarization microscopes in conjunction with the development of instruments for wafer-thin slicing of samples had made it possible to do exactly that. A pioneering paper in 1858 championed this use of a microscopic technique, for which the term (microscopic) 'petrography' was

---

[27]See Hofmann (1865*a*) pp. 416ff. On the further distribution of such chemical 3D models, see, e.g., Ramsey (1974) pp. 6ff., Meinel in de Chadarevian & Hopwood (eds. 2004) and further sources listed there. Meyer (1890) pp. 12f. reports about cheaper rubber models designed by the Karlsruhe chemist Paul Friedländer (1857–1923).

[28]First published in Dutch in 1874 and in French in 1875; cf. Ramberg & Somsen (2001) for a new, improved English translation. On van 't Hoff's stereochemistry, see Meyer (1890), Ramsey (1975), Snelders (1975), Paolini (1992), Meijer (2001), Root-Bernstein (2001), Meinel (2004) pp. 243, 265f. and Spek (2006).

[29]Fröbel had first studied architecture and then physics, chemistry and mineralogy, and he had worked as assistant to the crystallographer Christian Samuel Weiss (1780–1856) in Berlin before switching careers and turning to early child pedagogy and opening his first *Kindergarten* in 1837. See Meinel (2004) pp. 267ff. and (2009) p. 19. On Fröbel's intuitive training methods in German and Austrian kindergartens since 1844, from 1848 on also in British and US contexts, see Meinel (2004) p. 267, (2009) and Wollons (ed. 2000).

[30]That this is no mere rhetoric but an objection raised at the time becomes clear in the quote on p. 106 here.

minted and which later became an acknowledged subdiscipline of geology.[31]

Likewise, if in the year 1850 someone had foretold that it would soon become routine to examine steel samples by microscope, this visionary would have been considered just as insane. More so than rocks, steel and other metals are totally opaque, even in very thin slices, and, pray, what should one expect to see at such a high resolution, anyway? Nevertheless, by 1864, this very step had actually been taken. A Victorian gentleman scientist and naturalist who was familiar with the techniques of microcrystallographic examination had wondered why it could not also be used to examine metals. This section is thus the story of a twofold transfer of techniques, first from botany (and other areas of natural history where the microscope had long figured prominently) to crystallography and thence to what was soon to be called (microscopic) 'metallography.'[32] Within one generation, the odd practice of an innovative and queer individual had become accepted and canonized in various handbooks, atlases and institutions. The emergence of these two new subdisciplines of petrography and metallography was thus intimately linked with the formation of two new visual science cultures, both exhibiting many parallels with the formation of geology as a visual culture summarized in the foregoing subsection.

Various people played a part in these two transitions, but no one disputes that one researcher had a major role in both of them. Henry Clifton Sorby (1826–1906)[33] was born into a wealthy Victorian middle-class family living in Woodbourne near Sheffield. The son of a tool manufacturer was privately tutored then, after leaving the *Sheffield Collegiate School* at age 15, never attended college or university. With the comfortable resources of his private means, Sorby was able to maintain his own scientific laboratory and workshop and freely pursue his own lines of investigation as "unencumbered research," not having to justify himself or to report to anyone. This freedom of the self-educated amateur and gentleman scientist made it possible for Sorby to pursue ideas that would have been considered crazy or mind-boggling to his contemporaries.[34]

After a brief brush with crystallography, agricultural chemistry and botany,

---

[31]On the early history of petrography, see, e.g., Judd (1908), esp. pp. 199ff., Loewinson-Lessing (1954) and Humphries (1965).

[32]For surveys on the prehistory and early history of metallography, see Pasch (1979), Smith (1960, 1969, 1977), Piersig (2009) and further sources cited there. On crystallography, cf. here pp. 331ff.

[33]On Sorby's life and work, see, e.g., his two autobiographical notes Sorby (1876a, 1897). Sheppard (1906) and Smith (ed. 1965) pp. 43–58 have nearly complete bibliographies of Sorby's roughly 250 papers; see also Judd (1908), Higham (1963), Hammond (1989) and Piersig (2009).

[34]On his privilege of "unencumbered research [...] in complete immunity from routine employment," see especially Sorby (1876a) p. 149 and Judd (1908) p. 194.

Sorby turned his attention to physical geography and geology, particularly the mechanisms of rock denudation and the formation of river terraces. In his private laboratory, Sorby frequently used his microscope to examine all kinds of samples. For some reason, he was firmly convinced that this research instrument should also be used to examine the various rocks and crystals that he collected. In order to do so, he had to prepare very thin slices of these solid materials at thicknesses of typically only 1/1,000th of an inch (c. 0.025 mm), by no means an easy feat and one that took him until 1849 to achieve.[35] In a paper "On the Microscopical Structure of Crystals" that appeared in the *Quarterly Journal of the Geological Society of London*, Sorby summarized some of his important findings (cf. fig. 21). Overcoming initial apathy and opposition, Sorby eventually managed to convince his contemporaries that this idea was indeed worth pursuing further.[36] The conferral of the Wollaston prize by the *Geological Society* to Sorby in 1869 evidences gradual acceptance of this new visual technique, 12 years after the publication of his major research paper on his early microscopic findings. One decade later, Sorby even became president of this prestigious society. He published further petrographic results on the structure and origin of limestone and on noncalcareous stratified rocks.

Sorby's heavy reliance on the microscope as a generic research technology soon bore other fruit. For one, he became president of the *Royal Microscopical Society*, an acknowledgment of Sorby's ingenious use of microscopes for new classes of objects. In 1857, he was also elected a fellow of the prestigious *Royal Society of London*, an acknowledgment in general of his pioneering work in microscopic petrography, and in particular of his proof in 1853 of how so-called slaty cleavage is formed by mechanical pressure.[37] By 1865, he had developed a new type of spectrum microscope with which small samples of organic pigments, especially minute bloodstains, could be examined.[38]

Like many pioneers, Sorby left the consolidation and refinement of a field to others. – He himself found more satisfaction in opening up new fields. Quite in line with this general trait, he switched from the examination of rocks to their microcrystalline content, some of which also exhibited strange minute fluid

---

[35] According to Sorby (1897) p. 5 and Judd (1908) p. 195 and Higham (1963) pp. 36ff., Sorby was assisted here by William Crawford Williamson (1816–95), a local naturalist specializing in the preparation of thin sections of hard substances with the aid of diamond-hardened energy-wheels.

[36] On the early reception, in which skepticism dominated until 1861, when interest, especially among German geologists, began to mount, see Judd (1908) pp. 198ff., Higham (1963) chap. 2 and Humphries (1965) pp. 28f., 32ff.

[37] On this point, see Higham (1963) pp. 41ff.

[38] See Sorby (1865, 1867*b*, 1875*b*, 1876) and Higham (1963) pp. 78–86.

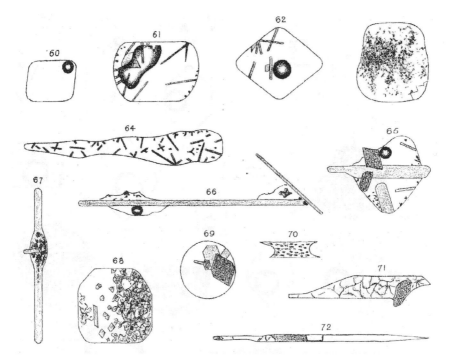

Fig. 21: The earliest micropetrographic plate. From Sorby (1858*a*) plate XVIII, nos. 60–72 (of altogether 120 samples), observed at magnifications between 60× and 1,600×; all plates drawn and lithographed by Sorby himself.

cavities. He next turned to the mechanisms by which rocks and stones might have formed millions of years ago. In 1862, he also started to compare terrestrial samples with meteorites. Metallic meteorites posed hitherto unknown problems for Sorby, since, no matter how thin he cut his slices, unlike rocks, they never became (semi)transparent, always remaining absolutely opaque. He was forced to modify his microscopic setup quite drastically, reverting from light transmission to light reflection.[39] The latter only made sense if he first prepared an absolutely smooth surface on his metallic specimen by several stages of polishing and then etching the polished surface, an involved procedure that could take up to five weeks per sample.[40] However, he managed to bring out very clearly the typical

[39]On the design of these light-reflection microscopes with parabolic mirrors and oblique illumination, see, e.g., Hammond (1989) p. 4 and Edyvean & Hammond (1997) pp. 55f. On precursors and on later progress in the design of these metallurgical microscopes, see Orcel (1972) and Piersig (2009).

[40]On the simple methods Sorby used in the early days for preparing samples, see Judd (1908)

Widmanstätten structures already known to experts, since they are clearly visible with the naked eye in some cut meteorite specimens. Suspecting that these crystal-like criss-cross patterns were characteristic of all ferrous materials, it was an obvious move for Sorby to next compare his micrometallographic samples of meteoric iron with hammered iron and steel samples from local suppliers in the vicinity of Sheffield. By 1863, again a new field had emerged, microscopic metallurgy, to which Sorby contributed some of the earliest insights. From 1864 on, Sorby cooperated with the photographer Charles Hoole in efforts to actually photograph select views through his microscope. But only in 1887 did it become technically feasible to print "in a satisfactory manner" ninefold enlargements of these early photographs by the Woodburytype process.[41]

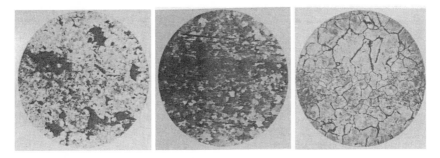

Fig. 22: Samples of Sorby's Woodburytypes of metallographic specimens, from left to right: hammered bloom, bowling bar iron (longitudinal section) and blister steel (also longitudinally), clearly exhibiting very different microstructure. From Sorby (1887) figs. 1, 2, 7.

Before this photographically enlarged reproduction became possible in 1887, Sorby resorted to a kind of direct printing from the specimen itself, called 'nature printing.' He explained the procedure followed in February 1864 as follows: "In preparing the blocks used for nature-printing, the specimens were cut type-height, one side made flat and roughly polished, and sufficiently well bitten with moderately strong acid to enable me to print with ink as from a wood block." These prints were first exhibited in natural size at the Bath meeting of the *British Association for the Advancement of Science* in September 1864. The earliest

p. 201 and Sorby in Beale (1868) pp. 181–3, extensively quoted in Humphries (1965) pp. 25–7. On the later progress, cf., e.g., Piersig (2009).

[41] See Sorby (1887) p. 255 (quote), likewise p. 259: on the "reproduction of these prints, some of the results are extremely satisfactory"; and Sorby's 1887 plates with figs. 1–17. On Hoole, see Croft (2006) p. 45, who mentions that Hoole was registered as a commercial photographer in Sheffield.

reproductions of Sorby's photomicrographs by the new heliotype process were distributed in a small print run at the annual conversazione of the *Sheffield Literary and Philosophical Society* on February 1, 1872.[42]

To gauge the progress of printing technology and resolution obtained since, I also reproduce two modern photographs taken from Sorby's samples in fig. 23.

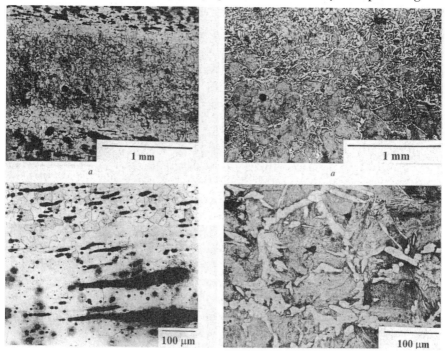

Fig. 23: Two samples of Sorby's metallographic specimens, each in two different resolutions as indicated below, labeled "Stuart's broken axle long," prepared in 1864. In modern enlargement. From Edyvean & Hammond (1997) p. 71, with permission of Maney Publishing.

In the early days of metallography, given the very limited quality of reproduction and enlargement, the collection of real samples exhibiting the major features was crucial. Sorby started collecting large numbers of these specimens (fig. 24), 69 of which are fortunately preserved at the Department of Metallurgy of the University of Sheffield.[43] The collections of photographs and lantern slides likewise soon grew to enormous proportions, requiring special cabinets to store them (fig. 25).

---

[42]Preceding quote from Sorby (1887) p. 259; cf. also Entwisle (1963) p. 321 for a facsimile of a surviving copy of the 1872 heliograph. It is not reproducible here because of the bad quality of reproduction in the journal *Metallography* of 1963!

[43]See Entwisle (1963) pp. 317ff. and the detailed survey by Edyvean & Hammond (1997).

Fig. 24: Photograph of Sorby's original display of microsections of iron and steel, 1863. From Edyvean & Hammond (1997) p. 66, with permission of Maney Publishing.

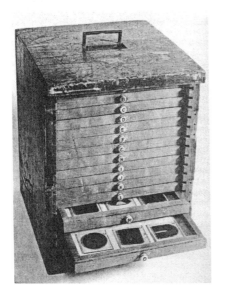 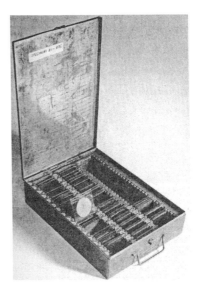

Fig. 25: Sorby's wooden specimen cabinet (left) and lantern-slide box (right). Both reproduced here from Edyvean & Hammond (1997) pp. 62, 64, with permission of Maney Publishing.

What in Sorby's case was still a very idiosyncratic practice of somewhat obsessive image taking and sample collecting became within roughly one generation the normal routine in materials testing laboratories all over the world.[44] The first handbook on materials science (*Handbuch der Materialkunde*, 1899) by Adolf Martens (1850–1914), the director of the Berlin materials testing laboratory, and the first richly illustrated atlas of typical metallurgical and metallographic images published in 1914 by Carl von Bach (1847–1931), professor of steam-engine mechanics, elasticity and machine components at the Stuttgart Polytechnic, indicate that this institutionalization process for metallography was provisionally complete. It had become an accepted subdiscipline of metallurgy and was being applied in materials testing laboratories, i.e., in state-funded institutions for quality control and research into materials improvement. A new visual culture of technology had formed.

To summarize this case: Sorby's retrospective statement from an address he delivered before the members of the *Sheffield Literary and Philosophical Society* at *Firth College* in 1897 shows that it had been anything but trivial or easy to establish two new applications of the microscope in petrography and metallurgy:

> In those early days people laughed at me. They quoted Saussure who had said that it was not a proper thing to examine mountains with microscopes, and ridiculed my action in every way. Most luckily I took no notice of them.

Likewise, about the later transition to metallography:

> In those early days, if a railway accident had occurred and I had suggested that the company should take up a rail and have it examined with the microscope, I should have been looked upon as a man to send to an asylum. But that is what is now being done.[45]

Why was it Sorby and not some other, equally independent, gentleman scientist with enough leisure also to afford to play around with microscopes? In my view, the reason is Sorby's specific combination of visual skills and interests, which gave him a certain advantage over his contemporaries: "He happened to become introduced to crystallography at an early age by his tutor, Rev. Walter Mitchell, and he labored to make himself proficient in drawing and the representation of objects in colour."[46] Following his naturalist inclinations, he prepared maps of

---

[44] On the further development of metallography and the growing institutionalization of materials testing laboratories c. 1900, see, e.g., Piersig (2009), Hentschel (2011*a*) and further sources cited there.

[45] Both preceding quotes from Sorby (1897) p. 5 et seq. On H. B. de Saussure, see here sec. 11.1.

[46] Both points and the quote are from Judd (1908) p. 194. Cf. also Humphries (1965) p. 18.

Yorkshire riverbeds in order to document changes in their courses, and adapted this mapping approach to the specific visuality of the microscope when conducting various other studies.[47] For his analyses of sedimentary rocks and cleavage, Sorby also experimented with compressed stacks of colored paper layered with wet pipe clay, thus essentially pursuing an experimental simulation approach.[48] This, combined with his use of 3D models, is another hallmark of visual cultures. His petrographic plates were often illustrated with lithographs designed and executed on stone by Sorby himself.[49] His later work on microscopic metallography as well as his final studies on algae incorporated extensive drawings and increasingly – wherever possible – also photographs. These are further indicators of a very visual type of person. His delicate drawings of observed micrographic structures required adequately refined reproduction. He used photomicrography and the newest printing technologies available, such as the Woodburytype process and 'nature prints' (discussed above). His yacht, the "Glimpse" – itself quite a visual metaphor! – was a fully equipped floating laboratory, in which he assembled and examined all kinds of marine animals. He also developed special techniques for mounting rather flat specimens onto a lantern slide in Canada balsam so that they could be projected onto a screen without too much distortion.[50] He furthermore investigated natural dyes in blossoms. Among his other hobbies and interests, we find architecture, archeology, palaeography, Egyptian hieroglyphics, medieval manuscripts and early maps; in his free moments, he also painted in watercolors. Undoubtedly, Sorby possessed several key skills and resources typical of visual cultures of his day. Scientific acceptance and institutional stabilization of the fields he had created came afterwards. Without his pioneering work, they might not have arisen; and without his strong visuality, he would not have become the pioneer in so many of them.

3.4 Wheeler and geometrodynamics

Some people might suspect that new visual science cultures were characteristic of the 19th century with its high dynamics of discipline formation and social differentiation.[51] In the face of such a proposition, I would like to discuss at least

---

[47] All preceding facts are from Sorby (1897) pp. 4ff. and Judd (1908).

[48] For details, see Judd (1908) pp. 196f.

[49] This is the case, for instance, with all five plates of Sorby (1858), signed "H. C. Sorby del. et lith." Cf. Higham (1963) p. 85 on a case in which Sorby even wanted to resort to chromolithography to depict absorption bands in blood spectra.

[50] See Entwisle (1963) p. 322 and Edyvean (1988).

[51] For a detailed account of differentiation in the sense of Luhmann, see Stichweh (1984).

one example clearly outside this temporal range, one from the second half of the 20th century: John Archibald Wheeler (1911–2008) and 'geometrodynamics,' as he called a specific approach within general relativity originating with his work.

Wheeler is generally mostly known today as the creator of amusing terms for weird objects and effects in relativity theory and quantum mechanics, such as geons, black holes and quantum foam. All of them have a striking visual connotation that makes them so appealing in the first place. Others might know that he was also a close collaborator of Niels Bohr, with whom he developed the so-called liquid-drop model of the atomic nucleus. This excursion into nuclear physics also led him to work for the Manhattan Project during World War II.[52]

Fig. 26: Cover of Misner, Thorne & Wheeler's textbook (1973). Courtesy of Charles W. Misner, Kip S. Thorne and John Archibald Wheeler.

Only serious students of general relativity might also know his masterful visual displays of things and theorems, otherwise totally incomprehensible or boringly formal. The textbook on *Gravitation* that he edited together with two of his most brilliant students, Charles W. Misner (*1932) and Kip S. Thorne (*1940), affectionately referred to as "the telephone book" or "the bible" because of its enormous dimensions – 1,279 quarto pages – is literally packed full of them:[53]

---

[52]For (auto)biographical accounts of Wheeler, see Wheeler & Ford (1998), Mills (2009) and Misner (2009) and the interviews by Kuhn & Heilbron (1962), Weiner & Lubkin (1967), Ford (1993) and Bičak (1978/2009).

[53]On MTW, as this huge textbook was also called, see Kaiser (2012) from the point of view of Wheeler's "experiments in genre" and Aaron Wright (2011) specifically on Wheeler's visual thinking with the aid of Penrose diagrams. On the publisher William H. Freeman, cf. also Heumann (2013) p. 322 and www.whfreeman.com/catalog/static/whf/college/datelinefreeman/news/history.htm (Feb. 2, 2014).

- perspectival tricks and artificial spatial foreshortenings
- cut-outs and extreme zooming
- displays of topological relations
- graphic transformations between coordinate systems or representations[54]
- combinations of different types of visual representations, etc.

Gravitation offers perspectivalism at its fullest; virtuosic switches between different types of representations and often ingenious combinations in unheard-of juxtapositioning. The image shown here in fig. 27, for instance, brings together graphics not normally encountered within a single figure on general relativity.

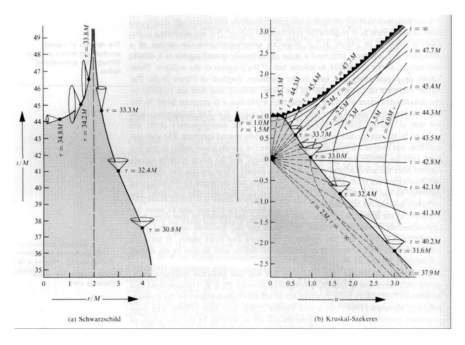

Fig. 27: The free-fall collapse of a star in Misner–Thorne–Wheeler mode, with the region of space–time inside the collapsing star marked gray, and the outside white. From Misner, Thorne & Wheeler (1973) fig. 32.1, p. 848. Courtesy of Charles W. Misner, Kip S. Thorne and John Archibald Wheeler.

[54] For instance, in Misner, Thorne & Wheeler (1973) fig. 31.4, p. 835, between Schwarzschild and Kruskal–Szekeres coordinates, or the one reproduced in fig. 27 here.

The graphs in fig. 27 combine at least four types of visual representation: perspectival cylinders and cones with shading $\oplus$ the world lines from the Minkowski diagram $\oplus$ light cones moving along a world line $\oplus$ Finkelstein coordinates. Instead of cluttering up the image too much, Wheeler managed to merge these four different types of representation into a surprisingly coherent image bringing across quite clearly the central idea: "the free-fall collapse is characterized by a constantly diminishing radius r [...] in a finite co-moving proper time interval $\Delta\tau = 35.1M$."[55]

Numerous photographs show Wheeler in front of a blackboard filled, not with equations like that of the archetypical physics teacher, but with neat, detailed and amusing cartoons, outdoing even those illustrating his textbooks.

Fig. 28: Wheeler at the blackboard. Photograph by Robert P. Matthews, 1971. Princeton University Archives, Special Collections, with permission of the Princeton University Library.

In general, one has to be careful not to overinterpret such photographs, which are often staged in critical awareness of the presence of the photographer. But the high number of such shots of Wheeler standing at a blackboard covered with his graphic sketches quickly dissolve any doubt in his case. Further sources corroborate this impression. I am thinking particularly of excerpts from notes Wheeler had written on sheets of ruled yellow paper to clarify certain points for his students. They are filled just as much with drawings of all kinds and only a

[55]Misner, Thorne & Wheeler (1973) p. 848.

minimal number of equations or other nonvisual elements.[56] For Wheeler, arguing and thinking in drawings was not showmanship but a part of life.

Where does Wheeler's compulsion to visualize weird effects come from? My hypothesis is: from Wheeler's early training. If we go through Wheeler's early training and courses, two things stick out as surprising and noteworthy:

(i) He did not start out as a physics student, but initially studied engineering and only later switched to physics, proceeding directly on to his PhD thesis without having first obtained his bachelor's or master's – a completely impossible feat nowadays. About this early training in engineering in Baltimore, where he was admitted as a freshman in 1927, he later remembered particularly the following: "Ever since my mechanical drawing courses as an undergraduate at Johns Hopkins University, I have enjoyed being my own artist and draftsman."[57]

(ii) The other unusual thing is his artistic mindset. After the war, in 1949, he bothered to take drawing lessons from a teacher who had been trained in Paris (at the *École des Beaux Arts*). Three decades later, he still specifically remembered them, when asked by a Czechoslovakian interviewer in 1978. Quoting this French teacher, he recalled:

> "He said that his fellow students there were so well trained in observing things carefully and accurately, to get the truth [...] This made a great impression on me – this concern for accuracy and truth.
>
> [...] in art you are trying to distill out of the situation some central thing and find out what that central thing really is and capture it in its naked essence, free of all complications. And that to me is what is so impressive in science. [...] So, to me there is a very great similarity between the two: the search for truth and the search for the absolutely central point."[58]

Wheeler's colleague Roger Penrose (*1931), to whom Wheeler owed a specific way of drawing space–time diagrams in the vicinity of black holes and other irregularities of space–time, likewise was embedded in more than one visual culture. The young Toronto-based historian and philosopher of physics Aaron Sidney Wright has recently shown that Penrose diagrams are a merger of two visual traditions: of Minkowski diagrams originating in 1909 as a new visual representation of space–time in special relativity and of artistic drawings of "impossible objects" originating in the work of Maurits Cornelis Escher (1898–1972) and then also

---

[56] For a few examples, see the published excerpts in Mills (2009) pp. 2254f.

[57] Wheeler in his autobiography (1998), coauthored with Kenneth Ford, who acted as a secretary to a man well into his 80s; quote from p. 31.

[58] Wheeler in an interview with Jiři Bičák (1978/2009) p. 688.

finding application in perceptual psychology. The latter was a subject of great fascination to both Roger Penrose and his father Lionel, a professor of genetics, with whom he had published amusing puzzles in 1958, only two years before the first publication of his new diagrams showing the "conformal structure of infinity."[59] The scopic domain of Penrose diagrams is thus a merger of two older traditions, with one of them – the endless Escher staircases – quite far away from physics and geometrodynamics: so we again have the feature of a remote transfer between formerly absolutely unrelated visual domains brought into contact by creative individuals who – for some biographically idiosyncratic reason – are well versed in both of them. That same structural feature of a fruitful intersection of two strikingly different visual cultures is also prevalent in Meghan Doherty's analysis of images in Robert Hooke's *Micrographia*, which she interprets as a merger of "visual conventions for portraiture," especially "the visual vocabulary developed by engravers for translating a three-dimensional world into a two-dimensional representation of it" with his specific techniques of "viewing of the microscopic world." Hooke's images were different from – and in fact far superior to – those of his contemporaries because his microscopic seeing and pattern recognition was "facilitated by his early training and lifelong interest in the arts."[60] Earlier on (cf. p. 50), we heard that William Playfair's own background was in engineering and that his brother John was a geologist. His creative innovations in visualizing economic or historical data might thus also be interpreted as a fusion or merger of older visual domains, transferring skills and representation modes into newly emerging domains such as economy.

Comparing several such well-studied examples like Wheeler, Penrose, Hooke, Playfair or Sorby made me realize that there might be a more general pattern behind such scattered and apparently idiosyncratic findings. The results of following up on this leads us to the next chapter, where I pursue this question more systematically on the basis of a prosopography of one particular field.

---

[59] On Penrose diagrams and their prehistory in the work of Roger and L. S. Penrose on "impossible objects," see Wright (2013) and primary sources quoted there.

[60] Doherty (2013) pp. 1–2 with several subsequent examples from Hooke (1665).

4

Pioneers of visual science cultures

What makes certain individuals into founders of remarkably visual subdisciplines or fields of science? Is there anything that these pioneers of visual cultures have in common, besides being highly creative, motivated and perhaps also a bit lucky to have been at the right place at the right time? This chapter will try to answer these question. We will first discuss a few exemplary visual pioneers in science and technology chosen from three different centuries and from very different national and cultural contexts. After tracing differences and a surprising number of similarities between these cases, we will reflect a bit in section 4.2 on the sense and nonsense of such bifurcations as iconophile vs. iconophobe or logocentric vs. iconocentric[1] (introduced in sec. 1.2). The remaining few exemplary cases elaborate these distinctions further. Then we are ready for a more systematic approach based on a detailed prosopography (or group biography) of a whole research collective (in sec. 4.3).

4.1   Some examples: Scheiner, Lambert, Young, Nasmyth

4.1.1   The visuality of a seventeenth-century Jesuit: Christoph Scheiner

Our first pioneer of visual science cultures is an unlikely candidate: a priest thoroughly embedded in the Jesuit theological tradition with its strong emphasis on argumentative skills and close textual exegesis of the bible. Although Christoph Scheiner SJ (1573–1650)[2] also mastered such verbal skills, he ventured far beyond this frame of reference to become a pioneer in several visual techniques within the drawing arts, astronomy, optics and anatomy.

---

[1]On the former distinction see, e.g., Hagner in Bredekamp, Werner & Fischel (eds. 2003) p. 105; on the latter, Hofmann (1999).

[2]On Scheiner's life and work, see Braunmühl (1891), Rösch (1959), Daxecker (2006) and further sources cited there.

Fig. 29: Vignette portrait of Christoph Scheiner SJ 1725. Note how this Jesuit astronomer is completely surrounded by objects relating to visual cultures: On the right he is holding a map of the Moon charted by the aid of his telescope, partially visible in the background. It is projecting the image of the Sun onto a screen tilted towards the onlooker. Below the telescope, a graduated circle stands for the many triangulation instruments used by astronomers at the time. Right next to the lunar map we see part of a right-angle gauge, a standard tool for both architects and geometers. Below the map a terrestrial globe represents the art of mapping the Earth, with which Scheiner was also familiar. The opened pages of the book at the bottom left display the titles of Scheiner's two most famous books on observational astronomy, *Rosa ursina* (1626–30) and *De maculis solaribus* (1612). Oil painting from 1725, kept at the Stadtarchiv Ingolstadt, online in Wikimedia.

Scheiner was born in late 1573 or 1575 in Wald, a small town near Mindelheim, Lower Allgäu, which at that time was in the Swabian part of the Habsburg empire. His family background is not known. By the age of 15, he had entered a Jesuit *Gymnasium* in Augsburg, joined the Jesuit order of St. Ignatius at Landsberg am Lech in 1595 and completed his two-year-long novitiate, taking his scholastic vows in 1597 and holy orders one year later. He was encouraged to complete his education with studies at the University of Ingolstadt, the seat of the first Jesuit College in the southern German lands. From 1603, the freshly qualified *Magister* Scheiner taught mathematics and Latin in the small town of Dillingen on the river Danube. It was there that he met an artist who gave him a glimpse at a device for copying drawings not only 1:1 but also on an enlarged or reduced scale.

Even though this artist, Georgius, did not let Scheiner take a closer look at the device, the priest had seen enough to start thinking about this interesting problem, and he soon came up with his own ingenious solution: the so-called pantograph (literally translated as the "everything-drawer").[3] Scheiner's published account of this device appeared as late as 1631, but news about it spread much faster in the tightly knit Jesuit order and reached well beyond Jesuit circles. The Paris natural philosopher Pierre Gassendi (1592–1655) enthusiastically called this simple but efficient instrument "the egg of Columbus."[4] Having heard about Scheiner's invention, the abdicated Duke of Bavaria, Wilhelm V, invited Scheiner to Munich in 1603 in order to be instructed in the art of using this drafting aid. Scheiner's original labeled woodcut (fig. 30) will serve to explain how this instrument functions.

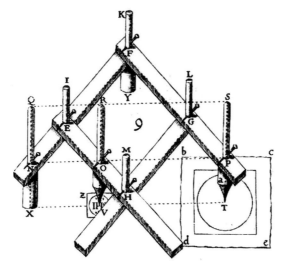

Fig. 30: Scheiner's pantograph for enlarging or miniaturizing drawings. From Scheiner (1631) p. 29; cf. also Stone (1753) and further commentary in Bryant & Sangwin (2008) pp. 51ff.

The slotted wooden frame, coupled in the shape of a parallelogram, is fixed at the support point X (in fig. 30). The draftsman traces along a figure IHS with a stylus (ROZ), thus setting the whole parallelogram in motion and causing another pen (SPa) at the other end of the construction to draw an enlarged, scaled copy (*bdec*)

[3]On the following, see Scheiner (1631), Wallace (1836) with a partial translation of Scheiner (1631), Braunmühl (1891) pp. 2–7 and Kemp (1990) pp. 180ff., Daxecker (2006) pp. 141–3.
[4]In a letter to Scheiner, April 13, 1632, quoted in Braunmühl (1891) p. 6 and Daxecker (2006) pp. 54–6.

of this line onto the blank page at T. The enlargement or reduction factor of the generated copy can be set at will by choosing the appropriate proportions PG:PF and GP:EN.

Later, Scheiner also invented a compass for drawing ellipses or parabolas instead of circles and an early version of a perspectograph.[5] His knack for instrument-making is also evident from the wooden sextant he built in order to observe a comet in 1607. Supplementary theological studies at the Jesuit college from 1605 onwards ended with a successful disputation in 1609. Scheiner then started teaching mathematics and Hebrew in Ingolstadt in 1610. His duties included lessons in practical optics, stereometry and astronomy as well as instruction in the handling of various instruments, ranging from simple solar dials to astrolabs and meridian circles. When news about the invention of the telescope spread throughout Europe in 1610,[6] Scheiner was one of its first enthusiasts for use as a new visual aid for astronomers. He quickly learned the art of grinding and polishing lenses and built his own telescopes, mainly to observe the Moon and the Sun. Together with his pupil Johann Baptist Cysat SJ (1585–1657), he also studied Galileo's *Sidereus Nuncius*. It contained the Italian pioneer of telescopy's description of the lunar surface as riddled with craters, among other equally striking observations. In the morning of March 6, 1611, Scheiner and Cysat climbed the tower of the Church of the Holy Cross in Ingolstadt in order to make use of the cloudy weather to determine with their new telescope the diameter of the Sun, which was just barely shimmering through the thin clouds. To their great surprise, they noticed dark spots on the solar surface: "either the Sun is weeping, or it has blemishes," Cysat is supposed to have said.[7] After further sittings had provided confirmation of this observation, also by other people whom they invited to look through their telescope under similar cloudy conditions or through heavily darkened glass, Scheiner converted the telescope into a projection instrument, thus inventing the so-called helioscope (cf. here fig. 31).

Rather than looking directly through the tube as all other observers had done, Scheiner mounted a telescope bf on fixtures TR and directed the image generated by the telescope $abcd$ onto a screen P set up orthogonally to the line of incoming light. This mounting had the additional advantage that a sheet of paper could be attached to the screen, and details such as those ominous dark spots observed

---

[5]See, e.g., Braunmühl (1891) pp. 42ff. and here further below on Lambert's perspectograph.

[6]On this invention and its huge impact on visual astronomy, see, e.g., van Helden (1977), Biagioli (2006) and Hentschel (2014*d*).

[7]See Rösch (1959) p. 190: "Entweder die Sonne weint oder sie hat Makeln." Cf. furthermore the introduction to Scheiner (1626–30) and Braunmühl (1891) pp. 11f., 30ff.

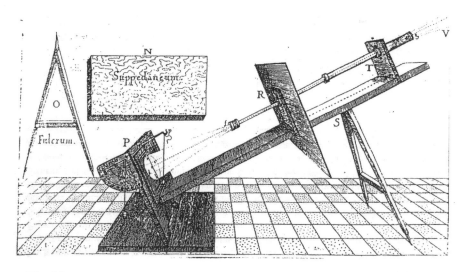

Fig. 31: Scheiner's immission 'telioscope' or 'helioscope'. From Scheiner (1626–30) p. 18.

on the solar surface could be outlined directly on the sheet. The helioscope thus allowed quick and easy recording of the exact positions, shapes and sizes of the dark spots. Scheiner soon noticed during observations over longer periods of time that the spots changed their locations on the solar surface. In three letters to the Augsburg patrician Marcus Welser (published anonymously under the pseudonym Apelles), Scheiner reported his findings in 1612 and 1613. (See here fig. 69 on p. 245 and fig. 32 below.) Compared against a later image by Scheiner (1626–30), the woodcut from 1613 shows this effect of apparent motion by sunspots larger and much more clearly.

Galileo Galilei, who had meanwhile also observed these spots, interpreted them as evidence for the absurdity of the Aristotelian and Christian belief in an immaculate Sun, as a perfect body in the supra-lunar realm of the heavens. Not surprisingly, the Jesuit Scheiner was more careful and conservative. To him, it was much more likely that these dark features were not spots on the solar surface but stars or other celestial objects partially occluding the Sun's surface when observed from the perspective of the Earth. That way, the Aristotelian doctrine could be preserved intact the way it had entered Scholastic doctrine.

These interpretational differences in solar observations led to an acrimonious debate between Galileo and Scheiner. Both sides resorted to all the available means of persuasion. Scheiner's last contribution to these debates was his extensive quarto volume *Rosa ursina*, published 1626–30. It assembles all his observations

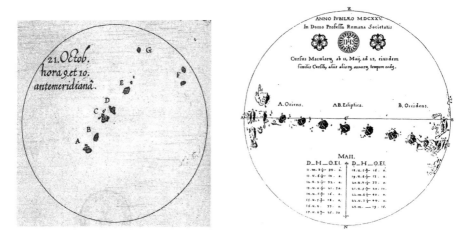

Fig. 32: Scheiner's observations of sunspots. Left: a group of sunspots observed on Oct. 21, 1611: from Scheiner (1613), copperplate by Alexander Mair. Right: One dark spot crossing the solar disk between May 11 and 23, 1625. From Scheiner (1626–30) p. 211. On this quasi-kinematographic display, in which a single object is depicted several times to indicate its displacement over time, see the main text (sec. 7.4 on p. 245).

so far on more than 784 folio pages, many of them with illustrations.[8] This book veered away from the former interpretation of solar spots being occluding planets to one of fires on a solid dark core in a strangely fluid mass. His precise mapping of the apparent trajectory of these spots across the solar disk showed that their paths are inclined with respect to the solar ecliptic by slightly more than 7°. This was something Galileo had overlooked in his first publications on solar spots. Scheiner also noted another phenomenon that is normally attributed to Alexander Wilson more than a century later: Near the solar rim, the solar spots seemed to retain some of their breadth well beyond what would be anticipated given perspectival foreshortening. Scheiner correctly interpreted this as evidence that solar spots are funnel-shaped.[9]

Aside from these astronomical studies, Scheiner also studied vision from an optical and physiological point of view. Scheiner considered the human eye a "natural tube" (*natürliches Tubus*), i.e., nature's analogue to the telescope. As fig. 33 shows, Scheiner developed this vague similarity into a full-fledged analogy. It is

---

[8]See Scheiner (1626–30), Daxecker (1996, 2006) pp. 37ff., 120–38 and Bredekamp (2007) esp. on Scheiner's illustrations.

[9]See Scheiner (1626–30) p. 506, and Braunmühl (1891) p. 63. Cf. Wilson (1774) on the 1769 rediscovery of this 3D effect, and www.strath.ac.uk/research/archive/news/solvingthesunspotpuzzle/ on the latest findings.

explained in great depth in his book bearing the programmatic title *Oculos hoc est fundamentum opticum*, published in 1619 and dedicated to Emperor Ferdinand II.[10]

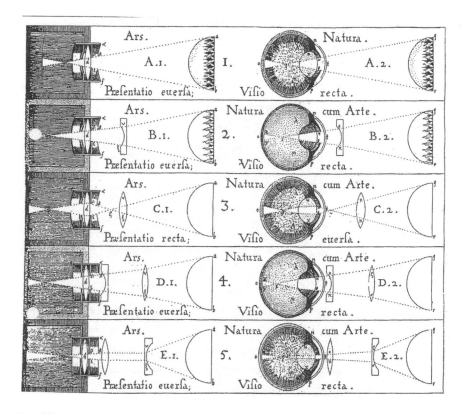

Fig. 33: The eye as a "natural tube," a *camera obscura*. On the left half of the figure, one sees a camera obscura in cross-section (*Ars*), on the right half, a human eye (*Natura*), in the top row unaided *visio recta*, below it with concave or convex glasses inserted between it and the light source (rows 2 and 3) and then with a telescopic lens combination in the mode of Galileo and Kepler (rows 4 and 5). From Scheiner (1626–30) p. 107 (excerpt).

Comparing the eye with a *camera obscura* was an idea that we already find in writings by Ibn al-Haitham, Della Porta and Kepler; but Scheiner was the first to subject this vague idea to anatomical test. Taking an eyeball from a butchered ox or sheep, Scheiner cut a hole into the upper wall of the sclerotica and removed all opaque media inside it, in order to be able to see the sharp images of luminescent objects

---

[10]For further details, see Braunmühl (1891) pp. 48ff., Rösch (1959) pp. 197ff. and Daxecker (2006) pp. 116–20.

inverted on its retina. In the second part of his book, which combined anatomical and optical insights, Scheiner tracked the path of light through all the layers and parts of the eye. To study the multiple refractions of light inside the eye at the interfaces between cornea, crystalline lens and vitreous humor, Scheiner compiled a detailed table of refraction angles – like Kepler a decade before, Scheiner did not yet know the law of refraction. It was discovered by Snellius before 1626, but was published only in 1648 by René Descartes. While he was conducting these extensive physiological experiments, Scheiner also noticed and examined the accommodation of the eye.[11] For popularization purposes, Scheiner built a transportable version of a *camera obscura* and seems even to have built a huge version that one could walk inside for the palace of the Archduke of Further Austria, Maximilian III, in Innsbruck.[12]

Father Scheiner thus already manifests nearly all the dimensions of visuality that figure throughout this book: He was a mediator between different visual cultures, ranging from astronomy to optics and from applied geometry to perspective theory; he exhibited a lifelong obsession with visual observations of various kinds, foremost telescopic ones; and he was an excellent draftsman, even developing several drawing instruments. He also applied these skills to architectural design, first and foremost in his contributions to the new Jesuit church in Innsbruck. The physiology of the eye and of vision was something of special interest and relevance to this keen observer; and he worked hard to train others in the visual skills he himself had already acquired through long practice.

4.1.2   The visuality of an 18th-century polymath: Johann Heinrich Lambert

Our exponent from the 18th century is Johann Heinrich Lambert (1728–77). Readers of this book, depending on their area of expertise, will perhaps have already come across this Swiss figure either as a philosopher or a logician, mathematician, astronomer or theoretical cartographer, perhaps also as a physicist and optician, a physiologist or as an expositor of perspectival drawing and various drawing utensils. In fact, he was all of these and much more still. This was a true polymath with an outright visuality at the core of many of his lasting contributions to science, philosophy and the practical arts. We will just focus on some of his activities in which this deep link between visuality and practice is most obvious (in roughly chronological order), referring the reader to further primary

---

[11]On the so-called Scheiner experiment with a card pierced with several pinholes held very close to the eye, see, e.g., Daxecker (1992, 1994, 2006) and Robinson (2006) p. 73

[12]On these gadgets and the connection with Archduke Maximilian, see Daxecker (2006) pp. 15f.

and secondary literature as we go along.[13]

Lambert was the son of a tailor in Mulhouse (Alsace). He was a good draftsman and calligraphist since his early youth, selling quite a few of his drawings to his wealthier classmates. Having merely completed primary schooling, the mainly self-trained young man assisted his father, various other employers and manufacturers of the area, before fortuitously becoming the private tutor to the grandson and the cousin of a count, Peter von Salis (1675–1749) in Chur. As a companion to his two private élèves on their travels through Europe in the mid-1750s, he met many important mathematicians and scholars. Leonhard Euler then arranged for his employment at the Prussian Academy of Sciences in Berlin, where he obtained a prestigious position as senior building officer (*Oberbaurat*). This drew him into contact with practices such as surveying, architecture and other applied arts; but the focus of his work up to his death in 1777 continued to be in natural philosophy and mathematical fields.

Lambert's book on free perspectival drawing was the first publication strongly tied into visual culture.[14] His interest in this topic may have arisen during his short assistantship in the iron works at Seppois during his youth.[15] Lambert enjoyed analyzing and drawing the complex machinery and furnaces, already presaging his lifelong fascination with instrumentation of all kinds.[16] As a case in point, Lambert's book on ichnography, as he called the theory of linear perspective and geometric drawing, contained various strands of a more practical, in fact instrumental, character. He spent much time describing all kinds of drawing tools, including not only Scheiner's pantograph, but also a proportion compass that he had designed himself. It culminates in a totally new invention, a so-called perspectograph for semi-automated mechanized drawing in perspective.[17]

The basic idea behind Lambert's perspectograph is depicted in fig. 34. A pair of straight rulers *BS* and *AO* are fixed with small nails at points *S* and *O* along a horizontal line at the height of the ideal observer above the ground (here represented by the line *EG*). The task is to make a perspectival image of the given

[13]For general surveys of Lambert's life and works, see, e.g., Fourney (1780), Huber (ed. 1829), E. Anding in Lambert (1760*b*) pp. 51ff. and Max Steck in Lambert (1959*c*). On his time in Chur, see esp. Humm (1972), for his diaries Bopp (ed. 1915), and on his unpublished papers Steck (1977). For Lambert's complete works online, see www.kuttaka.org/ JHL/Main.html.

[14]On the following, see Lambert (1759*c*), commentated upon and annotated by Max Steck.

[15]At age 15, according to Humm (1972) p. 18; cf. also ibid., pp. 39–40 for his vividly described impressions of Chur; these, too, imply a burgeoning visuality.

[16]There are no less than 106 different instruments in the inventory of his possessions compiled after his death. See Steck in Lambert (1759*c*) pp. 5, 27.

[17]On the following, see Lambert (1759*c*) pp. 5, 37, 45–9, 62ff., 164–76, 221–32.

object, here the star *HJNPK*. The two rulers are made to cross at each of this two-dimensional figure's five tips (here shown for *H*) so that it is just visible above ruler *AO* and to the right of ruler *BS*. Then a right-angled ruler *CLD* is shifted along the base line *EG* so that its left lower corner (at *L*) coincides with the left margin of the ruler *BS*. The rulers *CLD* and *AO* meet at *h*, allowing a reading of the length *Lh* (here 19 units). This length is then taken along the vertical scale *LC* of the right-angled ruler *CLD*, which indicates the height above the base line at which point *p*, representing point *H* of the object, has to be drawn in the perspectival image. The same procedure is repeated for each of the five tips. Then these points can be connected by straight lines, on the basis of another theorem in perspectival theory: lines are always depicted as lines.

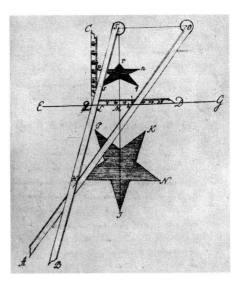

Fig. 34: The underlying principle of Lambert's perspectograph, designed in 1752. Published posthumously by Max Steck in 1943, in Lambert (1759*c*) pl. vii.

Executing such a drawing was possible in principle; but it was still quite impractical without any mechanical guides for the various rulers built into the construction. So Lambert took the next developmental step and designed a ready-to-use instrument with as many mechanical operations built into the design as possible (see fig. 35). Here, the base line (still labeled *EG*) is represented by a solid ruler *WP* fixed at *w* and *s* by means of screws, and the line *CD* likewise represents the height of the observer above ground level. The trick in this construction is that the two movable rulers are now laterally shiftable without losing contact with these two reference lines, because they are connected together by pivotable pegs that can move freely along the slots of both ruler systems, exactly above the two reference lines. As this example shows, a mechanical 'translation' of a given image (here of a baroque garden plan *KJVL*) to its perspectival image (here *rklp*) is thus obtainable much faster and more reliably than in the free-hand variant shown in fig. 34. This perspectograph was so carefully conceived and designed that it was actually manufactured and sold by the Augsburg instrument maker Georg Friedrich Brander (1713–83), with whom Lambert regularly corresponded.

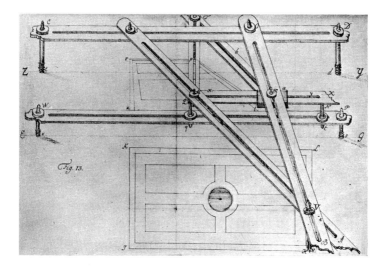

Fig. 35: Lambert's fully developed perspectograph as distributed by Brander. This pencil drawing by Lambert was first published in Lambert (1759*c*) pl. IX.

The main part of the book on free perspective was devoted to its strictly theoretical aspects as a mathematical technique. Lambert's work offered a rigorous introduction to its geometrical foundations, proceeding systematically from the simpler to the complex in a pedagogically well-conceived sequence, with many a sarcastic side remark about artists and draftsmen who still thought they could handle perspectival drawing merely intuitively. Like Gaspard Monge a few decades later (cf. here p. 269) – Lambert was often named as a last precursor to his *Géométrie descriptive* – Lambert also operated with a horizontal plane, the *Grundrißebene*, and with an orthogonal vertical plane (*Bildebene* or *Tafel*), on which the perspectival image was construed for the observer located far enough beyond these two planes. I defer a discussion of the details of perspectival drawing to sec. 5.1 of the present book. Here I just emphasize that, untypically for such perspectival treatises, Lambert repeatedly switched from a geometric approach to an algebraic approach to certain problems, even insisting on the mutual interrelations between these two fields in the manner of mathematicians following the tradition of Descartes, Viator or Desargues and strongly deviating from practitioners of perspectivalisms of the likes of Dürer or Leonardo da Vinci, whose pertinent writings he also knew. Lambert included a long chapter on the history of perspectivalism in the expanded second edition of his *Freye Perspective*, which also appeared in Zurich 15 years later. It was one of the very first historical accounts of this field (cf. here sec. 5.1).

A very original and creative part of this book was a section reversing the usual procedure and asking how to reconstruct the view point from which a given perspectival image is made, or how to find out the exact dimension of objects depicted in them. By this 'inverse' problem, compared with the usual task of perspective theory to construe images from objects with known dimensions, Lambert became the most important forerunner of photogrammetry. That discipline only began to take off during World War I with a proliferation of photographic images of fortifications and other military installations behind enemy lines taken from airships and airplanes. Coming from the pre-photographic era, Lambert picked other examples, of course, such as determining the size of buildings in the backgrounds of perspectival images.

Just one year after this first major publication, Lambert published the next milestone book. His *Photometria* handled aspects of illumination, shadows and the problem of gauging intensities of illuminated objects as a function of distance and the observer's viewing angle.[18] The Lambert law, also known as th Beer–Lambert law or Beer–Lambert–Bouguer law (named after August Beer, Johann Heinrich Lambert and Pierre Bouguer) relates the absorption of light to the absorption coefficient α of partially absorbing material through which the light passes. In general, the transmissivity T of light through a substance depends logarithmically on α and the length of the path l through the absorber $T = I/I_0 = 10^{-\alpha \cdot l}$, where I is the light intensity.

Lambert also studied the diffuse reflectance of light from diffusely reflecting surfaces, such as the surface of the Moon or some other 'Lambertian' surfaces. According to the cosine law, the radiant intensity observed from such a diffusely reflecting surface is directly proportional to the cosine of the angle between the observer's line of sight and the surface normal. In other words, the more obliquely one looks at such a surface, the darker it seems to be. Thus, the full-moon limb, for instance, seems to be darker than the central area of the disk. This law, which

---

[18]On the following, see Lambert (1760*b*), with detailed commentary by E. Anding. The invisible link between these two major works was very likely his interest in shadow casting – a topic treated in secs. 139–56 of Lambert (1759*a*).

is also known as the cosine emission law or Lambert's emission law, was already known to Kepler and Scheiner (see above), but only became more generally known after its publication in Lambert's *Photometria* in 1760.[19]

The same is true of the so-called distance law. According to this law, the light of any punctiform light source diminishes in intensity as the square of the distance to the observer. Four identical light sources shine equally brightly at a given location if they are twice as far away as does a single such light source. This law, which was also already known in Kepler's time but is often attributed to Lambert, was built into various photometers of the time. The distance of comparison light sources is adjustable so that the set distance at which they seem to be equally bright to the light source being measured serves as a gauge for its luminosity (relative to these comparison light sources).[20] Today, this law is used to gauge cosmic distances of a special class of stars whose absolute luminosity can safely be assumed to be comparable.

Lambert's studies of geometry are intimately related to perspective systems of projections. They led to his paper on parallel lines,[21] in which he foresaw the logical possibility of a non-Euclidean geometry, as well as to his work on theoretical cartography. Cartographic grids can be looked upon as solutions to the problem of projecting 3D objects onto 2D surfaces. His Notes and Comments on the Composition of Terrestrial and Celestial Maps[22] appeared as one part of an extensive work on mathematical applications and include seven new map projections. None of them was concisely labeled by Lambert, but in today's terminology[23] they are called the following:

1. Lambert conformal conic
2. Transverse Mercator
3. Lambert azimuthal equal area
4. Lagrange projection
5. Lambert cylindrical equal area
6. Transverse cylindrical equal area
7. Lambert conical equal area

---

[19] See Lambert (1760*b*) p. 22.

[20] Such photometers were used by Lambert (1760) himself, as well as by W. Herschel (1800) and by Fraunhofer (1815), for instance.

[21] See Lambert (1786), written around 1766 but published posthumously.

[22] This is the English translation of Lambert (1772*a*), published by Waldo R. Tobler in 1972.

[23] See, e.g., Snyder (1987, 1993) for further commentary.

I will limit my comments to the very first of these new projection systems. Also known as conical orthomorphic projection, it had been forgotten for more than a century before it was revived by the French and by the U.S. Coastal and Geodetic Survey for battle maps during World War I.[24] As fig. 36 illustrates, Lambert sought a compromise between the stereographic representation of a sphere and the Mercator nautical charts. The latter is famous for preserving angles, a crucial feature for navigation, and the former yields the greatest similarity in shape of plane figures with their mapped depictions. Lambert asked himself "whether this property occurs only in the two methods of representation mentioned or whether these two representations so different in appearance can be made to approach each other through intermediary stages."[25] Lambert obtained this conformal conic projection by making the intersecting angle between the meridians arbitrarily larger than its value on the surface of the Earth. He then applied it to a small map of Europe as an illustration (cf. fig. 36). In this mapping, two standard parallels have to be chosen, here between 20° and 70° North. The scale becomes too small above and too large beyond these latitudes. This in-built handicap of this system of projection has the complementary advantage, though, that there are no angular distortions along any parallel (except at the poles). While obviously impractical for geographical purposes, this projection system became quite important for regional and global aeronautical charts, for world maps and for the mapping of planets and their satellites.

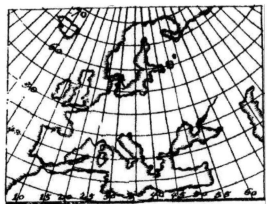

Fig. 36: Lambert's conformal conic projection, here applied to Europe, the Near-East and North Africa, with the latter appearing too large and Scandinavia too small. From Lambert (1772) pl. V, fig. XX.

While first visiting Göttingen, Lambert met Johann Tobias Mayer (1723–62), the professor of mathematics and director of the Göttingen Observatory. This versatile scholar shared many astronomical and mathematical interests with

---

[24]For these and other details, see Lambert (1772*a*) pl. IV–V and (1772*c*) pp. 28f. and Snyder (1987) pp. 104–11.

[25]Lambert (1772*a*).

Lambert, but their exchange about color systems was most fruitful. In a talk at the Göttingen Academy of Sciences in 1758, Mayer developed his way of representing color affinities in a system of triangular levels stacked on top of each other. The three primary colors – red, yellow and blue – are at the tips of the triangle in the middle of the stack, represented by the pigments cinnabar red, king's yellow (orpiment) and mountain blue (azurite). Eighty-eight intermediary shades are arranged in sub-triangles along the edges and in-between the tips and the gray center of the triangular level. Since each of these 88 colors can occur in various degrees of intensity, Mayer imagined a set of 10 such triangles differing from each other only by the amount of white or black mixed into these intermediary shades, the center of the uppermost level being pure white and the center of the lowermost triangular level pure black. Mayer could then simply read off the exact coordinates of a given sub-triangle color in his stack for a multiple of different nuances ($10 \times 91 = 910$), thus achieving an elegant alphanumeric classification system for colors: r^{12} was pure cinnabar red, whereas r^8bj^3, for instance, was a less intense red mixed with 3 units of yellow and one unit of blue.

Inspired by Mayer's system, Lambert developed one of his own. It was actually published before that of Mayer, who died suddenly in 1762, so his original ideas had to await a posthumous edition prepared in 1775 by Georg Christoph Lichtenberg (1742–99), another Göttingen luminary.[26] Lambert was able to publish his "color pyramid" three years before. It followed Mayer's idea of a triangular organization with one of the three primary colors at each corner of a triangular tray of hues and shades. The sizes of Lambert's trays differed, however. In his color system, the very top tray was a single white sub-triangle, with more and more nuance sub-triangles accumulating on the succeeding trays as the tones darken towards the bottom tray of his system, displaying the most saturated hues. It thus took on the shape of a regular pyramid rather than Mayer's triangular stack. To render it, Lambert chose vegetable colors in wax developed by the Berlin painter Benjamin Calau (1724–85), because these had excellent mixing qualities, so the in-between colors came out nicely (see here color pl. V). Coming from a tailor's household helped to remind Lambert to build a reality check into his system. Lambert himself expressed the hope that textile dealers, dyers and printers would consult his work in order to verify their color assortments and to denote specific colors more clearly. Descriptive terms otherwise always remained ambiguous.

[26] See Mayer (1758) and (1775) part IV and Lambert (1768, 1772); cf. Forbes (1849), Sherman (1981), Stromer (ed. 1998) chaps. 9–11 or Spillmann (2010) pp. 24–7. On their intense interaction and correspondence, see Humm (1972) p. 71.

It is tempting to continue on about his work on pyrometry and various other branches of the exact sciences and philosophy, etc. But I will stop here with reference to the secondary literature for further analyses.[27] What should have become clear with this abbreviated summary of Lambert's research is the strong interconnection between these various visual activities: his interest in perspectivalism, which also included attention to shadow casting, is closely interlinked with his work on photometry and geometry and his studies on cartographic map projection, which were also useful for astronomical mapping of stellar fields.

4.1.3 Drawing, writing and measuring: Thomas Young

The ophthalmologist, physicist and polymath Thomas Young (1773–1829) figures repeatedly in this book.[28] His name has already appeared in the context of the kymograph as a new class of self-recording instruments. Young demonstrated in his *Course of Lectures on Natural Philosophy and Mechanical Arts* a prototype he had built himself. Delivered in 1801–03 and published six years later with an extensive set of copperplates, the course contains many other indicators of an outstanding pioneer in visual science cultures. This subsection title derives from his tenth lecture, for instance. Hints interspersed among lectures on the motion of cohesive bodies and on statics informed his listeners and readers about all kinds of issues not normally part of a lecture on natural philosophy, even at that time. Young expanded on the art of drawing and its subtle transition into painting. He went into the history of sketching and writing, glancing back at the oldest cultures of mankind and spending extra time to discuss Egyptian hieroglyphics, which he had studied to the point of publishing several books, after succeeding in partially deciphering the system – that feat was soon accomplished by his French contemporary Jean-François Champollion (1790–1832).[29] Lecture 11 explained in detail the various printing techniques. The pantograph and proportional compass – just covered in the preceding subsections on Scheiner and Lambert – also found their place in Young's lectures, and likewise perspectival drawing (see here secs. 5.1 and 8.1). These lectures were delivered at the *Royal Institution*, founded in 1799 with the explicit aim of "diffusing the knowledge, and facilitating the general introduction, of useful mechanical inventions and improvements; and for

[27] Aside from the literature already mentioned in note 13, see Berger (1959) and Hentschel (2007) sec. 4.4.4 about Lambert's arguments on the ontological status of radiant heat.

[28] On his life and work, see, e.g., Arago (1854) vol. 1, pp. 241–94, Peacock (1855), Lees (1900), Robinson (2006) and Hentschel (2007) pp. 386–91 and further sources cited there.

[29] On this part of Young's and Champollion's work, see Young (1807b) pp. 74f. (with references), Lees (1900) or Robinson (2006) chaps. 10 and 15 and further sources cited there.

teaching, by courses of philosophical lectures and experiments, the application of science to the common purposes of life." A mixed bag of people attended them, ranging from members of the royal family to well-to-do artisans and citizens with enough leisure and keenness for learning. Experimental demonstrations accompanied many of Young's lectures in order to arouse and sustain the interest of such a varied audience – likewise those of his chemist colleague Humphry Davy (1778–1829) and Davy's assistant and successor Michael Faraday (1791–1867).

Prior to becoming a lecturer at the *Royal Institution* in 1801, Young had studied medicine, initially in London (1792–93), then in Edinburgh (1794), and finally at the prestigious Enlightened university in Göttingen, at that time part of the Kingdom of Hanover. There, he had obtained his doctorate in 1796 with a thesis on the anatomy and physiology of the eye. In 1797, he entered *Emmanuel College*, Cambridge. An inheritance granted him financial independence, so, even though he opened a practice in London, he never became a routine practicing physician, preferring rather to concentrate on his scientific research in the areas of physiology (sensory physiology of the eye), mechanics (the theory of elasticity) and optics (the wave theory of light, which he tried to promulgate against the then still prevalent Newtonian particulate theory of light).[30]

Young's work in physiological and physical optics predestined him to becoming a pioneer of visual approaches. Most notable with regard to the former are his studies on stereoscopic vision and on mechanisms of the eye, especially on accommodation, which he correctly explained as a slight change in shape of the crystalline lens, even though he got the eye's muscular anatomy wrong.[31] Young also experimented on retinal sensitivity and on after-images.[32] Issues such as chromatic aberration led him to the study of colors and their generation in the rainbow, optical refraction and diffraction. He subjected the optical spectrum generated by prisms or diffraction gratings to his intense analysis. Like Newton, he also experimented with generating white light as a superposition of various spectrum colors. Unlike Newton, Young claimed that only three primary colors were needed to recompose white light – red, green and violet. He illustrated this finding by means of color disks on which color combinations were painted

[30]On Young's arguments in favor of the undulatory system as a unifying theory of light, see Hentschel (2007) pp. 387f. and further sources mentioned there.

[31]On Young's anatomical and physiological work – which continued Scheiner's much earlier studies on accommodation – see, e.g., Young (1793, 1801); cf. Turner (1953), Sherman (1981) and Robinson (2006) chaps. 2–4 and 9 and further sources cited there.

[32]A research strand later taken up by Plateau: see here p. 156.

that merge into a composite color when rapidly rotated.[33] Most famously, Young also invented the concept of light interference by close analogy between light waves and waves forming when a stone is dropped in water. It is defined as a superposition of two positive or two negative undulations, and as an annihilation of positive and negative undulations by two interfering wave fronts. While this mental model allowed easy, intuitive comprehension of well-known phenomena such as diffraction along a sharp edge or Newtonian rings near the contact zone of two nearly parallel surfaces, Young could not yet describe these phenomena quantitatively. He still had to contend with a longitudinal conception of waves (like sound waves in air). The real breakthrough only came with Fresnel's and Arago's shift to a transversal wave theory of light one decade later.[34]

4.1.4 A Scottish engineer on the Moon: James Nasmyth

My last example in this section, taken from the domain of technology, will be James Nasmyth (1808–90), a British engineer best known for his invention of the steam hammer.[35] He was the son of the painter Alexander Nasmyth (1758–1840), well known for his portraits but famed as the founder of traditional Scottish landscape painting.[36] Some of his appreciation for the "beauty and majesty of old trees" or for old castles and mansions in idyllic scenery rubbed off onto his son, who accompanied him on many excursions in the fabulous Scottish countryside. James also learned many skills from his father, including visual thinking. For instance, Alexander designed 'bow-and-string bridges' which became an alternative to the standard pier-supported bridges. A substantial portion of James Nasmyth's *Autobiography* is devoted to his early years in Edinburgh, where he was raised in a well-to-do bourgeois setting with nursemaids and many caring relatives. Nasmyth particularly recalled the large illustrated folio editions of Vitruvius and Palladio among his grandfather's possessions: Michael Nasmyth had been a major builder in Edinburgh who had put great care and skill into finishing his buildings. Tactile memories, such as of the small fragments of basalt that his grandfather had used to

---

[33]On Young's trichromatic theory and his sensory-physiological experiments, see, e.g., Young (1807*b*), lectures 37–8. According to Robinson (2006) pp. 80ff., the anatomical correlates of Young's theory, three types of cones sensitive to the three primary colors, were only found in 1959.

[34]On the history of Young's concept of interference and the later path towards the fully developed wave theory of light, see Buchwald (1989) and further sources mentioned there.

[35]See his *Autobiography*: Nasmyth (1883), coauthored with Samuel Smiles (1812–1904), known for his "self-help" books. Cf. also Anon. (1883), Rowlandson (1870), Musson & Robinson (1969), pp. 66, 77, 98, 447, 459, 471 ff., Chapman (1997) and Robertson (2006).

[36]See Anon. (1882) and Nasmyth (1883) chap. II on A. Nasmyth's training, drawing and painting skills, local institutional context, clients and friends, and the *Dictionary of National Biography* on the illustrious Nasmyth family; James's brother Patrick (1787–1831) became a painter like his father.

fill the exposed joints of stone masonry to increase weather resistance, or James's pleasure in fine wood-working and modeling, come across vividly. He ascribed his artistic facility to an inheritance going back to his grandmother:

> I am fain to think that her delicate manipulation in some respects descended to her grandchildren, as all of them have been more or less distinguished for the delicate use of their fingers – which has so much to do with the effective transmission of the artistic faculty into visible forms. The power of transmitting to paper or canvas the artistic conceptions of the brain through the fingers, and out at the end of the needle, the pencil, the pen, the brush, or even the modeling tool or chisel, is that which, in practical fact, constitutes the true artist.[37]

Elsewhere he generalized this into a kind of Condillacian epistemology: The eye and the fingers were to him "two principal inlets to sound practical education. They are the chief sources of trustworthy knowledge as to all the materials and operations which the engineer has to deal with. No book knowledge can avail for that purpose." Sketching and drafting in pen and by brush remained practiced skills throughout Nasmyth's life. He was proud to have some of his later sketches printed as illustrations in his autobiography. Having experienced the versatility of the pen in expressing spatial constellations or in quickly recording complex structures, Nasmyth realized that "the language of the pencil is a truly universal one, especially in communicating ideas which have reference to material forms. And yet it is in a great measure neglected in our modern system of education." He also emphasized the role of sketches in supporting the memory: "Written words may be forgotten, but these slight pencil recollections imprint themselves on the mind with a force that can never be effaced."

His *Autobiography* allows us further glimpses into his own visual thinking as he was designing the steam hammer for which he justly became world-famous. The task he set for himself was to design a machine able to lift a heavy weight in order to be dropped onto the object being shaped or worked.

> Following up this idea, I got out my 'Scheme book' on the pages of which I generally thought out with the aid of pen and pencil, such mechanical adaptations as I had conceived in my mind, and was thereby enabled to render them visible. I then rapidly sketched out my Steam Hammer, having it all clearly before me in my mind's eye. In little more than half an hour [...] I had the whole contrivance in all its executant details, before me in a page of my Scheme Book [cf. here fig. 37]. The date of this first drawing was the 24th November, 1839.[38]

---

[37]Nasmyth (1883) p. 15; the following quotes are from this, pp. 95 and 56f.

[38]Nasmyth (1883) p. 231. On the history of this invention, cf. Rowlandson (1870). For an illustrated and commented chronological list of Nasmyth's many mechanical inventions and technical devices between 1825 and 1862, cf. Nasmyth (1883) pp. 387–431.

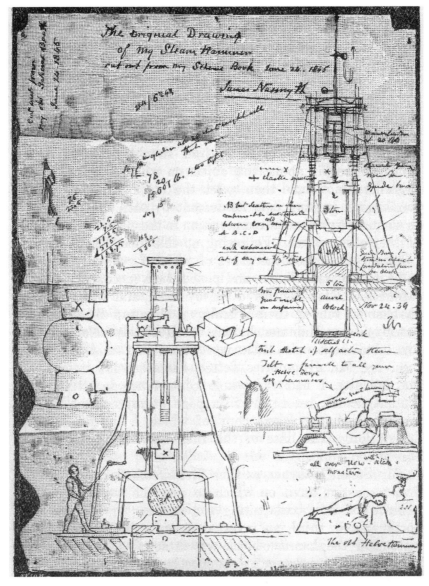

Fig. 37: Nasmyth's first drawing of his steam hammer, Nov. 24, 1839. At the bottom left, one sees a workman opening a valve to cause the hammer x to fall down upon the iron rod μ (drawn in hatched cross-section). A smaller drawing at the center of the page details the optimal shape of the hammer x, seen in side view, and in action on the left margin. At the top right, a variant is drawn with further design details, with various estimated dimensions and weights added. At the bottom right, there is a cartoon labeled "The old Helvetien," with a small demon labeled "J. N." hammering on his head; above it, transformed into an old-fashioned hammer works, labeled "all over now with sitch a monster"! From Nasmyth's 'Scheme book,' photomechanically reproduced in Nasmyth (1883) p. 232. The original is accessible at http://www.sciencephoto.com/image/362570/530wm/V2000027-Original_drawing_of_Nasmyth_s_steam_hammer-SPL.jpg (accessed June 15, 2011).

This efficient contrivance was patented on June 9, 1842 (British patent no. 9382) and soon became a standard tool. Steam hammers with automatic gearing (British patent no. 9850) were used since 1845 for heavy-duty jobs, such as dressing stones, pile-driving and iron-forging throughout the world.[39] The enormous power of steam hammers together with the adaptability of their hammering strength and pace made them versatile machine-tools without which the elegant steel constructions for suspension bridges or gigantic ships like Isambard Brunel's *Great Eastern* in the second half of the 19th century were unthinkable.[40]

Nasmyth earned a fortune from the production and international sale of these steam hammers. In 1856 he withdrew from business and focused his interest on areas of natural philosophy and the sciences, again picking strongly visually inclined areas, such as microscopy, geology and observational astronomy. He built various medium-size reflectors with high-quality mirrors he had made himself and observed the Sun and the Moon with these telescopes. His enlarged drawings of surface features of the Moon were exhibited at *conversazioni* in scientific societies and at the Great Exhibition of 1851 in London. They earned much admiration from other gentleman scientists, such as John Herschel (1792–1871) and aroused considerable interest also from outside astronomy. Nasmyth described his drawing techniques in great detail in his autobiography:

> I made careful drawings with black and white chalk on large sheets of grey-tinted paper, of such selected portions of the Moon as embodied the most characteristic and instructive features of her wonderful surface. I was thus enabled to graphically represent the details with due fidelity as to form, as well as with regard to the striking effect of the original in its masses of light and shade. I thus educated my eye to the special object by systematic and careful observation, and at the same time, practised my hand in no less careful delineation of all that was so distinctly presented to me by the telescope – at the side of which my sheet of paper was handily fixed. I became in a manner familiar with the vast variety of those distinct manifestations of volcanic action, which at some inconceivably remote period had produced these wonderful features and details of the moon's surface.[41]

---

[39]See Rowlandson (1870), Prosser (1900) and further sources cited there. Compare here also color plate XI for an oil painting by James Nasmyth depicting a large steam hammer in action in a foundry.

[40]On the patent history and a clone by Adolphe and Eugène Schneider, proprietors of the ironworks at Creuzot, see Nasmyth (1883) pp. 225ff., 331, Anon. (1890) and Prosser (1900).

[41]Nasmyth (1883) p. 316; on "practicing the hand," cf. Nasim (2012).

As the last sentence reveals, Nasmyth ventured beyond mere description. He tried to understand the causes of what he was observing by analogy and by mimetic experimentation with materials, in order to recreate the phenomena in his laboratory. The huge craters he observed on the lunar surface were to him clear indications of past volcanic activity. He therefore studied craters on the Earth during his visits in Italy and built plaster models of the lunar craters. These he then photographed under strongly skewed light to create fantastic landscapes that seemed to resemble what he had observed by telescope on the Moon. He compared the many parallel ridges to the wrinkles of his own hand or on a shrunken old apple, and the Moon's bright radial lines to crack-lines on a glass globe that had been put under pressure. All these spectacular, but highly hypothetical, findings were collected and lavishly illustrated in a book on *The Moon: Considered as a Planet, a World, and a Satellite*. Woodburytypes were used in the first edition of 1874, which counts as one of the very first books with photomechanical illustrations. They yielded beautiful images of high contrast and density. Less expensive techniques were used in later editions and the format also was diminished. Today's cheap facsimile editions do not give an adequate impression of the quality of the original images.[42]

Just as a caution to the reader: Nasmyth happened to be wrong about most of his causal claims made on the basis of his fascinating and intuitively so plausible visual analogies. The ability to reproduce certain features of a remote system by some process in the laboratory does *not* by any means guarantee that it indeed be the process causing the observed features in the remote system! Visual analogies and mimetic experimentation are extremely risky and uncertain forms of inference.

In the winter of 1865, right at the height of the debate on the proper conceptualization of these peculiar surface features, James Nasmyth visited the observatory at the *Collegio Romano* and surprised its director, Padre Angelo Secchi (1818–78), in the act of producing a 'representation' of Nasmyth's willow-leaf-shaped constituents of the solar surface. Nasmyth recounts in his *Autobiography*, how Secchi

> then pointed to a large black board, which he had daubed over with glue, and was sprinkling over (when we came in) with rice grains. "That," said he, "is what I feel to be a most excellent representation of your discovery *as I see it*, verified by the aid of my telescope." It appeared to Father Secchi so singular a circumstance that I should come upon him in this sudden manner, while he was for the first time engaged in representing what I had (on the spur of the moment when first seeing them) described as willow-leaf

---

[42]See Nasmyth & Carpenter (1874a–e); cf. also Nasmyth (1883) pp. 316ff., and Thomas (1997) pp. 202f., Utzt (2004) pp. 83ff. or Robertson (2006) on Woodburytype illustrations in the 1st edition.

shaped objects. I thought that his representation of them, by scattering rice grains over his glue-covered black board, was apt and admirable; and so did Otto Struve.[43]

Once again, a two-dimensional depiction of what was viewed appeared not to be enough – like Nasmyth and Carpenter with respect to the Moon, Secchi also sought a three-dimensional 'likeness.'[44]

4.2 Iconophile versus iconophobe types

It is tempting to neatly divide the communities of science and technology into two complementary halves, according to the criterion of whether the actors are visual types or not, i.e., whether they love or hate to work with images, whether they immediately search for an intuitive visualization or whether they stick to formulas, algorithms or other nonvisual aids of reasoning. The acrimonious debates and sharp polemics between, say, Kekulé and Kolbe then suddenly appear in a new light as deep psychological incommensurabilities between iconophile and iconophobe types. As we have seen in sec. 3.2, someone like August Kekulé was as much fixated on getting a spatial 3D image of what organic molecules look like as his ardent opponent Hermann Kolbe was averse to that idea of painting and drawing molecules on paper. For Kolbe, a former student of Friedrich Wöhler in Göttingen, who then became assistant and successor of Robert Wilhelm Bunsen in Marburg, this indulgence in wild speculations about 2D structures and 3D models of molecules seemed plainly absurd and against the positivist norms of refraining from all hypothetical, unproven elements in science.[45] Likewise, in June 1867, Sir Benjamin Collins Brodie (1817–80) defended his efforts to create a quasi-algebraic calculus of chemical operations against those who were so crazy as to believe "that the fundamental facts of chemical combination may advantageously be symbolized by balls and wires."[46] Charles Frédéric Gerhardt (1816–56) was also ridiculing "all attempts to express the grouping and arrangement of atoms."[47]

---

[43]Nasmyth & Smiles (1883) p. 391; cf. Hentschel in Hentschel & Wittmann (eds. 2000) pp. 23–5 and here sec. 13.1 on later photographs of solar granulation.

[44]On the Victorian distinction between instrumentalized, conventional 'model' and 'likeness' (aiming at verisimilitude), see FitzGerald (1902) and Kargon (1969).

[45]See pp. 92ff. above on architectural studies as the triggering moment in Kekulé's spatial thinking, and Kolbe (1881a) pp. 353ff., (1881b) pp. 489, 495 and (1877b) p. 474 against these "playful fantasies" (*Phantasie-Spielereien*), which he polemically linked with romanticism and even spiritism. For more on Kolbe, see Rocke (1993) and (2010) chap. 6.

[46]Brodie (1867) p. 296, arguing specifically against August Wilhelm Hofmann's "glyptic formulae," i.e., ball-and-stick models, which had been advertised by the editor of the chemist's journal *The Laboratory* a month before in May 1867, p. 78.

[47]See Brodie (1866) pp. 781, 784.

Kolbe, Brodie and Gerhardt were representatives of the older, analytic tradition of chemistry that can be traced back to the iatrochemistry of the 16th and 17th centuries, for which all that mattered were precise weights and weight-relations in chemical reactions as well as detailed chemical characteristics of substances and the mastery of instrumental procedures. In the 19th century, anti-romantic sentiments and Comte's positivism had lent this tradition additional anti-metaphysical impetus, henceforth banning all paraphernalia from chemistry beyond the immediately given. Consequently, hypothetical images known from alchemical scripts were banned from texts in this analytic tradition. Highly speculative 3D models of molecules, construed only on the basis of minimal information about chemical valences without any other experimental technique available (at the time) to test or verify such assumptions about the molecular architecture, were considered the worst possible crime against this positivist canon. Thus the only plates to illustrate the 524 pages of Kolbe's 1865 publication, bristling with chemical sum formulas, tables with weight relations quoted to 1 mg and lengthy descriptions of chemical procedures, are three floor plans of Kolbe's Marburg chemical laboratory.[48] To give three more examples of such research traditions averse to the use of images:

(i) Pliny the elder (A.D. 23–79) and Hieronymus Brunschwig (c.1450–1512) warning natural historians and pharmacists that "pictures are misleading" and "nothing more than a feast for the eyes."[49]

(ii) Joseph-Louis Lagrange (1736–1813), the mathematician and mathematical physicist, was proud to announce that he could write his two-volume oeuvre on *Mécanique analytique* (1787) without using a single diagram. To him it seems to have been an exercise in cleanliness to steer clear of diffusely visual material, in order not to contaminate his flawless differential equations and his rigid inferences based on his variational calculus.[50]

(iii) Biochemistry was a text- and formula-centered subdiscipline of chemistry until the mid-20th century. Textbooks before 1950 bristle with complicated chemical formulas but are nearly devoid of images. With the advent of x-ray crystallographic research technologies, with which protein structure and then also DNA structure were cracked from the mid-1950s onwards (cf. here sec. 6.2), this profile of a whole subdiscipline changed drastically and – it seems – irrevocably towards a heavily model- and image-infiltrated subject.

Not only I but also various other historians of science and technology could not resist the temptation to try out a bifurcation of scientific practices. Most

---

[48]See Kolbe (1865); similarly so for other chemical textbooks in this analytic tradition.

[49]For references and further commentary, see Nickelsen (2004*b*) p. 1.

[50]On the purist tradition of Lagrangian mathematics, see, e.g., Grattan-Guinness (1990) chaps. 4–5, (2005) pp. 234, 241–5.

famous is perhaps Peter Galison's (*1955) celebrated book on *Image and Logic* (1997). The supposed dualism of two modes of thinking and scientific practice even made it into the book's title. In particular, Galison describes a statistical mode of conducting research based on large samples of data, on the one hand, and on a search for the "golden" event, on the other hand. The latter mode will tend to seek the one best piece of evidence for the existence of a new particle, and in the end wield an image of this golden event as the ultimate irrefutable proof. The rival tradition will aim at a clear elucidation of its claims through statistics based on thousands of events, analyzed by semi-automatic electronic counters and a complicated logic of data evaluation. Both of these traditions have coexisted for a good part of the 20th century in such prestigious scientific fields as nuclear physics and elementary particles, so it is not possible to simply declare one of these two opposing strands as irrelevant or unscientific. Both have been practiced, and both have booked to their accounts big successes as well as misses and failures. Both have their pros and cons.[51]

In fact, there seems to be ample reason for a tentative bifurcation into a visual versus a nonvisual mode of scientific research. Sometimes, the actors themselves seem to push in this direction in their self-stylization as either totally averse to imagery or totally obsessed by it. As an example of this other extreme, witness William Thomson (Lord Kelvin), who ornamented his guest lectures on molecular dynamics and the wave theory of light, delivered in Baltimore in 1906, with the following statement: "I am never content until I have constructed a mechanical model of the subject I am studying. If I succeed in making one, I understand. Otherwise, I do not."[52] Or listen to theoretical physicist John Archibald Wheeler reflecting on his own style of thinking: "Well, I certainly feel today that if I can't make a picture, I don't understand what I'm talking about."[53] It is not clear to me whether Wheeler was consciously alluding to Lord Kelvin's famous statement or whether he just felt the same way. At any rate, Wheeler's and Kelvin's statements closely resonate with each other, both expressing an insatiable appetite for visual models so characteristic of strongly visual, i.e., iconophile personalities. In my own interactions with scientists and engineers from all kinds of fields, as well as in my studies on the history of science and technology, I have often come across cases easily and unambiguously relegatable under one of two categories: the iconophile or iconophobe.

---

[51] See Galison (1997) for a plethora of fascinating case studies and further references.

[52] Kelvin in his posthumously published notes of these Baltimore lectures.

[53] John Archibald Wheeler in the transcript of an oral history interview conducted by Kenneth Ford (1993), AIP. Cf. here sec. 3.4 for Wheeler's contributions to geometrodynamics.

But beware, since this classification of people remains a highly nontrivial task, as not all cases will be as clear-cut as the two above. It is possible to go completely amiss, especially if the available sources about an individual researcher or technician are limited in scope and depth. Let me give you just two examples of this, the second being a trap I myself had also fallen into in the past.

First example: Albert Einstein (1879–1955). Hailed as *the* most famous physicist of all time, he is certainly *not* renowned for particularly visual achievements, rather the contrary. Both the special and general theories of relativity, to which he contributed decisive ideas and insights, are known for their abstruse effects. Terms like time dilation and length contraction, light deflection and warped space might all sound reasonably intuitive and *anschaulich*, but the mathematical framework of tensor analysis and differential geometry is formal and repellent to most.

Thus it might be consoling to hear that Einstein himself had a rather troubled relationship with higher mathematics. He had done reasonably well during his schooldays in Ulm and Munich, but had left *Gymnasium* before graduation in order to avoid being drafted into the German army. After flunking the entrance examination at the Swiss Polytechnic, the *Eidgenössisches Polytechnicum* in Zurich, at age 15, he decided to finish off his secondary education at a Swiss reformist school in Aarau that followed the pedagogical maxims of Johann Heinrich Pestalozzi (1746–1827). There, Einstein was able to enjoy institutionalized learning because, for once, it did not rely on tiresome rote memorization so typical of German schooling. The focus was placed on nonverbal forms of thinking, such as drawing, 3D modeling and sculpture. After eventually passing the *Poly*'s entrance examination, he continued to work in this nonstandard style, skipping many of the maths courses, such as the lectures on differential geometry given by Hermann Minkowski (1864–1909). When the final examinations approached, he wrote in panic to his college mate and good friend Hermann Grossmann: "Grossmann, you have to help me!"[54]

Nevertheless, Einstein successfully completed his training at the Swiss Polytechnic and was then employed at the Swiss patent office until 1909, where he continued to refine this intuitive, *anschauliche* style of thinking. For seven years, Einstein was deeply immersed in engineering contexts. It was, indeed, part of his family background, making him better equipped to look at physics from an unusual angle. I argue that this unusual training and pronounced engineering background – his father and his uncle were employed at the cutting edge of the

---

[54]On the young Einstein's nonstandard education, see, e.g., Pyenson (1985) and Root-Bernstein (1989) p. 331 and (1999) pp. 16f., 63.

electrotechnics industry – helped Einstein develop a strong visuality – a relatively unknown facet even among experts. Consider Einstein's own words about visual thinking in his thought:

> The words or the language, as they are written or spoken, do not seem to play any role in my mechanism of thought. The psychic entities which seem to serve as elements in thought are certain signs and more or less clear images which can be "voluntarily" reproduced and combined [...] the above mentioned elements are, in my case, of visual and some of muscular type. Conventional words or other signs have to be sought for laboriously only in a secondary stage.[55]

His early biographer Carl Seelig, who knew Einstein well, wrote about his working routine: "he works more imaginatively and [...] never thought on formalistic lines. His powers of imagination are closely related to reality. He told me that he visualizes the gravitational waves with the help of an elastic body, and at the same time he made a movement with his fingers as though he were pressing an India rubber ball."[56] In this context, one immediately associates the well-known visualization of the bending of light near large masses: a heavy body resting on a piece of rubber stretched out horizontally with its imprinted square-grid pattern becoming curved as it nears where the heavy mass rests. This compelling visual analogy suggests that Euclidean space–time becomes non-Euclidean near heavy masses. I have no evidence, though, that this visual analogy is Einstein's – it seems to have been created in the context of geometrodynamics in the 1970s.[57]

My second example, showing how easy it is to err about whether a certain actor is visual or nonvisual in character is Paul Adrienne Maurice Dirac (1902–84). As one of the co-founders of quantum mechanics who later also prepared the way for quantum electrodynamics, quantum field theory and other even more formal approaches to problems in theoretical physics, he is immediately a strong candidate for an actor very much on the formal and algebraic side.[58] Like many other historians, I had also classified him accordingly in former work. This judgment was based on Dirac's publications. All of them are devoid of diagrams or figures, dry deserts of hundreds of stand-alone equations, linked together by as meager a prose conceivable. In the physics community, there was the running

---

[55] Einstein in response to a questionnaire by Jacques Hadamard (1945) pp. 142–3; cf. also Gardner (1985) p. 190.

[56] Seelig (1956) on Einstein, p. 155; cf. also Miller (2001).

[57] On Wheeler and geometrodynamics, see here pp. 107ff.

[58] On the contrast between algebraic and geometric thinking, see Root-Bernstein (1999) p. 62 and 68f.

joke about one Dirac unit being the lowest number of words utterable in a day.[59] This monosyllabic character of Dirac the man is reconfirmed by his interviews with Thomas S. Kuhn, Eugene Paul Wigner and Friedrich Hund.[60] Peter Galison had an opportunity in the 1990s to look at the Dirac Scientific Papers deposited, not at Cambridge University, England, where Dirac spent most of his career, but at Florida State University Libraries in the USA. He was very surprised to find hugh stacks of geometric drawings and personal notes, never published by either Dirac himself or anyone else who had been through these files.

Now, one might argue that this finding in and of itself does not mean very much, especially since Dirac himself never bothered to publish anything of the kind. But in these same files Dirac reflects on exactly the problem discussed above, i.e., how to classify people in a two-pronged manner. Here are his own words, taken from a lecture draft in 1972:

> There are basically two kinds of math[ematical]. thinking, algebraic and geometric. A good mathematician needs to be a master of both. But still he will have a preference for one rather or the other. I prefer the geometric method. Not mentioned in published work because it is not easy to print diagrams. With the algebraic method one deals with equ[ations]. between algebraic quantities. Even tho[ugh] I see the consistency and logical connections of the eq[uations], they do not mean very much to me. I prefer the relationships which I can visualize in geometric terms. Of course with complicated equations one may not be able to visualize the relationships, e.g., it may need too many dimensions. But with the simpler relationships one can often get help in understanding them by geometric pictures.[61]

Many of these drawings actually come from the context of projective geometry. This fits quite well with the following quote from another lecture draft, also from the early 1970s, in which Dirac goes somewhat more into detail about his practice of working out a problem for himself in geometrical terms, then reformulating

---

[59]For many anecdotes and stories about Dirac as the "purest soul," see, e.g., Kursunoglu & Wigner (eds. 1990) or Kragh (1990) and further sources cited there.

[60]Dirac's interviews with Thomas S. Kuhn and Eugene Wigner, conducted between April 1 and May 6, 1963, are kept in the Archive for the History of Quantum Physics, Oral History Interviews, Niels Bohr Library, American Institute of Physics, henceforth abbreviated AIP. The conversation with Friedrich Hund was recorded on film by the *Institut für den Wissenschaftlichen Film* (IWF) in Göttingen, Germany.

[61]P. A. M. Dirac: Draft of a lecture on the 'Use of Projective Geometry in Physical Theory,' Boston, Oct. 30, 1972, in the Paul A. M. Dirac Scientific Papers, Florida State University, Tallahassee, Florida, file: Lectures B4 F26); here quoted from Galison (2000) p. 146.

the result in analytic form and thus suppressing the geometric approach that had led him heuristically:

> My research work was based in pictures. I needed to visualize things and projective geometry was often most useful e.g. in figuring out how a particular quantity transforms under Lorentz transf[ormation]. When I came to publish the results, I suppressed the projective geometry as the results could be expressed more concisely in analytic form.[62]

Where did Dirac's surprising weakness for projective geometry come from? From the time of his early training after 1914 at the *Merchant Venturers' Technical College* in Bristol, where Dirac received his primary and secondary education from age 12 onwards, later being enrolled in electrical engineering. In Bristol – as in all technical colleges and polytechnic high schools of the time – geometry played a major role in the curriculum.[63] One Bristol mathematics teacher, Peter Fraser, seems to have been a particular fan of projective geometry, instilling in young Dirac the same fascination for this subject. When he then learned about relativity theory after Eddington's light eclipse expedition had made it suddenly world-famous in late 1919, Dirac of course also molded this new gospel into the paradigm of what he had learned in his Bristol mathematics classes. From 1923 on, with Dirac's switch to *St. John's College* at Cambridge, the acknowledged master of projective geometry at this college, Henry Frederick Baker, sparked more enthusiasm for this approach. Only later, from roughly 1925 on, did Dirac subdue his earlier allegiance to projective geometry and begin to couch all his results in the crystal clarity of his algebraic equations and his 'bra–ket' reformulation of quantum mechanics, for which he became justly famous.

Peter Galison fittingly concludes his paper on these findings with the assumption that "physicalized geometry – geometry grounded in spatial intuitions, visualizations, diagrammatics – collapsed under the language of an autonomous science. In a sense, Dirac's suppressed drawings were the hidden remnants of an infolded Victorian world. Public geometry became private reason."[64]

---

[62]P. A. M. Dirac: 'Recollections of an Exciting Era,' three lectures given at the Varenna Summer School, August 5, 1972. Quotations from the handwritten draft, Dirac's papers, file: Lectures B4 F3, here taken from Galison (2000) p. 147; on projective geometry cf. Joan Richards (1988).

[63]On the central place of descriptive geometry in the Parisian *École Polytechnique*, see here sec. 8.1, esp. p. 268. On the special role of geometry in curricula of Victorian England cf., e.g., Galison (2000), p. 152, Richards (1988), and further sources cited there, respectively.

[64]Galison (2000) p. 163 and pp. 155f. as well as Kragh (1990) chap. 1 on Dirac's education, as well as on his father's domineering role. While the second factor is truly idiosyncratic, the first, i.e., technical college training, is highly significant for my claims, spelled out further below.

4.3   A prosopography of spectroscopists

After these three, more or less impressionistic, samplings out of three different centuries, I would like to become a bit more systematic and check my list of characteristics of visual cultures in science and technology, spelled out in chapter 2, against an entire scientific community of a specific field: spectroscopy in the 19th century. This group will again feature some Victorians but will also include French, German, Italian and US-American researchers.   Any commonalities between them then ought to be related to something deeper than perhaps similar cultural backgrounds. You may now jump directly to the tabular survey of these findings (on page 151); but to maintain the discursive flow I will also briefly comment on each entry in sentential form, starting with the very last of the ten levels (listed here on page 84), i.e., with the entry of visual obsessions into the private lives of the protagonists.

As I have found confirmed among several dozens of the key 19th-century practitioners in spectroscopy, members of a visual science culture will often transgress the professional boundary to indulge their obsession with visual records at home as well. The Astronomer Royal for Scotland, Charles Piazzi Smyth (1819–1900), is a good example to illustrate this point.[65] He was born into a visual culture, his father William being a naval officer, a founding member of the *Royal Geographic Society* and a specialist in surveying and oceanographic mapping for the British *Hydrographic Office*. As amateur astronomer, Smyth Senior studied stellar colors.

Smyth Junior's extraordinary visuality might already be surmised from the sheer volume of visual material he produced throughout his life. The Piazzi Smyth collection of drawings, photographs, and manuscripts at the *Royal Observatory* in Edinburgh contains reams of sketches, watercolors and calotypes made during his extensive travels to South Africa, Tenerife, Russia, etc. All are solid documentation of his strong orientation toward the visual. He mastered painting in oils on canvas besides, as much as various photographic techniques, and even wrote a treatise comparing the advantages and disadvantages of various printing methods.[66]

For this "astronomer-artist," as he has been aptly described,[67] a natural attraction to the novel field of spectral analysis was a mere extension of his wider

---

[65]On Smyth's life and work see, e.g., M. T. & H. A. Brück (1988) as the most detailed overall biography, Copeland (1901), Warner (1983), Evans (1989) for shorter accounts, and Hentschel (2002*a–d*, 2008) on Smyth as a spectroscopist and on his general visuality.

[66]See Smyth (1843*b*), M.T. & H.A. Brück (1988) and Schaffer in Galison & Jones (eds. 1998).

[67]Such is the main title of the biographical study by Warner (1983). Nice samples are also reproduced in M. T. & H. A. Brück (1988). On Piazzi Smyth's photographic and stereophotographic work, see p. 22 above and Schaaf (1979*b*, 1980/81, 1984).

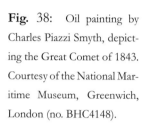

Fig. 38: Oil painting by Charles Piazzi Smyth, depicting the Great Comet of 1843. Courtesy of the National Maritime Museum, Greenwich, London (no. BHC4148).

search for visual patterns. He examined cloud formations, lunar craters, ferns and cultural relics of Ancient Egypt. Recording the treasures of the pharaohs, the fauna and flora of the Canary Islands or the ever-changing British skies was a lifelong passion that he shared with his wife Jessica.[68] Photography came in handy too, even though Piazzi Smyth was far from wanting in drawing skills. One obituary attributes to this scientist a gift of great artistic skill in committing to paper, canvas, and even frescoes, beautiful drawings, photographs, and paintings of the scenes he witnessed as a tourist, a gazer of the heavens and an optical experimenter. Even something as technical as drawing the solar spectrum was transformed by his mapping pen into an artistic drawing.

His explanatory texts to these spectra reveal his strong aesthetic fascination at seeing, for instance, how during a sunset off the coast of Spain: "the Fraunhofer line 'little a' had swollen up from the frog size to that of the bull, and had at last become positively elephantine in thickness and ponderosity, or it was even a case of a shrimp that had grown to be bigger than a whale." Likewise, the "awfully colossal proportions" of the Fraunhofer line capital A were to him "something for an intelligent man to have seen once before he dies."[69] But notice the strong

---

[68] On Smyth's photographs of "Cloud forms that have been," taken in retirement in the 1890s, see Thomas (1997) pp. 51, 87–91.

[69] Smyth (1877) p. 218; we are predating the age of political correctness here.

emotional bond to his subject! Expressions extolling the beauty of spectra abound in the literature.

Piazzi Smyth's Scottish colleague, the physicist David Brewster (1781–1868), was equally obsessed with visual perception and observation. He was an expert in physical optics, heavily involved in the debates on the physical nature of light, where he defended the Newtonian projectile theory of light against the adherents of the new wave theory, elaborating in particular on the multifarious applications of the polarization of light in crystallography, saccharimetry, etc.[70] He was the first to notice so-called 'axial images' generated by looking through crystals with polarized light.[71] In his *Treatise on New Philosophical Instruments* (1813), Brewster collected data on the refractive indices of nearly 200 substances in search of suitable material for an achromatic telescope. While attempting to improve colored eyeglasses to serve as a filter for telescopes and microscopes, he hit upon the selective absorption of highly specific colors, thereby founding the field of absorption spectroscopy.[72] Having entered spectroscopy, Brewster could not refrain from joining the debate on the number of primary colors in the spectrum.[73] Newton had famously opted for exactly seven, while others for less. William Hyde Wollaston (1766–1826), for instance, had chosen four. The outsider Johann Wolfgang von Goethe (1749–1832), whose polemics against Newton Brewster later bitterly criticized, even tried to cope with just two. Brewster settled on the number three and argued that the full spectrum was in fact a superposition of a red, a yellow and a blue spectrum. Following up leads from other experimentalists, such as Jean Senebier (1742–1809), William Herschel and his son John, Brewster also inquired into the chemical action of light. His researches on silver salts, then the subject of active research finally led to the invention of talbotype and daguerreotype in 1839.[74]

A fanciful instrument generating symmetrical optical patterns by multiple reflection and refraction off internal prisms, the so-called kaleidoscope, was invented by Brewster in 1817. This tubular viewing toy was soon pirated, thanks to a technical mistake in the patenting procedure, causing its rapid spread throughout the world, from the market fair to the nursery, and from the toy shop to the

---

[70]On Brewster's life and work, see Morrison-Low & Christie (eds. 1984), Buchwald (1989) and further references in Hentschel (2014d).

[71]On Brewster's techniques for generating and observing axial images, and for a comparison with Biot's and Goethe's approaches to these phenomena, see Nickol (2013) and primary references given there.

[72]For more on the latter, see Hentschel (2002a, 2014d) and further sources mentioned there.

[73]Cf. Sherman (1981) and Hentschel (2002a) pp. 36–9 and further sources mentioned there.

[74]Cf. Hentschel (2007) sec. 5.5.

designer's studio. Brewster's lenticular variant of the stereoscope, widely praised as far superior to Wheatstone's simpler design based on mirrors, also delighted innumerable customers enthralled by the three-dimensional glimpses it produced from a pair of two-dimensional photographs taken at slightly differing angles.[75] As a prime example of creators of visual cultures, Brewster also repeatedly dabbled in the physiology of vision, especially stereoscopy, color perception and contrast enhancement,[76] as well as photography. Early photo albums like David Brewster's, edited by Graham Smith in 1990 under the telling title *Disciples of Light*, reveal their obsession with fixing the moment on the photographic plate. They took shots of everything conceivable, long before our age of the omnipresent 'snap-shot' (a term coined by another Victorian spectroscopist already mentioned, John Herschel). But this pastime with light clearly predated the invention of photography and its subsequent improvements by several main actors in the history of spectroscopy (e.g., William Henry Fox Talbot and the aforementioned John Herschel, Hermann Wilhelm Vogel and William Abney). Various gadgets designed to produce entertaining visual impressions existed well before then. It is indicative that William Hyde Wollaston, the first to notice the dark lines in the solar spectrum in 1802, was much better known to his contemporaries for his invention of the *camera lucida*. Likewise, Josef Fraunhofer, who produced the first high-resolution map of the solar spectrum with its dark lines around 1814, also constructed a *camera obscura* in his summer house at Benediktbeuern as a local diversion.[77]

What inspired a young student to choose spectroscopy as his or her specialty? What made the some dozen leading researchers remain loyal to this field? I compared roughly fifty of the most prolific contributors to spectroscopic research throughout the 19th and early 20th centuries with regard to formative factors (such as family background and scientific training), and found that despite their totally different disciplinary embedding, they had more in common than might initially be surmised. Time and again, similar circumstances emerge that seem to have engendered a later preference for such a visual field within the sciences:

- A family background in the fine arts or artisanal crafts, such as engraving or lithography, weaving, or wood- or metal-working.
- Schooling at a polytechnic, a trade school or military academy (admitting civil engineers or architects), where drawing or mapping was emphasized

---

[75] See, e.g., Reynaud et al. (eds. 2000) pp. 44ff. and here p. 19 on the history of stereoscopy.

[76] Cf. Wade (1983).

[77] On the foregoing, see Rohr (1929), Fiorentini (2005), Hentschel (2002*a*) pp. 32ff. and further sources mentioned there.

in the curriculum.

- Employment as a teacher at such a polytechnic, often prior to any appointment in academia.
- Engagement as an instructor of perspectival drawing, descriptive or pure geometry.
- Application of specific visual skills (in drawing, engraving, lithography or photography), not only to the professional specialty, but also to other subjects as a form of recreation.

From among my prosopographic sampling, let us take Johann Heinrich Jakob Müller (1809–75), for instance. His father was a painter at the court of the prince of Waldeck, and later became director of the Darmstadt Art Gallery as well as the founding head of a graphic arts academy. Three brothers of our physicist received their initial training in the graphic arts there and later became renowned engravers or painters. This extraordinary enculturation within such a visually oriented, "artsy" environment might well illuminate the path leading to Müller's cooperation with court photographers to achieve his scientific goals in ultraviolet spectrography and to invent a graphic method for converting a prismatic spectrum into a normal (wavelength-proportionate) spectrum map. His skill in drawing likewise underpins his motivation to transform a rather dry textbook by Pouillet into a richly illustrated free adaptation that underwent eight editions during Müller's lifetime, each one incorporating even more wood engravings based on Müller's own drawings. Müller had a strong affinity for visual aids in teaching as well. In his lectures he made regular use of demonstration drawings (executed by the university's drawing teacher, Lerch). These innovative classroom posters included the solar spectrum several years before Bunsen and Kirchhoff's breakthrough in spectrum analysis, which led to the proliferation of their spectrum charts in poster format. And his perfectionism with regard to the quality of the reproductions explains why he was the first to publish an ultraviolet photograph of the solar spectrum. The available techniques of photomechanical reproduction were not yet sufficiently developed, so he decided to have prints of his wet collodion photograph on albumen paper pasted into every single copy of the 6th and 7th editions of his celebrated textbook.

Several cases of such artsy or artisanal family backgrounds exist. The father of the astronomer Karl Friedrich Zöllner (1834–71) was a woodcut engraver and cotton printer; and the pioneer in celestial photography Warren de la Rue (1815–89) was the son of the founder of a stationery printing firm. Another striking example is the French spectroscopist Pierre Jules César Janssen (1824–1907), who came from "une famille bien connue dans les Arts."[78] His grandfather, in particular, had been the famous architect Paul-Guillaume Le Moyne, and his father a well-known clarinetist. Janssen exhibited a talent for drawing since the age of 5, and paid frequent visits to painter ateliers in his youth, from ages 14 to 16. His travel diaries are filled with all kinds of fine sketches (as are Piazzi Smyth's or Listing's). Before he became the founding director of the Meudon Solar Observatory, he had been professor of physics at the *École Speciale d'Architecture*, where he gave courses for architects, for instance, on effective illumination. After 1873, Janssen acquired a deep interest in the new medium of photography and its multifarious uses, first and foremost for astronomical purposes, but also for application in meteorology, geology and other fields of the natural sciences.[79] This fascination induced him to preside over various photographic societies during the course of his career in astrophysics. And despite all of his success in science, this son and grandson of artists never ceased to speak about the "beauty" of the phenomena that he studied, indeed their "sublimity."[80]

Intense lifelong appreciation for the visual arts is documentable in the cases of various other spectroscopists, who are not by any means confined to the 19th century. We might even just be seeing the tip of the iceberg here, because of the unfortunate tendency of traditional biographers and archivists to eschew nonscientific aspects of scientists' lives as purportedly irrelevant for an understanding of their science. I would argue quite the opposite. In order to grasp the personal styles and modes of thinking of our figures, their preferences for, or abhorrences of visuality have to be taken into account, without arbitrary exclusion of their pastimes and general predilections. To take another example from the 20th century, the mastermind of the MK stellar spectra classification scheme, William W. Morgan (1906–94), went beyond the usual hobby of art collecting and photography. One of his obituaries mentions: "In the local primary school, during his later years, he thoroughly enjoyed serving as Picture Lady, which involved taking a work of great art to one of the classrooms, showing the children how to see its patterns, and helping them to understand its greatness."[81] Seeing patterns – or

---

[78] On Janssen's vita, see, e.g., Bigourdan (1908), Baume Pluvinel (1908), Levy (1973) and Launey (2001) for a list of his correspondents. Cf. here pp. 247 and 379.

[79] See, e.g., Janssen (1887*a*, *b*) on astronomical applications of photography, which he predicted to be "a revolution no less productive than the one marking the introduction of lenses in astronomy" ("une révolution qui ne sera pas moins féconde que celle qui a signalé l'introduction des lunettes en astronomie," p. 1068), and (1888*a*) on meteorology on the summit of Pic du Midi.

[80] Quotes from Janssen's opening speech at the international congress for photography, accompanying the 1889 World Exhibition; cf. Sicard (1998) p. 60 for a longer excerpt.

[81] See Garrison (1995) p. 507 and Hentschel (2002*a*) pp. 357ff., 429.

knowing how to search for them – was also one of the most important skills in the search for series formulas (Balmer, Rydberg) or for structures in band spectra (Piazzi Smyth, Alexander Herschel, Deslandres, Bjerrum).

Johann Benedikt Listing (1808–82) is another example from among our spectroscopists noted for their outstanding drawing or printing skills. This physicist, son of a brushmaker, had obtained a grant from the *Städelsche Stiftung* in Frankfurt to study architecture and the arts before he switched to the sciences. Aside from his work on physiological optics and spectroscopy, he is best known today for his topological studies, which contain striking visual analogies. He illustrated the rather arcane subject of linear complexions in space with various types of knots found in ordinary daily life and spiral structures found in botanical and zoological specimens. Such pronounced visual thinking may be linked to his early training in architecture, which certainly included a hefty serving of drawing and other graphic arts. Later this enabled him to lithograph his own plates for his textbook on the physiology of the eye. His choice of iris print to illustrate his study of the color ranges in the spectrum also reflects his fastidiousness in printing matters.

The astronomer Hermann Carl Vogel (1841–1907) and his assistant Wilhelm Oswald Lohse (1845–1915) did not have to rely entirely on lithographers either. They were able to draft the plates for their atlas of the solar spectrum from 1879 themselves and also experimented with various photomechanical reproduction techniques such as albertype (*Lichtdruck*) and a self-made method of autography based on photomechanical zink etchings. The backgrounds of these two Potsdam astronomers are revealing. Both Vogel, son of a Leipzig school director, and Lohse, son of a master tailor, had attended the Royal Polytechnic in Dresden before embarking on studies in the natural sciences at the University of Leipzig. Polytechnic schools, with their emphasis on drawing lessons, or military colleges, with their field-mapping courses, form the background of surprisingly many 19th-century spectroscopists (roughly 50% of my prosopographic sample in that century; cf. here table 1 on pp. 151f.). By contrast, in other fields, such a migration between the very different worlds of polytechnics and universities was quite unlikely. Thus we begin to see why spectrum mappers differed in their visuality from other science cultures, such as chemistry, mechanics or theoretical astronomy, which all were taught at universities according to a traditional, i.e., at that time decidedly nonvisual, curriculum.

Lohse was clearly a perfectionist obsessed with visual representations. During the 40 years he worked in the Bothkamp and Potsdam Observatories, Lohse experimented with photographic emulsions, comparing different kinds of commercially available types, as well as testing developers and fixers. He was among the first to use the new method of gelatine dry-plate photography for astronom-ical purposes.[82] Aside from photographs of the Sun and its spectrum, Lohse also published extremely refined sunspot drawings (see fig. 39), together with recommendations on how best to reproduce them in print.

Fig. 39: Lithograph of a sunspot observed in Potsdam in 1875. Based on a drawing by Oswald Lohse, and printed by the *Lithographische Anstalt* of J.G. Bach in Leipzig. From Lohse (1883) pl. 41.

Lohse's superb sunspot drawings had been executed by a technique he had devised for himself, graphite stumping. After making a rough sketch in charcoal at the telescope, Lohse transferred the outlines in graphite pencil onto sturdy Bristol board. The finer details of the sunspots were made by stumping with leather, cork, paper or sponge, and finally the superposed filigree bright parts were rendered by erasing with a rubber, using a few transparent gelatine templates cut in frequently recurring shapes. In his article on astronomical drawings from 1883, Lohse emphasized that he had taught himself this method, also known as *dessin à l'estompe*. A beautiful original drawing of a sunspot by Lohse employing the skillful technique of white heightening hangs on the workroom wall at the Einstein Tower in Potsdam.

Speaking of solar-spot drawings, another name springs to mind: Samuel Pier-pont Langley (1834–1906), who is known for his work on infrared spectroscopy.

---

[82] On Vogel's and Lohse's vitas, see Kempf (1915) and Hentschel (2002*a*) 82ff., 187ff., 421ff. On Lohse's huge photographic plate archive for the period 1879–89 preserved in Potsdam, see Tsvetkov et al. (1999). On astronomical plate archives more generally, cf. Kroll & Bräuer in Hentschel & Wittmann (eds. 2000).

Among astronomers, he was probably better known for his meticulous drawings of sunspots dating from 1873, which showed minute details in the penumbra that became visible during moments of exceptionally good seeing but were not photographically recordable for another hundred years. It will not come as a surprise that this son of a wholesale merchant also had undergone unusual training for a researcher, having studied not science but engineering and architecture. Incidentally, the same is true of Kirchhoff's student assistant Karl Hofmann (1839–91), who had studied at the polytechnics in Vienna and Carlsruhe as well as at the Freiberg Mining Academy in Saxony before adding mineralogy, chemistry and physics among his courses at the Ruperto-Carolina University in Heidelberg. The Swiss spectroscopist Walther Ritz (1878–1909), who joined the search for series formulas, was doubly preconditioned, being the son of a landscape painter and having enrolled at the Zurich Polytechnic.[83] Both Frank McClean (1837–1904) and Victor Schumann (1841–1913), two obsessive photographic spectrum mappers, had only concentrated on this specialty after long-time employment as chief engineers: Schumann had previously worked for a number of engine factories and McClean held a position in his father's engineering firm, *McClean & Stileman*. In Schumann's letters, these commercial jobs only feature as a source of irritation; nevertheless, the engineering skills involved explain why it was Schumann as opposed to someone else who successfully constructed the highly complex vacuum spectrographs and devised the appropriate gelatine-free emulsions for photography in the ultraviolet.[84]

Besides these examples, I could also have referred to many other spectroscopists. Their science education deviated from the standard, with them having obtained a thorough training in programs emphasizing visual skills. Indeed, some had not only been educated in the applied graphic arts during their youth, but later also actually taught at institutions where this visual style was promoted. Henry Augustus Rowland (1848–1901), for instance, initially taught physics at the *Rensselaer Polytechnic Institute* in Troy, New York, whence he himself had graduated in civil engineering in 1870. Hermann Wilhelm Vogel (1834–98), who conducted sensitization experiments on photographic emulsions, taught photochemistry, spectrum analysis and applied optics at the Royal Prussian Commercial Academy (*kgl. Gewerbeakademie*), which later merged with the Architectural Academy (*Bauakademie*) to become the Charlottenburg Polytechnic in 1879. Likewise, the doyen of spectroscopy in Germany, Heinrich Kayser (1853–1940), taught at the Hannover

---

[83] On Ritz, see Pont (ed. 2012), with reproductions of pencil drawings by this talented draftsman.
[84] On McClean and Schumann, see Hentschel (2002*a*) and further sources cited there.

Polytechnic before accepting a call to assume the physics professorship at Bonn University in 1894. And Friedrich Paschen (1865–1947) was Kayser's assistant at the Hannover Polytechnic before he later received a call to Tübingen University, where he became known for his experiments on the Zeeman effect and other spectroscopic precision measurements. Listing taught mechanical engineering at the *Höhere Gewerbeschule*, the local precursor institution to the Hannover Polytechnic, where Kayser later taught, and Robert Wilhelm Bunsen (1811–99), codiscoverer of spectrum analysis, taught at the Kassel *Höhere Gewerbeschule* from 1836 to 1839. Johann H. J. Müller, too, offered courses in geometrical drawing several times during his long tenure as professor of experimental physics at Freiburg University.

The cases mentioned thus far exemplify personal links to civil engineering. There are links to its military counterpart as well, foremost, of course, the faculty at the *École Polytechnique*, which had been founded as a training institution for French military officers. Aside from Alfred Cornu (1841–1902), Eleuthère Mascart (1837–1908) and Charles Fabry (1867–1945), we also have a few British examples: The pioneer of infrared spectrum photography, William de Wiveleslie Abney (1843–1920), taught photography at the *Chatham School of Military Engineering*. In 1883, Abney began to offer lectures in photography at the *South Kensington Science and Art Department*, where Joseph Norman Lockyer (1836–1920) was another member of the teaching staff.

Tab. 1: Next two pages: Tabular summary of the prosopography of 30 spectroscopists active prior to 1900. From Hentschel (2002*d*), pp. 598–9.

Name	Pertinent family background	Training or practice in architecture or engineering	Other types of tutoring, private lessons	Teaching in visual fields	Active interest in physiology	Photography	Visual pastimes
Abney, W. de Wiveleslie		Royal Military Academy; Royal Engineers		Photography: Royal School of Military Engineering, Chatham; South Kensington		XX	
Balmer	Brother, son and grandson: painters; brother-in-law: Burckhardt	Berlin Bauakademie	Lessons in physics and spectroscopy by A. Hagenbach (University of Basel)	Geometry, perspectival drawing at Höhere Töchterschule & University of Basel			Architectural design of churches, bridges, worker houses
Brewster				Optics, astronomy, Edinburgh University	Color perception, kaleidoscope	X	first director of Scottish Society of Arts
Bunsen				1836–39 at Kassel Höhere Gewerbeschule	Photometry; Fettfleck-photography		Geological excursions
Cornu		École Polytechnique & École des Mines		Optics, spectroscopy: École Polytechnique		X	Cornu spiral as graphic technique
Fabry		École Polytechnique		Optics, spectroscopy: École Polytechnique + interferometry	Color intensity, light diffraction	X	
Fraunhofer	(Orphan)	Autodidact (reads Klügel etc.)	Artisanal training	Practical training of glass workers & techniques			Camera lucida
Herschel, J.	Astronomer family		Drawing lessons			X	
Janssen	Grandfather: architect P. G. LeMoyne		Drawing and painting in various Parisian ateliers	1865–71: École spéciale d'architecture (e.g. lectures on effective illumination)	IR absorption of the eye; reaction time (granulation)	X	Drawings in travel journals. ...
Kayser	Brother: Emanuel Kayser; geologist			Before 1894: Hannover Polytechnic			
Langley	Father: wholesale merchant	X				X	Travel photography, art collection
Listing	Father: master brushmaker	Study grant in architecture ...	Drawing lessons at Städelsche Kunstakademie, Frankfurt	Hannover Höhere Gewerbeschule, later Göttingen Univ.; geometry, topology (knots)	Color perception, eye physiology	X	Drawing, e.g., travel journals
Lockyer	Father: medical practitioner		Editor of Nature (many illustrations)	South Kensington, Science and Art Department		X	
Lohse	Son of master tailor	Königliches Polytechnikum, Dresden	Graphite stumping (autodidact)			XX	Sunspot drawings
McClean	Father: engineer	Various engineering companies				XX	Sunspot drawings
Michelson	(Emigré child)	US Naval Academy		Optics (Cleveland & University of Chicago)			Drawing

Name	Pertinent family background	Training or practice in architecture or engineering	Other types of tutoring, private lessons	Teaching in visual fields	Active interest in physiology	Photo-graphy	Visual pastimes
Müller, J. H. J.	Father: court painter, later head of art academy; brothers: painters		Drawing and painting lessons from father and brothers	Drawing, geometry, frequently with posters (Lerch)		X	Woodcuts for his text books
Pickering, E. C.				Graphic data analysis at Massachusetts Institute of Technology (before 1876)		XX	
Ritz	Father: landscape painter	Cours techniques at Lycée cantonal; Eidgenössisch-Technische Hochschule, Zurich		(Died as *Privatdozent*)		X	IR-sensitive emulsions
Roscoe		Liverpool Institute		Member of Royal Commission on Technical Education	Photo-chemistry with Bunsen		Laboratory architecture
Rowland		Rensselaer Polytechnic				XX	
Schumann	Father: engineer	Königliche Gewerbeschule, Chemnitz; engineer at engine factory				XX	
Smyth	Godfather: astronomer Piazzi	Geodetic survey, South Africa	Tutoring by father, Maclear, Herschel	Astronomy at University of Edinburgh & Society of Arts	Photometry & color perception	X	Drawing, pyramidology
Talbot						XX	Botany, arts
Trowbridge				Superintendent of drawing at Massachusetts Institute of Technology			
Vogel, H. C.		Königliches Polytechnikum, Dresden				X	
Vogel, H. W.		Königliche Gewerbe-akademie, Berlin				XX	
Wollaston	Father: astronomer				Binocular vision		Camera lucida
Young		Professor at Royal Institution		Optics; professor at Royal Institution 1801–03	Medical PhD; eye physiology		Deciphering of hieroglyphs
Zöllner	Father: pattern-maker and cotton printer				Light intensity; color perception		

As is verifiable in the sixth column of Table 1, many spectroscopists not only attended but also taught at polytechnics or vocational schools, often before getting their professorship appointments at university – at a time when the invisible gulf between low-prestige technical schools and privileged universities was still quite daunting. Such experience as Kayser's, Paschen's and Listing's at the Hannover Polytechnic, or Jules Janssen's six years of lecturing at the *École Spéciale d'Architecture* instilled a strong sense for techniques of visual representation. One could also quote the example of the *Massachusetts Institute of Technology* (MIT), where Edward Charles Pickering (1846–1919) and John Trowbridge (1843–1923) played a crucial role in establishing 'graphic methods' in physics courses before they joined the Harvard faculty. Trowbridge even taught courses in technical drawing. Pickering's first female student, Sarah Frances Whiting (1847–1929), then transferred this pedagogic approach to *Wellesley College*, where many of *Harvard College Observatory*'s female stellar spectrum classifiers were trained.[85]

4.4   Generalizability of these claims

All of the above examples and a plethora of further published case studies show that special scopic domains cannot be studied in isolation from the broader culture of vision and representation into which they are embedded. To research and to document the latter isn't easy, since information on such nonscientific aspects of biographies has often been tucked away as irrelevant or marginal. For instance: What kind of paintings hung on the walls in their private dwellings? Which hobbies did they cultivate? Only when we know the visual worlds in which our actors moved, and the cultures of perception and representation in which these were embedded, will we be able to gauge the specificity of their own visual representations and their respective visual preferences or aversions.

For intensely researched figures such as Charles Darwin (1809–82), these biographical visual resonances are easily traceable, whereas for the great majority of scientists, physicians and technicians, we first have to uncover them from the myriad of side-remarks in obituaries, correspondence and other such sources happening to say something about the visuality of our actors. Concerning Darwin, we know that he maintained a lively interest in the fine arts, photography and printing techniques.[86] As a student at Cambridge University, he frequently visited the *Fitzwilliam Galleries*, discussing the paintings on display with the curatorial staff. Under the tutelage of Charles Thomas Whitley (1808–95), Darwin even started to

---

[85] For more on this, see Hentschel (2002*a*) sec. 9.5 and further references there.

[86] E.g., Prodger (1998), Smith (2006) and www.darwinproject.ac.uk/ (accessed Aug. 28, 2013).

collect engravings. Later he also traveled to the *National Gallery* in London, where he was particularly impressed with the Renaissance masters. Among his close friends we can list the amateur portrait photographer Julia Margaret Cameron and the pioneers of photography John Herschel and Fox Talbot. For his book on the *Expression of Emotions in Man and Animals* (1872), he established contacts with specialists in medical and psychiatric photography, such as Guillaume Duchenne de Boulogne (1806–75) and George Charles Wallich (1815–99) and the professional photographer Oscar Rejlander (c.1813–75), who contributed photographs of insane patients or of electrophysiologically contorted human faces.[87] All of these photographs were either transformed into wood or steel engravings or reproduced as heliotype, using the brand-new photomechanical reproduction technique invented by Ernest Edwards (1837–1903). For other books, Darwin sought out the most skilled draftsmen and copper or wood engravers of the Victorian age, even when that meant risking the publication's economic success.

Darwin's prime advocate in Great Britain, Thomas Henry Huxley (1825–95), was equally famous for his profusely illustrated books for the furtherance of the evolutionary cause. He had inherited artistic and engineering skills from his father. Throughout his life, Huxley habitually recorded and documented in pen, pencil and watercolor.[88] In fact, his drawing talent was so marked that Huxley for a while seriously considered pursuing a career in this field, or in technical drawing in engineering, to which he also always felt a close affinity. In the end, though, he opted for the study of medicine. Staying at *Charing Cross Hospital* until 1845, he then became assistant surgeon for the *Royal Navy*, and accompanied Captain Owen Stephens aboard HMS Rattlesnake on a long expedition to North Australia, New Guinea and the Louisian Archipelago (1846–50). Even though an official artist also participated in this expedition, the official report was later illustrated by lithographs and wood-engravings based on Huxley's far-superior drawings. In a talk on "Science and art in relation to education" and later as governor of *Eton College*, Huxley deemed drawing lessons imperative for all students, on the argument that "it gives you the means of training the young in attention and accuracy."[89] His lectures were remarkable not only for his famous rhetorical skills.

---

[87] For more on this context, see, e.g., Prodger (1998) and Smith (2006) chaps. 5–6; cf. also http://special.lib.gla.ac.uk/exhibns/month/nov2009.html

[88] For the archival collection of Huxley's preserved zoological, anthropological and ethnographic drawings and photographs at Imperial College, see www.imperial.ac.uk/recordsandarchives/huxley papers/HUXS021.htm (accessed Sept. 20, 2013).

[89] Huxley 1882, quoted from his *Collected Essays* from 1892, vol. 3, pp. 160ff.; cf. also Newth & Turlington (1956), Bodmer (1997) and Jarrell (1998) for further details on Huxley's visual culture,

Huxley frequently resorted to skillful zoological and zoological sketching on the chalkboard. With a piece of chalk in one hand and a wiper in the other, he could vividly show his students how a few changes in shape can transform one species into another. This was an exercise in visual thinking that also came ready to hand when he was pondering how a certain species might move, just on the basis of a skeleton or drawings. When dry emulsion plates made photography portable in 1870, Huxley also started to build up an anthropological and zoological photograph collection. Like Darwin, he hired some of the best engravers and lithographers available to transform his drawings and photographs into print, even though he often remained discontent with the result of their efforts. In all these respects, Huxley fits our prosopographic grid quite perfectly.

To give an example from a quite different field: The Belgian physiologist and physicist Joseph Plateau (1801–84), whom we have already met (in sec. 1.6) as inventor of an important pre-cinematographic device, the *phénokistiscope* (cf. here also sec. 7.4), was the son of a floral painter, who wanted his son to become an artist, too, and sent him to the *Académie de Dessein* at Brussels. The artistic milieu in which Plateau grew up turned him into an ardent, highly skilled draftsman and watercolorist as well as an enthusiastic collector of insects. In 1822, Plateau began studying arts at the University of Liège. Under the influence of the Belgian statistician Quetelet, the young naturalist eventually decided to switch to the natural sciences. In 1828, he submitted his PhD thesis on human visual perception, presenting the first exact measurements of the effects of colors on the human retina. From then on, he pursued path-breaking research in sense physiology and on afterimages, irradiation, color contrasts, colored shadows and other topics all related to visual issues. In the 1830s, he constructed instruments to study afterimages and the apparent merging of images only briefly seen. This makes him into a pioneer of film and cinema. In 1840, his attention turned to the phenomenon of surface tension in fluids, which he analyzed with precision experiments on 3D models of surfaces of revolution based on his experimental observations and stereo-photographs.[90] Self-experimentation, including looking directly into the bright sun for 27 seconds, tragically ruined his good eyesight and eventually made him blind.[91]

---

including his drive to teach drawing, modelling and singing to working-class children as "the best possible preparation for technical schools."

[90] On these experiments, models and photographs, see Wautier, Jonckheere & Segers (2012) pp. 271–4.

[91] For detailed surveys of Plateau's life and work, see, e.g., Mensbrugghe (1885), esp. pp. 390ff., Joseph Wachelder in Dürbeck et al. (eds. 2001) and Wautier, Jonckheere & Segers (2012).

Self-experimentation was also something practiced by Christian Ernst Wünsch (1744–1828), a professor of mathematics and physics in Frankfurt on the Oder. Wünsch was the son of a simple weaver, who spent a miserable youth in utter poverty. To finance his studies of medicine and philosophy, later also mathematics and physics at the University of Leipzig, he made portraits of his fellow students and copied their college notebooks. He also built planetary and cometary orreries, which finally got him a fixed position and a fellowship to support his further academic studies. He became best known for his color theory published in 1792.[92] His claim that all colors could be generated on the basis of only the three primary colors green, red and violet was criticized and ridiculed by Johann Wolfgang von Goethe, no less, who parodied it in one of his *Xenien*:

> Yellow-red and green make yellow, green and violet-blue make blue! Pickle
> salad does make vinegar, so it really must be true. [93]

That Wünsch could liberate himself from the dictate of the Newtonian seven-color theory is due in no small measure to his artisanal background and his mostly autodidactic upbringing, which spared him exposure to the dogmatism in optics of the university curriculum at that time.[94] In close parallel to Listing and Wünsch, the famous crystallographer René-Just Haüy (1743–1822) was also born into the household of a poor weaver, in his case a linen weaver in the small town of Saint-Just-en-Chaussée in the Beauvaisis region north of Paris.[95] Without the support of attentive teachers – in Haüy's case, Premonstratensian canons in the cloister of St. Just, who arranged a fellowship for their highly talented poor élève – neither Listing nor Wünsch nor Haüy would have been able to study to become foremost scientists of their day. Their family backgrounds in artisanal contexts certainly trained their powers and acuteness of observation.

One very important factor in characterizing individual visuality is the kind of education such actors received, possibly also in addition to their main – or bread-and-butter – courses of study. Surprisingly many among my prosopographic set of spectroscopists received less of a scientific than technical training, such as in

---

[92]On Wünsch's *Versuche und Beobachtungen über die Farben des Lichtes* see, e.g., Frank (1898).

[93]"Gelbroth und grün macht das Gelbe, grün und violblau das Blaue!
So wird aus Gurkensalat wirklich der Essig erzeugt." Goethe and Schiller: Xenion 175; transl. by A.M. Hentschel. Cf. also his letter to Schiller, January 13, 1798.

[94]On this Newtonian color theory and its dogmatic entrenchment in the late 18th century, see Hentschel (2002*a*) pp. 27ff. and (2006*c*).

[95]On Haüy and his pioneering role in the proper visualization of crystallographic structures, see below pp. 334ff.

engineering or architecture; both these fields offer a high percentage of instruction in drawing or drafting as part of the regular curriculum.

Let me remind you here that these spectroscopists share this unusual exposure to architecture and/or technical drawing with various other scientists who became famous for breakthroughs associated with visual thinking: the chemist August Kekulé, the materials scientist Henry C. Sorby, or the physicist John A. Wheeler, for instance (see above secs. 3.2–3.4). August Kekulé had studied architecture before opting for chemistry and later envisioning the structure of the benzene ring. In the end, architecture wasn't just a "passionate hobby" for Kekulé. In a sense, he always remained an architect. His compulsion for visualizations was simply transposed onto another field. He became the "architect of chemistry," modeling the spatial arrangement of atoms and their chemical bonding. During his youth, he also received regular drawing and painting lessons, on Sundays at an engraver's studio. Even as an academic student, he took instruction in modeling clay, as well as in woodcarving and woodturning.[96] Likewise, the astronomer Giovanni Schiaparelli, whom we will meet later (in sec. 9.2) as the "discoverer" of Martian canals, had also obtained a thorough training in engineering and architecture at the University of Turin before he turned to astronomy, where he could use his outstanding drawing skills to the fullest in his planetary observations.[97] That his father was an artisan, a furnace-maker to be precise, strengthens our prosopographic correlation between artisanal family background and a likelihood for a later connection with a visual culture of science or technology.

That training in civil engineering or architecture may almost inevitably elicit a pronounced visual acuity is already well documented by Antoine Picon's case study on the training of French architects and engineers in the Age of Enlightenment. Eugene Ferguson and Brooke Hindle have advanced similar claims about 'thinking with pictures' or models as an essential part of engineering culture.

> Many features and qualities of the objects that a technologist thinks about cannot be reduced to unambiguous verbal descriptions; therefore, they are dealt with in the mind by a visual, nonverbal process. The mind's eye is a well-developed organ that not only reviews the contents of a visual memory but also forms such new or modified images as the mind's thoughts require. As one thinks about a machine, reasoning through successive steps in a dynamic process, one can turn it over in one's mind. The engineering designer, who brings elements together in new combinations, is able to

---

[96] See Anschütz (1929) vol. 1, pp. 7–13, 657–58 and Rocke (2010) pp. 39f., 65f.
[97] On Schiaparelli's training and family background, see MacPherson (1910) p. 468, Knobel (1911) p. 282 and here p. 299.

assemble and manipulate in his or her mind devices that as yet do not exist.[98]

It is thus part of the engineer's mind-set to be able to solve design problems by a creative play with spatial configurations in machines, mentally altering arrangements, sequences, dimensions and thus applying visual thinking in its full glory. Engineering drawings not only exhibit a special kind of graphic language that needs to be learned in courses and practical exercises; they also encapsulate a particular style of visual thinking in images and processes. As an aside, Ferguson also deplored the recent demise of mechanical drawing or design courses in engineering curricula, fearing a consequent loss of nonverbal imagination and visual thinking. This catastrophe foreseen by Ferguson has not taken place so far, which might be due to a substitution of these traditional drawing techniques by more modern techniques, such as CAD drawing on screen. Skills of visual thinking might shift and alter but by no means become totally erased.

My thesis is that individuals initially trained in drawing and other graphic techniques as part of their education in engineering, architecture or the applied arts were later able to transfer their specific visuality into the natural sciences and hence became instrumental in the formation of visual science cultures such as spectroscopy, radiology or x-ray crystallography. The astonishing regularity of so many prominent spectrum mappers having undergone some kind of technical training invites this tentative link between training in the applied and graphic arts with a propensity for visual cultures within the sciences. The rapid rise of visual cultures within science during the 19th century may thus be closely linked with the incorporation of drawing lessons in European school curricula after 1800, as well as with the inclusion of descriptive geometry in engineering curricula, as emphasized in Eugene Ferguson's perceptive study on the engineer's eye.[99]

Electron microscopist Heinz Bethge's (1919–2001) double qualification as physicist *and* engineer fits smack into this picture; likewise Robert Hooke's (1635–1702) balancing act between (instrument) technology, architecture and natural philosophy of his time.[100]

To the best of my knowledge, there are only very few precursors to our prosopographic approach as far as the examination of visual science cultures

---

[98] Ferguson (1992*a*) p. xi; cf. Hindle (1983).

[99] On the foregoing, see Picon (1992), Ferguson (1977, 1992) and Hindle (1983).

[100] Good insights into Robert Hooke's visuality, ranging from architecture, the art of portraiture and city planning to microscopy and astronomy, are offered by Harwood (1989), Inwood (2002) and Doherty (2013). Cf. also Robinson (1948) and, especially on microscopy, Dennis (1989).

is concerned. The earliest hint at a systematic connection between scientific creativity and skills far outside the realm of science and technology is found in studies by Francis Galton (1822–1911). In a prosopographic study on *English Men of Science* that first appeared in 1874, Galton studied the social and psychological "nature and nurture" of fellows of the *Royal Society* in the 19th century and noted a statistically significant affluency of musical and mechanical skills as well as a surprisingly broad range of interests.[101] The neurologist Paul Julius Möbius (1853–1907) and the physical chemist Wilhelm Ostwald (1853–1932) found similar character traits in some eminent mathematicians and physico-chemical scientists, respectively. In 1951, Anne Roe published her study on members of the US *National Academy of Sciences*, documenting a strong correlation between outstanding mathematical abilities and visual thinking skills.[102] Stimulated by these findings, Robert Root-Bernstein (*1953) started his study of "correlative talents."[103] While working on his PhD thesis on the origins of physical chemistry, in general, and on the cofounder of stereochemistry Van 't Hoff, in particular, he had become attuned to the problem of explaining why some scientists differ in their motivations and creativity. In a prosopographic study of some of the "most creative scientists," Root-Bernstein checked how many of them also showed strong talent and interest in the arts. He claimed to find a correlation between creativity and "artsy" skills, or between mental imaging skills and the importance attached to the arts by his actors:

> The ability to imagine new realities is correlated with what are tradition-
> ally thought to be nonscientific skills – skills such as playing, modeling,
> abstracting, idealizing, harmonizing, analogizing, pattern forming, approx-
> imating, extrapolating, and imagining the as-yet unseen – in short, skills
> usually associated with the arts, music and literature.[104]

In particular, Root-Bernstein found that for his set of 40 scientists, spread evenly over the natural sciences, the "visual arts, particularly painting and drawing, although among the least common hobbies reported by the scientists, are significantly correlated with both high-impact and high publication citations cluster

---

[101] See Galton (1874). Cf. also Hilts (1975) on the Darwinian and eugenic context of Galton's writings, which were most concerned with the tricky issue of hereditary transfer of talent and "genius."

[102] See Roe (1951).

[103] See Root-Bernstein (1985, 1999, 2001) and Root-Bernstein et al. (1995) pp. 117ff. for further references to prosopographic studies on skills and creativity in science.

[104] Root-Bernstein (1985) p. 51. On the use of imagery by scientists, cf. also Roe (1951) and A. I. Miller (1996).

status."[105]   By contrast, other hobbies such as poetry, singing, electronics or record collecting yielded no such correlation with the impact ratio. But why would avocations in the fine or applied arts advance the creativity of scientists and technicians working in spheres so vastly different in kind? Root-Bernstein's explanation is temptingly simple:

> People who practice making things – poems, paintings, weavings, musical instruments, etc., – learn the process of creating. In order to create something, one must first image it in one's mind and then acquire the skills necessary to translate one's vision into reality. [...] despite the very real differences between disciplinary products, the experience of the process itself is transferable.[106]

The problem I have with Root-Bernstein's claims is methodological. First of all, he does not clearly define "creativity," nor is it easy to do so – perhaps it is impossible – on the high level of generality in which he positions his claims. That means, though, that his social demarcation criterion to define the study group is too loosely circumscribed. Who exactly is a creative scientist? And who – by contrast – only a Joe Blow regular scientist or technologist? The other side of the correlation is not defined any better: talent for playing piano counts as much as for drawing or painting, reciting poetry or theatrical acting on stage. With such a grab bag of interests, on the one hand, it is easy to demonstrate positive correlations with all kinds of reference groups, especially if they are as loosely defined, as well.

My set of markers are much more limited and sharply defined than Root-Bernstein's. What I take to be indicators of the likelihood for someone to move into a visual science culture or even to form a new one are:

- former training in architecture or engineering prior to switching to science;
- and/or an artisanal family background.

Both of these markers are very likely to lead to exposure to formal drawing instruction, which in turn encourages a talent for draftsmanship and better schooling of the eye for discerning subtle, inconspicuous features of objects or processes. This leads to a higher likelihood for developing extraordinary skills in pattern recognition, and abstraction, which in turn increases the probability that this person will end up as a conspicuous exponent of a visual culture in science or technology. The strong correlation already suggested by my systematic prosopography of spectroscopists, and which I aim to corroborate by further examples taken from

---

[105]See Root-Bernstein et al. (1995) pp. 120ff. for the quote and the statistical data.
[106]Root-Bernstein (2001) p. 301.

across the board in the present study, is generated by plausible causal chains that must explain *why* these correlations are so strong – something that is also missing in Root-Bernstein's broader claim. In fact, it even seems possible to extract data for my more specific claim out of his data.[107]

The surprising frequency of artisanal backgrounds can tentatively be linked to findings by the social historian Peter Burke. His analysis of *The Culture and Society of Renaissance Italy* (1972) examined 600 actors from all sectors of the cultural elite. These encompass painting, music, literature and other arts, as well as architecture, technology and the sciences of the day. In terms of family background, the painters, architects and sculptors, on the one hand, differed strongly from the humanists, writers and natural philosophers, on the other. The former, i.e., the representatives of visual cultures in Renaissance Italy, had a high percentage of artisans among their parents (90 of them to be precise, whereas only 40 of them claimed aristocrats, merchants or lawyers as their immediate forefathers). The corresponding figures are quite the opposite for the nonvisual branches of Renaissance culture, i.e., the writers, natural philosophers and humanists: their fathers' professions are (only) 7 artisans but 95 aristocrats, merchants or lawyers. So here again we do establish a strong correlation between (i) belonging to a visual culture and (ii) having an artisanal family background.[108] Growing up within an artisanal context must have trained the eye and aroused interest in issues of visual perception, such as gauging proportions, distinguishing color nuances or aesthetics. The other elites were more refined in their language, rhetoric, and forms of argumentation, for instance. Thus even someone like Leonardo da Vinci humbly spoke of himself as an "uomo senza lettere," an unlettered man, while he was staying at the Milanese court in 1487.[109] What seemed to him, an illegitimate child of a lawyer and a country girl, to be a mortifying personal fault that he tried to compensate for by studying vocabulary and reading whatever he could lay his hands on, was, in other respects, actually an advantage. It liberated him from some of the cultural fixtures that university-trained humanists were bound to. Thus it was Leonardo and his fellow artist-engineers and not humanists who became the pioneers of visual science cultures, such as perspectival theory and practice (cf. here sec. 5.1). Comparable statistical analyses of roughly 200 illustrators active in 1987

---

[107] See Root-Bernstein (1989) pp. 318–27 for a long table listing "artistic proclivities among eminent scientists and inventors," with pp. 318–21 limited to visual skills, such as, painting, etching and sketching, sculpting, drafting, photography and training in architecture.

[108] On these data, see Burke (1972), esp. his section on "the creative elite."

[109] See, e.g., Chastel (ed. 2002) p. 228. Cf. Edgar Zilsel (2003) on the "artist-engineers" of the Renaissance.

shows that successful illustrators tend to be nonconformist and unconventional, with male illustrators often stemming from families with incomes well below the average of their community, whereas female illustrators having statistically significantly higher parental income and education levels.[110]

What can we learn from such more focused prosopographic surveys? As with all efforts at inductive generalization, nothing can be proven on the basis of repeated singular occurrences. But in my opinion it would be wrong to ignore the hint at a systematic pattern behind these findings. Therefore I would like to end this chapter with a general appeal to future researchers on visual science cultures: look behind and beneath the surface. More specifically, look at the backgrounds and training of pioneers in these fields, in order to understand their visuality! Why did they, and not others, come up with a visual technique for representation that proved to be so fruitful that a whole visual domain could be formed around such an innovative practice? What triggered the birth of visual cultures in science and technology? In my view, it was quite often an import from another field that was already visual in character, not necessarily a neighboring one – actually often a quite remote one. This transfer of visuality is achieved by individuals who for some reason have a foot in both fields, thus facilitating this highly creative and often daring transfer (sometimes making it in fact psychologically hard to miss, even). The general scheme as well as the technicalities of these transfers will be the topic of the next chapter.

[110]See Stenstrom (1991) p. 9.

5

Transfer of visual techniques

Examples of transfers from an already-visual field to one still lacking in that particular resource abound in the literature on visual cultures. H. Floris Cohen (*1946), who studied the global migration of ideas and practices in the early-modern period from a comparative point of view, prefers to speak of "transplantation." The architect and historian Tim Peters distinguishes 'transformation' from 'translation,' the latter here understood as "moving information between professional fields."[1] Brooke Hindle has shown, for instance, how the American painter Samuel Finley Breese Morse (1791–1872) became one of the key inventors of the electrical telegraph by systematically transferring images and mental models from visual domains he was already familiar with into the technical world.[2] Equally daring transfers of visual techniques had been achieved in earlier periods. The Chemnitz physician Georg Bauer (1494–1555), better known under his Latinized name Agricola, imported the cut-away style that he was familiar with from Vesalius's anatomical plates (1543) into technological treatises. Vesalius, Leonardo and other Renaissance anatomists offered views into the interior of the human body by artfully omitting some outer layers. By analogical transfer, Agricola instructed the woodcutters for his *De re metallica* (published posthumously in 1556) on how to skip part of the terrestrial surface to allow views into the interior of a mine shaft or pumping utensils.[3] Similarly, Sir Christopher Wren (1632–1723) acted as a mediator between architecture and anatomy when, transferring into brain physiology his expertise in high-quality three-dimensional renderings of complicated spatial structures, he illustrated Thomas Willis's (1621–75) celebrated treatise on *Cerebri Anatome* (published in 1666).[4] One century later, the French anatomist Félix Vicq d'Azyr (1748–94) applied advanced crystallographic research principles to the visualization of the brain. His teacher Haüy at the *Cabinet du Jardin des Plantes* (precursor to the Parisian Natural History Museum), modeled crystals as composed of regular quasi-atomic building blocks. Analogously, Vicq d'Azyr advocated conceptualizing the brain as a composite of successive layers of tissue structures

---

[1] See Cohen (2007, 2011) and Peters (1998), who also refers to "border-crossing thinking."

[2] See Hindle (1983) chaps. 4–5. Morse's prototype telegraph was mounted on scaffolding!

[3] On this fascinating transfer across widely separated domains, see Hans Holländer in Holländer (ed. 2000) pp. 643–72.

[4] On this example and its context, see Cavalcanti et al. (2009) pp. 16–18 and Jardine (2003).

that could, in fact, also be stripped away one by one in anatomical preparations.[5] The concept of symmetry has roots in the architecture and anatomy of antiquity. It was revived and creatively reinterpreted in the Renaissance, transferred into crystallography in the early 19th century and imported from there into physics in the late 19th and early 20th centuries.[6]

Among the most fascinating 20th-century examples of such *transfers* – as I will continue to call them – are those presented by James Elkins in various publications since the 1990s.[7] One of them concerns the population geneticist Sewall Wright (1889–1988), who introduced into his subdiscipline a new type of diagram that strongly resembles topographic mapping. Wright's maps displayed altitude lines, like height isolines indicating locations at exactly the same altitude above sea level (typically rendered in steps of 10, 25 or 50 m height, depending on the scale of the map). However, his quasi-landscapes are a "hypothetical multidimensional field of gene combinations" broken down into a two-dimensional plot with the peaks (carrying + signs) indicating the maxima of genetic "fitness" and, analogously, the valleys (denoted by −) local minima.[8] By making full use of the well-known conventions of cartography, Wright can assume that his readers are immediately able to understand his diagrams without too much explanation or prior training or revision. As the many arrows in his images indicate, these diagrams not only allow one to "walk through his landscapes" as Elkins puts it, but also to think in them and with them and thus to 'understand' the mechanisms of gene combination more intuitively.

A similar importation of landscape-viewing conventions was achieved by Gerd Binnig (*1947) and Heinrich Rohrer (*1933) at the IBM research labs in Zurich in 1982. They transformed their height profiles of silicon surfaces scanned by their new scanning tunneling microscope (STM) into a quasi-landscape. They exaggerated the minute height differences (of only 15 Å) by enhancing them by 55%; then they cut out the individual printed profiles on thick paper, glued these slices together and obliquely illuminated them, to add three-dimensional depth to

---

[5]See again Cavalcanti et al. (2009) p. 18 on Vicq d'Azyr and here pp. 334ff. on Haüy.

[6]See Hon & Goldstine (2008) and the widely dispersed primary sources cited there. On symmetry considerations in crystallography, see here pp. 331ff.

[7]On the following, see especially Elkins (1995) pp. 565ff., (2007).

[8]Both quotes are from Wright (1977) pp. 446 and 452. Incidentally, his name is consistently misspelled 'Sewell' by Elkins (1995) pp. 565f., who is less interested in the protagonists of his examples. For a detailed finding aid of Sewall's papers, kept at the *American Philosophical Society* in Philadelphia (also including a concise survey of his vita), see www.amphilsoc.org/mole/view?docId=ead/Mss.Ms.Coll.60-ead.xml (accessed Apr. 28, 2011).

the image from the shadows they cast.[9] As Jochen Hennig has pointed out in his brilliant historical analysis of this episode, the lamp with which these shadows were cast on the artificial 'landscape' is just visible at the upper left margin of figure 40, taken from the article from 1983 by these two STM pioneers.

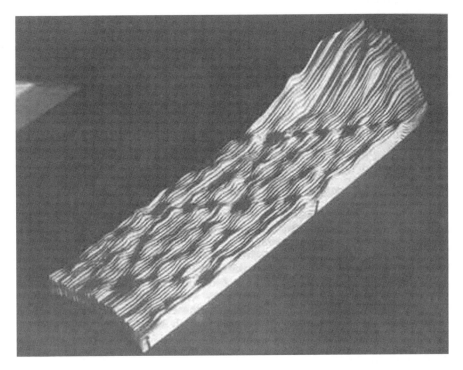

Fig. 40: Silicon surface explored by STM, cut out on paper and illuminated from the side to simulate three-dimensionality. The rhomboidal 7 × 7 unit cell with 12 height maxima is clearly discernible from the deeper corners at its edges. Its diagonals (of c. 50 Å and 27 Å) were in excellent agreement with crystallographic values already known for the 111 unit cell. Reprinted with kind permission. From Binnig, Rohrer, Gerber & Weibel (1983) p. 120, fig. 1. © 1983 American Physical Society.

Hennig's 2011 dissertation on image practice in early nanotechnology reconstructs a long line of representations, a whole image tradition (in Stefan Ditzen's meaning) that ranges from a late 20th-century paper cutout model of a silicon surface from scanning tunneling microscopy to coin and medal-relief copying techniques in the

---

[9]On this episode, see Gerd Binnig, Heinrich Rohrer, Christoph E. Gerber & Edmund Weibel (1983) and Jochen Hennig (2005) and (2011) sec. III.1, pp. 140 ff.

19th century.[10] Do we describe such image traditions as stable scopic domains with changing groups of practitioners? Or as visual practices stable only for a short time that get their longer-term stability by slow adaptation and amalgamation in everyday practices? Hennig's examples rather speak in favor of the latter option. The later Nobel prizewinner Gerd Binnig had started out by adding a few shadows by pencil using a ruler, in order to make these images more compelling and easier to 'read.' His later 3D paper models (in fig. 40) are simply an extension of this line of representation. The choice of a differential representation by his IBM colleagues Donald M. Eigler and Erhart K. Schweizer also sought to make an unfamiliar structure discernible by suggesting it in relief, using a device already familiar from other contexts. Likewise, Charles Thomson Rees Wilson (1869–1959) with his pioneering cloud-chamber images. He artificially suppressed the background condensation in order to make his images 'clearer.' By calling the minute traces of the paths of ionized particles "thin threads," he consciously chose a macroscopic metaphor to familiarize a wholly new scopic domain by means of a well-known phenomenon from our everyday world.[11] The same pattern recurs in earlier historical episodes to the one demonstrated by Ditzen in his fine analysis of 17th- and 18th-century microscopy. It is strongly reminiscent of the challenges posed to European draftsmen in accurately drawing morphologically entirely new species brought back from ethnological and natural historic expeditions of the period. In their first attempts to draw and paint the kangaroo, Cook's artists inadvertently produced a strange cross between a deer and a rabbit (cf. here fig. 7 on p. 38). Microscopists likewise saw familiar patterns in their novel motifs and, for instance, portrayed a larva as the absurd grimace of a satyr.[12] Because this *Gestalt* was very familiar, it was adopted in various other atlases and became part of a microscopic image tradition lasting for a surprisingly long time. Modifying a *bon mot* by Schopenhauer, one could say: We are able to see what we want, but we only recognize what we already know beforehand – that is, what we have already internalized as a *Gestalt* and can thus discriminate from its background noise. Or – according to the advertisement by the Cologne art publishing house DuMont: One only sees what one knows! (*Man sieht nur, was man weiß.*) More recent authors in science studies have emphasized this same point. The Zurich historian and sociologist of medicine Barbara Orland wrote:

[10] See J. Hennig in Hessler (ed. 2006) and Hennig (2005) and (2011) pp. 133ff., 221ff.

[11] For further details on this example, see Galison & Assmus (1989), Chaloner (1997) and Wolfgang Engels in Hessler (ed. 2006) pp. 57–74.

[12] On this misinterpretation of *Hydrachnid* larvae in Joblot (1718), see S. Ditzen in Hessler (ed. 2006) pp. 41–56.

In order for the invisible-made-visible also to acquire evidential status, it must be inscribed into habitual custom. Many forms of perception usual to us today, for example, the x-ray view into the body's interior, first had to be normalized, routinized, matter-of-factualized before – as introduced and no-longer-questionable traditions of seeing – they could become referents of other, newer image technologies.[13]

What is the common element in these above examples? Is this a generalizable finding? Delving into more and more idiosyncratic examples won't help answer these questions. Therefore, we will start our in-depth discussion with one paradigmatic secondary text that has carefully studied this transfer of techniques for visual representation in great detail. Our earliest instance of such a transfer of visual techniques is the gradual adoption of perspectival drawing methods employed by architecture and painting in mechanical drafting. We will then stay within the engineering context and look at the migration of so-called indicator diagrams developed in the late 18th century, which at first were nothing more than a clever way to scale the efficiency of steam engines but became standard for conceptualizing thermodynamic processes in the mid-19th century. Another canonical example for this historical study of transfer processes is Carsten Reinhardt's book on *Shifting Boundaries*. Reinhardt analyzes the emergence of NMR in physics after 1945 and its gradual shift into chemistry, keeping a close eye on the concomitant changes in practical usage and social context.[14] In sec. 5.3, we will summarize his findings and those of other authors such as Stuart Blume, Lisa Cartwright and Regula Burri, who have traced the further migration of this NMR technique to diagnostic medicine (where it was redubbed 'magnetic resonance imaging,' or MRI for short). In doing so, we will devote particular attention to structural features concerning the transfer of visual techniques from one culture to another.

---

[13]See Orland in Gugerli & Orland (eds. 2002); transl. by A. M. Hentschel. See also Henderson (1988); cf. Rasmussen (1996, 1997) on a similar 'domestication' of electron microscopy imagery.

[14]See Reinhardt (2006).

5.1 The gradual diffusion of perspectival drawing

The history of perspectival drawing is an ideal place to start because various transfers between different visual domains are involved. Euclid's *Elements of Geometry* (c. 300 B.C.) can be regarded as a first example of such a transfer because it systematized long ongoing practices such as land surveying and estimations of container volumes. Euclid's much less well-known and shorter treatise on optics was likewise a compilation of the contemporary knowledge about lines of vision. His extramission theory of light was that rays issue from the eye and not from the object, as in competing intromission theories. Surveying and optics have been closely linked since antiquity, because of the widespread practice of optical leveling with surveying instruments such as the diopter or the chorobates.

Aside from these archetypical tasks of applied geometry, another early area of application was scenography, i.e., the art of making theater backdrops. From brief allusions in writings by Plato and Aristotle, we know that by 500 B.C. illusionistic stage backdrops making use of perspectival shortening must have already been pretty convincing. Unfortunately, none of these antique scenes have been preserved, but frescoes in various buildings of late antiquity indicate that the art of perspectival drawing was already fairly well developed in Roman times. It had reached the so-called fishbone perspective stage. The receding edges of depicted objects, if continued, all meet along a common median line. They do not yet converge at a single central vanishing point.[15]

For Euclid, the apparent size of an object seen with the naked eye (monocularly) was only a function of the angle subtending the rays of vision issuing from the eye that touch the extreme edges of the object. The closer an object is, the larger this angle gets and the larger the object seems to be. Euclid thought there was a minimal angular distance between two neighboring rays of vision. If any object fell between those two limiting rays of light, it was *per definitionem* no longer visible to the observer emitting those rays of vision. This assumption seemed to explain why all objects cease to be visible beyond a certain limiting distance.[16] The

---

[15]Excellent examples are found, e.g., in the ruins of Pompeii, all dated before A.D. 79 (cf., e.g., Little (1937), Beyen (1939), White (1956) pl. 7–12, Edgerton (1975), Tobin (1990) pl. I, and here color pl. I. On the link between perspective and scenography, see also Little (1937) and White (1956), pp. 45ff. with excerpts from Vitruvius.

[16]See Euclid (1945) for an English translation. Better commentary on the Greek terminology is found in the edition by Heiberg 1899, in French translation by ver Eecke (1959), and in Kheirandish (ed. 1999) esp. on Arabic versions of Euclid's optics. For further details on Euclid's so-called angle perspective in contrast to modern linear perspective, cf., e.g., Little (1937) pp. 491ff., White (1949) pp. 58–61, (1956) pp. 43ff., Wheelock (1977) chap. 2, Veltman (1980), Brownson (1981), Theisen

masterful treatment of several projection techniques in Ptolemy's *Optica* and *Geographia* shows that this knowledge was around until the second century A.D. From Vitruvius, we learn that the hybrid field of *perspectiva*, combining basic knowledge of geometry, optics and anatomy, also formed part of the architect's training.[17] The optical core of this theoretical knowledge of Greek $O\pi\tau\iota\kappa\eta$ survived in a few Islamic treatises on optics and in scholarly texts of the Latin *perspectiva* tradition in the Middle Ages, but there was no contact between scholastic authors such as Roger Bacon or Robert Grosseteste and the contemporary painter Jan van Eyck.

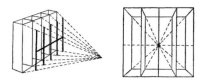

Fig. 41: Differences between Renaissance linear perspective (left) and Euclid's angular perspective. In linear perspective, straight edges of objects are always mapped onto straight image lines, whereas in natural perspective curvilinear distortions occur for off-center lines. From White (1949) p. 59.

Medieval paintings exhibit simpler forms of depth representation, though, with a mainly symbolic indication of locality, a multitude of viewing points and fragmented spatial organization. However, the early Italian Renaissance led to a rediscovery and systematization of these antique techniques of visual projection and perspective construction.[18] The first, still incomplete, traces are visible in paintings by Giotto di Bondone (1266–1337), whereas Masaccio's (1401–28) Trinity in Santa Maria Novella in Florence (c. 1425–27) is already perspectivally perfect. The Renaissance artisan-engineer, architect and painter Filippo Brunelleschi (1337–1446) devised perspectival demonstrations using a mirror to show the perfect fit between his painted depiction of the Florentine Baptisterium and the original, when viewed from the same point at which he had constructed his perspectival representation of it. By that time, four motives had converged: (i) the new idea to paint an

---

(1982), Simon (1988), Tobin (1990), Knorr (1991) and Andersen (2007) appendix I.

[17]See Vitruvius's *De Architectura*, book I, chap. 2, sec. 2, book VII, chap. 5, preface and secs. 1–3; cf. also www.ancientsites.com/aw/Article/420727 (accessed July 4, 2011) for online excerpts and an English translation, and Little (1937) p. 488.

[18]Good surveys on the history of perspective are White (1949/51), Carter (1970), Edgerton (1975, 1991), Janowitz (1986), Field (1988), Kemp (1984) and (1990) part I, and Field (1997).

instantaneous view rather than visually narrating something, (ii) some inspiration from the theory of optics, (iii) a strong fascination for mathematics in general and geometry in particular, and (iv) experiments on depicting complicated but geometrically regular objects.[19] In 1435, the architect, art-theoretician and humanist Leon Battista Alberti (1404–72) canonized the rules of linear perspective in *De Pictura* (first published in 1440). It offered the first comprehensive theory of linear (or 'artificial') perspective (as opposed to the 'natural' or 'synthetic' perspective of antiquity; cf. fig. 41). This new skill is executed in a perfect manner in paintings and drawings by Piero della Francesca (c.1420–92), who also wrote a treatise on the *prospectiva pingendi*,[20] as well as by Leonardo da Vinci.[21] Within one generation, the set of practical rules for painters and architects on how to achieve perfectly illusory spatial depth on a flat surface was transformed into a thorough system. This mathematically deepened perspectivalism became not only a new mixed science in the scholastic classification of knowledge,[22] but at the same time a new 'world view' or – as the art historian Erwin Panofsky and the philosopher Ernst Cassirer put it – a new "symbolic form." It determined not only how to represent the world but also what to select for this representation. Geometrically regular bodies, for which strict perspectivalizing rules existed, were clearly preferred. In this sense, 'scenography' was indeed a visual culture *par excellence*, to redefine in Bruno Latour's sense "what it is to see and what there is to see."[23] During the Middle Ages, paintings were understood as material surfaces onto which symbolic signs or icons were placed in order to narrate a story visually, with the size of each

[19]See Wittkower (1953). According to Field (1987) p. 31, Brunelleschi was motivated to study perspective because he was concerned that a building's proportions be seen properly by a viewer from street level. Field (1997) and Andersen (2007) discuss the underlying mathematical theories. See esp. Andersen (2007) pp. 3ff. on these four stimuli. Cf. Manetti (1970) on Brunelleschi's vita, and Veltman (1979) p. 330 on Brunelleschi's and Alberti's prior work in surveying ancient ruins; Lynes (1980) claims a connection to Brunelleschi's expertise in the shadow-geometry of sundials.

[20]See Alberti (1435), Piero della Francesca (before 1492), Elkins (1987), Field (1997) chaps. 4–5 and Andersen (2007) chap. II, esp. p. 72f. on Piero's construction of a foreshortened human head.

[21]Edgerton (1975), Kemp (1978) and Andersen (2007) chap. I on Brunelleschi; Panofsky (1940), White (1949) pp. 70–9, Pedretti (1963), Kemp (1977), Elkins (1988), Veltman (1986), Turner (1992) pp. 139ff., Grafton (2000) chaps. 3–4 and Andersen (2007) chap. III on Leonardo, who is dealt with here on pp. 17f. and 242f below.

[22]On issues of disciplinary location of *perspectiva* and remaining differences between a mathematician's and an artist's version of perspective, see Veltman (1979). According to Field (1987) p. 6, "artists had no reason to be deeply interested in the mathematical rules"; Hamou (1995) and Peiffer (2002).

[23]See Panofsky (1927) and Latour (1986) pp. 9–12; cf. Henderson (1995) pp. 198–200 and (1999) pp. 9 and 28.

object in the image reflecting its importance rather than its physical extension. The backdrops were often blocked out of view, being either filled in entirely in gold or heavily stylized. Perspectival paintings of the Renaissance open up the backgrounds, and the whole painting is understood as a transparent window – opening onto a view of a potentially infinite world. Objects and figures were painted proportionately smaller the further they were supposed to recede into the background, and parallel edges of objects or houses all converge on one central vanishing point.[24] The art historian Michael Baxandall (1972) has pointed out the repercussions that the new craze for perspectivalism had on the motifs of paintings in 15th-century Italy: checkerboard pavement, equidistant columns, circular cylinders and geometric cones suddenly appeared everywhere, because these pictorial elements allowed rigorous geometric construction of the image according to the recipes given by Alberti, Brunelleschi and Piero della Francesca. Even forms that were not intrinsically geometric were tendentially regularized to fit into the perspective mold. Thus, when the artists Erhard Schön (c.1491–1542) and Hans Sebald Beham (1500–50) wrote treatises on proportion (cf. fig. 42), they both started to geometrize even organic bodies in order to be able to represent their foreshortening perspectivally correctly when regarded from an acute angle. In a similar manner, letter type, fortifications, gardens, ornaments and many other elements of everyday life fell under the spell of geometry.

With Albrecht Dürer (1471–1528), a painter and draftsman born in Nuremberg, who sojourned in Italy 1494–95 and 1505–06, this knowledge also migrated northwards.[25] While Dürer and his Italian contacts are best known as painters, it would be wrong to confine these illustrious figures to the art world. All of the above artists were very interested in mathematics and its application to the real world. In 1507, Dürer bought the first printed edition of Euclid's *Elements of Geometry*, which had just been translated into Latin by Zamberti in 1505, and studied it intensely. Later, Dürer himself wrote separate treatises on fortification and on metrology. He also had cartographer contacts, such as Johannes Stöberer (1460–1522), with whom he worked on projection techniques for terrestrial mapping, and with astronomers such as Johann Werner (1468–1528) in Nuremberg,

---

[24] According to Panofsky (1943*b*) p. 329, this rule was adopted from 1340 onwards in Italy, and somewhat later in Central and Northern Europe, reaching the British Isles only in the 17th century; cf. Carter (1970) §11 and Andersen (2007) pp. 489ff. on the late reception in Britain and on Brook Taylor's (1685–1731) important contributions to the theory and practice of linear perspective.

[25] See Panofsky (1943*b*) pp. 11, 330f. on Dürer's decision to extend his second Italian stay and his trip to Bologna to acquire the then still secret theory of perspectivalism, possibly from the mathematician Luca de Pacioli (c. 1445–c. 1514).

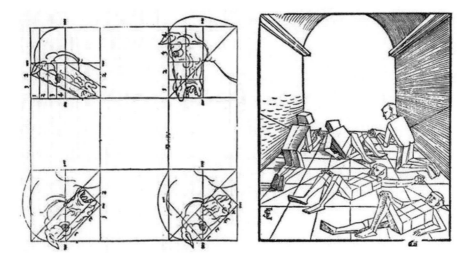

Fig. 42: Perspectival shortening of geometrized bodies. Left: from Beham's *Mass oder Proporcion der Ross*, Nürnberg 1528. Right: from Schön's *Underweisung der Proporzion*, Nürnberg 1538, fol. cii.

with whom he worked on a stellar map and discussed the intricacies of geometrical techniques. In 1522, Werner wrote the first occidental textbook on conic sections. Three years later, Dürer followed with his *Unterweysung der Messung mit dem Zirckel und Richtscheyt in Linien, Ebenen, und gantzen Corporen* (1525), and in the same year he also made the woodcuts for the 1525 edition of Ptolemy's *Geography*.[26] We will pick a few examples from this treatise, as it is another perfect case of this transfer or migration of ideas from one context into another. Whereas his compatriot Werner still avoided the use of projective techniques out of a lack of familiarity, Dürer tried his best to clarify the basic ideas behind them for his readers. He made use of this intuitively clear manner of showing a three-dimensional object in various projections at the same time (cf., e.g., here fig. 43, right). In general, he approached geometrical problems from the point of view of a practitioner. He wrote in German and used practitioner's terms rather than highfalutin scholarly or Latin expressions; for instance: 'boar's teeth' (*Eberzähne*) for acute angles, 'cutting line' (*Ortstrich*) for diagonal, 'air bladder' (*Fischblase*) for circle segment or 'burning line' (*Brennlinie*) for parabola.[27] Even his translation of *geometria* as *Messkunst*, i.e.,

---

[26] See Veltman (1980*a*) pp. 403, 406f. and Edgerton (1975) on the link from Ptolemy's cartographic projection techniques to linear perspective; furthermore Field (1997) pp. 150ff. on links to the theory of the astronomical planisphere.

[27] The last term implicitly uses the knowledge that all rays from a faraway light source can be focused onto one point by means of a perfect parabola, thus achieving the perfect focus used in

as the art of measurement, is a case in point. Often he mentioned original ways of constructing geometric figures, such as a special form of conchoid (literally translated: shell line) or of a limacon (now called Pascal's limacon, even though Dürer had described its construction nearly a hundred years earlier). Ellipses were practically constructed by him by the aid of a wire wound between two sticks acting as the two foci of the ellipse (today often called the gardener's construction for its ready application in the circumscription of ellipses, so often needed in baroque gardens). Following Werner, Dürer demonstrated how ellipses, parabolas and hyperbolas can be obtained as sections through a cone, thereby anticipating projective techniques later exploited in Gaspard Monge's descriptive geometry.[28]

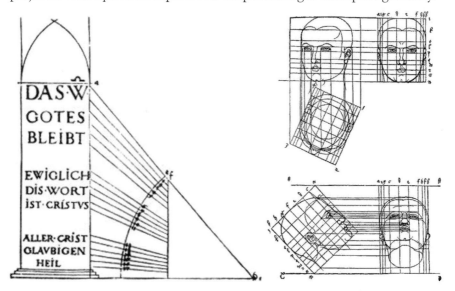

Fig. 43: Left: Foreshortening of text displayed high above the viewer must be compensated for by increasing the size; from Dürer (1525). Right: geometrized treatment of perspectival foreshortening of the human head; from Dürer (1532). This kind of representation, in which the head is shown perspectively and in projection onto plan and in profile, already prefigures Monge's descriptive geometry (cf. here sec. 7.4 and Filippo Camerota in Lefèvre (ed. 2004) pp. 198ff.).

The practical problem of finding the right size of letter type for several lines of

---

so-called burning glasses to light a fire; cf. also Panofsky (1943*b*) pp. 327, 337 on Dürer's intended readership encompassing artisans, and Olschki on Dürer's importance for the German language.

    [28]In the *Unterweysung*, he still misinterprets ellipses as an (asymmetric) egg line *Eierlinie*), though. For further details on Dürer's applied geometry, see Taton (1951) pp. 63–6, Panofsky (1943*b*) chap. VIII, esp. pp. 338ff., Veltman (1979) p. 341, Peiffer in Lefèvre (ed. 2004) and literature cited there.

text situated far above the heads of its readers motivated him to make a graphic representation of the tangent function (cf. fig. 43, left). He also described a pretty good approximation of angular division by a factor of 3. He was always aware of whether a given construction was exact (*demonstrative*, as he put it), though, or only approximate (*mechanice*); and he also knew the difference between a mathematical figure and its concrete physical representation.

Most interesting within the context of this chapter are Dürer's remarks and supporting drawings about how to obtain perspectively correct images. As fig. 44 shows, he had found an ingenious way to demonstrate the clue to perspectival drawing, namely to construe any perspectival image as a two-dimensional section cut across the three-dimensional visual cone. In the figure, this section is represented by the wooden frame in the middle right of the image, whereas the visual cone is represented by the spanned wire ending at a single point, the location of the eye of the image's virtual viewer. Each point of the image can thus be found as the intersection of this wire with the section plane. The person on the right-hand side of the image is reading off these intersection points as coordinates along the *x*- and *y*-axes marked on the wooden frame fixed in the section plane.

Fig. 44: One possible device for point-by-point construction of perspectival drawings. Woodcut from Dürer (1525), fig. 64 on his second-to-last page (unpaginated).

The procedure shown in Dürer's woodcut is not supposed to be followed in practice – the point of his illustration is rather figurative. He demonstrates that on the canvas, each point represents an exact location in space, and that this point-by-point mapping is an algorithmically closed, exact procedure.[29] *Perspectiva* – in the Middle Ages the abbreviation for the science of optics – had quite literally become the art and science of "through-vision." To construct a perspectival drawing meant imagining the image as a geometric surface situated between the object and the viewer, with each point of the object projected onto an image point on this plane. In other words: any perspectivally correct image was an intersection of the canvas through the viewer's visual cone. According to Alberti, for whom the human eye was a live mirror, each perspectivally correct painting was an intersection with the visual cone.[30] It is no accident that this very image by Dürer became the icon for perspectivalism, as it very clearly transmits the central idea of the technique even though this procedure is by no means the easiest to follow for practical purposes. But it is more direct and easier to understand than, say, Brunelleschi's tricky usage of mirrors or Alberti's and Piero della Francesca's complicated *costruzione legittima* in which the perspectival representation was constructed on the basis of a planar and profile view in the architect's manner, now interpreted as projections onto two orthogonal planes.[31] Even Viator's *perspectiva cornuta* (literally horned perspective, i.e., a perspective with more than one vanishing point), explained in his *Artificiali perspectiva* of 1505, was described much more clearly and comprehensively by Dürer in his *Unterweysung*, there called *näherer Weg*. In this "easier construction," diagonal lines AO, BO, CO, ... are drawn from the front pavement row (cf. fig. 45) to the point O whose position is defined by the assumed height and lateral distance of the ideal observer of the image from the depicted scene. Each of these diagonals defines a point on the rightmost vertical, marked by the series of primed smaller letters A/, B/, C/, ... in the figure, and the horizontal lines through these points then give the correct spacing of traverse parallels as seen in perspective.

Dürer's *Unterweysung der Messung* was the first text about these issues not published in Latin or Greek but in a vernacular language. This greatly helped spread the gospel into far-flung circles outside the scholarly world. Practitioners in architecture or engineering learned their geometry and perspective drawing from hands-on introductions like Dürer's rather than from high-brow deductive textbooks in the Euclidean tradition, which continued to dominate mathematics

---

[29]See Turner (1992) pp. 147f. on literal versus figurative meanings of perspectival images.

[30]*intersegazione della piramide visiva*; cf. Dürer: "ein ebene durch sichtige abschneydung aller der streym linien, die auß dem aug fallen auf die ding, die es sieht." Cf. Panofsky (1943*b*) pp. 332ff. and Edgerton (1975) pp. 174ff. for the sources of these two quotes and further commentary.

[31]For full details on all these constructions and further historical background, see, e.g., Rohr (1905), Panofsky (1943*b*) pp. 332ff. and Janowitz (1986). On Jean Pélerin (1445–1524, latinized name Viator), see Ivins (1973).

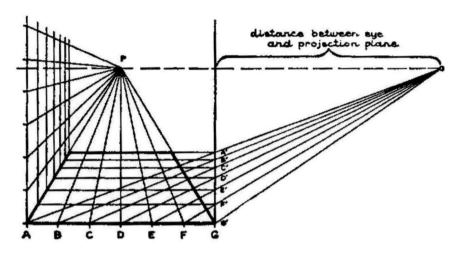

Fig. 45: Dürer's simplified perspective construction. Adapted from Dürer (1525) by Panofsky (1943*b*) p. 335.

at universities. Dürer thereby became a transmitter of ideas and practices from the abstract world of mathematics into the real world of engineers and artisans. Perspectival drawing became the standard technique of representing buildings, including the virtuosic switches between the older representation forms of flat-on and elevation views (left half of fig. 46) and the perspectival view created from them (right half of fig. 46) by strictly following the rules spelled out by Brunelleschi, Alberti, etc. Dürer's step-by-step treatment of the construction of perspectival views of a cube on a pedestal actually prefigured the rise of descriptive geometry with Gaspard Monge and the *École polytechnique*.[32]

In machine drawings, however, the so-called cavalier perspective or the orthonormal projection onto one or two sides of the object remained standard until the 19th century. Both of these simpler representational forms rendered parallel edges of an object as strictly parallel lines on paper. Thus both dispensed with the fixed point of convergence and the unequal foreshortening of edges seen from different angles, which require some experience with perspectival representation to decode the spatial relations for proper depiction. How much conventionality is hidden in perspectival constructions, or how natural they are is still a very

---

[32]See Dürer (1525) fols. 52–63. On the dominance of orthographic projection in engineering, see, e.g., Lefèbvre in Lefèbvre (ed.2004) or Henderson (1999). On the connection to Monge, see, e.g., Taton (1951), Nedoluha (1960) chaps. 3–4, Booker (1963), Field (1987) p. 36, Andersen (2007) chap. XIII and here sec. 7.4.

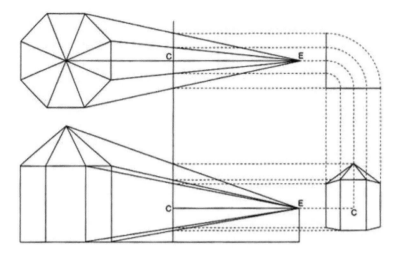

Fig. 46: Perspectival view created from plan and elevation. From Carter (1970) fig. 5. By kind permission of Oxford University Press.

controversial point. Ethnographic experiments with various African trial groups of different tribal origin, educational level and degree of urbanization showed that illiterate and isolated subgroups tend to have a flat perception of pictorial material. Schematic depth cues such as object size, superposition and perspective were present in these images, but how to construe spatial relations from these cues has to be learned (in our civilization, this already happens at preschool age). Some of these experiments were later also confirmed by other ethnic groups. Furthermore, studies of cultural transfers from, e.g., Europe to early-modern China show how difficult it was for Chinese illustrators just to copy plates with perspectival views.[33] All of this seems to confirm Nelson Goodman's strong thesis about the deep cultural inculcation of systems of representation. But there are some contrary indications, too. Why did Alberti's and Brunelleschi's contemporaries so readily accept the claim to superior fidelity for these new perspectival representations even though none of them were raised in this visual tradition? And why can even seven-month-old infants or chimpanzees recognize depth and height in (linear-perspective) photographs?[34]

Altogether, this first section of the current chapter already contains several

[33]See, e.g., Hudson (1960), Deregowski (1972), Leach (1975) on Africa and Edgerton (1991) chap. 8 on China.

[34]On these debates, see, e.g., Derksen (2005) and further cross-cultural literature cited there, as well as here note 24 on p. 15.

telling examples of remote transfers between different visual fields: from the pretheoretical contexts of practical surveying to Euclid's mathematical context of geometry and the equally theoretical and scholarly *perspectiva* tradition, reverting back to real-world applications in architecture (Alberti and Brunelleschi) and painting (Piero della Francesca and Leonardo). The humanist Alberti bridged the occupations of architect and city-planner with those of mathematician and theorist of the arts. The mathematization of perspectival painting led to a rise in status in Renaissance Italy. Likewise, the artist and polymath Albrecht Dürer was an ideal intermediary between these worlds of painting, engraving, woodcutting and detailed drawing, on the one hand, and of architecture, fortification, scenography, surveying, cartography, astronomy, engineering and practical draftsmanship, on the other. A technique like shadow-projection linked fields as varied as astronomy, geography, optics and topography; and intellectual mergers of these range from Ptolemy to Alhazen, Witelo and Levi ben Gerson to Leonardo da Vinci, Alberti and Dürer.[35] Aside from these cognitive transfers and mergers of practical and theoretical traditions, we also have various examples of geographic transfer, most notably from Greek and Roman antiquity via the Islamic world to the medieval centers of learning, thence to Venice, Florence and the centers of the early Italian Renaissance, and then onward to Nuremberg and other important trading towns north of the Alps, again with Dürer as the transmitter of Italian theory and know-how of perspectival representation to Central and Northern Europe. Aside from these various geographic transfers, we also have cognitive and disciplinary shifts: first from architecture to geometry and mathematics, then back into the arts world, and from there into anatomy and medicine. In the 18th and 19th centuries, "geometrical drawing" was applied in the new subdisciplines of anthropology and craniology to gauge the absolute size and form of human skulls,[36] with the aid of drawing apparatus with grid-frames (cf. fig. 47) stemming directly from Dürer's famous drawing utensils depicted in his *Unterweysung der Messung*.

Another case in point for such transfers will be the line of development from the 15th-century artist's practice into the new projective geometry of Girard Desargues (1591–1661) and others in the 17th century,[37] to which we will return in chap. 8 discussing the further development of descriptive geometry.

---

[35] On these invisible and often overlooked links, see Edgerton (1975), Kaufmann (1975), Kemp (1977, 1990), Veltman (1979, 1980*a*), Field (1987, 1997), Grafton (2000) and Peiffer (2002).

[36] On this medical branch of perspectival drawing, see, e.g., Mann (1964) p. 9 and Kinkelin (1882).

[37] More on this in Field (1987) p. 36, (1997) p. 224, Andersen (2007), and here in sec. 8.1.

Fig. 47: A cranioscope by the Frankfurt anatomist J. Ch. G. Lucae (1814–85) from 1844 and a craniograph with the ensuing visual record by Ltd. Karl August von Cohausen (1812–94) from 1873. Both images taken from Kinkelin (1882) pp. 106 and 115.

5.2   Indicator diagrams: from industrial secret to thermodynamics

In the 18th century, steadily rising demand for iron and coke for smelting led to a real impasse. Many coal and iron mines in Britain and on the continent had been dug so deep down into the ground that they were on the verge of flooding. Motive power was urgently needed to pump water out of the mine shafts and to drive the bellows and rolling mills of the nearby iron works. British ironmongers and engineers produced a long stream of technical inventions and improvements in the new sector of heat engines, which were steam-driven for the generation of power. The first wave of developments was dominated by practitioners like Thomas Newcomen (1664–1719), who created the first practical steam engine for pumping water. Few traces are left of this research because such inventors tended not to have had any formal training whatsoever.[38] In the last quarter of the 18th century, these sporadic developments became more systematic and better documented. The epitome is the *Boulton & Watt* manufactory founded in Birmingham in 1773. It produced modified Newcomen steam engines based on the first patent issued in 1769.[39] Watt's engine generates steam slightly above atmospheric pressure. The steam condenses inside a small metallic cylinder immersed in cold water, causing rarefaction and thus leading to a down-stroke of

---

[38]On the history of steam engines, see, e.g., Matschoss (1901, 1908), Kanefsky & Robey (1980) and Musson & Robinson (1969) chap. XII.

[39]On James Watt, his partner Matthew Boulton (1728–1809) and their context, see, e.g., Muirhead (1854), Dickinson & Jenkins (1927), Hills (1989, 2002) and Miller (2008).

the piston inside the cylinder. The expansion of the steam upon heating causes an up-stroke of the piston. Watt's design was thus the first double-stroke engine, working on both the up- and down-strokes.[40] Another crucial improvement over earlier designs was Watt's invention of a separate condensation chamber into which the steam is injected. This condenser remains cold while the cylinder remains hot, thus significantly reducing heat loss inside the engine and improving the overall efficiency with which heat is converted into mechanical motion.

Various developmental strands intertwine in the person and career of James Watt (1736–1819), giving him a considerable advantage over his competitors lacking one or several of these strands. On one hand, Watt was an ingenious developer and inventor, and a virtuosic visual thinker. For instance, his solution to the problem of connecting the piston rod of the sealed cylinder to the beams of the pump driven by his steam engine is brilliant. While the piston sealed inside the cylinder moves vertically up and down, the beam of his double-stroke engine is necessarily supported at its center of mass. This implies that both ends of the beam move along arcs of a circle rather than straight up and down in a vertical line. In 1784, Watt found as a solution to this mechanical power transmission problem a construction with four bars, coupled like the arms of a pantograph (see here sec. 4.1 on p. 115). A pantograph reproduces on a larger scale the motions of retracing a smaller drawing. Similarly, Watt's construction copies the vertical up-and-down motions of the piston at the other end of his clever beam construction. As an aside: Watt was quite justly proud of this 'parallel motion' construction. It was brilliant visual thinking. More specifically, it was a creative mental transfer of a device (the pantograph) used in one context that he happened to be familiar with (i.e., technical and perspective drawing) into a new context where it provided an elegant solution to a difficult problem.[41]

On the other hand, Watt also had contacts in the academic world with its growing number of conceptualizations of heat and steam engines. Watt had been university mechanic and a kind of laboratory assistant to the chemist Joseph Black (1728–99) at the University of Glasgow from 1757 to 1766. Black's commission in 1765 to repair a model of a Newcomen engine in his large collection of

---

[40] See the description in the *Encyclopædia Britannica* of 1797, based on Watt's patent specification, and quoted in Carnot (1824*d*) caption to plate 1 verso p. 54; cf. also Muirhead (1854) vol. 3.

[41] For further technical commentary and good animations of the resulting beam motion, see, e.g., Bryant & Sangwin (2008), http://en.wikipedia.org/wiki/Parallel_motion and .../Watt_steam_engine, http://staff.science.uva.nl/~leo/lego/linkages.html and http://de.academic .ru/dic.nsf/dewiki/1492536, and http://web.mat.bham.ac.uk/C.J.Sangwin/howroundcom/linka ges/index.html.

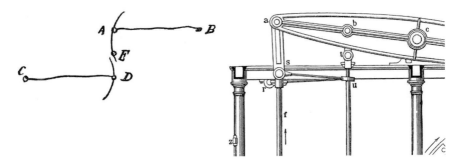

Fig. 48: Left: James Watt's sketch of how he conceived his solution to parallel motion. From a letter to his son, Nov. 1808, published in Muirhead (1854) vol. II, p. 88, with further commentary in Reuleaux (1875*b*) p. 4. Right: Watt's parallelogram *in situ*; from http://de.academic.ru/dic.nsf/dewiki/1492536.

models actually started Watt thinking about how such engines could possibly be improved.[42] One specific device that a *Boulton & Watt* employee, namely John Southern (1758–1815), had invented towards the end of the 18th century will concern us further throughout this section: the so-called 'indicator diagram.'[43]

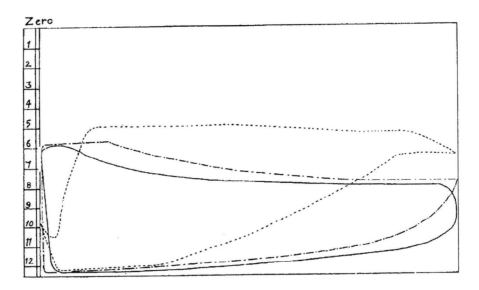

[42] On Watt's relationship with Black, see Robinson & McKie (eds. 1970) for their correspondence, and Miller (2008) pp. 48–52 with further references.

[43] On the history of this device, see, e.g., H.H. Jun. (1822), Shields (1938), Baird (1989), and Hankins (1999). The first published account appeared in 1820.

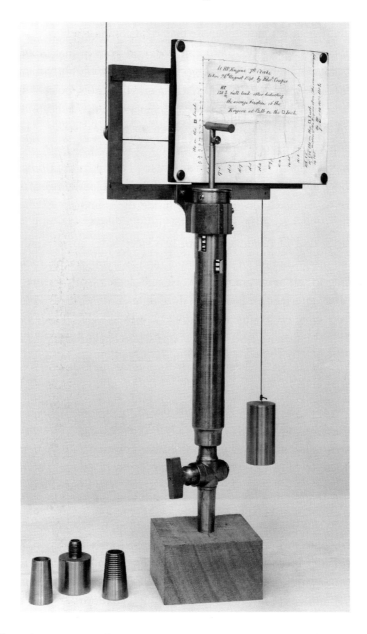

Fig. 49: Left: Museum model of the indicator diagram apparatus as used by the Boulton & Watt company to gauge the power or 'effect' of their steam engines, c.1803. Reproduced with permission from the Science Museum/Science & Society Picture Library (no. 10281385). Facing page: Indicator diagram as recorded in 1803 in Manchester on Mr. Houldsworth's steam engine, from the Boulton & Watt collection, redrawn by Richard L. Hills (1989) p. 93 (here turned 90°). Reprinted by permission of Cambridge University Press.

Originally, this device was nothing more than a rough-and-ready way to gauge the power of a steam engine. Southern and Watt had noticed that more powerful machines maintained a higher pressure inside the cylinder and had a higher lift (the difference in height of the piston at its highest and lowest positions). Around 1796, Southern had the brilliant idea of combining these two indications into one gadget permitting quick measurement of a steam engine's power. He affixed a pen at the top of the piston, somewhat extending it by a stable steel rod, so that the pen jumped up and down with each stroke of the engine. He had the pen move orthogonally to this motion on a recording scroll in a horizontal direction proportional to the variation in pressure inside the cylinder. As the left part of fig. 49 shows, during a full cycle of the steam engine's operation, the pen is guided along a kind of quadrangle, with two steep sides corresponding to the quick up-and-down motion of the cylinder, and two less steeply inclined sides corresponding to the more gradual increase and decrease in pressure inside the piston in the two intermediate phases. In the right part of the figure, various dotted lines show the same cycle after the settings of various valves have been slightly changed. Thus it illustrates how the power of steam engines could be optimized (in this case, up to 9.4 horse power) using indicator diagrams as intermediary gauges of the effect of each mechanical alteration.

This invention remained a strictly guarded company secret for roughly two decades, and only employees of *Boulton & Watt* were taught how to use these diagrams as a quick way to gauge the relative power of, as Watt put it, the 'effect' of a steam engine (estimated in units of horse power, i.e., how many horses would be needed to perform the same work). The trick with the indicator diagram was simply to measure the area of the quadrangle: the larger this area, the more powerful the engine is. At this point, there was no underlying theory whatsoever behind this practice. It was just a good working recipe that *Boulton & Watt* mechanics had happened to chance upon and follow for several decades from c. 1795 onwards. Since this company exported its steam engines in large numbers to countries of every description, from Russia to America, with company mechanics often sent along to explain their workings to the local users, it was unavoidable that eventually the secret would leak. While in active service in a civil engineering corps, the young Frenchman Sadi Carnot (1796–1832) may have seen the diagram and realized that it was actually a depiction of four phases of internal combustion inside a steam engine. This son of the engineer-officer and politician Lazare Carnot (1753–1823) had been educated at the Parisian *École Polytechnique* (since 1812) and at the *École du Génie* in Mezière (1814). So he was much more theoretically inclined than the *Boulton & Watt* mechanics. Looking for a general description of heat engines, he rephrased engineering practice in a form compatible with the caloric theory of heat prevalent at the time. Heat was considered to be a weightless fluid flowing in and out of these machines like water flowing in and out of a water mill. Whereas the overshot water mill takes its power from the difference in heights of the water entering and leaving the mill, heat engines were supposed to tap their power from the difference in temperature of the heat fluid entering and leaving the engine.[44]

Carnot reconceptualized the working of steam engines as a four-stage cycle, with two phases of adiabatic compression and expansion, during which the overall quantity of heat inside the engine does not change, and two phases where heat flows in or out of the engine without a change in pressure or temperature because the changes in volume are synchronous. During these so-called isothermal phases, the density of heat per unit volume, interpreted as temperature in the caloric theory of heat, remains constant inside the piston, whereas in the other two phases the temperature does change. Carnot presented these considerations nearly purely in textual form; it was still far from complete or easy to understand. The cycle (later dubbed the Carnot cycle in his honor) was described in words as an enumerated list of stages of the process.[45] The only visual support provided for this reasoning in Carnot's text is the occasional very schematic engraving of a piston inside a cylinder in various positions. This shows that Carnot's reasoning was not at all visual but abstract and formal.

Carnot died early. His abstract reconceptualization of a steam engine's operation would possibly have been forgotten again if Émile Clapeyron (1799–1864), another *polytechnicien*, had not taken it up. After finishing his studies at the *École Polytechnique*, Clapeyron worked on various engineering projects in Russia in the 1820s,[46] and later, back in France, on the construction of the first railway line connecting Paris to Versailles and Saint-Germain. While still in Russia, he may have come into contact with mechanics from *Boulton & Watt*. There he may have also obtained a copy of Carnot's *Refléxion sur la force motrice du feu*. Able to draw

---

[44]On this so-called waterfall analogy, see Carnot (1824*d*) p. 15; *chute de calorique* analogous to *chute d'eau* through a mill. Cf., e.g., Cardwell (1989); on the caloric theory of heat, see, e.g., Callendar (1911), Heilbron (1993) and Miller (2008) pp. 53ff. On Carnot, see also the introductions to Picard (1872) and to Carnot (1824*e*).

[45]See Carnot (1824*d*) pp. 10 (only 3 stages, incomplete), 17–21 (7 stages, with the seventh indicating the cyclic repetition of stages 3–6).

[46]On Franco-Russian engineering links and Clapeyron's vita, see Bradley (1981), Emmerson (1973) p. 202 and Kurrer (2008) pp. 310ff.

a link between these two worlds, Clapeyron realized the importance of Carnot's insights and made them more communicable in both mathematical and graphic form. Clapeyron thus 'translated' Carnot's insights into the language of the modern theory of heat, which had meanwhile dispensed with the caloric concept.[47]

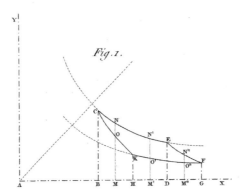

Fig. 50: Idealized heat cycle of a steam engine, with the *x*-axis representing volume, and the *y*-axis pressure. The two dashed hyperbola segments define the isothermal curves pV = constant, the sections EF and KC are the two adiabatic phases. From Clapeyron (1834*a*) fig. 1.

It was in Clapeyron's version that the indicator diagram became known to the academic world under the new labels: Carnot(–Clapeyron) diagram or pV diagram, because pressure and volume are its two axes; and the curves traced by the engine's motion are actually describable by simple formulas, such as pV = constant for the two adiabatic curves. Unlike the indicator diagram of Watt and Southern, the Carnot–Clapeyron diagram, depicting an ideal work cycle of a heat engine, was thus highly theoretical. It is charged with theoretical background knowledge from the rapidly evolving science of heat, soon to be rechristened 'thermodynamics' by William Thomson (1824–1907).[48]

The four-stage Carnot cycle remains to this day an integral part of all courses in thermodynamics (for students of physics, chemistry and applied sciences) and in the theory of heat engines (for engineering students). For an example of a very

---

[47] From the annotated edition of various unpublished notes found among Carnot's papers – also published in Picard (1872) as well as in a selected English translation in Carnot (1824*d*) pp. 60–9 – it becomes clear that towards the end of his life Carnot himself had struggled to free himself from the constraints of caloric theory, but this was unknown at the time.

[48] On the history of thermodynamics, see, e.g., Truesdell (1980) and Müller (2007).

recent effort to bridge the gap between these two worlds of practice and theory by synchronizing the motions of a piston with the corresponding lines of the Carnot diagram, see fig. 51, taken from a modern online lecture course.[49]

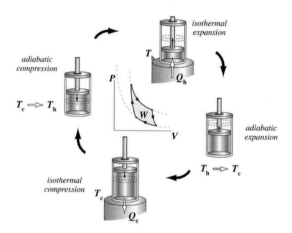

Fig. 51: A modern version of the Carnot cycle. Image by John Wetzel, from www.wikipremed.com/image.php?img=010304_68zzzz130800_30301_68.jpg& image_id=130800

What this episode tells us is how a graphic device that was initially nothing more than a working recipe developed in the rough-and-tumble world of engineering practice was transformed into a theoretically highly sophisticated diagram to explain the working of such engines in fairly abstract terms. Carnot and Clapeyron acted as mediators between the two worlds of engineering practice, on the one hand, and abstract theory, on the other, since they were to some degree in contact with both worlds: with the latter as former students of the highbrow *École polytechnique*; with the former insofar as both Carnot and Clapeyron were employed in practical contexts for part of their careers. This episode is also a case in point for those who argue that often science is only the afterthought of technological practice rather than technology being but applied science (as many still seem to believe). At any rate, thermodynamics certainly owes more to the steam engine than the steam engine owes to thermodynamics.[50]

---

[49] See www.vectorsite.net/tpecp_10.html and whs.wsd.wednet.edu/Faculty/Busse/MathHomePage/MathHomePage/busseclasses/apphysics/studyguides/chapter12_2008/Chapter12StudyGuide2008 (the latter last accessed June 7, 2011, no longer online in Feb. 2014).

[50] That science owes more to the steam engine than vice versa is an often-repeated slogan,

Since the overall theme of this chapter is transfer, it is appropriate to mention that the story did not stop with the transfer from the worlds of practical technology to theoretical thermodynamics. Various historians of science and medicine have pointed out that the great success of indicator diagrams in the production of steam engines led to further transfers of the underlying idea of complicated systems recording their own internal motions. By the middle of the 19th century, appropriately modified apparatus for self-registration of dynamic changes cropped up in various other fields, most prominently in physiology. Carl Ludwig's kymograph, for instance, registered blood pressure, muscle contractions, or nerve activity, and, by the turn of the century, the 'graphical method' had become a veritable craze.[51]

5.3   NMR: from physics to chemistry and medicine (MRI)

Nuclear magnetic resonance (NMR) is a physical effect in which atomic nuclei, either in individual atoms or in molecules, absorb external radiofrequency waves and then re-emit them in a magnetic environment.[52] NMR only involves nuclei with a magnetic moment and is highest for light nuclei such as hydrogen (protons). The spin of the atoms temporarily becomes aligned with the direction of the applied external magnetic field. During the 'relaxation interval,' the atoms, having been forced out of equilibrium to artificially align themselves parallel to the external field, will return to their equilibrium distribution. It is during these few microseconds of readjustment that they emit the nuclear resonance radiation that is experimentally recordable by a pick-up coil surrounding the specimen under examination. This phenomenon was first observed by Isidor Rabi (1898–1988) in 1938 in the context of atomic beam experiments in Chicago. A focused stream of heated atoms escaping through a small opening in an oven was directed through a magnetic field and their electromagnetic radiation recorded.[53] This initial context in which NMR first appeared was limited to high-precision measurements of

allegedly first coined by the Harvard physiologist and sociologist Lawrence Joseph Henderson in a Harvard lecture course in 1917. On the unclear provenance of this phrase, see Miller (2008) p. 64, note 47. For more on this controversial science–technology interplay, see, e.g., Kerker (1961).

[51]On this second transfer phase from engineering into physiology, see Braun (1992), Brain (1996), Brain & Wise (1994) and Miller (2008) pp. 70f.

[52]For general surveys on the history of NMR, see Andrew (1984), Partain (eds. 1988) chaps. 1–4, Mattson & Simon (1996) on seven key actors, and Reinhardt (2006) on disciplinary trajectories and instrumentation, all with extensive references to the literature. Further links and citations are found at www.ebyte.it/library/refs/Refs_NMR_AboutHistory.html

[53]On Rabi's pioneering work, for which he received the 1944 Nobel Prize in Physics, see, e.g., Mattson & Simon (1996) chap. 1 and further references there.

magnetic moments in atomic and nuclear physics. It was not yet a visual culture, but primarily an experimental culture yielding as final results a few numerical values for atomic and subatomic quantities.

In 1946, Felix Bloch (1905–83) at Stanford University and Edward M. Purcell (1912–97) at Harvard University made the NMR technique applicable to experiments with liquids and solids, earning them their shared Nobel Prize in Physics in 1952. Both Bloch and Purcell had backgrounds in radar research, which is essentially also a study of resonances, namely of microwaves in cavities; and both were familiar with advanced electromagnetic technology and electronics, which facilitated their transition to the study of NMR. But, as with Rabi's and Ramsey's work, Bloch's and Purcell's did not yet transform NMR into any particularly visual field, either. Both remained very much focused on atomic and nuclear precision measurements.[54] However, Bloch performed the first biological experiment with NMR when he placed his finger in the test coil and obtained a strong proton NMR signal. Although it did not yet contain spatial information, the signals from blood, tissue, fat, bone marrow and other components of the finger were integrated.[55] Measurements other than resonant frequencies since 1950 involved spin echoes to pulsed electromagnetic signals. The relaxation times granted insight into the spin–lattice and spin–spin interactions inside the specimens. As Edward Raymond Andrew (1921–2001) wrote in a historical review of NMR, "in the first decade of its activity, NMR was largely the province of the physicist and the physical chemist."[56] In a two-volume textbook on NMR imaging that appeared in the 1980s, Purcell reflected retrospectively on the reasons why a transmission of NMR techniques into other, more visual domains was not achieved much earlier:

> By 1950, plus or minus a year or two, the basic physics that underlies NMR imaging was for practical purposes completely understood. That includes the magnetic dipole moments and electric quadrupole moments of relevant nuclei; the relaxation times [...] and their dependence on molecular viscosity, the dynamic behavior of spins of all sorts in oscillating fields; both continuous and pulsed; the chemical shifts that were soon to open up an immense field of application in organic chemistry. No physicist working with NMR at that time would have been surprised to see a proton resonance

---

[54] On Bloch's and Purcell's experiments and their theories of nuclear magnetic absorption and nuclear induction, respectively, see Rigden (1990), Mattson & Simon (1996) chaps. 3–4 and further sources cited there.

[55] See Bloch, Hansen & Packard (1946) and Andrew (1980) p. 472.

[56] Andrew (1984) p. 116; cf. also Andrew (1955) for the first textbook referencing all 400 pioneering papers, and Mattson & Simon (1996) chap. 6 on the discovery of spin echoes by Erwin L. Hahn (*1921) in 1950.

with a mouse or a human finger, in the coil. Its amplitude [...] would have been quite predictable. Yet with all this knowledge ready to apply, the realization of medically useful NMR images lay more than 20 years in the future. What essential ingredients were lacking?[57]

In Purcell's opinion, the reasons for this intellectual and practical blockage were a too low sensitivity, a lack of sufficient computing capacity to process such signals and a missing drive to obtain useful images of the interior of objects with NMR.

The resonance frequency of NMR depends on the strength of the external magnetic field as well as on the magnetic properties of the isotopes involved to allow a determination of the magnetic moments of the nuclei under examination. Aside from this resonance frequency, one can measure the relaxation time, which depends on the atomic or molecular weight and on the inner-molecular binding forces. The NMR signal is thus like a fingerprint of the substance under examination. Scientists like Norman F. Ramsey (*1915) took an interest in this technique and started to develop it into a research technology for chemical analysis. Ramsey and his colleagues from physics and chemistry soon noted that certain molecular groups and radicals of organic chemistry all had their own clear signature in NMR spectra. This allowed a quick and reliable analysis of organic molecules (cf. fig. 52).

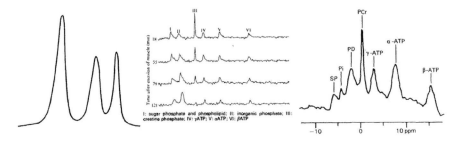

Fig. 52: NMR spectra from 1951, 1974 and 1983 show the progress made in spectral resolution and the sophistication with which the peaks could be correlated with the presence of specific substances. Left: A proton NMR spectrum from ethyl alcohol, showing peaks due to CH_3, CH_2 and OH groups (Arnold et al., 1951). Center: An NMR spectrum, recorded from ^{31}P nuclei, of various phosphates and ATPs in an intact, freshly excised rat leg muscle (Hoult et al., 1974). Right: A ^{31}P-NMR spectrum of these same phosphates and ATPs in a living human head, examined in a 1.5 T magnetic field with a signal-receiver coil placed over the patient's temple (Bottomley et al., 1983). From Andrew (1984) pp. 116f. Reprinted by permission of Oxford University Press.

---

[57]Purcell in Partain et al. (eds. 1988) p. xxvi.

The mammoth research projects to correlate all kinds of molecular components with their characteristic NMR spectra certainly count as an extensive and very successful research program in pattern recognition. But I would not say that this was a visual science culture yet, because the NMR spectra were identified and catalogued numerically by their most striking resonance frequency peaks.[58]

It took another two decades for the transfer from physics and chemistry into medicine to finally take place. In the 1970s, medical imaging technologies were enhanced substantially by what was later renamed magnetic resonance imaging (MRI). The historical details of this transfer are still somewhat controversial, but I think it is fair to say that the lion's share of this transfer can be credited to Raymond V. Damadian (*1936). He had been trained as a medical doctor at the *Albert Einstein College of Medicine* in New York City (MD 1960) and had actually attended lectures by Purcell on quantum mechanics during a postdoctoral fellowship at *Harvard Medical School* in 1963.[59] In the mid-1960s, Damadian was a medical researcher affiliated with the *Downstate Medical Center* in Brooklyn, New York. In September 1969, after his efforts to obtain an NMR spectrometer through Purcell failed, Damadian wrote an application for funding to purchase a high-field NMR spectrometer. In it he certified: "I will make every effort myself and through collaborators, to establish that all tumors can be recognized by their potassium relaxation times or H_2O-proton spectra and proceed with the development on instrumentation and probes that can be used to scan the human body externally for early signs of malignancy."[60]

Experimenting with NMR signals from diseased and cancerous tissues, Damadian noticed in 1970 that these samples differed from normal tissue by having prolonged relaxation times. He immediately realized that this would open up NMR techniques to medical diagnostics, and he published these findings in 1971. He submitted his first patent application for a medical scanner based on these differences between cancerous and noncancerous tissue in March 1972; after careful examination, the patent was granted in 1974.[61]

---

[58] On Ramsey's pioneering work on magnetic resonance spectroscopy and his chemical shift theory, which tried to correlate the number and position of NMR emission and absorption peaks with electron configuration, molecular structure and binding energies, see Mattson & Simon (1996) chap. 2, Reinhardt (2006) pp. 55ff. and further sources cited there.

[59] On Damadian, see his company webpage www.fonar.com/fonar_timeline.htm and Partain et al. (eds. 1988) chap. 4 and Mattson & Simon (1996) chap. 8. On his early link to Purcell and to Ramsey's daughter, see ibid., pp. 218f.

[60] Letter to the scientific director of the *Health-Research Council of the City of New York*, dated Sep. 17, 1969, and available online at www.fonar.com/.

[61] US Patent No. 3,789,832, dated Feb. 5, 1974, upheld and enforced by the US Supreme Court

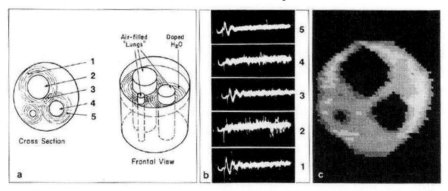

Fig. 53: Experimental NMR images of a 3D test-object shown in part a, comprising two cylinders (2 and 4) filled with air, representing the two lungs, and a smaller one to test resolution, all immersed in doped water (1 and 5) representing the other parts of a human chest. The NMR signals for each of the numbered sections are shown in part b to diverge strongly enough to be differentiable. Part c shows a black-and-white version of Damadian's rough image of the phantom chest, reconstructed later on the basis of the data and their spatial distribution by a field-focusing imaging technique, originally obtained as a 14-color video display (color pl. XV top). From Damadian et al. (1977) figs. 1 and 2 by permission of FONAR Corp. & Physiological Chemistry + Physics + Medical NMR.

Up to this point, Damadian's experiments had been conducted in the manner of a physicist, as mere recording of relaxation times and diffusion coefficients, without any imaging technique apart from the oscilloscope graphs of the NMR signals (see part b of fig. 53). In order to obtain some sort of 2D image, Damadian developed the field focusing technique. Every point in a coarse-grained 2D array was iteratively measured by superimposing the NMR signal with a parabolically increasing magnetic field. This left just one point of the lowest magnetic field, the 'sweet spot,' which gave a spatially uniquely locatable signal. Data-taking thus meant moving all the volume elements of the sample stepwise across the sweet spot to complete one full scan. It was a tedious and time-consuming procedure that took hours to complete and was thus not suitable for clinical applications (cf. the caption of fig. 55 below for examples of data-taking times).

A more elegant way to achieve such nuclear magnetic resonance imaging was conceived by Paul Lauterbur (1926–2007) from the *State University of New York* at Stony Brook. Initially in a notebook entry in early September 1971, later also in a paper that appeared in 1973, Lauterbur suggested that linear gradients of magnetic field strength could be superimposed on the resonance frequency signal.

against *General Electric Corp.*, Oct. 6, 1997; see Damadian (1971) and www.fonar.com/timeline.htm for scanned versions of all pertinent patents and other documents relating to Damadian's R & D.

The resulting measurable resonance frequency was slightly altered, depending on the spatial location of the resonating nucleus in the sample. If one worked with materials with known resonance frequency, one could in principle correlate the strength of a signal with its physical location along this gradient field (cf. the left part of fig. 54). This became a key idea of today's nuclear magnetic resonance imaging,[62] replacing Damadian's field focusing technique (fig. 54, right).

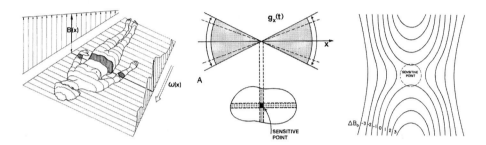

Fig. 54: Three different imaging technologies for NMR from the 1970s. Left: Lauterbur's linear gradient $B(x)$ superimposed on the signal $\omega(x)$ to identify the spatial slice from which a signal issues. Middle: Ernst's sensitive point technique using a sequence of radiofrequency pulses with alternating phase. Right: Damadian's field focusing technique with a local minimum in a static magnetic field to uniquely identify the point of measurement. All images are taken from Richard R. Ernst: A Survey of MRI techniques, in Partain (eds. 1988) pp. 23, 27 and 28. © Elsevier Ltd. by permission.

Independently of Lauterbur, Peter Mansfield (*1933) had also experimented with NMR techniques at the University of Nottingham, England, initially with the goal of imaging crystals. Motivated by Damadian's and Lauterbur's publications, Mansfield switched to examining biological tissue and published his first successful results of medical imaging by NMR in the mid-1970s. His key contribution to NMR imaging was the idea of sequencing and pulsing the gradients so as to obtain a clearer signal to allow localization of the signal-emitting unit cell along three orthogonal gradients. Thus fully 3D reconstruction and imaging was possible. Lauterbur suggested the term 'zeugmatography' for NMR imaging (derived from the Greek term *zeugma* for joining two things together – here,

---

[62]See Lauterbur (1973) and (2003) with a precursor in Gabillard (1951) for the one-dimensional case; cf. also Mattson & Simon (1996) chap. 9 on Lauterbur, esp. pp. 713f., and ibid., p. 249 on two MRI precursor experiments by Purcell and his PhD student Herman Carr in 1952 and 1954, however, which were not followed up.

the two superimposed magnetic fields).[63] But neither this scholarly term nor other suggestions, such as spin mapping or spin imaging, found acceptance in the scientific community. The scary word 'nuclear' was silently dropped in the early 1980s,[64] and today the technique is universally called magnetic resonance imaging (or MRI for short). MRI is a fusion of various key ideas and technologies that had all been around for quite a while already. But their fusion triggered the sudden boost in this image-making technology:

- the basic technology of NMR, originating in physics (Rabi, Bloch, Purcell)
- the idea to use NMR for chemical analysis (Ramsey)
- the construction of large solenoid magnets generating homogeneous fields
- the realization that cancerous and noncancerous tissue have different NMR signals (Damadian)
- the superposition of a linear magnetic field gradient (Gabillard, Lauterbur), or nonlinearly increasing fields, with a local minimum (Damadian's field focusing technique)
- back-projection techniques (Radon, Oldendorf, Hounsfield, Lauterbur)
- robust computer applications of these mathematical techniques

The construction of experimental prototype MRI scanners, in which all of the above ideas were implemented, turned into a race between Damadian, Lauterbur and Mansfield. Damadian was first to produce the experimental proof of NMR distinguishability of cancerous and noncancerous tissue in 1970. In the face of disclaimers by some experts, such as "any further discussion of scanning the human body by MR is visionary nonsense," Damadian and Lauterbur both started building prototype MR scanners. Lauterbur succeeded in spatially resolving a simple phantom composed of two small tubes of water immersed in a larger tube in 1973, and Damadian managed to discriminate and spatially resolve air-filled cylinders immersed in water in the same year. Lauterbur then started to experiment with NMR examinations of clams and earthworms, nuts and tree branches, and (after obtaining somewhat larger NMR spectrometers allowing larger specimens of up to 3 cm diameter) a live mouse. Peter Mansfield started with an excised chicken leg

---

[63]See Lauterbur (1973, 1986) and Andrew (1980).

[64]I.e., even before the Chernobyl disaster in 1986; see, e.g., Anon. (1981) p. 9: "Dr. Ross prefers the term ... MRI ... to NMR ... to underline the fact that it does not involve nuclear medicine radioisotopes [like PET, for instance, see the following section], and to avoid the possible controversial connotation of the word 'nuclear' as it relates to power generation and weapons."

in 1974, then proceeded to examine a human finger.[65] In March 1976, Damadian took his first NMR image of a mouse thorax in cross-section, and in July 1977 he obtained the first NMR cross-section of a human chest (see here fig. 55 and color pl. XV). On October 26, 1977, E. M. Purcell wrote to R. V. Damadian: "Thank you very much for sending me a copy of your historic first picture. I congratulate you and shall keep the picture as a perpetual reminder of how little one can foresee the fruitful application of any new physics."[66] In 1978, Mansfield followed suit with the second human whole-body image. It was of his own abdomen and took only 40 minutes data-taking time compared with Damadian's and Minkoff's 4.45-hour ordeal the year before. Around that time, skepticism among radiologists began to wane and NMR began to be accepted as an alternative imaging technology in medicine. There were further improvements in magnet design, moving on from finicky superconducting magnets to iron-cased permanent magnets of a more open design less apt to elicit claustrophobic anxiety in patients. As faster computers and increasingly sophisticated programs yielded better-quality images, it became an increasingly important and indispensable research and diagnostic technique. In 1978, Damadian founded his *FONAR* company and he marketed the world's first commercial MRI scanner from 1980, the Fonar QED 80. This whole-body scanner still used Damadian's field-focusing nuclear resonance technique – from which the company acronym is derived. Lauterbur and Mansfield followed suit with their first model, manufactured by the British company *EMI*; and other providers of medical instruments, such as *Varian* and *General Electric*, soon also began to flood the rapidly expanding market with their NMR variants. Image quality continued to improve over the next two decades (cf. figs. 55 and 56); the average times to scan a patient and device prices fell as a consequence.

The first NMR 'image' (see fig. 55) quite clearly is a matrix of 12 columns and 16 rows filled with 116 measurements – most of the zeros are outside the test body, and the highest numbers indicate solid bone in the lower chest, with the left and right lungs also yielding very low numbers. This 'image' – like all NMR images – is thus nothing but a construct of surfaces enveloping areas of roughly equal relaxation times, indicating comparable materials. More modern MRI has gradually increased the number of volume elements (also called voxels) to now 512×512 per slice. Each increment in progress was closely linked to rapid advances in computing power and speed. The much finer image resolution

---

[65] See Mansfield & Maudsley (1977) – the first human NMR image obtained; cf. Mattson & Simon (1996) pp. 726f.

[66] Quoted in Mattson & Simon (1996) p. 252, posted on www.fonar.com (access Aug. 22, 2008).

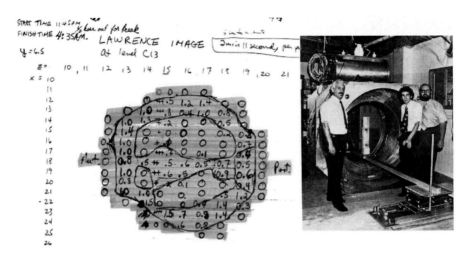

Fig. 55: Left: Laboratory sketch of oscilloscope measurement data of the first MR scan of a human chest, obtained on July 3, 1977. Damadian's assistant, Lawrence Minkoff, was the subject because he was the smallest in his team and barely fit inside the tight NMR scanner (right). It consists of a circular 53-inch superconducting coil magnet cooled by the liquid helium and liquid nitrogen tanks above. The subject is repositioned iteratively, horizontally and vertically, by means of the motorized construction at the front, after each NMR measurement in the focus of the field (the 'sweet spot'), until all the data points have been recorded. Note the rather long scanning time needed: start time 11:45 PM, finish time 4:35 AM, with a half-hour break and an average of 2 minutes and 11 seconds per image point, and thus roughly 4 hours data-taking time in all. Left: from Michael Goldsmith's and Raymond Damadian's laboratory notebook. Courtesy of Fonar Corporation; cf. Damadian (1978) and here color pl. XV for a 14-color computer display of the same data. Right: from Damadian (1980) pl. 1, fig. 7, by permission of the Royal Society, London.

that results sometimes leads one to forget that these are not simply photographs. Each NMR parameter range is dynamically assigned a certain gray or color value. This free adaptability of the gray or color scale to the data range allows fine tuning of the display areas where there are only very subtle differences in the measured values. It also adds a certain conventionality to the images that are clearly actively constructed and not just passively registered.

Considering the rapid improvements in image quality and designs of MRI scanners affecting all the functional parts, more and more hospitals and medical research institutions decided to purchase MRI scanners, despite their exorbitant cost of between 1 and 5 million U.S. dollars. It is one of the most expensive medical imaging technologies.[67]

---

[67] For data on the USA, Canada, and Britain, see Eden (1984) p. 60, Kevles (1997) pp. 173ff.

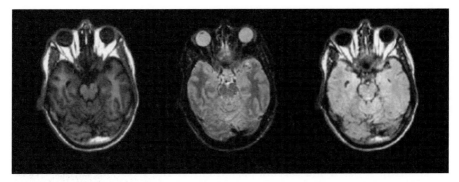

Fig. 56: Three modern MRI images of the brain, taken around the year 2000. Compare the improved resolution and quality with fig. 55. All three images show the same axial section of the same human head, but the conventions chosen to represent gray and white brain parts, fat and tissue differ widely. It illustrates the large play available to the radiologist in choosing the display and contrast most suitable to what he or she wants to examine. From Hennig (2001). © Universität Freiburg im Breisgau. On MRI image optimization, cf., e.g., Partain (eds. 1988) chaps. 6–8.

Regula Burri has conducted one of the most thorough national surveys of the diffusion of this high-tech into the medical world, for Switzerland. A first pilot plant was bought for the Zurich University Hospital in 1982. At that time, Swiss medical care planners were still assuming that the total demand would be less than 10 such MRI plants for the whole of Switzerland. De facto, the demand increased far beyond this level, with the total number of MRI scanners installed doubling between 1995 and 2000. By that year, Switzerland already had 14 MRI scanners per million inhabitants, i.e., altogether c. 100. It is still the country with the highest density of MRI scanners per number of inhabitants in the world.[68] By 2002 approximately 22,000 MRI scanners had been installed worldwide, which performed more than 60 million MRI examinations in that year.[69] The trend is still rapidly rising, since the price for simple, low-field MRI scanners has sunk below the million-dollar threshold.

Because of its versatility, with virtually no hazards for most patients, and its ability to yield additional diagnostic insights beyond high-resolution morphological information from the relaxation parameters, MRI has advanced to a reliability

[68]See Burri (2000) pp. 17ff. and (2008); cf. also Dussauge (2008) on Sweden and Prasad (2005*a*) on the different trajectory of NMR and MRI in India.

[69]Both data according to Dussauge (2008) p. 14; on current practice, cf. Semczyszyn (2010) pp. 113ff.

of something like the "gold standard" of imaging technology within medicine.[70] It had become "an invaluable aid in the whole healthcare chain" and a routine component rather than a luxury.[71] Despite the fact that MRI and magnetic resonance tomography (MRT) have become increasingly routine in modern medical diagnostics, the images still evoke "a sense of wonder and excitement." Even MRI practitioners are known to still express amazement when looking at an image. An ethnomethodological interview of radiologists and medical technicians working regularly with the scanners recorded the statement: "I always go 'Wow' [when I see an MRI]. It's as if you sliced a person in half and looked at them."[72] The sociologist Kelly Joyce has examined tropes in the discourse on these medical imaging technologies in clinical routine user contexts, in laboratory shop-talk and in popular science texts. She found three discursive modes, each one of them at variance with what a careful semiotic analysis of MRI technology would yield:

1. MRI images and the patient's body are treated as "interchangeable."
2. MRI images are seen as "superior" to other images of the human body.
3. MRI "technology itself is portrayed as an agent."[73]

Point (1) suppresses the strongly convention-laden character of the MRI images as an artificial construct based on nothing but strings of data. Point (2) neglects the complementarity of various medical imaging technologies. Each has its advantages and disadvantages, so intermedial comparisons and hybrid representations are frequent.[74] Point (3) commits the fallacy of regarding machines and instruments as acting on their own and images themselves "doing the talking," whereas it is the actors handling these gadgets and explaining the images that are communicating. Thereby MRI acquires the misleading status of a perfectly transparent, "unpremediated slice of the world"[75] rather than of its very indirect re-presentation. At the

---

[70]See the quotes to this effect in Prasad (2005), Burri (2000) p. 21, where 94.9% of all Swiss radiologists in a survey concurred, and Joyce (2005), quote on p. 439.

[71]Quote from Hans Ringertz, chairman of the Nobel Committee in 2003 – see Dussauge (2008) pp. 13f. and nobelprize.org/nobel_prizes/medicine/laureates/2003/ (access Aug. 3, 2011).

[72]Anonymous quote in Joyce (2005) p. 437; the images of the *Visible Human Project* (VHP) undertaken at the *National Library of Medicine* (NLM) are freely accessible on the Internet. The fact that the male subject was an executed criminal adds a peculiar touch to this virtual slicing of human life; cf. Cartwright (1995) p. 25 and Prasad (2005*b*) p. 302.

[73]Ibid., pp. 438ff. and 457, with many examples from the literature and interview excerpts, such as "MRI telling the story" headlining an exhibition (quoted on p. 443).

[74]Cf., e.g., Carusi (2011) pp. 311ff. and here fig. **??** for a recent examples.

[75]Joyce (2005) p. 457 quoting Susan Sontag (1990) p. 69 on an equally widespread misreading of photographs.

same time, the physician's role – and more so that of the technicians operating the machines, with their extraordinary skill at enhancing otherwise hidden features inside a patient's body – is rendered invisible, as the imaging technology is set up as the front-stage hero. The MRI scanner or the fantastic images it produces are exhibited in science museums and in medical exhibitions accompanying large congresses, etc., not the medical technical assistants (MTAs) handling it or the interpreting radiologist.

What is obscured are the skills that radiologists and MTAs only acquire after many years of training and practice in generating, handling and interpreting MRIs. These include the recognition and suppression of artifacts, such as 'cross talk' (i.e., effects induced by overlapping neighboring voxel slices, creating little white dots in the image), 'UBOs' (i.e., "unidentified bright objects") and "old friends" (i.e., anatomical variations from person to person) that are easily mistaken for pathological findings if the examiner disregards such variations.[76] Extensive practice in MRI leads to the acquisition of special skills and visual knowledge that has been described as a "technomedical" or "radiological gaze." I would prefer to call it a particular visual domain within a field of related medical imaging technologies, such as CT, PET, x-raying, ultrasound and more conventional anatomical representations. All of these imaging technologies work towards deconstructing corporeal opacity. There are many switches back and forth between them in current medical practice. So we do indeed have a kind of "kaleidoscope of technomedical gazes."[77]

Altogether, this episode of NMR and MRI shows that it took several decades for a technology that had emerged in the context of Cold War physical research to migrate first into chemistry and much later also into medicine. Even after Damadian had shown the possibility of applying NMR tools to the analysis of biological specimens and to the discrimination of carcinogenic from normal tissue, it took another full decade for the first NMR scanner to become available on the market. Thus, "the development of MRI as a clinical tool was [...] not a straightforward translation of an idea into a machine. The first MRI for clinical use were developed in the first half of the 1980s after a decade of complex sociotechnical tuning."[78] The large magnets, smaller coils and signal receivers had to be combined with the mathematical algorithms and the associated software to

---

[76]For examples and further discussion of these potential sources of error in MRI, see Joyce (2005) pp. 447–52 and Dussauge (2008) chaps. 2, 6–7.

[77]Quotes in the preceding from Dussauge (2008) pp. 19ff., 26ff., 199ff., who rightly insists that "different gazes have co-existed, and co-exist, in medical practice" (ibid., p. 29).

[78]Prasad (2005a) p. 465. Blume (1992) covers the exploratory phase of NMR from 1973 to 1977.

construe the NMR images. Then the intended user groups had to be won over and patients had to be convinced to subject themselves to examinations inside a confining tube. Finally, with the award of the 2003 Nobel Prize in physiology or medicine to Lauterbur and Mansfield, MRI was ushered in as a cultural icon for the virtuosity of modern imaging technologies, for the "medical art of seeing with blind technologies of the invisible" and for our current craving to make the human body transparent.[79]

5.4   CT and PET scanners in medicine

For a while, NMR was known by its nickname: "No More Röntgen." However, x-ray technology had also already developed further into true 3D imaging shortly before NMR was introduced into medicine under the new marketing label MRI, alternatively – and somewhat more poetically – characterized as "the art of see-ing with protons." Thus x-raying became yet another example of a new 'digital anatomy.' As will be detailed further (in sec. 8.3), conventional radiology had thus far been a virtuosic art of shadow-reading (hence the older term skiagraphy). In computed axial tomography (CAT), soon abbreviated to computed tomography (CT), an object is x-rayed from many different angles, by rotating either the object by a defined angle after each exposure or the x-ray generating tube and recording unit around the stationary specimen. These 'images' are not exposed on photo-graphic film, but are electronically stored as long sequences of digits related to the degree of transparency at each coordinate of the targeted object. While each of these digital traces of the amount of radiation recorded behind the object is nothing more than a digitized skiagraph of the object, equivalent to a normal x-ray image, the full set of these data allows one to calculate the spatial distribution of x-ray-absorbing matter inside the volume as a whole. This feat, analogous to reconstructing a forest from various side views into it, is made possible mathe-matically by means of the 'Radon transform' and its inverse. This transform had been invented by the Bohemian mathematician Johann Radon (1887–1963) in 1917 for other purposes. Similar procedures were also used decades later in radio astronomy and electron microscopy.[80] It became useful for the calculation of CT images only from the early 1970s, when more powerful, faster computers became available. From 1967 on, Godfrey Newbold Hounsfield (1919–2004) at *Electric*

---

[79]Dussauge (2008) p. 18; on the contested transparency of medical images, cf. Semczyszyn (2010) pp. 194ff.

[80]See Radon (1917), Bracewell (1956, 1967), Oldendorf (1961), Gordon, Bender & Herman (1970), Gilbert (1972), Cormack (1983, 1992), Kalender (2006) and Lauterbur et al. (2003) p. 246, who only retrospectively learned about these precursors to his own 'back-projection' algorithm.

and Musical Industries (*EMI*) in Hayes, Middlesex, and James Ambrose (1923–2006), radiologist at *Atkinson Morley's Hospital,* Wimbledon, England, cooperated on experimental tomographic imaging. They first worked with simple anorganic test objects. Around 1968, they sampled organic material from cattle and pigs, and from the early 1970s also human tissue.

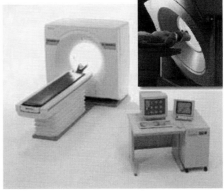

Fig. 57: The evolution of CT scanners from 1969 to 2000. Left: Hounsfield's first experimental system with a motor-driven carriage system to reposition or rotate the square-shaped water-tight glass tray containing the test object. Right: a CT scanner from around 2000. www.epileptologie-bonn.de/cms/upload/homepage/lehnertz/CT1.pdf (p. 5, July 24, 2011).

The first *in vivo* CT scan of a human cerebral cyst was done in 1971, but the *EMI* team continued into the mid-1970s with their development of a first prototype.[81] The brain scanner evolved into the whole-body scanner we know today. As can be seen in the left half of fig. 57, the very first experiments still rotated the sample, with the x-ray tube remaining stationary. This is obviously impractical for medical applications. Therefore, modern CT scanners, such as the full-body scanner displayed in the right half of fig. 57, have the patient lie motionlessly while the recording unit revolves around him or her inside a closed ring-shaped box unit, scanning slice by slice at a flabbergasting speed. The first CT scan of a pig brain in 1968 with the experimental system on the left half of fig. 57 took nine days, followed by a two-and-a-half-hour period of computer calculation of the some 28,000 measurement data – obviously both of these stages were far too long for applications with human subjects. Commercial computer tomographs with much shorter measurement spans of less than 4 minutes and

---

[81] See, e.g., Hounsfield & Ambrose (1973) and Hounsfield (1980).

much faster calculations of the images became available only from 1973. *EMI* dominated the few hundred annual sales until the mid-1970s, when *General Electric, Ohio Nuclear* (later renamed *Technicare*), *Pfizer* and a few other suppliers also entered the market, charging around a quarter million pounds per unit with running costs of about £ 50,000 per year.[82] The rapid development of increasingly powerful computers and efficient hardware also led to continual improvements in image quality, from a poor 80 × 80 image matrix in 1974 to a higher-resolution 512 × 512 matrix of modern spiral CT scanners in 2000, to the current 1024 × 1024 voxel high-resolution systems.

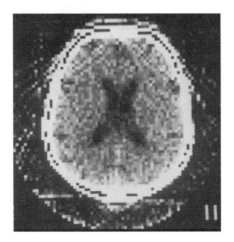 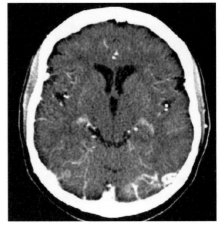

Fig. 58: A comparison of the first CT scans of the human brain (1974) and a more recent image from 1994. From Kalender (2011) by permission of W. A. Kalender. For the further progress in resolution up to 2010, see fig. 59, as well as Kalender (2005, 2006, 2011) on various later technical improvements.

Such images might suggest that CT scans are just an improved form of x-ray images. But MRI, CT and PET differ from older medical imaging technologies much more radically in that the specific angle at which the images must be taken is less constraining. A plate in an atlas, a photograph or an x-ray image all display their object from one specific point of view, presumably chosen by the artist, photographer or radiologist as suited to the purpose at hand. But each such perspectival view occludes certain features: the rear of the depicted object in

---

[82] See, e.g., Banta (1984) p. 73, Stocking & Morrison (1978) pp. 13, 29, Süsskind (1981), Barley (1986), Blume (1992), Kevles (1997) 155ff. and Kalender (2005, 2006).

engravings or photographs, or the bones situated behind other bones seen from a given perspective in x-ray images. Taking several images from different angles certainly helps to remedy this in-built weakness. But each such image remains perspectively bound. MRI, CT and PET images are not dependent on a single point of view, because they are calculated digitally after the recording apparatus has registered signals from various angles (modern CT scanners are typically programmed for 800–1,500 different directions). These body scanners capture cross-sectional views of the interior of the scanned object in a 360° all-round 'view.' Imaging programs later synthesize these 'impressions' into thin slices of, say, a patient's body. From various such slices, various body parts or – if needed – the whole body can be recomposed digitally. The orientation of these slices in the later data display can be changed arbitrarily. One and the same CT scan, stored electronically, can later be used to produce sections along any plane of interest, whether axial (dividing transversally from front to back), sagittal (dividing left from right vertically) or coronal (dividing front from rear vertically) or even nearly tangentially to the surface. The digital data also substantially simplifies image processing and alterations, for instance, subsequent color enhancement or filtering, setting threshold values or shading. High-speed modern computers make it possible to calculate 'fly-throughs' through the human body. In these digital sequences of images – either as sequential peeling, layer by layer, along one axis of the body or as a virtual flight along this axis – selected parts of the human body such as the interior of a vein or the colon are digitally displayed as would be seen by a virtual observer navigating inside the body in a tiny spaceship.[83]

The option of selectively screening out specific materials allows modern medical imaging technologies, for instance, to map the diffusion of water in the nervous fibers of the human brain or to visualize a rising oxygen concentration in tumor tissue. The morphological resolution has increased enormously since the earliest days of these technologies (compare figs. 58 and 59). The transition has also been made from merely morphological to functional imaging. Neural activity and its changes over time are now mappable. In positron emission tomography (PET), a short-lived, positron-emitting radioactive tracer incorporated into a biologically active molecule is injected into the bloodstream. The active molecule is taken up by the tissue of interest, where the positrons from the radioactive decay of the

[83] For examples, see www.nlm.nih.gov/archive/20120612/research/visible/vhp_conf/le/haole .htm, www.youtube.com/watch?v=h5ZUmlET-nI or www.youtube.com/watch?v=mjrUaCEFco (all last accessed July 24, 2011). For further analyses, cf. Cartwright (1995), Gugerli (1998), Waldby (2000), Gugerli in Gugerli & Orland (eds. 2002) and further sources cited there.

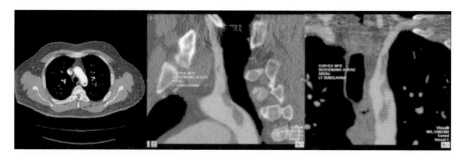

Fig. 59: Axial, sagittal, and coronal thoracic angiographic CT images of a "thrombus involving the proximal 3 cm of the left subclavian artery and extending 3 cm caudally into the thoracic aorta" of a 42-year-old man. From Roche-Nagle et al. (2010) fig. 1. © SAGE Publications Ltd.

tracer annihilate with electrons to produce gamma rays, which are then detected by the PET scanner. The amount of tracer taken up by the tissue, and thus the gamma rays detected by the scanner, will depend on the biological (e.g., metabolic) activity of the tissue. It is therefore possible, using PET, to trace and determine the levels of processes such as brain metabolism, neural activity or the action of a particular drug. On the other hand, PET images lack the clarity of anatomical detail, so, often both imaging technologies are combined (cf. here color pl. XVI).

Full digitization of medical imaging since the mid-1990s reconnected with scientific progress not only the generation of images but also their interpretation. A new subdiscipline, medical image computing, was formed. It led to the design of amazing image enhancement techniques and computer-aided detection (also abbreviated CAD, but not to be confused with the older acronym for computer-aided design used by engineers) for medical diagnostics. These modern medical imaging technologies thus doubtlessly constitute a radically new visual regime, which not unfittingly has been called "cyborg visuality."[84] Like a cyborg – defined as a being breaching the boundaries between the natural and social or between the nonliving and the living – the medical images of this new visual regime have lost their materiality and have become a virtual, digitized medium. The human body had already lost most of its opacity to x-ray exposures, Röntgen-cinematography, fluoroscopy and other such older medical imaging technologies. Now it is completely digitized and thus virtualized. Various parts of it can be recomposed at will, at any resolution. Color or other features are predeterminable by the observer rather than by the object of his or her perspicacious observation.

---

[84]See, e.g., Cartwright (1995) or Prasad (2005*b*) pp. 292, 298, 310, tracing this concept back to Donna Haraway 1991 and Sarah Kember 1998.

Another characteristic of this new cyborg visuality is its rapid development, unlike the older medical imaging technologies, such as anatomical atlases, which remained stable or only evolved slowly over entire generations. Constant reconfigurations and hybrids of old and new occur, forcing all practitioners involved to steadily update their knowledge base. Discriminating the normal from the pathological and locating the latter has become a task in differential and comparative analysis, involving frequent switches between media and imaging technologies, and dynamic interactions between scientists, technicians or radiologists and their image data. Despite the many new possibilities of medical imaging, which is essentially data-based image computing, the interpretation of these images is still considered the domain of the highly skilled radiologist's discerning eye.[85] "Computer-aided diagnosis" is in the offing, though. It will lead to another deep transformation of medical practice in the near future within new infrastructures of "picture archiving and communication systems" in which each fresh image obtained from a patient can be semiautomatically compared against a huge diagnostic database of digitally stored normal and abnormal features.[86]

[85]On the risks involved, and on possible alternatives, see here sec. 9.4 and Peitgen et al. (2011). On the hybrid status of image-data, "once or even twice removed from reality," see Kassirer (1992) p. 829, Prasad (2005b) p. 292 or Adelmann et al. (2009).
[86]On CAD and PACS, see, e.g., Doi (2006).

6

Support by illustrators and image technicians

We have already had occasion to look beyond the scientist or technologist inventing or utilizing visual representations in his or her practice. Behind or around this
figure, other, often little-known, people repeatedly appear: draftsmen, illustrators,
photographers, web-specialists or other image technicians. The specific choice
between these options, of course, depends on the historical period and disciplinary
context of the episode we are looking at. This chapter deals with the broader
context of specialized printing establishments or experts on visualization, be they
woodcutters, lithographers, photomechanical printers or plaster-model makers.
Rather than isolating the contributions by naturalists, their draftsmen, engravers
or photographers, etc., it is more helpful to look at the full set of people involved
in the conception, production and dissemination of a scientific image. In his
Philosophia Botanica of 1751, Carl von Linné (Linnaeus) put it thus: "A painter, an
engraver and a botanist are all equally necessary to produce a good illustration;
if one of them goes wrong, the illustration will be wrong in some respect."[1] In
her study on botanical illustrations in the 18th century, historian of science Kärin
Nickelsen spoke of 'work collectives' to denote these highly organized and differentiated teams of naturalists, their draftsmen, engravers, colorists, printers and
publishers. Together they produced the marvelous atlases, *Kräuterbücher* and illustrated compendia.[2] This term appropriately highlights the collaborative character
of the enterprise, yet it is important to keep an eye on who exactly is contributing
what. In his masterful study of *Art and the Scientist*, Geoffrey Lapage distinguishes
four different scenarios (which he then supports by cases mostly from the fields
of anatomy and zoology):[3]

- scientists who left no information about how (if at all) they interacted with
  illustrators;
- scientists who employed artists but did not supervise their work;

[1] I came across this quote in a webpage by the 'botanical artist' and illustrator Juan L. Castillo
(his translation into English): see www.juanluiscastillo.com/english/about/education.html (access
Nov. 1, 2012).

[2] See Nickelsen (2004*a*) for various examples, and esp. p. 327 on *Arbeitskollektive*.

[3] See Lapage (1961) chap. II: The kinds of scientific illustration, and chap. V: Drawings done by
scientists.

- scientists who did supervise their artists' work; and finally
- scientists who made their own illustrations (e.g., drawings or photographs that were either preserved as such or later reproduced photomechanically).

It seemed clear to Linné that "botanists who have practiced the arts of painting and engraving along with botany have left us the most outstanding illustrations."[4] The following examples will show, though, that various other schemes could also be followed in order to arrive at high-quality plates – each with its own advantages and problems. Before the invention of photography, drawing or painting was the only way quickly to record an observation of an object if preservation of the specimen was not possible. With plants, for instance, the old tradition of storing dried samples in so-called herbaria was one way around the problem of their perishability or further ripening.[5] Drawing a complex organism like a plant is by no means a trivial matter of just recording what is before one. Simple experiments have shown how much other factors affect the drawing of even much less complicated objects. Important factors determining the result were the artist's

- age and experience,
- previous training,
- degree of familiarity with other drawings of the same type of object, and
- inculcation in cultural drawing styles and visual representation.

The following series of copies of the letter W drawn by test persons of various ages and training (fig. 60) are a case in point.

Depending on the observers' artistic experience and acuity, these test letters capture either the minimal attributes (as in case b), where only the four strokes and their approximate relative orientations are drawn, or further details, such as relative widths and shading of the lines (in c and d), or the precise numerical dimensions needed for cutting type or for digital reproduction as electronic font (in e). Drawing hence depends not only on accurate observation or 'seeing' the object to be drawn, but equally on the weighting of sensory elements and their culturally encoded interpretation. What is consciously noticed as a relevant feature, and what is consequently selected as an important feature to be included in the drawing, crucially depend on the information you already have on the object

---

[4]Same source as in footnote 1.

[5]On herbaria and herbals (*Kräuterbücher*), see Blunt (1979), Brandes (1987) and Müller-Jahncke (1995); cf. here sec. 6.1.

Fig. 60: The letter W in a modern serif font and copies of it by hand, done by a very young child (b), a ten-year-old (c) and an untrained adult (d), together with a professional typesetter's analysis of it (e). From Booker (1963) p. 9. © Institution of Mechanical Engineers, London.

and on your goals. Ultimately, it will also depend on the given iconic tradition of other images of this same object – or, as art historians would put it, on its iconography.[6] Drawing is thus not a purely or mostly receptive regurgitation of sensory impulses, but a highly subjective and interpretative activity. Therefore, drawings are often relatively easy to date – sometimes it is even possible for experts to attribute to a specific draftsman or artist an unsigned drawing or sketch lacking other direct hints as to its authorship, simply because of the highly idiosyncratic style of drawings. Aside from the trained eye, it is also the practiced hand of the observer and the intricate interplay of hand-to-eye coordination that leads to the masterfully drawn record.[7] Even if a drawing is only supposed to be copied by hand without any further alteration, each hand-drawn copy will smuggle in certain idiosyncrasies of the draftsman and subtle variations in motif. This permits experts to distinguish a real Rembrandt from a fake, and it also causes a subtle evolution in images if one follows chains of representations of a particular object over time (see here sec. 7.1 on copy relations). And drawing – ending in a unique, non-multipliable re-presentation of the object under study on paper or canvas – was only the first stage of a sometimes longer sequence of transformations of this object before its visual representation could finally be inspected as a printed plate or in some other disseminable format. Our first example will be woodcuts, but later other techniques as well, such as copper or steel engraving, etching, mezzotint, lithography, photography, heliogravure and autotype. Finally, photomechanical printing and other modern techniques of mass

---

[6]On this iterative process of: seeing of, seeing as (*Gestalt* interpretation), instructions to the executing hand, verification of what is drawn, and finishing up, see Booker (1963) pp. 9ff., Arnheim (1969) chap. 14, Gombrich (1960) chap. II and Semczyszyn (2010) pp. 47ff.

[7]On the example of nebulae drawings, see the introduction to Nasim (2012), who provocatively called his book "Observing by hand."

production were added to the spectrum of options available for making multiple 'offprints' of one 'original.' Each has its own strengths and weaknesses in terms of cost, quality and adequacy of representation.[8] Available monetary and personal resources, representational goals, and contextual factors had to be brought into accord; and as we will see in the following, this was by no means always easy. Descriptions of a few famous successful collaborations between naturalists or later scientists and their artisanal aids (in secs. 6.1–2) will be followed by cases generating stress and personal frictions (sec. 6.3).

6.1   Leonhart Fuchs and his team of artisans

Guides to the bewildering variety of plants and herbs found in our habitat were in high demand already during the Middle Ages. One could assemble dried and carefully pressed samples in a herbarium; but after a while they lose their pigmentation and often crumble to pieces. Alternatively, these plants could be drawn on paper, parchment or vellum, thereby producing a more permanent record of their shapes and perhaps also of their colors.[9] 'Herbals' or *Kräuterbücher* are among the earliest printed books since the invention of printing, alongside the Bible and other religious texts. This fact indicates how widespread the desire was to gain an overview of plants and their uses in cooking and pharmaceutics. But information about their possible toxicity was equally crucial. Such herbals fulfilled important functions as guides to the local flora, as aids in determining and classifying unknown plants, and as sources of information about their potential uses.[10]

Leonhart Fuchs (1501–66) was one of the three "fathers of botany" together with Otto Brunfels (1488–1534) and Hieronymus Bock (1498–1554).[11] Fuchs was born in the Bavarian town of Wemding, the youngest son of the local mayor. Obtaining a basic education at the grammar school in Heilbronn and at the *Marien-Schule* in Erfurt, where he learned not only Latin but also Greek, he subsequently matriculated at the University of Erfurt, where he graduated with his *baccalaureus artium* in 1517. Continuing his studies at the University of Ingolstadt, he became

---

[8]See Gascoigne (1986), Bridson & Wendel (1986) and Bridson & White (1990) for examples of each of these printing techniques; cf. Bridson & Wakeman (1984) for an extensive bibliography.

[9]For a 14th-century example, see Pfister (ed. 1961). On the tradition of *herbaria picta*, cf. Müller-Jahncke (1995), and on herbaria Brandes (1987).

[10]Good surveys of illustrated botanical works include, e.g., Blunt (1950), Nissen (1966) and Lack (2001). For techniques on how to draw plants, cf. West (1988).

[11]On Fuchs, see, e.g., Hizler's *Leichenrede*, first published in German translation in Fuchs (2001), the foreword to Fuchs (1543*b*), Brinkhus (2001), P. H. Smith (2006) and Kusukawa (1997, 2012).

Magister in 1521 and *doctor medicinae* in 1524. After working for two years as a practicing physician in Munich, Fuchs accepted a call to the chair for medicine at the University in Ingolstadt. Religious tensions with the dominant Jesuit clergy there about Fuchs's sympathy for Lutheran trends soon motivated him to leave Ingolstadt again and become the personal physician of the Protestant margrave, Georg of Brandenburg, in Ansbach. In 1533, another appointment brought him back to Ingolstadt; but it was not long before he returned to his Ansbach position because, unfortunately for him, nothing had changed in that strongly Catholic town and old animosities had flared up again. Hopes of founding a new university in Ansbach were dashed, so Fuchs decided to accept an appointment to the University of Tübingen, where he became professor of medicine in 1535, soon advancing to the position of dean; he also repeatedly held the highest office, as rector of the university. As dean and rector, Fuchs was intensely engaged in modernizing the university curriculum at Tübingen. For instance, he introduced into medicine Vesalius's anatomical atlas as the most reliable and reasonably complete anatomical introductory text. It was to replace Mondino dei Luzzi's *Anathomia* from 1316, which he dismissed as "full of errors." To further the first-hand visual knowledge of his students, he also arranged the purchase of the first human skeleton for teaching purposes. Aspiring physicians were supposed to "see with their own eyes, how the bones of the human body are shaped and set or oriented, one upon another."[12] As far as botany was concerned, Fuchs introduced the tradition of regular botanical excursions into the environs of Tübingen. He and his students then practiced their skill at determining and classifying plants and picking out suitable, fresh specimens for their extensive illustration endeavors. Fuchs transformed the garden of the former nunnery where he was living during his Tübingen period into one of the first university botanical gardens.[13]

Some 497 plants are described, together with more than 500 superb woodcuts based on first-hand observations, in Fuchs's *Historia stirpium* from 1542 and his *New Kreüterbuch* of 1543. Roughly 400 of these plants were of European origin; the others were exotic and obtained by exchange. All the known names and synonyms were given for each plant, followed by details about its size, shape and color, where it typically grew, when it bloomed, its nature and complexion, and finally any pharmaceutical effects it might have, which had figured so prominently in the earlier literature. The *Historia* was originally issued in two volumes and

---

[12]See the introduction by Klaus Dobat to Fuchs (1543*c*) p. 11: "wie die gebain im Menschen geartet und ains uf das ander sazet oder gericht ist."

[13]On this episode, see Fuchs (1543*c*) pp. 19f. and further sources cited there.

illustrated with 511 woodcuts; the *Kreüterbuch* was a single folio of more than 900 pages and 517 full-size plates. Fuchs called his second book *New Kreüterbuch* in conscious distinction from Brunfels's *Contrafayt Kreüterbuch*, which had appeared in two volumes in 1532 and 1537 in Augsburg, and Bock's *Kreütter Buch* of 1539 printed in Strassburg. The latter was a thorough inventory of plants compiled by a Lutheran preacher, scholar and physician on his extended travels from the Ardennes into the Swiss Alps. Bock focused on their healing properties. Only since its second edition in 1546 did it contain a substantial number of woodcuts carved by David Kandel (1520–90), who already took Fuchs's 1543 publication as his model.

A comparison between the woodcuts of the teasel plant (fig. 61) in Brunfels (1532) and in Fuchs (1543) shows that the humanist scholar Brunfels had obviously decided to depict real samples with all their defects. His woodcutter Hans Weiditz (c. 1500–36), a pupil of Dürer, reinforced this tendency. He depicted a teasel in its specific, obviously already rather sad wilted state, with drooping leaves and very few flowers. Fuchs, on the other hand, chose the complementary strategy of depicting a highly idealized burgeoning sample without any visible faults.

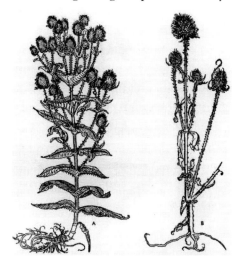

Fig. 61: The teasel (*dipsacus*) in Fuchs (1543), labeled A, compared against Brunfels (1531), labeled B; juxtaposition by Blunt (1950*c*) p. 66.

The contrast could not be stronger between the withered autumnal specimen taken by Brunfels and the bountifully blossoming teasel found in Fuchs. Whereas all the leaves in Fuchs's plant appear pairwise in perfect symmetry, the few leaves shown by Brunfels have different lengths and look decrepit. There is no doubt about which image is "prettier." But it is equally clear that we are far more likely to come across a sample closer to Brunfels's depiction than to Fuchs's. In this sense, Brunfels is more realistic, whereas Fuchs offers more memorizable and recognizable details for plant identification. Thus both images, different as they are, have their pros and cons. Fuchs's main argument against his predecessor was that Brunfels (or perhaps just his printer?) had sometimes used the same woodcut for two or three different plants – a practice not uncommon in early-modern publications. By contrast, there was a one-to-one correspondence between images and plants in Fuchs's *New Kreüterbuch*.[14]

Fuchs's *Historia stirpium commentarii insignes* was published in Latin, like all of its forerunner illustrated herbals. These "notable commentaries on the history of plants" partly still followed the ideal of traditional natural history by presenting long-winded scholarly accounts of ancient botanical knowledge, with the current knowledge provided as a mere appendix. Fuchs's *New Kreüterbuch* was aimed at a much broader array of potential readers. Even though future physicians could also profit by its study, its main purpose was to help the layman in tracing and identifying plants and cultivating and applying them pharmaceutically.[15] It was immediately published in German, with English and Dutch translations following in subsequent years. Various abbreviated versions appeared after 1545. It was part of the trend toward publishing in the vernacular also noticeable in Vannoccio Biringuccio's *Pirotechnia*, which appeared first in Italian in 1540 and was translated into Latin only later, or Andreas Vesalius's *De humani corporis fabrica* (1543) and Agricola's *De re metallica* (1556), to name just three other lavishly illustrated books of the same period. Fuchs continued to collect plant specimens after the *New Kreüterbuch* had appeared in 1543. At the end of his life, a set of over 1,500 of his plant watercolors had accumulated, replete with full botanical identification and commentary on each image. All these materials are preserved at the Austrian National Library.[16] Quite a few of them already existed also as woodcuts made at Fuchs's own expense; but his efforts to find a publisher for this vastly expanded compendium had been in vain. In the following, we will focus on Fuchs's *New*

[14]On this point and its implications, cf. Kusukawa (1997) pp. 406f., (2012) pp. 15–19.

[15]Fuchs (1543) in his unpaginated foreword (*Vorred*: "dem gemeinen mann zu der erkanntniß der kreüter … nützlich und füeglich zu sein."

[16]They were finally published as *Kräuterbuchhandschrift* in Fuchs (2001).

Kreüterbuch simply because for contingent reasons Fuchs's personally hand-colored copy has been preserved and is now kept at the *Stadtbibliothek Ulm*. Altogether, the documentation on the context of its production is extraordinarily good for an early-modern book.

Fuchs did not publish his *Kreüterbuch* single-handedly. To produce such a voluminous work of more than 500 high-quality woodcuts, he needed a whole team of highly qualified and specialized experts. This artisan team included the draftsman Albrecht Meyer (c. 1510–after 1561); the transfer artist Heinrich Füllmaurer (c. 1500–47/48), who transferred the drawings onto the wooden blocks; the woodcutter Veyt Rudolff Speckle (c. 1505–50); and the printer Michael Isengrin (fl. 1538–1560), who ran the printing establishment *Officina Isingrimiana* in Basle, one of the centers of early-modern book printing aside from Nuremberg and Augsburg, Amsterdam or London.[17]

That Meyer and Füllmaurer were depicted executing specific tasks does not mean that their contributions were strictly limited to drafting for Meyer or to carving or transferring the image onto the block for Füllmaurer – quite the contrary. First of all, we know that Füllmaurer was also a talented painter. For instance, in 1541, he made the most famous oil portrait that we have of Fuchs.[18] Füllmaurer and Meyer also collaborated on painting church altars in Gotha and Mömpelgard. Most importantly, the preserved draft drawings for Fuchs's *Kreüterbuch* do not bear any monograms by these two artists. As was frequently the case in so-called atelier production, each of them just did whatever there was to be done, as needed. We also know that Fuchs closely supervised the production of each plate, because many of these preliminary drawings include his corrections and/or supplementary drawings, at least some of them presumably drawn by himself.[19] It is very remarkable that Fuchs lavished such praise on his artisans' work, especially on Veyt Rudolff Speckle, "who so skillfully expressed the outlines of each picture by carving [*sculpendo*] that he seems to compete with the painter for glory and victory."[20] In compensation for their skillful renditions, such superb woodcutters as Speckle could expect to get the lion's share of the

[17] On the history of early-modern book production, see Johns (1998) and Kusukawa (2012) part I and further references cited there. On the interplay between Fuchs and these artisans, see Baumann-Schleihauf (2001) and pertinent chapters in Fuchs (2001).

[18] It is now preserved in the *Württembergische Landesmuseum* in Stuttgart, Inv. No. 1933-622. Cf. upload.wikimedia.org/wikipedia/commons/4/40/Renaissance_C14_Füllmaurer_Leonhart_Fuchs.jpg

[19] For examples, see again Baumann-Schleihauf (2001) and Fuchs (2001) pp. 30ff.

[20] Fuchs (1542) as translated with further commentary by Kusukawa (1997) pp. 404f. and (2012) 45ff., 89f., 132ff.

Fig. 62:   Portraits of two artisans who contributed to Leonhart Fuchs's *De historia stir-pium* (1542) and *New Kreüterbuch* (1543):   On the right, the draftsman Albrecht Meyer; on the left, the transfer artist Heinrich Füllmaurer.   Not depicted are the wood-block cutter, the printer and the publisher.   A hand-colored version of this plate is accessible online at http://special.lib.gla.ac.uk/exhibns/month/oct2002.html (last accessed Feb. 14, 2014).

wages paid out for the production – up to 15 times as much as the artist, who had drawn the images in the first place.[21] The actual cutting into hard sycamore or pear wood was done by Speckle. But since there is no evidence that Speckle ever stayed in Ingolstadt or Tübingen for longer periods of time, we have to assume that these woodcuts were done in Strassburg where Speckle resided until his early death in 1550.[22]

As far as the more extensive herbal manuscript that Fuchs started to compile after the publication of his *Kreüterbuch* in 1543 is concerned, we know that Speckle was no longer involved. Apart from Meyer and Füllmaurer, a third artisan contributed high-quality watercolor sketches that all bear the initials IZ, standing for Jerg Ziegler (post-1500–74/75). Close analysis of the new and rare specimens

[21] See ibid., p. 406 and further sources cited there, on which this estimate is based.

[22] For the few sources we have on the work of Speckle, whose name is sometimes also spelled as Veit Rudolph Speckel or Specklin, see ibid., the Stuttgart database DSI or www.idref.fr/059216565.

depicted, in conjunction with information about when Fuchs had obtained them, indicates that Ziegler worked for Fuchs between 1555 and 1664.[23]

The woodcut technique had the disadvantage of yielding only black-and-white images without much shading in gray apart from very rudimentary hachures. They could merely describe a plant's basic contours. Therefore, such plates were occasionally colored in by hand after the printing had been completed. Such hand-coloring jobs were usually contracted out to the publisher, who had to pay for these commissions in addition to the printing costs. On the one hand, such colored plates certainly made the finished product much more attractive; but on the other hand, they increased the costs prior to sale and hence also the financial risk. Bankruptcy was a frequent occurrence for publishers including, most famously, Gutenberg himself. Publishers usually tried to minimize such illumination costs by hiring women or even children to do the job for low pay. Such thriftiness, in turn, generated other problems, as the quality of colored plates crucially depended on precision in the choice of color for each segment and on the neatness with which the paint was applied.[24]

Like other botanical illustrators, Fuchs and his artisan team aimed at a faithful representation of the plant specimen. At the same time, however, they had to bear in mind other goals of equal importance – for instance, maintaining a sufficiently clear distinction between the various species depicted, so that the user of the compendium could easily correlate a given sample with the range of species selected for illustration. They also had to abide by the representation technique then available. Woodcuts demanded a certain reduction in a given motif. The emphasis is placed on outlines and a relatively coarse-grained overall resolution. That made it difficult – if not impossible – to represent very small details of the plants. One representational strategy was to select developmental stages considered particularly typical or characteristic of the plant at hand. In practice, several such developmental stages, temporally separated from each other by weeks or even months, were often depicted counter-factually as if they could occur synchronously. This was very much in the interest of condensing information onto a single plate but ran counter to the goal of fidelity to reality. All too often, such stages never are observable together on any one specimen. Another illustrative strategy often employed in botanical plates is visual emphasis. Either the size is artificially increased or the positioning is deviated from what real

---

[23]See Baumann-Schleihauf (2001), Fuchs (2001) pp. 35ff. and www.unimuseum.uni-tuebingen.de/38dinge/dinge29.html

[24]For examples, see Nickelsen (2004) pp. 77, 202 on botany, or Hentschel (2002) pp. 122ff. on spectroscopy, or Blum (1993) and Jackson (2011) on zoology.

plants manifest, in order to guide the user's attention to certain features deemed particularly important, for instance, to the plant's sexual organs, such as the fruit or pollen anthers, so crucial in their classification within the Linnean system.

Fuchs became one of the fathers of scientific botany and defined the standards for botanical illustrations for many decades. He owed this privilege to no small degree to the first-class artisans he had found partly in his immediate academic environment at Tübingen and partly searched out elsewhere (from Strasbourg) whenever the high standards he had in mind could not be met locally. Fuchs's success was thus due to the achievements of the entire team. The same can be said of other multi-volume endeavors in early-modern natural history, be these Gesner's *Historia animalium* or Buffon's *Histoire naturelle générale et particulière*, for which more than 80 draftsmen and women, engravers and printers have been identified.[25]

Conrad Gesner (1516–65), for instance, was an ardent naturalist and a prolific author of fat volumes heavily illustrated with high-quality woodcuts. His *Historia animalium* appeared in four volumes with altogether 4,500 pages and large woodcuts on virtually every second page; a fifth volume on snakes appeared posthumously in 1587.[26] Furthermore, he published *Opera botanica* (with 1,500 illustrations), a handbook of mineralogy, *De omni rerum fossilium* (1565), and (posthumously) a *Historia plantarum* (1587). Many of the images in his books were based on samples from his own collections, obtained as a field collector of botanical specimens, during Alpine mountain climbing tours (described in a little book by himself), but also by exchange and gifts from his vast network of friends and correspondents, who were pleased to enrich his collections.[27] In order to better survey the work of his illustrators, he regularly housed at least one draughtsman and one wood-cutter in his own spacious living quarters as Zurich's city physician. According to Wendell E. Wilson's *History of Mineral Collecting*, "he had his illustrators render

---

[25] On Buffon's illustrators for the 36 volumes that appeared between 1749 and 1788 (with 8 more to follow posthumously), see Hoquet (2007); cf. also the online Stuttgart Database of Scientific Illustrators (DSI) – simply search for "Buffon" in the "worked for" search field.

[26] See Conrad Gesner (also spelled Konrad Gessner): *Quadrupedes vivipares* 1551; *Quadrupedes ovipares* 1554; *Avium natura* 1555; *Piscium et aquatilium animantium natura* 1558; in German transl.: *Thierbuch* 1565. See Holthuis (1996) and Kusukawa (2010) on illustrative contributions to Ges(s)ner's *Historia Animalium* (1551–58).

[27] His mineralogical collection, for instance, was later acquired by the Swiss anatomist and naturalist Felix Platter (1536–1614) in Basle, carefully preserved by his offspring, and ended up as part of the *Naturhistorisches Museum* in Basle: see Wilson (1994) pp. 24, 172; cf. also here sec. 12.1 on early-modern mineralogical cabinets and collections.

several of his mineral specimens, but was not well pleased with the results."[28]

6.2   Linus Pauling and Roger Hayward

Linus Pauling (1901–94) was one of the most popular scientists in the USA.[29] One reason for his popularity certainly is that he was one of the very few to get not *one* but *two* Nobel Prizes, the first in 1954 in Chemistry for "research into the nature of the chemical bond [...] and its application to the elucidation of complex substances." The second in 1963 was the Nobel Peace Prize in recognition of his effective activism against atomic bomb testing. Another reason for his popularity was his outgoing personality and his intense engagement with the interested public. He gave inspiring presentations for laymen or lectures for college freshmen as well as specialized courses for PhD students. He also wrote many books and articles at all these levels. A common characteristic of his interactions with these audiences was his heavy use of visual representations, whether 3D models (see here fig. 63 on p. 220), drawings and plates, slides, or celluloid films. It is important to note that his use of visual representations was not limited to didactic purposes, even though it certainly encompassed them. They also served as heuristic devices. Just as so many of our other protagonists in this book, Pauling thought visually and seems to have always tried to visualize processes and structures. He was thus continuing the research style of early structure and stereochemistry that we already met (in sec. 3.2) among 19th-century examples, such as Kekulé, Crum Brown or van 't Hoff. Because of the importance of spatial thinking in Pauling's oeuvre, he – like Kekulé – was also often called an "architect of molecules."

Another parallel to our prosopographic findings (in secs. 4.3–4.4) – besides this allusion to architecture – is Pauling's training. He also started his education within a polytechnic context, as a student of chemical engineering at *Oregon Agricultural College*. From 1927, he taught chemistry at the *California Institute of Technology (Caltech)*, where he had also submitted his PhD thesis and where he was to pursue most of his career. In the mid-1920s, Pauling became preoccupied with research on quantum chemistry. At that time, it was – and to some extent still is to this day – not a visual science culture but a highly formalized and heavily mathematical subbranch of chemistry. In those early days, it was still dominated

---

[28]Wilson (1994) p. 24.

[29]On his life and career see, for instance, Perutz (1994), Hager (1995) and Nye et al. (eds. 1996), and online http://scarc.library.oregonstate.edu/coll/pauling/bond/index.html and paulingblog.wordpress.com/2012/11/28/mary-jo-nye-on-paulings-models/ (Febr. 14, 2014).

by theoretical physicists.[30] A Guggenheim fellowship allowed Pauling to spend a year in Europe, where he met some of the pioneers of quantum mechanics in Munich, Copenhagen and Zurich. Even though he was very impressed with recent advances in theoretical physics, he did not want to narrow himself down to this discipline. He continued to search for a niche in need of his specific talents, less in the direction of exacting calculation and more toward intuitive model-building.

In the 1930s, Pauling took interest in the larger molecules of biochemistry. They had hitherto been left aside as far too complex for rigorous treatment by the new methods of quantum mechanics. They were of crucial importance in biochemical research, which was figuring prominently on campus at *Caltech*. Therefore it was very easy for Pauling to obtain funding for this kind of research. In both his research and teaching, Pauling made extensive use of 3D models of these large molecules. This his colleagues from quantum chemistry would never have done at that time. He also combined model-building with examinations of substances by x-ray diffraction. This technique had already become standard in crystallographic analyses, because equally spaced layers along the symmetry axes of crystals give strong signals that are relatively easy to interpret in terms of interatomic distances and angles between the layers. But it was a new and quite daring idea to apply this technique to large organic molecules. Pauling pioneered this research strand and masterfully combined it with his knowledge of the theory of chemical bonding. Thus he founded a research school of structural chemistry at *Caltech* specializing in proteins and other very large organic molecules. His discovery of the α-helix in amino acids in 1953 was the culmination of x-ray analysis at the very limits of what was then possible, based on meager diffraction data from protein fibers, such as human hair, fingernails and muscle tissue or from frozen proteins.[31] Interpreted in the light of our findings in chapter 3, Pauling transferred visual techniques of x-ray crystallography into structural chemistry and merged them with other components of general and quantum chemistry to create a new visual science culture.

The art of chemical model-building had already developed somewhat beyond the state where we had left it (in sec. 3.2) in the 19th century, but essentially 3D modeling of molecules in chemistry still basically meant ball-and-stick structures like those of Hofmann or Kekulé (here depicted in fig. 20). These ball-and-stick models were all well and good for inorganic chemistry and crystallography,

---

[30]On the early history of quantum chemistry, cf. Gavroglu & Simões (2011).

[31]See Pauling & Corey (1953, 1954). Cf. Perutz (1994) pp. 669f. on Pauling's reasoning style, and Cambrosio (2005) and Elkins (2007) pp. 62f. on his visual improvements of structural formulae.

where the purpose often was to recognize symmetries in crystals composed of one molecule. But using these kinds of models for the large and messy organic molecules that Pauling and his team were targeting was a hopeless enterprise. In particular, they were of no help for one of his most pressing problems in structure resolution, namely, the question of where a certain part would 'fit' best. Unlike what is suggested by simple ball-and-stick models, in which all atoms are represented by balls differing only in color to denote specific elements, *de facto* not all atoms have the same effective radius. Some, such as potassium, have very large effective radii, whereas others, such as hydrogen, have a quite small radius, thus allowing a much higher packing density.[32] Chemical combinations of these atoms thus impose further constraints on how they can be built, apart from simple pairwise coupling of free valences. This element of visual thinking was not incorporated in the ball-and-stick model variant then in use. It is nonetheless of crucial importance in modeling large molecules. Various chunks of molecular subgroups all have to be placed next to each other under very specific cramped conditions. Pauling therefore developed a totally different type of chemical model, called the 'space-filling model.' He later realized that he was not the only one to do this, nor the first. Prior to him, a German physical chemist by the name of Herbert Arthur Stuart (1899–1974) had arrived at basically the same insight.[33] But because Stuart became an ardent advocate of Nazi ideology when the National Socialist Party took power in Germany in 1933, his thoughts were not heard outside Germany, and after the breakdown of the Nazi regime in 1945, Stuart was quite understandably stonewalled. As a consequence, his '*Kalottenmodelle*' continued not to exert much influence beyond Germany. But Pauling's career during the hostilities, at one of the international centers for scientific research on the other side, was unbroken. His impact on contemporaries and succeeding generations was deep. So nowadays space-filling models are more or less automatically connected with Pauling's name and are sometimes simply called Pauling models or CPK models (for Corey, Pauling and Koltun).[34]

---

[32]This effective van der Waals radius should not be imagined as the atom's actual extent, which is much smaller, but rather as half the distance at which a chemical bond with another atom would occur: see, e.g., Pauling & Hayward (1964) appendix 5 for tables of van der Waals radii.

[33]On Stuart, his models and political misconduct in 1934, see Stuart (1948), Hentschel (ed. 1996), Johnson (2013) and http://en.wikipedia.org/wiki/Herbert_Arthur_Stuart with further refs.

[34]The biochemist Robert Brainard Corey (1897–1971) was a member of Pauling's team. Walter Lang Koltun (1928–2010), MIT, headed the Biophysics and Biophysical Chemistry Study Session of the National Institutes of Health. He developed Pauling's models into a cheaper and lighter variant that was patented in 1965 and then marketed internationally: see Corey & Pauling (1953) p. 627. There they thank Roger Hayward for "ingenious solutions of many of the problems of

Fig. 63: Linus Pauling giving a lecture amidst an array of CPK models of α- and double-helix DNA. From Oregon State University Archives, Pauling papers, no. 196-i.060, by permission.

From 1947 on, the laboratory workshop of Pauling's department in fact became a mini factory for his own type of space-filling molecule models. He presented himself as the biggest and most important protagonist of this new style of chemical modeling. The CPK models came in two variants, one at a scale of 1.25 cm/Å with wooden calottes and metal screws linking them, the other at 2 cm/Å with plastic calottes and snap fasteners. These models were not just a teaching device but an important heuristic aid for Pauling's own research. "If you have a model, you know what the permissible structures are. [...] The models themselves permit you to throw out a large number of structures that might otherwise be thought

design and construction" of these models. Cf. Koltun (1965), Gurd (1974), Francoeur (1997) and Francoeur & Segal (2004). On the haptic qualities of these models and their use in modern research, cf. also Francoeur (2000), Hennig (2011) pp. 228ff. and www.hps.cam.ac.uk/whipple/ explore/models/modellingchemistry/spacefillingmodels/ (accessed Sep. 10, 2011).

impossible."[35] For instance, in the early 1950s, Pauling and his team were busy building 3D models of the α-helix of relatively simple proteins such as keratin, collagen and gelatin; these models crucially helped to find plausible configurations of polypeptide chains and the possible packaging of these chains, of course always used in conjunction with the latest x-ray diffraction data of proteins. An eye-witness, Max Perutz (1914–2002), a chemistry graduate from Vienna who later stayed in Cambridge, England, as a postdoc, reported that Pauling occasionally amused himself by building cut-out paper chains of planar peptides to toy with in search of a suitable way of folding them up into a spatially more compact form – this was the way he came up with the helical structure.[36] This pioneering research thus sovereignly combined x-ray crystallographic analysis with a broad range of physico-chemical data and chemical constraints; and ultimately also opened up the path to structural analysis of DNA. The steps taken by Rosalind Franklin and Maurice Wilkins in London, and by James Watson and Francis Crick in Cambridge, England, toward the discovery of the *double*-helix structure of DNA followed similar trajectories and were strongly inspired by Pauling and Corey's paper on the α-helix from 1953 as a *single* right-handed screw.[37]

Pauling's impact on students was extraordinary. Even those not having the privilege of hearing Pauling's exalted lectures seem to have felt this quite strongly, just from reading his textbooks. Pauling's first classic, *The Nature of the Chemical Bond*, fascinated young Max Perutz. He wrote soon after its publication in 1939:

> His book transformed the chemical flatland of my earlier textbooks into a world of three-dimensional structures. It stated that "the properties of a substance depend in part upon the type of bonds between its atoms and in part on the atomic arrangement and the distribution of bonds," and it proceeded to illustrate this theme with many striking examples.[38]

Pauling seems to have found Roger Hayward (1899–1979) during the late 1930s to help him illustrate this book. Hayward was a part-time watercolor painter and pastel artist.[39] Too late for the first edition, its later editions as well as several

---

[35] Pauling, as quoted in Root-Bernstein (1999) p. 258; cf. ibid., chap. 12 on modeling in general.

[36] On the preceding, see, e.g., Corey & Pauling (1953), Pauling & Corey (1953, 1954), Perutz (1994) p. 670 and Nye in Klein (ed. 2001).

[37] See, e.g., Watson (1964), Olby (1974, 1994) and Keller (1996) 107ff.; cf. here p. 339.

[38] Perutz (1994) p. 668.

[39] On Hayward, see the online documentation compiled by Miriam Kramer (Hay-ward's niece) and John Benjamin (his cousin) and published by Oregon State University: *Roger Hayward: Architect, Artist, Illustrator, Inventor, Scientist*, at scarc.library.oregonstate.edu/coll/pauling/bond/people/hayward.html; on his optical work, cf. Bell (2007); on his coopera-tion with Pauling and publisher W. H. Freeman, cf. Heumann (2013).

other textbooks and articles by Pauling were heavily illustrated with images deriving from Hayward's pencil or crayon.[40] Because many samples of his graphic art are easily found online, let us just look at two examples from Pauling's *Architecture of Molecules* (in fig. 64) that reproduce well in black and white. They clearly show Hayward's characteristic soft and very plastic style of drawing in pastels, a far cry from air-brush 'realism' that appeared in the 1960s and later took over the market.

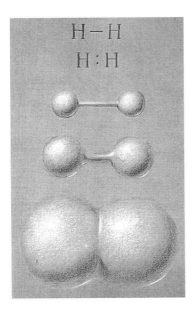
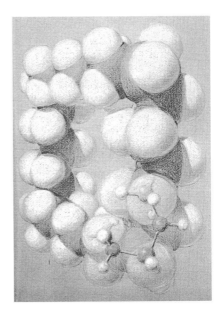

Fig. 64: Two pastel illustrations by Hayward for Pauling. Left: Five different ways to represent the chemical bonding between two hydrogen molecules. The first two lines reflect two different chemical symbolisms; the next two are graphic versions of ball-and-stick modeling; and the final line is Pauling's type of space-filling model. Right: A large carbohydrate ring of a $C_{24}H_{48}$ molecule, again depicted by a space-filling model. It is much smaller than a ball-and-stick model. While the former conveys the compactness of the molecule as a whole, the latter is more clear and precise in depicting atomic distances and valence angles. Thus both forms of visual representation have their pros and cons. From Pauling & Hayward (1964) figs. 5, 23.

Pauling and Corey's article on the structure of proteins for the semipopular journal *Scientific American* in 1954 initiated close cooperation between Hayward and this highly influential science magazine, for which he worked both as illustrator

[40]See, e.g., Pauling (1948, 1957), Pauling, Corey & Hayward (1954) and, most importantly, Pauling & Hayward (1964); cf. again Heumann (2013).

and as contributor to its amateur scientist section. When he formally ended this cooperation in 1973 because of failing eyesight, the editor thanked him profusely "for the tremendous contribution you have made over the years. You know that you have contributed more than the illustrations themselves. You have set the whole style for this kind of illustration."[41] In 1958, the publishing house that ran this journal, W. H. Freeman in San Francisco, offered Hayward a ten-year contract for exclusive publication rights to all his illustrations at this press. Hayward gladly accepted because it offered him financial stability and a guaranteed influx of illustration jobs ranging from chemistry to mineralogy, and mechanics to electronics, in addition to general science books. This way, Hayward was indeed able to define the overall style of scientific illustrations during this period.[42]

Hayward was fastidious about accuracy. In order to prepare for an illustration in an area of chemistry he had never heard of before, Hayward first took private tutoring from R. M. Langer at *Caltech*, which he paid for with various watercolors. Hayward also worked as illustrator for John D. Strong (1905–92) at the *Caltech* physics department, for the architect and astronomer Russell W. Porter (1871–1949), and for a handful of other such clients. But Hayward was more than just an illustrator. He had studied architecture at MIT, hoping that this subject would give him a "balanced diet of aesthetics and mechanico-science."[43] He graduated with honors in 1922 and intermittently worked as architect and architectural consultant. He held 10 patents for various mechanical and optical inventions, and also worked as technical consultant for the *Mt. Wilson Observatory* and for Beckman Instruments. He designed mounted telescopes, for instance a Schmidt–Cassegrain telescope. For the *Griffith Planetarium*, he built a hugh plaster model of the lunar surface (on a scale of 50 feet for the diameter; cf. here fig. 65, and later also smaller-size versions (6-ft diameter) for the *Adler Planetarium* in Chicago and for *Disney World*.[44]

Particularly for Pauling, Hayward was much more than a regular illustrator. Their preserved correspondence shows that the latter's status steadily grew from a mere executor of Pauling's wishes and visions, to that of a consultant, with whom every little detail of each plate was intensely discussed, to that of a co-author and

---

[41] Dennis Flanagan to Hayward, Dec. 18, 1973, quoted in Kramer & Benjamin (undated) part II.

[42] Ibid., part III and again Heumann (2013) for excerpts from Hayward's and Pauling's correspondence with each other and with publisher William H. Freeman. Wheeler's books in geometrodynamics were also mostly published by Freeman; cf. here sec. 3.4.

[43] From a letter by Hayward written in 1950, quoted in Kramer & Benjamin (undated) part I.

[44] See http://miehana.blogspot.com/2009/07/mr.html and Kramer & Benjamin (undated) part II. Cf. also his 32-page pamphlet on *Molding and Casting* (undated), teaching the essential processes of molding and casting with molten metal, wax, and other materials.

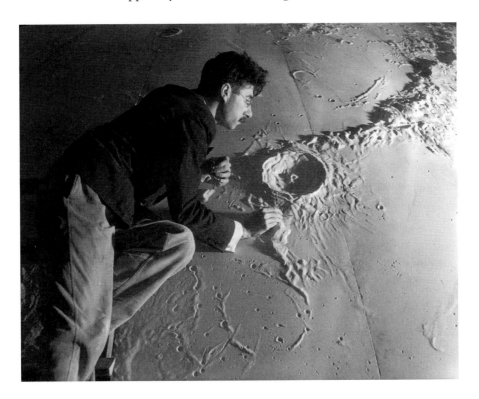

Fig. 65: Hayward working on his model of the lunar surface for the Griffith Planetarium, 1934. Los Angeles Public Library Photos, Security Pacific National Bank Collection, no. 00010364, by permission; see also http://miehana.blogspot.com/2009/07/mr.html (accessed Sept. 3, 2011).

illustrator colleague. That the flow of ideas and arguments really went both ways in a cooperative manner is neatly shown by the following exchange – taken from their correspondence in 1951. On July 19, the ever-impatient and extremely busy Pauling wrote to "Dear Roger":[45]

> Could you during the next few days make some drawings. We need to get a paper off for publication immediately, because I have learned that someone else (a Swede) is doing some closely similar work, and I think that we might as well publish our results as obtained so far.
>
> I would like to have drawings made of the structure of Na_2Cd_{11}, closely similar to those that you have already made for me in pastel. If you need to pick these up, please drop by the laboratory and get them. There is,

---

[45]Some (not all) excerpts from this correspondence are found in Kramer & Benjamin (undated) part IV. The full text of this letter is found at http://scarc.library.oregonstate.edu/coll/pauling/ bond/corr/corr152.9-lp-hayward-19510719.html (last accessed Feb. 14, 2014).

however, one difference, which I shall describe below – this involves an interchange of the six larger atoms and six of the smaller atoms.

Also, the drawings to be made now should be line drawings from which cuts can be made – that is, black and white, line drawings, with stippling or some other type of shading of the atoms.

Note how Pauling uses the first person plural to speak of their – joint – work ("We need to get a paper off"). In fact, only Pauling felt pressured about priority concerns with the unnamed Swedish competitor. One week later, on July 27, 1951, Hayward responded with a modified sketch and added the following slightly annoyed commentary:[46]

I believe that a review of the enclosed sketch for the revised figure E discloses that the octahedron cannot be placed where required if the radii and spacings are consistent with my interpretation of your directions. Furthermore if all the triacontahedra are shown completely surrounding the octahedron, the figure will be unintelligible [...]. I will reiterate that I do these figures for the pleasure involved. Such a catalogue of criticisms of drawings which you requested me to do in a hurry is not pleasant.

Pauling quickly replied on August 2. The following passage (ibid.) shows that, in this case, he really did take Hayward's suggestion.

I return the sketches on the Na_2Cd_{11} structure with my apology. I have just discovered, a couple of days ago while going over calculations with Dr. Ewing, that I had placed the large atoms in the wrong positions in the rhomb [...]. I think that this will take care of the steric difficulty that you have pointed out [...].

A more visual trace of this interplay of visual and verbal thoughts can also be provided by the following preparatory sketch for a plate that ultimately appeared in Pauling's book on *The Architecture of Molecules* in slightly altered form – proof that Hayward's handwritten suggestions were implemented.

Pauling, the "architect of molecules," and Hayward, the architect–inventor–illustrator, were a great team, no less so than Leonhart Fuchs and his artisans in Nuremberg and Strassburg had been 400 years earlier. These are certainly not singular cases. Parallels to other fruitful collaborations between scientists and illustrators do exist, though few cases have been documented as thoroughly. For anatomy, marvelous work on the interplay of great physicians such as Samuel Thomas von Soemmerring (1753–1830) and Friedrich von Esmarch (1823–1908)

[46] See http://paulingblog.wordpress.com/2008/05/01/roger-hayward-and-linus-pauling/.

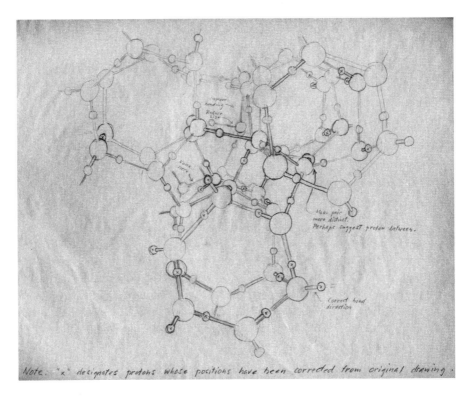

Fig. 66: Hayward's annotated pencil sketch of the structure of ice, 1964. Pauling papers, Special Collections & Archives, Oregon State University, no. 1964b4.1-ice, by permission.

with their illustrators has been published.[47]   The Kiel surgeon Friedrich von Esmarch, for instance, for decades intensely cooperated with his student, then private assistant and illustrator Ernst Kowalzig (1863–?).[48] Once von Esmarch had an interesting medical case in his hospital, he quietly asked his assistant to please make a drawing before any treatment was begun so that the change of symptoms (hopefully an improvement) afterwards could be documented. This visual documentation was also used in later medical courses.[49] He used to refer to these pictures as "his images," but the joint authorship was clear to the audience

[47] On Soemmerring and his special training of the copper engraver Christian Köck (1758–1818), see Geus (1985); on Esmarch, see Wolf & Härle (eds. 1994).

[48] On this collaboration, which led to many joint publications, see Jörn Henning Wolf in Wolf & Härle (eds. 1994), pp. 9-25.

[49] Ibid., p. 16: "wemm nal ein besonders schöner [sic] Fall vorlag, fragte mich Esmarch leise: Können Sie davon wohl eine Skizze machen, und wenn er [der Fall] malerisch war, machte ich sogar ein Gemälde in Farben."

and clear attributions in von Esmarch's many publications specified that Kowalzig had produced the plates.[50] Similar case studies on the intricate interaction of illustrators with their clients have also been published for natural history, especially botany and zoology. The versatile Italian naturalist Giovanni Antonio Scopoli (1723–88), for instance, convened intensely with his artists, who provided him with hundreds of drawings and guache paintings for his *Deliciae florae et faunae insubricae*, an ambitious inventory of plants, animals and minerals found in Northern Italy. A few of Scopoli's annotations to semi-finished drawings from the production process have been preserved and document the meticulousness with which he inspected and criticized the artists' efforts: "The outlines of the leaves and the primary lines of the calyces not yet opened must be more exact and less pitturesque. The section of the fruit and the seeds are missing."[51] When insects or other small animals had to be depicted, Scopoli was equally exacting in his instructions on how to paint these entomological plates: "Each insect must be drawn twice, seen from above and below, outlining the antennae and the mouthparts separately with a magnifying glass, together with their mandibles. Furthermore, it is necessary to draw exactly the outline of the body and of the feet alone seen from above, without colours or shadows."[52] The Nuremberg physician Christoph Jacob Trew (1695–1769), a most ardent naturalist, also heavily invested in the training of his illustrators in order to get anatomical and botanical plates of the highest quality.[53] He had met the gardener and flower painter Georg Dionysos Ehret (1708–70) in botanical courses. He regularly commissioned flower aquarels, often one or two per week, each at a compensation of one gulder – a lot of money in those days.[54] He instructed Ehret to depict the plants "according to nature," since he did not care about ornamentation, seeking veritable scientific documentation and adequacy: "Cousin, sir, please be so kind as to remind the artist that I very much look upon it as that is all be made according to Nature, for I need it not for decoration alone, but also for utility."[55] Aside from shipping to Ehret loads of high-quality

---

[50] Ibid., p. 17; for an overview, see www.worldcat.org/identities/lccn-n2008-183371.

[51] Handwritten note by Scopoli on a drawing of *Physalis peruviana*, attributed to Giuseppe Lanfranchi (1737–1800), publ. & transl. in Siviero & Violani (2006) pp. 221f. On Lanfranchi, cf. the DSI online entry.

[52] From a letter to Count Castiglioni, publ. & transl. with further textual examples, images and commentary in Siviero & Violani (2006) pp. 222f.

[53] On Trew, see Schnalke (ed. 1995), Ludwig (1998) chap. V, Nickelsen (2004) and further sources mentioned there.

[54] See Calmann (1977) or Nickelsen (2004a) pp. 34, 38ff. on Ehret and his interaction with Trew, ibid., pp. 79ff. on Trew's interaction with Haid; Charmantier (2011) pp. 394ff. and the Linnean correspondence online at linnaeus.c18.net/Letters on the Ehret-Linné friendship.

[55] See Trew to his relative, the naturalist Beurer, Dec. 22, 1731, transl. by Ann M. Hentschel and

paper, he also gave him concrete instructions on how to execute his aquarels and drawings showing all the systemic botanical markers, including the flower, fruit and pollen. "In all this I would, however, like to have bid again that Nature be most clearly brought out and wherever possible to add each time the fruit or the seed."[56] Apart from Ehret, a whole army of other trained illustrators, draftsmen and engravers also worked for Trew.[57] Like Trew, Soemmering and von Esmarch before him, the Jena-based zoologist Ernst Haeckel (1834–1919) also personally trained his illustrator, in his case the lithographer Adolf Giltsch (1852–1911), before they jointly illustrated books on skeletons of radiolaria, medusae and other see animals in the 1870s.[58] The sophisticated stylization – i.e., a move away from pedantic scientific accuracy in Giltsch's color lithographs – turned these images into veritable "artforms of nature," influential not just in marine biology but also in the visuality of the art nouveau movement which was just then gathering speed. Haeckel's radiolaria, first published in his 1887 *Report of the Exploring Voyage of the H.M.S. Challenger* and later reproduced in various semipopular books and articles, eventually even found their way into René Binet's celebrated entrance building to the Paris World's Fair of 1900.[59]

In the 20th century, another well-known example of a close cooperation between a chemist and an illustrator is Richard Earl Dickerson's joint work with Irving Geis (1908–97) on such chemical classics as Dickerson's textbooks on *The Structure and Action of Proteins* (1969), on *Hemoglobin* (1983) and on *Chemistry, Matter and the Universe* (1978). All were generously illustrated by Geis,[60] who – like Hayward – also worked for *Scientific American* and other influential journals. The advent of affordable desktop publishing programs with digital image manipulation,

---

quoted from Schnalke (ed. 1995), p. 103: "Der Herr Vetter sey so gütig und erinnere den Künstler, daß ich gar sehr darauf sehe, daß alles der Natur gemäß komme, denn ich es nicht alleine zur Zierrath, sondern auch zum Nutzen verlange."

[56]Trew to Beurer, Feb. 1732, quoted from Schnalke (ed. 1995), p. 104: "Bey allem aber will ich nochmals gebeten haben, die Natur auf das deutlichste zu exprimieren, und, wo es möglich, jedesmal die Frucht oder den Saamen beyzusetzen."

[57]On their vitae, see the biographical appendix in Ludwig (1998) and the Stuttgart database DSI. On the training of draftsmen involving the copying of masterful illustrations by senior illustrators, see, e.g., Nickelsen (2004a) p. 52. The Spanish botanist and expedition leader Gómez Ortega (1741-1818) provided detailed oral and written instructions to his illustrators: Bleichmar (2012) p. 83.

[58]On the collaboration between Haeckel and Giltsch, see Lapage (1961) pp. 29–32, and Breidbach (2006). Other fascinating examples of close interplay between zoologists and their illustrators are found in Lapage (1949, 1951, 1961), Knight (1977) and Blum (1993).

[59]On these broader cultural resonances, see Breidbach (2006) and Richards (2008).

[60]See Dickerson (1969, 1983, 1997) and Anon. (1993).

such as Adobe's *Photoshop*, has certainly increased the temptation for researchers today to produce their own graphics. The modern visualization programs *RasMol, Molscript* and *Raster3D* offer chemists sophisticated options for creating covalent bond diagrams, space-filling representations or ribbon-style diagrams of complex molecules with just a few mouse-clicks.[61] Even though science departments now frequently offer their own courses on scientific visualization and on the handling of the latest software for 3D imagery, animation or simulation, there is still an enormous need for professional help and expertise at the high end of the sector.[62] In the image-saturated discipline of medicine, the first *Department for Art in Medicine* was founded by Max Brödel (1870–1941) at *Johns Hopkins University* in Baltimore as early as 1911.[63] A second such institution for biomedical visualization was founded at the *University of Illinois* in Chicago in 1921, and dozens of others followed elsewhere.[64] An international *Association of Medical Illustrators* was founded in 1945. It issues an annual *Source Book* with examples and contact information for top-notch illustrators.[65] Numerous centers for visualization have been founded since the 1990s worldwide,[66] and experts on medical animation now obtain prestigious grants formerly reserved to scientists,[67] a sure sign of maturity of the currently booming field of interactive visualization.

6.3   Friction between scientist and illustrator

After these few examples of extraordinarily fruitful and constructive interactions between scientists and their illustrators, we cannot close this chapter without a brief look at a few less than harmonious cases. Obviously, there are in-built tensions between the two groups. Problems start with the issue of timing. It takes time to

---

[61] See Goodsell (2003*a, b*) and Francoeur & Segal (2004).

[62] See Goodsell (2003*b*) pp. 1295ff. or Lok (2011) for very recent examples of fruitful "collaboration with an artist" in chemistry. On the "combined eye of surgeon and artist" in medicine, cf. Johnson & Sainsbury (2012).

[63] See Crosby (1991) and www.hopkinsmedicine.org/medart/HistoryArchives.htm (accessed March 3, 2012) on Brödel and Baltimore.

[64] See www.ami.org/medical-illustration/graduate-programs.html (accessed March 3, 2012) and www.campusexplorer.com/colleges/major/F4193881/Medical-Communications/   66C74576/Medical-Illustration-Medical-Illustrator/.

[65] The *Medical Illustration Source Book* has been issued annually since 1987. On the current situation, cf., e.g., http://en.wikipedia.org/wiki/Medical_illustrator (accessed March 3, 2012).

[66] For instance VISUS at the University of Stuttgart in 2002: See Ertl (ed. 2010). On the global trend, cf. Friedhoff & Benzon (1989).

[67] See, e.g., Lok (2011) on the case of the biomedical animator Drew Berry at the Institute of Medical Research in Melbourne, Australia, who won a MacArthur grant in 2010, and on other interesting cases of cooperation between scientists and illustrators in the 21st century.

engrave, lithograph or carefully photograph – usually more time than is scheduled by their impatient clients. Once a plate was finished for printing, it often did not quite meet the specifications, necessitating major or minor changes, causing further delay and mounting stress. In order to minimize this friction, experienced illustrators – such as Julius Geissler (1822–1904), who headed the lithographic establishment of J. G. Bach in Leipzig after the death of its founder Johann Gottlob Bach (1809–47) – even issued leaflets of instructions for scientific authors, to ensure that their initial drawings and associated commentaries were as clear and unambiguous as possible.[68] Illustrators were likewise thoroughly instructed and provided with plenty of ready-made examples in "How to Draw" manuals.[69] Engravers likewise had handbooks that canonized and standardized standards of representation and accuracy.[70] However, the most conscientious naturalists actually hired illustrators at their own cost and carefully checked their work. Conrad Gesner (1516–65) and Charles l'Écluse (1526–1609), for instance, are both known for constantly supervising and frequently correcting their drawing and hand-colouring.[71] For the botanical, geographical and ethnological expeditions to the Spanish Indies, detailed instructions and constant supervision were not enough: unhappy with the work of the painters and draughtsmen educated in the Spanish Academy of Fine Arts, José Celestino Mutis (1732–1808) hired about 60 mostly native American illustrators who had been trained in local workshops.[72]

Even if all of these recipes are followed, things could still go amiss, though. Before the advent of chromo-lithography and other color-printing processes, plates printed with a monochrome technique were often hand-colored. Poorly paid and uneducated illuminators, often female and not infrequently children, were hired to perform these routine tasks, which nevertheless required extraordinary manual skill in the delineation and in the finding of the right color, regardless of how detailed the instructions were or the sometimes accompanying color tables. A sloppy job at this stage could still spoil an otherwise-perfect plate. As a case

---

[68]See, e.g., Geissler (1889); on Geissler and Bach, see the Stuttgart database DSI.

[69]One example for many is Keith West's guide on *How to Draw Plants*, published in 1988 (with references to similar works published earlier). Check your regional online book catalogues for titles, such as *How to draw x*, with *x* being anything from animals and landscapes to cars and wheels.

[70]See, e.g., Doherty (2012b) on William Faithorne's *The Art of Graving and Etching* from 1662.

[71]For these and other examples of micro-management of artists by naturalists, see Wilson (1994) p. 24, Jackson (2011) and Kusukawa (2012) p. 145. Despite all such efforts, anatomist Felix Platter bitterly complained about the "malevolentia" of his illustrators.

[72]See Bleichmar (2012) chap. 3, Puig-Samper (2012) and further references given there, esp. the 49 vols. *Flora de la Real Expedición Botánica del Nuevo Reino de Granada*, publ. 1783–1816, now partly online at www.rjb.csic.es/icones/mutis/paginas/ (accessed Dec. 22, 2013).

in point, the botanist Joseph Dalton Hooker (1817–1911) complained about an "incompetent colorist" who "utterly ruined" his plates of floras that had been illustrated by Walter Hood Fitch (1817–92).[73] Fitch had been trained in botanical illustration by Hooker, and when the bottleneck of hand-coloring was later closed by Fitch turning to chromo-lithography, he became one of the most prolific Victorian botanical artist.[74] Other examples of high-quality plates being ruined by bad coloring include John Gerrard Keuleman's plates for Richard Bowdler Sharpe's. *Monograph of the Alcedinidae, or Family of Kingfishers* (1868), some of which were – so the author complained in his introduction – "marred by the incapacity of the colourists."[75] For the roughly 100,000 copies needed for the 435 plates in John James Audubon's *Birds of America* (published as a series of small subscription sets 1827–38), famously printed in double elephant size (c. 26 inch by 39 inch) in order to be able to represent all birds in "the size of life," the publisher William Home Lizars (1788–1859) in Edinburgh had to hire a team of more than 50 artists who were naturally not all on the same skill level. Unwanted variations in the coloring of the plates thus occurred, and complaints about their inequality resulted. When these criticisms were forwarded to the artists and hand-colorers, the inherent tensions between illustrators and their employers exploded: in 1828, all of them went on strike because they refused Audubon's wish to have one of them fired who had been doing a "miserable daubing." In order to resolve the impasse, Lizars decided to dismiss the whole team and to assemble a new team: "a set of women [...] who I hope will manage it even better than these radical skamps."[76] Soon thereafter, Audubon shifted the whole project over to the engraver Robert Havell Jr. (1793–1878), who managed to finish the mammoth project with roughly 175 complete sets.

Quality evidently reigned over practicality where illustrations were concerned. Surveys of the botanical literature have confirmed that a surprisingly low number of high-quality illustrators served the high demand for excellent illustrators Europe-wide and dominated the market for such jobs in the early-modern period and even into the early 19th century.[77] In search of an illustrator for his anatomical (and later also botanical) plates, Albrecht von Haller (1708–77) in Berne turned to

---

[73] See Secord (2002) p. 41.

[74] See the Stuttgart database DSI and sources on Fitch mentioned therein.

[75] For more on this example, see Jackson (1975) chap. 7 and (2011) pp. 37f.

[76] For these quotes from Audubon's journals and a letter from Lizars's workshop, see Jackson (2011) pp. 42–5 and www.aradergalleries.com/works.php?id=25& wks=53 (accessed Sept. 1, 2013).

[77] See the surveys by Blunt (1950), Nissen (1966) and Nickelsen (2004a) p. 55 for specific examples of a handful of botanical illustrators working for clients all across Europe.

Christoph Jacob Trew in Nuremberg, who recommended a local copper engraver, Georg C. Lichtensteger (1700–81), who eventually did carry out several contracts for Haller.[78] It is indicative that the Swiss naturalist much preferred to deal with this Nuremberg illustrator rather than with the Berne competition, even though it was much more involved and impractical.[79]

Some of this source of friction was avoidable. The illustrator Elaine R. S. Hodges advised her collaborators that a good "working relationship between the scientist and the artist" is establishable if the scientist bears in mind to:[80]

- provide the illustrator with a detailed listing of what is to be drawn;
- indicate the size of the drawing and the technique by which it is to be depicted and later reproduced;
- obtain the technical specifications from the publisher so that the right technique for the job is chosen, to avoid later reworking;
- allow the illustrator to experiment with various angles, shadings and colors, in order to optimize the image in terms of its clarity, density of information and interpretability;
- consult with the illustrator at this stage without hesitating to provide input, yet leaving some leeway for the illustrator to optimize the job;
- frequently check for accuracy and suitability of each illustration together with the illustrator, starting with the preliminary sketches to avoid prohibitively labor-intensive revisions of finished renderings.

After all, scientific illustration – as opposed to popular or commercial artwork – "should be beautiful, and the best quality as art, but accuracy comes first. A beautiful but inaccurate drawing is useless to science. [... however,] the scientific illustrator interprets what is presented, reconstructs broken or missing parts, eliminates artifacts such as dirt, and shows layers of anatomy. No machine can replace the mind of the artist or scientist."[81] Even nowadays, when paper-based representations have yielded way to computer-aided forms of visualization on the computer screen, the basic message of this quote still rings true.

---

[78] On Lichtensteger and roughly a dozen other draftsmen and engravers working for Haller, see Gloor (1958); on Trew see above, p. 227

[79] On the importance of Nuremberg illustrators in the 17th and 18th centuries, see Ludwig (1998).

[80] These points are raised in Hodges (1989); cf. also Geissler (1889).

[81] Ibid., p. 104. Neither the 'mind' nor the 'eye' of either, I would add. On the "combined eye of surgeon and artist" in medical illustration, cf. Johnson & Sainsbury (2012); on "natural historical duets" of illustrator and naturalist in Spanish expeditions, see Bleichmar (2012) pp. 90ff.

7

One image rarely comes alone

The historiography of visual studies has shown in a plethora of cases that a historical reconstruction of research practice involving imaging techniques cannot focus on just one representational technique or on isolated images, as important as either may be for the case at hand. We inevitably find images embedded in larger sequences or more complex chains of representations of variable media and type. These chains of representations are, on the one hand, a diachronic succession of different types of images and modes. Their technical reproductions in print media, on the other hand, are intricately coupled with each other as a synchronous side-by-side of different forms of representation. Images in science are thus continual provocations – to contemporaries in terms of potential gaps, defects or possibly alternative renderings; to successors as something to overcome, possibly to improve or comment upon. In order to circumvent this constant provocation by his own images and paintings, to escape the recurrent urge to rework or repaint them, Leonardo da Vinci is known to have carefully packed up many of them, labeling one such package – tongue in cheek – thus: "Don't unpack me, if you value your freedom."[1]

The flea in a plate of the *Micrographia curiosa adiuncta observationibus circa viventia* (1691) by Filippo Bonanni (1638–1725) can be directly linked to the famous fold-out plate in Robert Hooke's *Micrographia* (1665). This becomes clear beyond doubt when you notice that Bonanni took over the idea of depicting the reflection of a window in the facetted flea eye. However, he added flea eggs not depicted in Hooke's plate, not to speak of many other major and minor variations in style. More generally speaking, typically only certain aspects of predecessor images are copied, whereas others are varied, optimized or given changed emphasis, according to the pictorial styles, expectations, goals, and image traditions of the microscopist and his audience.[2] It would thus be a mistake to analyze any scientific or technological image in artificial isolation from its context of foregoing images and from the cultural tradition in which that image arose. We will begin

---

[1] Quote from Bredekamp (2005) p. 158; cf. also Fehrenbach (1997) p. 325.

[2] On this particular example within microscopy, see Stefan Ditzen in Hessler (ed. 2006), who speaks of "stable microscopic motifs within image traditions" (*Bildtraditionen gleichbleibender mikroskopischer Bilder*). What remains the same is the subject, with strong variations in style and detailed content.

this exploration in sec. 7.1 with a short resumé of Kärin Nickelsen's surprising findings on copy relations in 18th-century botany, and then delve further into diachronic and quasi-synchronic chains of visual representations.

7.1   Nickelsen on copy relations in botanical illustrations

In a *Diplom* thesis as well as in a PhD dissertation submitted to the University of Berne in 2004, the historian and philosopher of biology Kärin Nickelsen (*1972) carefully analyzed a complex set of 137 illustrations of seven plant species taken from various botanical publications of the 18th and early 19th centuries.[3] She was able to document a total of 197 copy relations between motifs of 101 of these plates (i.e., roughly 75%), often with more than one motif per image copied from plate to plate. In the overwhelming majority of cases (165), this borrowing from old images in new ones was only partial. Merely 7 (less than 5%) were more or less complete copies of a whole plate (disregarding possible minor variations in insignificant features). The remaining 26 were classified as sharing a vague inspiration in motif, without there being any specific similarity. More particularly, among these 165 cases of more than fleeting resemblance, 17 corrected the older images, whereas 5 introduced new errors. Pragmatic reasons lay behind the remaining 143 partial copy relations. The denotations used were merely improved (in 59 instances) and the informative content was optimized (in 49 instances). In 38 of them, simplifications were made in order to reduce the necessary cognitive work for the user of the plates.[4]

All in all, these statistics and examples show that the quality, adequacy and utility of these botanical atlases and plates rather increased over the course of time. Nickelsen's finding thus explicitly contradicts frequently made complaints about the corruption of image quality by repeated copying. Such were often made in the context of textual or mathematical diagrams in codices copied during the Middle Ages, or by scholars deploring the mindless copying of Newton's figures by lesser contemporaries.[5] In 18th-century botany, it was consequently certainly *not* laziness or incompetence that motivated so many illustrators – the very best being among them – to take into account how their predecessors had depicted a

---

[3]See Nickelsen (2000) and Nickelsen & Grasshoff (2001) chap. V for the detailed comparisons, and Nickelsen (2004) for more on fig. 67, further literature and other examples. Cf. also http://penelope.unibe.ch/docuserver/compago/home_bot.html (last accessed Aug. 7, 2011).

[4]See Nickelsen (2004a) pp. 270, 279, 317 for the above quantitative data, and Nickelsen (2000) or Nickelsen & Grasshoff (2001) chap. V for examples.

[5]See Lohne (1968); cf. Nickelsen (2004a) p. 17 for her starkly contrasting findings implying an overall qualitative improvement over time.

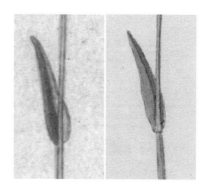

Fig. 67: Two simple examples of copy relations in botanical illustrations by Jacob Sturm (left) and Johann Philipp Sandberger (right). The left pair, depicting the fruit of the meadow saffron, is classified as a full copy because all essential details were taken over (despite the minor variation in placement of the two details at the bottom right). The right pair, detailing the leaf of *Anthoxanthum odoratum*, commonly known as sweet vernal grass, constitutes a partial copy because Sandberger drew a closed sheath at the node, with a narrower blade right above this point. Both examples are from Nickelsen (2006) pp. 209–10. By permission of Springer Science + Business Media, B.V.

plant species that they were supposed to illustrate. The aim for many of them was the very best image achievable by the technical and pecuniary means available to them. This legitimate goal actually even required them to take into account older efforts to do so, as their starting platform. Rather than blindly copying, these experienced illustrators very carefully analyzed older plates, and often compared several of them, only taking over whatever they deemed suitable and correct. At the same time, they more or less freely altered whatever they thought was inappropriate, suboptimal and, of course, rejected or suppressed whatever was plainly wrong. Because these new images, in turn, became the comparative basis for later efforts to depict these same plants, a complicated network of predecessor–successor relations is formed, which can be depicted like a *stemma* of philological dependence relations between different texts. See the diagram in fig. 68 for one example relating to a toxic plant commonly known as autumn crocus, meadow saffron or naked lady (in German also known as *Herbstzeitlose*) as depicted from Tournefort 1700 to Hayne 1817.

　　This finding of such intricate nets of dependencies induced by copy relations is so surprising at first because it goes against the explicit Enlightenment rhetoric of *sapere aude* in so many of these illustrated works. Time and again, the author and his illustrators assured that "Here everything is from nature only."[6] This rhetoric

---

[6]Thus, for instance, John Hill in the unpaginated preface to his *Vegetable System* (1750–75), quoted

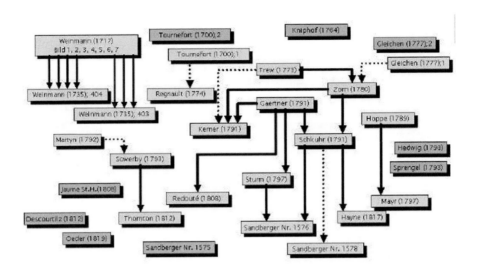

Fig. 68: The network of copy relations for different plates depicting the species *colchicum autumnale* (meadow saffron), 1700–1817. The dotted lines stand for uncertain cases. Isolated boxes signify botanical plates depicting the species for which no copy relations could be traced. From Nickelsen (2004*a*) p. 273, by courtesy. Also available online with further commentary at http://philosci24.unibe.ch/compago/tree_bot/tree_2.html (accessed Aug. 6, 2011).

of monomaniac authenticity was engraved so deeply in the mind of the reader and user of these plates that it was simply ignored how strongly the scientific practice deviated from this ideal world of isolated geniuses supposedly never influenced by anything happening around them. Whereas users and readers were left under the impression that each of these plates had been produced directly from the sample depicted, possibly even on location where they grew, *de facto* these plates were all done in a studio as the final result of lengthy, active image construals. Detailed on-the-spot examinations (wherever possible) were combined with a comparative study of older representations, intense discussions between illustrator and author (see chap. 6), and repeated efforts to come to grips with the species (often generating many different drafts and versions prior to the printing stage).

  Because this issue of copy relations is of such obvious importance in scientific and technological practice, a small group of researchers at the University of Berne, where Nickelsen was assistant professor, developed some software, called COMPAGO, to conduct a systematic comparison of images from various sources

from Nickelsen (2004*a*) p. 259. However, the anatomical plates of Diderot's and d'Alembert's *Encyclopédie* appearing from 1777 were all drawn from the best available precursors.

and hands.[7] The head of the group, historian and philosopher of science Gerd Grasshoff (*1957), contributed general ideas about the epistemology of scientific images interpreted as models; and the research student Hans-Christoph Liess (*1972) added findings on copy relations in medieval astronomical diagrams. Copy relations between those images and diagrams are found to be even more frequently reducible to a very small number of basic types.[8] I hope that this software and the attendant comparative historiographic approach will continue to be applied to many other fields in which illustrations figure prominently, so that we will eventually get broader comparisons of such copy relations.

7.2   Diachronic succession of printing techniques

Most of the examples treated by Nickelsen concern minute variations in visual representation by one and the same medium: hand-colored drawings and engravings. New techniques of visual representation were repeatedly invented, and each led to drastic changes in the resulting images.[9] As Martin Rudwick argues: "Technical advances in illustration might be said to have played a part in the history of palaeontology similar to that of improvements in instrumentation in the physical sciences."[10]

Engravers speak of "translating" a motif taken from a drawing, a painting or a photograph into a line engraving. This metaphor is well chosen insofar as something is inevitably lost in a literal textual translation from one language to another. Likewise for the 'translation' of a gray-tone shading into a line pattern. Not even cross-hatching or other tricks of the engraver can evoke a visual equivalent of a certain shading. On top of this merely technical limitation there is the subjective element, for instance the style of the engraver somehow worked into the image. Sometimes the user of such a plate will not be satisfied with it. We have many examples of scientists drawing into printed plates, in an

---

[7]On the COMPAGO software, developed with funds by the Swiss Nationalfond and freely downloadable, see Graßhoff, Liess & Nickelsen (2001) and philosci24.unibe.ch/botany/index.html on the program, as well as http://philosci24.unibe.ch/botany/documents.html on the digitized botanical source materials (both webpages last accessed Aug. 6, 2011).

[8]Grasshoff and Nickelsen have since left Berne. Image types and copy relations in astronomical texts and manuscripts 1450–1650 are currently being studied at the *Department of History and Philosophy of Science at the University of Cambridge*, cf. www.hps.cam.ac.uk/diagrams/.

[9]For literature on these techniques of printmaking and picture printing, see Bridson & Wakeman (1984); for examples, cf. Gascoigne (1986); on this interplay in botany, see Bridson & Wendel (1986); in zoology and anatomy, Blum (1993), Bridson & White (1990), Choulant (1920), Herrlinger & Putscher (1972) and Thornton & Reeves (1983).

[10]Rudwick (1976) p. 9.

effort to improve them, or to record what is wrong with them.[11] In other cases, the lack of satisfaction with existing re-presentations leads naturalists, scientists and technicians alike repeatedly to try to capture it better. Charles Piazzi Smyth is a good example of such perfectionism. He often switched the recording technique (e.g., from line drawing to watercolor to pastel painting to photography) to record the phenomenon at hand, be it the rain spectrum or clouds.[12] Nor is photography, or, in our day, a digital image obtained by a charge-coupled device (CCD) or by a black-boxed digital camera, an ultimate, inalterable end in this endless chain of re-re-presentations.[13] For photographs, either manual retouching or modern image-processing programs are used to improve their overall quality. For astronomical CCD images, Michael Lynch and Samuel Y. Edgerton, Jr., documented the following options:[14]

- removing electronic bias causing an uneven background;
- patching up out-of-focus areas;
- improving the overall contrast and adapting the gray scale;
- removing whole rows of burnt-out pixels where the CCDs of the camera did not work at all, thus producing a white stripe;
- removing cosmic-ray traces (correctly recorded but unwanted background noise for the investigator's purposes);
- adding small markers, arrows or cursor boxes to highlight important segments of an image, etc.

All of the above steps would be considered totally unproblematic by astrophysicists, as none of them really changes the informative content of the images involved. These routine steps all just help to bring out the latter most clearly and unambiguously. It would be considered comparable to amplifying a weak acoustic signal, perhaps also as "perfecting reality," but not as creating or faking it.[15]

7.3   Near-synchronous chains of representation

All chains of visual representations in science and technology looked at so far in this chapter are diachronous – various types of visual representation, i.e., rough

---

[11]See Kusukawa (2012) pp. 156ff. on Gessner improving upon Mattioli by drawing in his copy of Mattioli's book, or Rudwick (1997, 2005) pp. 287f. on Cuvier's creative re-use.

[12]See, e.g., Warner (1983), Brück (1988) and Hentschel (2002).

[13]See Lynch & Edgerton (1988) on the clean-up of raw data in astrophysics, using CCD images.

[14]See Lynch & Edgerton (1988); cf. also Elkins (1995) pp. 557ff.

[15]The last quote is from Elkins (1995) p. 558, who links this to a Kantian meaning of aesthetics.

sketches, initial and reworked drawings, final drawings, lithographs or copper plates, finally prints and possibly later revised versions of them follow in steps separated by weeks, sometimes months or years. Their sequence could thus be construed as examples of scientific progress. That visual representations are eagerly improved as soon as new and better techniques of representation become available does not come as a surprise. But it is not yet the full story about Latour's cascade of scientific images. Even within very short timespans, we find scientists and technologists producing much more than just one (allegedly optimal) image. They produce several, often to be used quasi-simultaneously, with frequent alternating back and forth between the different modes of visual representation. Each of them might have certain advantages, but at the cost of unavoidable disadvantages elsewhere, so constant virtuosic switching is the only way to avoid the muddles one would get into by following only one of these representation modes. Examples abound.[16] To give only one example here: STS scholar Tom Schilling has interviewed and accompanied geologists, exploration company officials and geostatistics experts in Alaska, Wyoming and British Columbia during their field trips for mineral exploration. He observed plenty of different techniques or recording, visualizing and simulating used side by side, and with goals ranging from exploration and orientation to finally convincing themselves and others whether systematic mining is economically feasible or not at a specific site. Among the visual techniques used in this highly specialized area of geological exploration, we find traditional geological mapping conjointly with schematic modeling, serial aeromagnetic scanning, geostatistical analysis in three-dimensional units, or voxels, whose size depends on the area to be screened, the available computer space and the required spatial resolution.[17]

Pragmatic philosophers of science, such as Ronald Giere, Mauricio Suárez and William Wimsatt, have pointed out that scientists frequently depict the objects of their study in more than one representation. None of these depictions can grasp everything, all of them are perspectival not only in the literal sense of only showing one side of the object, but even more so in terms of the visual conventions of what to depict, what to leave out, what to emphasize and what to suppress. Nelson Goodman (1906–98) and Catherine Z. Elgin have spoken of 'exemplification,' i.e., of emphasizing the highlighting of certain features of scientific objects at

---

[16]See, e.g., Starikova (2010) on different ways of presenting mathematical objects, Kaiser (2000) on Feynman graphs as an alternative to algebraic quantum electrodynamics or Crosland (1962), Klein (2001) and Rocke (2010) on alternative graphs in chemistry.

[17]On the details of this example see Schilling (2013) and references given there.

the price of others that are downplayed or completely ignored.[18]   A case in
point: The biologist Stephen Harrison uses at least ten different forms of visual
representation to answer the question he posed in his Harvey Lecture: "What
do viruses look like?"[19] There is accordingly no unique answer to this question.
– Science is pluralistic, nowhere more so than in the "thicket of representation"
(Wimsatt) that it provides of its objects. As a flip-side, there is no 'truth' in
scientific images, but rather pragmatic adequacy or iconic accuracy, i.e., a greater
or lesser degree of captured properties or *relata*. Scientific re-presentation is a
(careful) selection, a partial homomorphism, not an all-or-nothing isomorphism.
I thus deviate from Laura Perini and other philosophers who try to construe
scientific images as "having the capacity to bear truth" by representing properties
of the real world by virtue of an allegedly "isomorphic relation between symbol
and referent."[20]

In his studies on electron microscopy in Halle, Falk Müller has shown us that
actors such as Heinz Bethge (1919–2001) practiced two fundamentally different
processes of preparing samples for electron microscopy side by side throughout
the 1960s and 1970s.[21] Both continued to yield new insights into those indefati-
gable NaCl crystals and their growth as examined by the researchers in Halle.

For images produced by the advanced research technique of raster tunnel
microscopy (RTM), Jochen Hennig has shown how all of these 'images' – in
fact artificially generated visual constructs based on complicated atomic measurements – have to be seen against a two-dimensional grid of image types and seeing
traditions in order to "make sense." At all times, all such generated images are
construed against a background of available representation types, and they only
work within a horizon of ways of seeing (and interpreting) objects, patterns and
processes that is more or less accepted, often not even consciously registered at
the time.[22] For cell physiology, bacteriology and biochemistry, Jane Maienschein,
Thomas Schlich and Alberto Cambrosio have examined the intricate interplay of
photography and drawing.[23] Maienschein analyses the use of images by the Amer-

---

[18]See, resp., Giere (2010), Suarez (2003), Wimsatt (1990) and Elgin (2004) on "true enough." Cf.
Goodman (1969), Files (1996) and Perini (2004) pp. 38ff., (2005) on the hotly contested relevance
of resemblance as a precondition for representation.

[19]See Harrison (1991); similarly Wimsatt (1990) on six different representations of genes.

[20]Perini (2004) p. 46 and (2005) p. 262 and Wimsatt (1990) on the "thicket of representation."

[21]For details of the processes of imprinting and of gold-covering (*Abdruck- und Golddekorationsver-
fahren*), see Falk Müller in Hessler (ed. 2006) and Müller's habilitation thesis (in prep.).

[22]See Hennig (2005, 2011) and Hennig in Hessler (ed. 2006).

[23]See Maienschein (1991), Schlich (2000) as well as Schlich in Rheinberger et al. (ed. 1997)
pp. 165ff., and Cambrosio, Jacobi & Keating (2005), respectively.

ican zoologist and geneticist Edmund Beecher Wilson (1856–1939) as a sequence of visualizations that starts out from photographs that are more or less retouched in order to enhance contrast, then leading to simplified drawings in gradually increasing diagram style. Altogether, we thus get a shift from the initial presentation of data to their more schematic re-presentation, re-re-presentation, etc. While this sounds like a very linear shift again, the majority of case studies show an intricate and highly nonlinear back and forth between all kinds of representational forms, devices and techniques. For instance, the historian of medicine Thomas Schlich has shown how the bacteriologist Robert Koch managed to link cause and disease in the laboratory by systematically superimposing iconic and functional representations of bacteria.[24] Scientists, technologists and medical researchers are opportunists and happily switch back and forth between sketches, diagrams and photographs, between drawings, plates and atlases, in our day also happily incorporating 3D animations, computer simulations, films (cf. sec. 7.4).

What I recommend be done much more often than currently is a systematic way of protocolling the sequences of different visual representations in their cognitive context as they follow each other in research practice. David Gooding has attempted to do that for Faraday's research in electromagnetism; Alberto Cambrosio and his co-workers as well as Annamaria Carusi have followed this up in various fields of biomedical research,[25] but we would need far more examples perhaps to see patterns in this maze of long chains of representations. Only then could we say whether the sequencing as described by Jane Maienschein is actually typical (a) for cell physiology at the time of E. B. Wilson, (b) at other times as well, (c) for biomedical research altogether or (d) even more broadly for other areas of research. Only then could we hope to see similarities and/or differences between these various fields, and only then would we begin to understand whether specific types of visual representations are typical for certain stages of research.

7.4   Cinematographic images and science films

Video, celluloid film, multimedia and digital simulation are at the very heart of visual studies.[26] As we will see, moving images have been *desiderata* of many scientists and engineers, well before the invention of the kinematograph. It has been on the wish-list of observers and experimenters for a long time to escape the limitations of static representations, depicting only one stage in an ongoing

---

[24]See Schlich (2000) and here pp. 53 and color pl. XIII for further details.

[25]See Gooding (1990, 2004, 2006); Cambrosio et al. (1993, 2005) and Carusi (2008, 2011, 2012).

[26]See the references mentioned in the introduction on pp. 61ff.

process. This section will start with a brief discussion of examples of this urge to make scientific iconography dynamic and to develop representations conveying processes. Our first inventor of film *avant la lettre* will be Leonardo da Vinci, who repeatedly felt this urge and invented graphic means to satisfy it. We will then fast-forward – to use a film metaphor – to so-called chronophotography, i.e., early efforts to capture dynamic processes by sequences of photographs taken at very short, regular intervals of time. This part of the story already exhibits a clear transfer of techniques: from astronomy (Janssen) to physiology (Marey). With ultrafast photography, invented to record extremely rapid processes, we have another transfer into areas such as ballistics, aerodynamics, and hydrodynamics. By then, turning to early cinematography in fields such as cell biology (Comandon), Brownian motion (Perrin), convective cells (Bénard) and capillarity (Ollivier), we will transgress the area of clever substitutes for films in areas of research where actual filming was not possible and enter the world of the Lumière brothers and early cinema. Film cameras and their associated projection apparatus[27] made it possible, of course, to use the new medium of film directly, both as a research instrument and as a new way to present science and technology.

7.4.1   Some early precursors to motion pictures

The direct visual representation of motion occurred relatively late in the history of humanity, even though this desire may well have been there quite early on. For a long time, there was simply no direct technique of visual representation and recording to capture fleeting moments of rapid processes or quickly changing objects. Pharaonic reliefs and medieval paintings often spatially combined tem-porally sequential events into a single big image. It was thus taken to be a kind of narrative, chronicling stages in a pharaoh's life, for instance, or of Jesus – a valu-able device especially in ages when the majority of people on Earth were illiterate. Such images might be counted as efforts to represent time, or temporal sequence, translated into relations of spatial proximity, but not yet of motion itself. Drawings by the artist-engineer Leonardo da Vinci are among the first clear attempts to do so. The art historian Martin Kemp has pointed out that Leonardo's notebooks are "filled with sketches of small figures in motion."[28] In his marvelously illustrated book on Leonardo's *Experience, Experiment and Design*, Kemp has assembled some

---

[27] On Edison's invention of the kinetoscope, first patented in 1888, see Carlson & Gorman (1990) and Braun (1992) p. 189. On the later cinematograph and early film history, cf., e.g., Marey (1902) pp. 326ff., Weiser (1919) pp. 16ff., Martinet (ed. 1994), Curtis (2005, 2009) and Wellmann (2011).

[28] Kemp (2007) pp. 90f.

of them into strings of sequential drawings in the manner of a thumbnail movie. The illusion of motion is created by quickly flipping through a succession of images capturing different stages of rapid activities such as stepping up and down from a pedestal or driving a stick into the ground with a mallet. One can likewise consider Leonardo's studies of surface ripples on water disturbed by a stone cast into it, or his studies of the flapping of bird's wings in the air, horses and other animals in rapid motion, or the trajectories of grenade fire from mortars.[29] Kemp convincingly argued that such images were "theory machines" for Leonardo. They helped him understand the process at hand, be it the free fall of a cannonball or the flow of water around an obstacle. This also means that they were indeed much more than mere depictions or illustrations, since they carried along rather implicit classifications of their parts and prototheories of the operating forces. Thus, for instance, his trajectories of mortar grenades are no longer triangular, as they were sometimes schematically depicted in impetus-theory-inflected texts. In Leonardo's drawing of a bombardment from c. 1504,[30] the trajectories are continuously curved, thus intuitively prefiguring insights into ballistics that we only connect with Galileo in the 17th century. Among the visual techniques of representation used by Leonardo to represent rapid motion, we find (i) series of overlapping contours indicating the position of an object at various moments superimposed in one image, (ii) the use of dashed lines to suggest rapid displacement, familiar to us from comic strips, and (iii) cartoon-style sequential imaging in quasi-cinematographic mode. Leonardo was also the first person to notice that rapidly moving colored objects (like red-hot coals in the dark, for instance) seem to color their path, since the human eye reacts too slowly to separate this trajectory into its constituent images. Johannes Segner (1704–77) estimated the minimal time needed for a sense impression to register in the human mind to be 0.1 s; Chevalier Patrice d'Arcy (1725–79) estimated 0.15 s, which was considered too high by the Belgian physiologist and physicist Joseph Plateau.[31]

Aside from the representational techniques already mentioned, there were other options as well:

[29] See again Kemp (2007) pp. 117, 145ff., 161. I refrain from reproducing any of these well-known images here.

[30] Kemp (2007) p. 161, Windsor Castle Royal Collection, available at www.royalcollection. org.uk/search?Search=12275.

[31] See Plateau (1829/30); on the latter's life and work, see Mensbrugghe (1885) and here p. 156; cf. also http://mhsgent.ugent.be/engl-plat3.html. Thomas Young (1807*b*) vol. 1, p. 357 quoted a broader range of 0.01–0.5 s, depending on the physical conditions of seeing.

(i)   the blurred image (especially of rotating wheels or disks in motion);

(ii) a superposition of several images of one and the same object in various stages of motion, typically only drawn in outline so that the changes in place or configuration become clear;

(iii) the choice of bodily poses indicating fast dynamics (like pulling strings or pointing fingers); and, finally,

(iv) sequential images drawn either side by side (typically from left to right or from top to bottom in Occidental sources) or printed on subsequent pages, sometimes even designed to be speedily thumb-flipped.

The last two suboptions in (iv) were chosen, for instance, by Christoph Scheiner and Galileo Galilei, respectively, in their publications on sunspots soon after the invention of the astronomical telescope. Galileo's book on sunspots had 48 full-size plates showing his observations between June and July 1612, for the attentive reader to run his thumb through the page edges of this part of the book and virtually 'see' the sunspots moving across the solar disk.[32] His adversary, the Jesuit astronomer Scheiner, preferred much smaller depictions of these sunspots, all of which he deemed to be but "apparent," since no marks were supposed to exist on a perfect supralunar body like the Sun. Consequently, he squeezed all 40 depictions of the solar disk as observed on various days between October 23 and December 20, 1611 into the lower half of one large plate (cf. fig. 69).[33]

The other options listed above, especially (iii), were much simpler and were consequently preferred for applications where the details of the object depicted did not matter too much and rough outlines or indications of directions of motion sufficed. Thus, the two Weber brothers Ernst Heinrich and Wilhelm Eduard, for instance, chose superpositions of profile drawings in their book on human locomotion of 1836[34] – incidentally, the first strictly calculated scientific images, not only in human physiology.

When Daguerreotypes and Talbotypes came into usage in 1839, the situation did not automatically improve. On one of Daguerre's early plates showing a Parisian street, all houses and other static objects are seen quite clearly, but no

[32]See Galilei (1613) pp. 57–96. In the appendix to this book (pp. 151–9), Galileo inserted a similarly cinematographic sequence of Jupiter's satellites as observed in March and April of 1613.

[33]See Scheiner (1612). On Scheiner's life, work and visuality, see here p. 113. On the contemporary controversy over the interpretation of sunspots, and on the differences in representational styles, see van Helden & Reeves (eds. 2010) and Bredekamp (2007) chaps. 7–9.

[34]See Weber & Weber (1836). I owe this reference to Canales (2011), who also gives a nice survey of other early efforts to display motion directly.

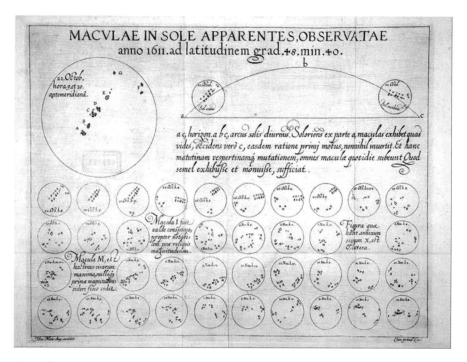

Fig. 69: "Apparent sunspots" observed between October and December 1611 by the Jesuit Christoph Scheiner (note Scheiner's reservations!). Copper engraving by Alexander Mair based on Scheiner's drawings, from Scheiner (1612) unnumbered plate.

moving objects are visible because the exposure time for such plates was initially several minutes. The only human recognizable on this plate is a gentleman who had stopped to have his shoes cleaned and therefore had stayed stationary at one spot long enough to be photographically registered.[35] Of course, one could mount several such static photographs on devices such as the phenakistiscope, invented in 1832 by the Belgian physiologist and physicist Joseph Plateau (1801–84) or the zoetrope, invented in 1834 by the British schoolmaster William George Horner (1786–1837).[36] But for several decades these gadgets were only used for entertainment and amusement, and most notably *never* with photographs, only with redrawn or printed versions.[37]

[35] For this and other examples of early photography and kinetic visualization, see Newhall (1944), Canales (2006) and Chitra Ramalingam in Brusius et al. (eds. 2013) pp. 245ff.

[36] On Plateau's life and visuality, see here p. 156; on these two gadgets, see here p. 20.

[37] On the failure of these devices when used with contemporary photographs, see Canales (2011). Excellent animated examples are at the following website (accessed Feb. 14, 2014): courses.ncssm.edu/gallery/collections/toys/opticaltoys.htm

It is interesting to note here that when the French philosopher Henri Bergson (1859–1941) coined the term *méthode cinématographique*, he was well aware of these century-long precursors to cinema. He was acquainted with that urge among researchers to go beyond static depiction of rapid processes and changing objects. For Bergson, the cinematographic method included all practices in which sequential images were supposed to represent movement or change over time. Ironically, he believed that this aim was illusory. So, against the trend of the early 20th century of upcoming motion pictures, he advised contemporary scientists and technologists to "set the cinematographic method aside" and to search for a "second kind of [intuitive] knowledge" in a "philosophy of life" and "creative evolution" – to no avail, as we now know.[38]

7.4.2   Transit of Venus and the photographic revolver

In 1874, a rare event occurred – a transit of Venus across the solar disk as seen from the Earth. The precise timing of this transit, which occurs only twice every 121 years, was important, as it allows an exact determination of the distance between the Sun and Earth. Astronomers from all over the world set out to determine this transit time. Various expeditions were organized to remote places on Earth where this Venus transit was observable in 1874. A French astronomer, Pierre Jules César Janssen (1824–1907), was discontent with conventional techniques, since they all depended on subjective estimates of the precise moment when Venus entered the field in front of the solar disk, something not easy to determine given the extreme differences in brightness and size between these two objects. Wanting to objectify and standardize these observations, Janssen invented what he dubbed a *revolver astronomique*. Like a regular revolver with a magazine carrying several (typically six or more) bullets that can be fired in short intervals because the magazine quickly rotates on after each shot, Janssen's photographic revolver had a circular photographic plate driven by clockwork that rotated to the next position after each photographic image was taken. This way, Janssen could take 48 exposures in 1.5-second intervals within the 72 seconds during the crucial phase of Venus's passage on December 8, 1874 in Japan (see fig. 70 and color pl. XII). The exact instant of contact could thus be determined more precisely than by observation, since finely resolved daguerreotypes allow close microscopic inspection of the contact area long after the fleeting event.[39]

---

[38]I owe this indirect link to Bergson and his philosophy, a prime example of an icono-

Pl. 1: Detail of a Roman mural from the villa of P. Fannius Synistor at Boscoreale (near Pompeii), c. 30–40 B.C., now preserved at the Metropolitan Museum of Art, New York. The buildings are depicted in the so-called second style, exhibiting perspectival overlaps and occlusions, foreshortening, oblique shading, but no central vanishing point. Photograph by Alethe 2007, Wikimedia Commons.

Pl. 2: Uroscopic analysis. Hand-colored woodcut from Pinder (1506), Wellcome Library, London M0007286, depicting a physician demonstrating the method to a student. They are surrounded by 20 urine glasses (each painted in a different tint) with abbreviated captions for the different diagnoses. On this iconographic tradition rooted in 15th c. manuscripts, see Sudhoff & Singer in Ketham (1491*b*), pp. 90–2, 100 and pls. I–II. For a variant in a manuscript by Gill de Corbeil (1165–1213), royal physician to Philippe-Auguste, King of France, see Armstrong (2007) p. 386.

Pl. 3 (opposite): Technical drawings in English shipwrightry. Top: two Elisabethan shipwrights at work, copying dimensions of a ship to be built from a floor and elevation plan by means of a compass. Notice the careful effort by the maker of this drawing to obey the perspectival rules; he even marked the vanishing points of the floor, side walls and ceiling in the center of his image. Middle left: visual analogy of a ship's hull with a fish. Both from the *Fragments of Ancient English Shipwrightry*, a collection of drawings, texts and tables partly attributed to Matthew Baker of c. 1586; Pepys Library, Magdalene College, Cambridge MS 2820, fols. 8 and 24, and reproduced with further commentary in Baynes & Pugh (1981) pp. 70ff. Cf. also Johnston (1994) chap. 2 on Baker and his artisanal and mathematical contexts. Middle right & bottom: plans of the 14-gun sloop HMS Atalanta, which had a maximum breadth of 26 feet and 9 inches and was launched in August 1775. Black lines are chosen for the outline, red for inside fittings and stabilizers. The superimposed lines show the change in the ship's profile over its length and were used for making enlarged templates in 1:1 format. Courtesy of the National Maritime Museum, Greenwich, London (no. J4428).

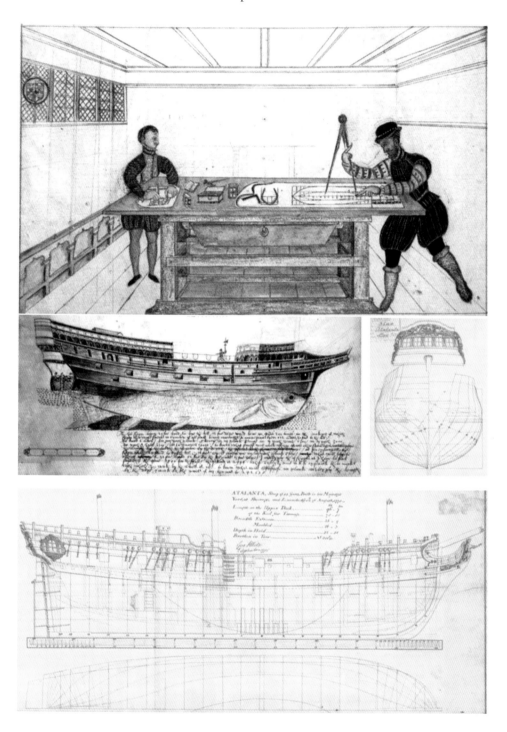

Pl. 3: Technical drawings in English shipwrightry, c. 1586 and 1775.

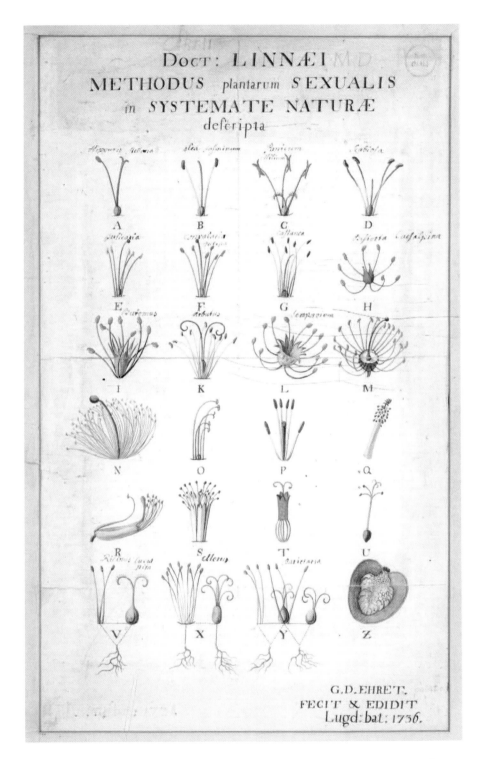

Pl. 4: Georg Ehret's illustration (1770–80) for Linné's sexual classificatory system in botany. Cryptogams comprised the last of 24 classes. It was first published in Ehret (1736) and republished in Linné's *Genera Plantarum*, 1737. This original hand-colored proof with penciled corrections is kept in the Natural History Museum, London. Cf. Calmann (1977) pp. 44ff. and www.nhm.ac.uk/natureplus/community/library/blog/2011/04/01/item-of-the-month-no-8-april-2011–georg-ehrets-original-drawing-to-illustrate-linnaeus-sexual-system-of-plants.

Pl. 5: Color pyramid by Johann Heinrich Lambert. Blue, yellow and red are at the corners of a triangular base, with nuances of whiteness rising toward the tip of the pyramid. Cf. here sec. 4.1.2 for further commentary. From Lambert (1772d) pl.

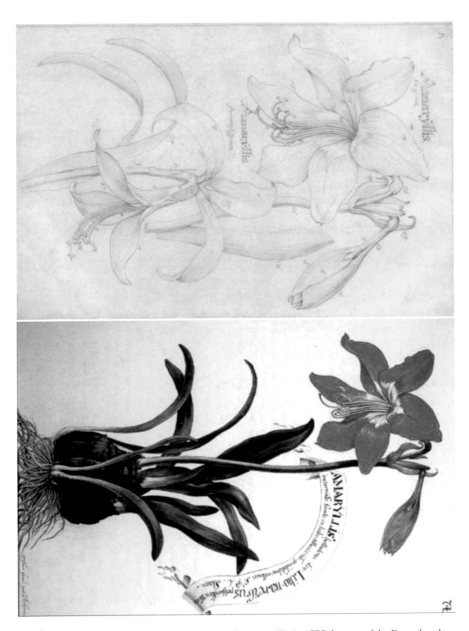

Pl. 6: Graphite sketch and hand-colored print of an amaryllis (c. 1775) by one of the Bauer brothers Joseph, Franz or Ferdinand. Naturhistorisches Museum, Vienna, by permission. The numbers on the sketch refer to the color scale (see here color pl. 7) that all the Bauer brothers used to decode colors. Austrian National Library, Liechtenstein: Princely Collections, Vaduz-Vienna; cf. Lack & Ibáñez (1997) p. 93 & pl. 318.

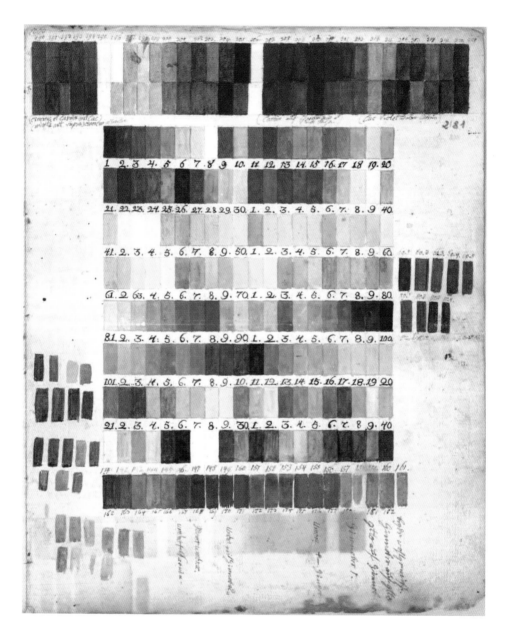

Pl. 7: Color chart used by the botanical illustrators Josef, Franz and Ferdinand Bauer, hand-painted by Thaddäus Haenke, Czech naturalist and member of Malaspina's expedition, in watercolors, c. 1775. ©Archivo del Real Jardín Botánico (Div. VI-H, 3, 2, f. 281), CSIC, Madrid. When drawing a specimen on a botanical excursion, they only noted the color code and postponed the actual painting for later. Cf. also Lack & Ibáñez (1997) pl. 317 and Mabberley & San Pino (2012).

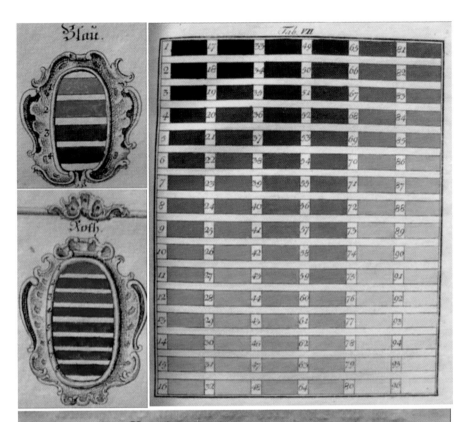

Pl. 8: Color coding schemes. Top and middle left: the color fields blue and red from Jacob Christian Schäffer's *Entwurf einer allgemeinen Farbenverein* (1769). Württembergische Landesbibliothek. Top right: a rational color code by Christian Friedrich Prange, charted horizontally for variations in hue and vertically with increasing brightness. Below: application of Prange's color nomenclature in the description of insects, from Christian Friedrich Prange's *Farbenlexicon* (1782) p. 411 and pl. VII.

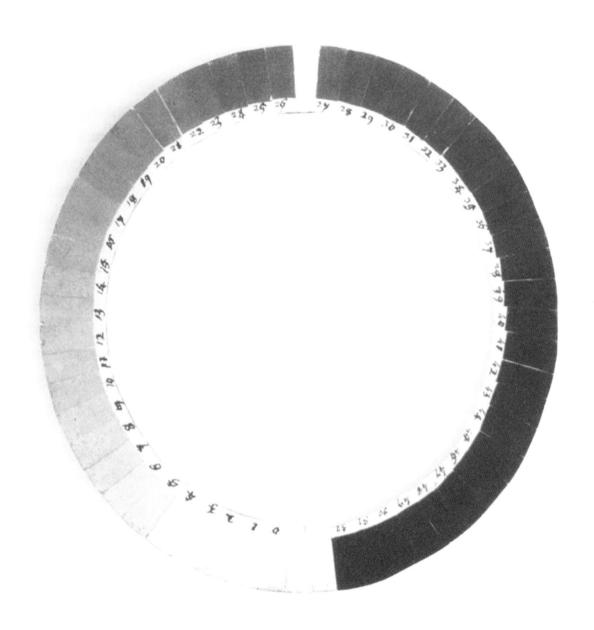

Pl. 9: Horace Bénédict de Saussure's most refined cyanometer, c. 1789. This original cyanometer with 52 shades of blue is preserved at the Musée d'Histoire des Sciences, Geneva. From the top of Mont Blanc, de Saussure recorded grade no. 39 (cf. here sec. 11.1).

Pl. 10: *Trompe l'oeil* painting of the mineral cabinet of Alexandre Isidore Leroy de Barde (1777–1828), watercolor and gouache 1813, exhibited in the 1817 Paris Salon, now in the Paris Louvre. ⓒ bpk – Bildagentur für Kunst, Kultur und Geschichte, Berlin.

Pl. 11: Nasmyth's large steam hammer at work on a large iron bar inside a foundry works, 1871. Oil painting by James Nasmyth on wood, image size 400 mm × 500 mm, possibly exhibited at a Conversazione of the *Institution of Civil Engineers* on June 6, 1871, as item 46. If so, it was probably painted at their request. Reproduced with permission from the Science Museum/Science & Society Picture Library (no. 10276175).

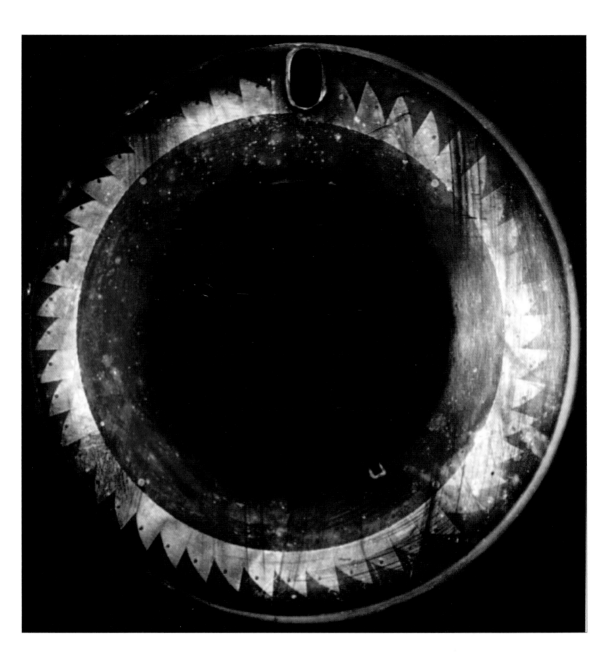

Pl. 12: Janssen's serial daguerreotype recording of the transit of Venus, showing 48 exposures (each 1.5 × 1 cm in size) made within the 72 seconds of the crucial phase of Venus's passage on December 8, 1874 in Japan. Cf. here sec. 7.4.2 for details on the photographic revolver with which this image was taken.

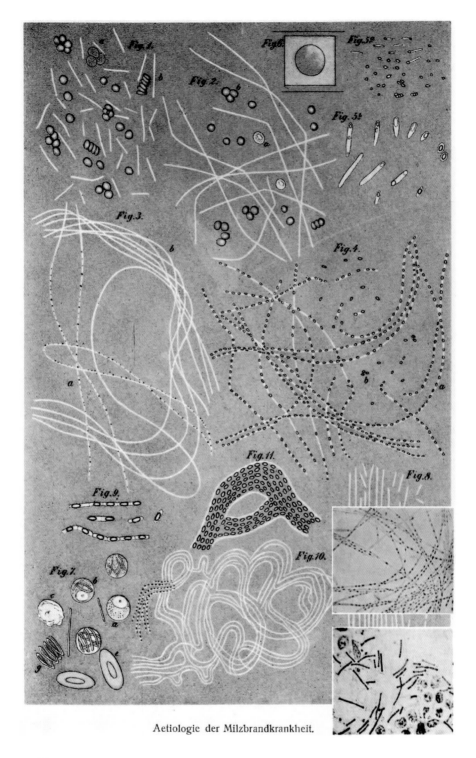

Aetiologie der Milzbrandkrankheit.

Pl. 13: Chromolithograph from Robert Koch (1876) pl. 11, drawn by the bacteriologist, depicting the life cycle of anthrax bacillae, compared against two inset photographs of the same bacterium (at the bottom right), from Koch (1877) pl. 3, figs. 7 and 9.

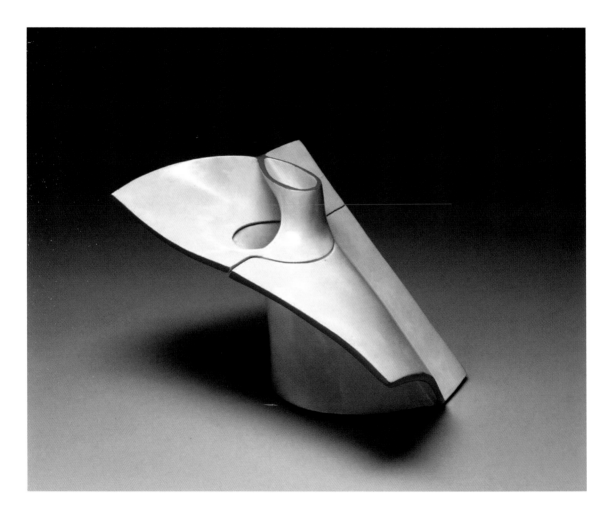

Pl. 14: Model by Alexander Crum Brown of a half-twist mathematical surface, featuring a non-Euclidean so-called Klein bottle, c. 1900. Reproduced with permission from the Science Museum/Science & Society Picture Library (no. 10314737).

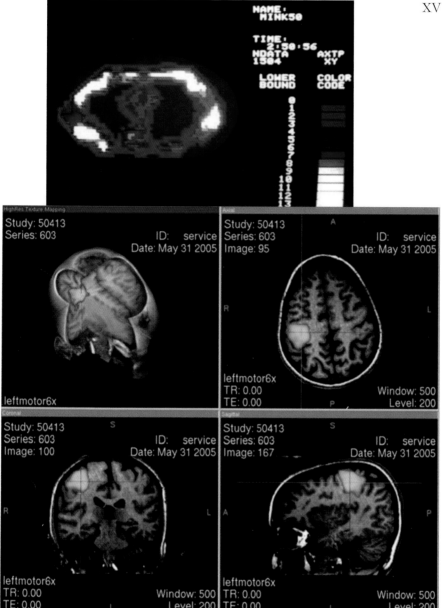

Pl. 15: Top: Computer color display of the earliest NMR of the chest of Damadian's research assistant, Lawrence Minkoff, taken in July 1977. White and red indicate bones, dark blue the air-filled lungs and body fat. The image is quite pixeled because of the low resolution of this prototype NMR scanner. From Damadian, Goldsmith & Minkoff (1977) pl., by permission of Physiological Chemistry + Physics + Medical NMR and FONAR Corp., Melville. Below: a functional magnetic resonance image (FMRI) of a human brain as the patient is tapping his fingers. The one region in bright red and yellow is functionally related to the kinaesthetic task of finger-tapping and functionally identifiable because of higher oxygen intake during this localized brain activity. Taken by Martin Wittle in 2005, from http://de.wikipedia.org/wiki/FMRI.

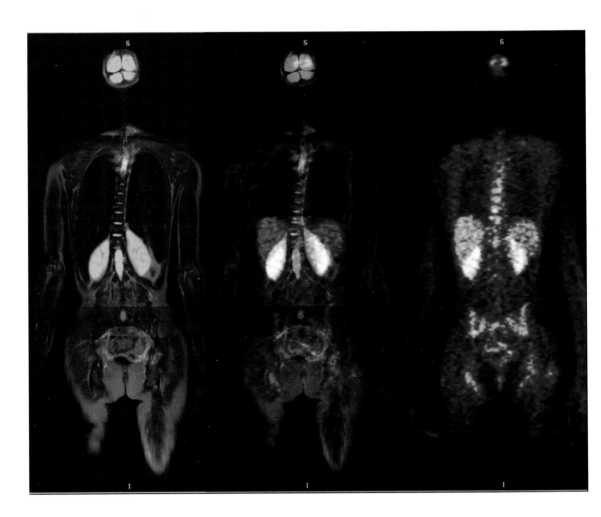

Pl. 16: Image fusion of MRT and PET scans. Left: a pure MRT T1-scan of an adult woman with high-resolution anatomical information. Right: a pure PET image of the same patient, showing the breasts, lungs, kidneys and lymph nodes as the metabolically most active zones, but in poor anatomical resolution. Center: a merging of these two images yields the optimal combination of anatomical and metabolic information. Reproduced by permission of Prof. Dr Reto Bale, Section for Microinvasive Therapy, Medical University of Innsbruck, Austria. See http://sip.uki.at/bildfusion.htm (accessed July 24, 2011).

Fig. 70: Janssen's photographic revolver to observe the transit of Venus, 1874. Left: Exploded views. The tooth-wheel cog JK rotates the daguerreotype plate P and shutters C and T' with the aperture F. Center: Photograph of these components; the aperture at the bottom left is periodically opened and shut by the shutter disk, here placed outside the apparatus for demonstration. The disk inside the box is an exposed daguerreotype plate (see color pl. XII for a closeup). Right: The actual set-up during observation. Sunlight is projected into the observer's hut by a heliostat mirror; the observer monitors the timing of the start and end of each serial recording. From Flammarion (1875) pp. 356f. and Janssen (1876) p. 104.

7.4.3 Muybridge's and Marey's chronophotography

When Janssen showed his results to the *Société Française de Photographie* in 1875 and to the *Académie des Sciences* in April 1876, he pointed out that his apparatus might also be used for the study of animal movements, especially those of birds, once materials sensitive enough to permit the very brief exposures necessary became available. This suggestion to apply chronophotography to studies of animal motion was followed, both in France and in the USA. In 1878, the photographer Eadweard James Muybridge (1830–1904) succeeded in taking serial photographs of a galloping horse. He had strung 24 equidistantly parallel threads across the riding track. Each of these threads was connected to a high-speed 1/500-s camera shutter that was electrically released when the horse struck the thread. He thus obtained 24 photographs documenting the various stages of a full gallop, one of them depicting all four hooves off the ground (cf. fig. 71, left). Two popular science journals, *Scientific American* (October 19) and *La Nature* (September 7, 1878), published engraved reproductions of these photographs. London-based *Nature* followed suit on April 3, 1879. Subscribers could cut them out and place

phobeenvironment, to Canales (2002) pp. 589, 612 and (2006).

[39] See Flammarion (1875) and Janssen (1876), who emphasized that this instrument was an "automoteur [...] sans aucune intervention de l'opérateur" (p. 101). Cf., e.g., Forbes (1874), Weiser (1919) p. 13 and Sicard (1998). Schaefer (2001) and Canales (2002) also deal with contemporary alternatives and remaining problems of Janssen's method. For more on this and on Janssen's later work on solar granulation, see here sec. 13.1, p. 379.

them in their zoetrope to view the still-motion photographs in rapid succession (cf. fig. 71, right, with a later version for Muybridge's zoopraxiscope). By replacing the threads with an electrical commutator device for releasing the shutters at precise intervals, Muybridge was then able to take photographic series of other moving beings.[40]

Fig. 71: Left: Engravings made from Muybridge's photographs of a galloping horse. From *Scientific American*, Oct. 19. 1878. Right: A later version of such photographs mounted on a disk for viewing in Muybridge's zoopraxiscope.

Muybridge's ingenious but costly solution of simply multiplying the number of cameras did not work for most other applications. So Jules Janssen's alternative of using just one camera lens but rotating the photographic plate was developed further. The French photographer and physiologist Étienne-Jules Marey (1830–1904) produced a so-called chronophotographic gun (*fusil photographique*) in 1882.[41] Marey had been conducting endless photographic trials on the locomotion of animals and humans since 1879. Like Janssen's photographic revolver, Marey's chronophotographic gun also took several images on a single photographic plate. In-between, the plate was covered by a variable-speed circular-disk shutter that had been designed in cooperation with the watchmaker Charles Dessoudeix. Marey's

---

[40]On Muybridge and his impact, see Newhall (1944) p. 43, Darius (1984) pp. 34f., www.stephenherbert.co.uk/muybCOMPLEAT.htm and his archive at www.archives.upenn.edu/faids/upt/upt50/muybridgee.html (both accessed June 18, 2011).

[41]See Marey (1882*a, f*) and (1902) pp. 322f. and Janssen (1882) on Marey's contact with Janssen and his priority. On Marey, see Weiser (1919) pp. 8–15, Darius (1984) pp. 38f., Martinet (ed. 1994) pp. 34–47, and Douard (1995); Braun (1992), esp. pp. 170ff. on the cinematographic method; pp. 228ff. on Marey's legacy; Cartwright (1995), Borrell (1986) and Brain (1996) on Marey's 'graphic method.' Cf. also www.cinematheque.fr/marey/expo-ligne/.

earliest photographs still had a defect, though. There was considerable overlap between the various phases of motion (cf. fig. 72, top). For this reason, from 1883 on he had his test models wear black clothes with a bright stripe highlighting the limbs and joints (cf. fig. 72, bottom). Because the visible surface of the photographed subject – often Marey's assistant Georges Demeny (1850–1917) – was thereby substantially diminished, the number of photographs per second could be greatly augmented, from approximately 10 to roughly 100 per second.

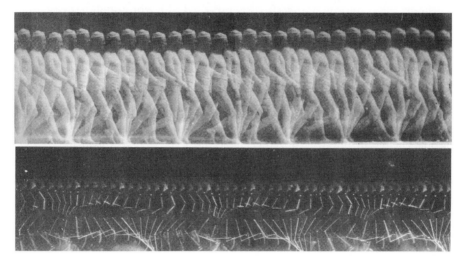

Fig. 72: Marey's chromophotographs of humans in motion. Top: Marey's assistant Demeny walking. Bottom: A running Joinville soldier, dressed in black with a white stripe attached to the sides of his arms and legs. Both taken in 1883 from negative dry gelatino-bromide plates kept at Musée Marey de Beaune, Collège de France Collection. Reproduced in Braun (1992) pp. 81 & 84.

As the historian of photography Marta Braun put it in her analysis of Marey's early "photochronographic" practices:

> Marey had cut apart the pictures that made up these earliest film bands and recomposed their movements in slow motion in an electric zoetrope to confirm his analysis in real time [...] slowing down some movements and speeding up others. He was not after a machine that would replicate the continuity of perceived movement: such an apparatus would have been no use to him in his work.[42]

There is an interesting parallel to this: The earliest experiments in x-ray cinematography were likewise "not a true film, but synthetic, being composed of a large

---

[42]Braun (1992) pp. 170, 173f.; for an interesting link between Marey's practices and later cinematographic loop techniques, see Curtis (2004) pp. 241f.

number of ordinary roentgenograms of the leg in different positions, which were
then arranged so that they seemed to show the movements of the leg."[43] The visuality of these actors was still very much channeled by older gadgets like Plateau's
phenakistiscope or Muybridge's zoopraxiscope, in which quick successions of single images created the illusion of a moving object.[44] This synthetic recomposition
of motion able to recreate the impression of motion actually became an important
way to cross-check the truthfulness of the photographs.[45]

From 1888 onwards, Marey also experimented with strips of photographically
sensitive paper shifted frame by frame through the camera. To protect the paper
against stray light, it was rolled up into a scroll inside a cylinder that was unwound
mechanically with a crank. This was the prototype of a kinematographic camera.
In 1890, Marey presented his 'chronophotographe' at the *Académie des Sciences* in
Paris. He could record series of up to 120 images per second. Five years later,
Louis Lumière publicly presented his first 'cinématographe.' From 1898 onwards,
flexible chemically photosensitized celluloid replaced paper as the image-bearing
material. It was perforated along both margins so that it could be mechanically
transported through the camera by a toothed-cog system. Further improvements
in the recording and projection later reduced the flickering for which early films
are so famous. However, such mechanical transport of film was limited to a
maximum of about 300 exposures per second. This was reached around 1912 by
Pierre Nogues (1878–1971); so, for very fast processes, other solutions had to be
found. Lucien Bull (1876–1972), who had been working in Marey's photographic
laboratory since 1895, abandoned the idea of mechanical film transport for high-
speed applications such as photographing a bullet penetrating plywood or the
rapidly beating wings of a dragonfly. He instead switched to a system in which a
rotating prism reflected the light from the camera onto a static film strip. With
an electrically generated intermittent flash illuminating the rapidly moving object
every 1/20,000 s and a sufficiently quickly rotating prism, he could achieve a
separation of 20 mm between the images on the film strip.[46] In the USA, similar
experiments on high-speed photography and cinematography were done by A. M.
Worthington (1852–1916) at the *Royal Naval Engineering Corps* in Devonport, Great
Britain, and by Harold Eugene Edgerton (1903–90) at MIT, whose photographs

[43] See Holm (1944) p. 163 on these pioneering studies by John MacIntyre in 1897.

[44] On these precinematographic devices, which were mostly used for entertainment purposes,
see, e.g., Füsslin (1993), Crary (1995) and here p. 20.

[45] On this interesting interplay of synthesis and analysis and the related technical division of
recording and projecting apparatus in early chronophotography, see Canales (2011) pp. 341–8.

[46] On Lucien Bull, see, e.g., Martinet (ed. 1994) pp. 107–9.

of droplet splashes in milk became an icon for ultrafast photography in the 20th century.[47]

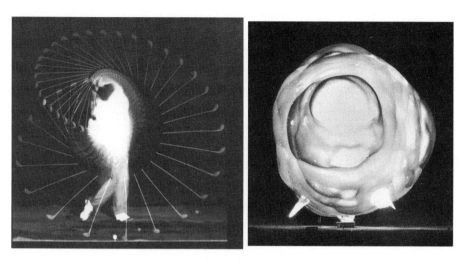

Fig. 73: Ultrafast photography by Harold E. Edgerton. Left: Multiple flash photograph of a golfer hitting a golf ball at a rate of 100 flashes per second for half a second, taken in 1938. Notice the slight flexing of the shaft after the club hits the ball. Right: Ultrashort exposure of an exploding atomic bomb, 1952. Automatic equipment engages the shutter one microsecond after ignition at a distance of seven miles from the epicenter. The tower there is already nearly fully vaporized, with a somewhat faster expansion of the cloud along the stabilizing cables of the tower. Both photographs by Harald E. Edgerton. From Jussim & Kayafas (eds. 2000) pp. 43 and 145. Courtesy of MIT Museum, Cambridge, Mass., Edgerton Digital Collections, nos. HEE-NC-38001 & 52010).

[47] See Darius (1984) pp. 54f., 96ff., Douglas & Joyce (1994), Edgerton (2000) and Jussim & Kayafas (eds. 2000) on Edgerton's high-speed photographs. The close link between serial photography and early cinematography is also confirmed by early handbooks such as Liesegang (1920).

7.4.4 Microcinematography

For the early history of motion pictures, various authors have already pointed out repeated instances of an intermeshing of artistic concerns (close-ups, zooming, and artificial speeding up and slowing down of motion) with exploratory experimentation on quasi-film documentation of strange phenomena (such as the formation of Bénard cells in a heated fluid or Brownian motion).[48] Early French microcinematography is particularly noteworthy in this regard. It was pioneered by Victor Henri (1872–1940) and Jean Comandon (1877–1970). We will discuss briefly Comandon's efforts to display blood flow inside the vessels of a living body, more specifically in the tiny capillaries of a tadpole's tail, each measuring no more than 1/100 mm in diameter. Because in normal microscopy, the greater the magnification the less illumination of the sample is possible, Comandon had to resort to a trick: ultramicroscopy or dark-ground illumination. He illuminated his specimen from the side, so that the blood is seen against a dark background, with the light coming obliquely from the scattered rays. The resulting strong contrast and very clear illumination of the outlines were the technical precondition for seeing anything on the film despite very short exposure times of 1/32 s, necessary for taking 16 photographs per second in order to create the impression of a continuous motion for the film viewer. Comandon's light source was a powerful electric arc lamp fed with a current of 30 amps. His Zeiss microscope had a parabolic Siedentopf condensor for side illumination, and the cinematographic camera came from the company *Pathé-Frères*, who were also responsible for public projections of the film.[49] As fig. 74 shows, the normal red blood cells manifested as brilliant rings and the white blood cells as somewhat cloudier, slightly larger white granules with shaded nuclei.

---

[48]See, e.g., Lefebvre (1993), Martinet (ed. 1994) pp. 69–85, Fieschi (2000) p. 234, Gaycken (2002), Landecker (2005) pp. 909–21, Curtis (2005, 2009), Aubin (2008) and Bigg (2008, 2011).

[49]For these and further technical details, see Comandon (1909) and Anon. (1911). On the marketing and dissemination of these films in France, cf. Lefebvre (1993) and Gaycken (2002).

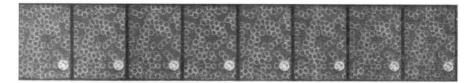

Fig. 74: A short segment of Comandon's 1909 kinematographic film of normal blood flow, recording the rapid motion of red blood cells (top and center), and the slower motion of one white blood cell (bottom right). From Anon. (1911) p. 214.

The advantages of this portrayal on film of circulating blood were manifold: The viewer got an idea of the rapid kinetic motions of the suspended corpuscles. Even more exciting was that one could also see processes such as phagocytosis ("the gradual surrounding and ingestion of a red corpuscle by a white cell"), the multiplication of white cells in a feverous organism, or blood infected by a trypanosome. A further increase in magnification allowed Comandon to show spirocheta pallida, indicators of syphilis. These are the worm-like spirals in the dark zone of fig. 75 (right).

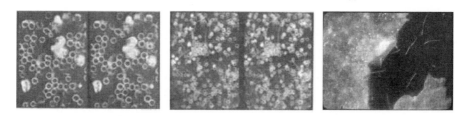

Fig. 75: A few detail stills from Comandon's kinematography, showing trypanosomes in the blood (left), relapsing fever with its higher percentage of white blood cells (middle), and spirocheta pallida in syphilitic blood (far right). From Anon. (1911) p. 214.

As this last example already suggests, the purpose of these films was twofold: On the one hand, it was a pioneering effort to extend human vision and insight into processes normally invisible to the naked eye. Whereas only one person can look through a microscope at a time, films permit projection onto a larger screen. Several colleagues and students being able all to see the same thing allowed effective sharing of observations and usage in the classroom. In particular, the new options afforded by this medium – of intense magnification and in fast-forward mode for very slow evolutions,[50] or conversely playback in slow motion for quickly

[50] A superb example from geology is provided in Brandstetter (2011). A short film of model

moving objects – opened up new horizons for scientific investigation. It became possible to count and to trace individual cells. Their irregular and rapid motions had previously made this practically impossible under a normal microscope.[51] Émile Reynaud's praxinoscope and Edison's kinetoscope – precursors of modern film projectors – were eventually improved by the insertion of a heat-absorbing water cell between the hot light source and the celluloid, making 'stop motion' feasible without running the risk of inflaming the film material. Thereby, it became possible to analyze individual frames picked out of longer sequences.[52] Figure 76 depicts the relative motions of cells in the endoplasm of an immersed amoeba, recorded with an optical lens combination yielding 1,500 × magnification onto a strip of film. The individual exposures measured 18 mm × 24 mm (the standard film format at this time). The five subsequent exposures of the film negative selected for analysis were enlarged by a factor of 11× for photographic printing. We thus arrive at a total magnification of c. 5,000×. Each of these film stills was carefully redrawn, as shown in the accompanying strip, concentrating on the outer contour lines and relative motions. Another analysis technique also practiced was to superpose several still frames to accentuate the relative motions of cells. By the way, fig. 76 is also a good example of the fruitful interplay between photography and drawing in scientific practice.

It also became possible to create so-called 'cine loops' for cyclic motions, i.e., repeat sequences of images that could be used to analyze rhythmic body cycles such as the heartbeat, respiration and the bending of joints.[53] At the same time, these films were instruments of knowledge dissemination far beyond the close-knit circle of specialists and could reach far broader audiences. Given a choice between a dry textbook and such a film, showing "the whole blood history of an attack [...] from the interval between the crises when no organisms are present, through the period of multiplication to the termination of the attack with the tendency of

experiments simulating geophysical folding already in the 1930s allowed the visualization of even geological time scales.

[51] For applications of this, see Comandon (1909) p. 940 or Comandon et al. (1913) p. 465 on pulsations of cells from a chicken embryo *in vitro*. Henri's (1908) tracing of individual particles in Brownian motion or Fortner's (1933) tracing of erratic motions would not have been possible otherwise.

[52] On this invention by Albert E. Smith in *Vitagraph Studios* around 1912 and the impact of this transition on cinematic visuality from frozen to unfrozen images, see Lefebvre (1993) p. 145, Gunning (1999) pp. 822ff.

[53] See, e.g., Janker (1931, 1949). For further refs., see Curtis (2004) pp. 241ff., who draws interesting links to older photographic practices of serial photographs that likewise focused on rhythmic events.

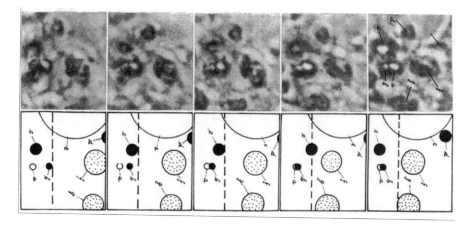

Fig. 76: Tracing of cells in the endoplasm of an amoeba, based on 11× enlarged prints of selected stills (top row) as the source for outline drawings (bottom row). From Fortner (1933) p. 15.

the spirals to aggregate together and eventually disappear," who would not rather chose the film?

> The latter [...] is not the least of the functions which such moving pictures can fulfil. There are thousands of people in this country who are intimately acquainted with the cellular constituents of the blood, and their various shapes and functions, thousands who have seen the ordinary bacterial preparations, [...] but of these thousands not one-tenth have actually seen the amœboid movements of a leucocyte or a spirillum wriggling its way between the corpuscles [...]. Yet these are things which it concerns them to understand, and no amount of imagination can supply the clearness and comprehension which actual seeing can give. The kinematograph might well become a most efficient aid to the teaching of very many biological, and especially medical, subjects.[54]

7.4.5   Later science films

These pioneering efforts in French medical cinematography were taken up in Germany and in the USA to some extent,[55] most notably by radiologists, who

---

[54] All previous quotes from Anon. (1911) p. 215. Similar arguments are used in Weiser (1919); cf. also Curtis (2009) p. 88.

[55] See, e.g., Braun (1898), who advertised cinematography for the examination of heart mechanics at a meeting of the *Deutsche Naturforscher und Ärzte* in 1897, Stein (1912) and Weiser (1919) for a survey of medical cinematography, Boon (2008) on the UK and Curtis (2009) pp. 90 and 96 on the enthusiastic reception especially in Germany.

Fig. 77: Dr. Jean Doyen, the son of Eugène-Louis Doyen, pioneer in surgical cinematography, presenting autochrome anatomical sections in Paris c. 1914. Courtesy Prof. Thierry Lefebvre, Paris.

saw it as a means of objectifying and documenting their examination of dynamic processes, otherwise only observable on a fluorescent screen.[56] Medical cinematography flourished. Films proved to be ideal for training physicians and medical personnel. Other sectors in which cinematography became popular as early as in the 1920s were the biosciences: zoology, embryology and anthropology, for instance, as well as psychology. In these fields, film demonstrations became a routine element of professional conferences. By the 1930s, they also became a broadly accepted research tool, especially in the behavioral sciences.[57] The ethologists Konrad Lorenz (1903–89) and Niko Tinbergen (1907–88) both used the film medium intensely in their examinations and documentation of animal behavior. Polls held among students showed that more than 90% appreciated such films as informative *and* entertaining.[58] In National Socialist Germany, the government – very conscious of the importance of public media – even founded a central state institution for visual media in science and education, the *Reichsanstalt für Film und Bild in Wissenschaft und Unterricht* (RWU). After World War

[56]On medical x-ray cinematography, see Weiser (1919) pp. 68–74, Janker (1931), James (1935), Mitchell & Cole (1935), Holm (1944) and Holmgren (1945).

[57]See Michaelis (1955) and Mitman (1999) pp. 60f. on the general acceptance in the biosciences.

[58]See Mitman (1999) chap. 3, especially on ethology in Germany and Austria, Great Britain and the USA, and www.humanetho.orn.mpg.de/en/eindex.html for the "film archive of human ethology."

II, a UNESCO commission essentially acquitted the RWU of charges of having been an instigator of Nazi propaganda, classifying only 10% of the scientific films produced in Nazi Germany as "tendentious." The RWU was succeeded by a Göttingen-based *Institut für den Wissenschaftlichen Film* (IWF) and an international collection of 16 mm scientific films, the *Encyclopædia Cinematographica* (EC).[59] Other major archives for such scientific productions are the *Haus des Dokumentarfilms* in Stuttgart, Germany. In France, there is *La Villette* (Paris) and the *Cinémathèque de la Ville de Paris*, as well as the *Institut Lumière* in Lyon. Holland has the *Nederlands Filmmuseum* in Amsterdam; the UK, the *Science Museum* in London; and the USA, the *Public Moving Image Archives and Research Centers* at the *Library of Congress Packard Campus for Audio Visual Conservation*.

Altogether, during its first 50 years, only some branches of science were interested in the new medium. A survey of the early "scientia productions" by *Éclair*, France's third-largest movie company, between 1911 and 1914, shows that 72% treated zoology and 14% botany, whereas physics and chemistry represented only 12% of the entire film corpus of that period.[60] One possible reason is the limited resolution and sensitivity of the film material. But many scientists and engineers did not yet realize the new potential. Occasional enthusiasm about motion pictures among a few of its protagonists was countered by widespread skepticism. The concern was their tendency toward mere spectatorship and sensationalism. Sometimes the response was even downright condemnation of the new mass culture.[61] With the broad diffusion of television sets in American, Asian and Western European households from the 1960s, the film medium became ubiquitous. Science journalism then also increasingly made use of the new visual media. Television and video clips, for instance, were used to record and broadcast interviews with important scientists, show off laboratories or experimental setups, or popularize science in the form of simplified sequences.[62] Especially in the USA, the rise of film as a medium of science communication is strongly con-

[59]On these institutions and developments, see Mitman (1999) pp. 70ff. On the IWF, now liquidated, which housed one of the world's largest collections of documentary science films and continued to receive funding from the German federal government until 2007, see also http://de.wikipedia.org/wiki/IWF_Wissen_und_Medien and http://de.wikipedia.org/wiki/Encyclopaedia_Cinematographica and literature mentioned there.

[60]See Lefebvre (1993) p. 141 on these numbers, and Gaycken (2002) on the target groups (predominantly young children) and style of scientific "vernacularization" in France at that time.

[61]On these conflicting views on movies, see Curtis (2009), esp. on Germany, and Canales (2011) pp. 349ff. on France. For a thorough survey of early research films in the biosciences, see Michaelis (1955). On the importance of wildlife films for the American image of nature, see Mitman (1999).

[62]See, e.g., Gattegno (1969), still full of pedagogic optimism with regard to the new medium.

nected – some even say "coincident" – with the boom in mass entertainment in Hollywood.[63] The televised manned flight to the Moon was followed by millions of viewers at home. It is a prime example of the dual use of the film medium both as an indispensable technical means of communication and as an instrument of information and mass propaganda.[64] Two obvious later examples of intense and bidirectional interplay between science and technology, on the one hand, and popular images of science and technology, on the other, are the interplay between science-fiction movies and research, development and design in the aerospace industry and between scientific simulation and commercial computer games in the 1990s and 2000s.[65] With the advent of high-speed graphic processors, animated renditions of research results and simulations, ranging from cosmic developments, such as the big bang, to car-crash analysis, has become routine in many branches of science and technology.

7.5   The drift of scientific images into the public sphere

With chronophotography, film recording and interactive simulation, we have reached media types strongly related to the issue of the popularization of science and technology. It is noteworthy that *some* of Muybridge's and Marey's series received quite broad publicity. The same is true of science documentaries such as Heinz Haber's (1913–90) in the 1960s and 1970s, and even of Volker Springel's millennium simulation.[66] Such transfers are also demonstrable in certain earlier episodes where scientific images originally produced for research migrated into fairly popular contexts, such as into Camille Flammarion's (1842–1925) ubiquitous 19th-century popularizations. Such shifts only work with some images, and often they coincide with subtle changes in meaning and a loss of sophistication. Sometimes this image simplification was not so subtle, but fairly drastic: intentional shifts in meaning, for instance, frequently by the use of scientific icons in advertisements. It is often just a metaphorical transfer onto a commercial product of desirable attributes such as high reliability or laboratory precision. For this

---

[63]Exemplary nature films, including the popular TV series 'Flipper'; Mitman (1999), quote from p. 60; ibid., chaps. 5–6 on "Disney's true-life adventures," and "domesticating nature on the television set."

[64]On Apollo flights and the media, see Kauffman (2009) and further references given there.

[65]On the impact of films on the visual culture in medicine, see, e.g., Cartwright (1995) and Curtis (2004). On the image of science and scientists in films up to the 1990s and interesting statistics, see Martinet (ed. 1994). According to Jacques Jouhaneau, ibid., pp. 248–57, medicine and the biosciences continued to dominate the scene.

[66]See, e.g., Utzt (2004), Jochen Hennig in Adelmann et al. (2009) and Hessler et al. (2004).

purpose, only well-known icons of science are chosen – the stereotypical scientist (a wizened male, sometimes with the unkempt Einstein hairdo and wearing protective goggles) functions as guarantor of proven quality. Images from scientific practice are only rarely selected, and, if so, only if the public at large is sufficiently familiar with them. For instance, an ultrasound scan of a fetus was chosen for the advertisement by a Swedish car manufacturer shown in fig. 78, published around 1990 in various popular journals, such as *Harper's Magazine*.

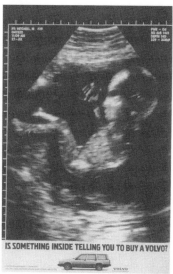

Fig. 78: Advertisement for Volvo with a retouched ultrasound image of a fetus, 1991. From Sturken & Cartwright (2001) p. 293.

This advertisement became the focus of a heated debate in the subsequent years, as it resonated strongly with the contemporary debates about the abortion controversy.[67] What interests me here is the actual usage of the 'scientific' image. Compared with 'real' ultrasounds of fetuses from the 1980s, this image strikes one as being of far too good quality. All the fingers of the apparently waving hand are discernible. It is undoubtedly a strongly retouched image to ease immediate recognition of the unborn child. Otherwise the ad would have lost its punch. This feature is typical of such usage of scientific images in advertisements. The fuzziness and sophistication of the original is streamlined in order to make the point clearer. Often the only thing that really counts in such contexts is the metaphorical transfer of the attributes onto the product. In our example, the

---

[67]Those political, feminist and anti-feminist facets are irrelevant here, but I refer the reader to the pertinent literature: Yoxen (1989), Sturken & Cartwright (2001) pp. 291–4, Taylor (1992) and further references there.

fetus is supposed to be as safe inside the Volvo as it is in the mother's womb.

While this car manufacturer's advertisement does not show the various stages of modification of the original image, other examples allow one to reconstruct a whole chain of 'translations' of a particular image, originally meant to be scientific, into stages further and further removed from its original context. A nice example is the (notorious) march of human progress: a row of primates walking from left to right, starting with a gibbon (*Pliopithecus*) at the left and ending with a modern human at the head (see fig. 79).

Fig. 79: The "march of progress," originally designed by Rudolph Zallinger in 1965, displayed 15 different bipeds. This simplified version from 2007 by José-Manuel Benitos and M. Garde reduced the number to six. Wikimedia Commons. The original fold-out color plate is also accessible at http://madartlab.com/2011/03/14/ai-have-we-made-progress/

This image was originally designed by the natural history painter and muralist Rudolph Zallinger (1919–95) for a special issue of *Time Life Books* on "The early Man," which appeared in 1965. The author, the anthropologist F. Clark Howell (1925–2007), later remarked that neither he nor the illustrator intended "to reduce the evolution of man to a linear sequence, but it was read that way by viewers. [...] The graphic overwhelmed the text. It was so powerful and emotional."[68] This was an image with a veritable life of its own, at odds with the intention of its creator. Its widespread distribution and intuitive, clever design, impressed the image deeply into the iconography of Darwinism.[69] Even though experts like

---

[68] Howell in Barringer (2006); cf. also http://en.wikipedia.org/wiki/March_of_Progress (accessed May 31, 2011) for further discussion on the reception of this image.

[69] On the repercussions of Darwin and Darwinism in Victorian visual culture, see Prodger (1998) and Smith (2006). For the broader impact on other nations and time periods, cf. Kort et al. (2009) and Larson & Brauer (eds. 2009).

Stephen Jay Gould have long tried to debunk this simplistic picture of evolution, it is here to stay. Amazingly, this very same image appeared on the cover of a Dutch translation of Gould's classic *Ever since Darwin*, certainly not at his own instigation. It is an ironic comment on the efforts of professionals to control the iconography of their field. Some images have an impact on the public arena that is simply beyond control. Allusions to this image in particular appear in all kinds of other contexts, from political cartoons to campaigns for and against creationism or feminism.[70] Creationists have exploited the sloppy use of this misleading image in their uphill battle against Darwinism,[71] thus conflating inaccurate image usage in some popular texts with the tenability of the underlying theory. What these heated debates show, though, is how much impact visual icons of science and technology have. It reveals the urgency of improving "visual literacy" among the general public. This is one of the main tenets of visual studies in curricula and educational programs – as only visual literacy of this kind will lead to a more critical handling of such images and block the formation of myths and misunderstandings.

A sidelong glance at the example of Haeckel's embryology will further illuminate why and how a few science images succeed and become icons.[72] It is well known that the Jena-based biologist Ernst Haeckel, whom we have met before as a protagonist of chromolithographic illustration (see p. 228), used images to full effect in his propaganda efforts for Darwinism in Germany. It is somewhat upsetting that the embryological series of schematic images with which Haeckel tried to convince his audiences of the evolutionary scheme were partially faked in order to bring out the essential similarity in the ontogenesis of humans and apes, dogs, etc. in their early stages of development even more strikingly. What was really outragious to Stephen Jay Gould, though, is that those images still crop up in semipopular journals and schoolbooks long after serious doubts arose as to their trustworthiness soon after their initial publication. To him, it was their "mindless recycling" that led to the persistence of these (incorrect) drawings in the press.

---

[70] A host of amusing examples are given in Gould (1989) pp. 30–5, including an Indian poster on children's education, an advertisement for cigarettes as the highest stage of human advancement and an advertisement for Granada's TV rentals (with a naked human on the far right proudly carrying a TV set in his arms). Further examples are listed in http://en.wikipedia.org/wiki/March_of_Progress.

[71] See, for instance, Wells (2000) pp. 108f. Wells, polemicizing against these icons of evolution, argues that "much of what we teach about evolution is wrong." For a survey of the critical scientific response to this book, see http://en.wikipedia.org/wiki/Icons_of_Evolution.

[72] On the following, see Gould (2000), Hopwood (2006) and Hopwood (in prep.).

"At this point, a relatively straightforward factual story, blessed with a simple moral story as well, becomes considerably more complex, given the foils and practices of the oddest primate of all. Haeckel's drawings, despite their noted inaccuracies, entered into the most impenetrable and permanent of all quasi-scientific literatures: standard student textbooks of biology. [...] We should therefore not be surprised that Haeckel's drawings entered 19th century textbooks. But we do, I think, have the right to be both astonished and ashamed by the century of mindless recycling that has led to the persistence of these drawings in a large number, if not a majority, of modern textbooks!"[73]

7.6   Viscourse on top of discourse

A dense network of interrelations is formed between images and their contexts of presentation, usage and critique. It is formed out of diachronic chains of stepwise improvements and synchronic rival options, as well as out of an interlocking between different forms and types of representation. Analogously to Foucault's concept of 'discourse,' which aims at a coherent whole of all *textual* sources of a certain era or field, the term 'viscourses' (in German: *Viskurse*) has been proposed by the sociologist of science Karin Knorr-Cetina to denote this elaborate lattice of *visual* interconnections.[74]

Viscourses do not replace discourses, but are rather intertwined with them. The quasi-ethnological observations of laboratory settings pioneered by Knorr-Cetina, Latour and Lynch[75] have shown that the process of making an image is a rather complicated and often controversial affair in which many people take part, often interactively, to optimize the final output for publication as conference overheads, as PowerPoint slides, or perhaps as images in scientific journals or textbooks. These are just the final products in a very long chain of earlier versions, and if one compares succeeding editions of such textbooks, even they at times still change. When it comes to scientific images, the call for improvement in quality practically never stops. Mike Peyton, the long-standing illustrator of the semipopular journal *New Scientist*, has captured this idea of a long production line in science quite well in cartoon shown in fig. 80:

[73] Gould (2000) p. 42; cf. Hopwood (2006) and (in prep.) for an in-depth study of the heated debates surrounding Ernst Haeckel's embryo images.

[74] See, e.g., Knorr-Cetina (1999) on 'viscourse,' Jay (1993) for a critique of the neglect of images in Foucault's structuralism and Frank (2006) p. 46 on the parallelism in the theoretical developments leading to these two concepts.

[75] See, e.g., Latour & Woolgar (1979), Amann & Knorr-Cetina (1988), Lynch (1985), Lynch & Edgerton (1988) and Lynch & Woolgar (eds. 1990).

Fig. 80: An endless chain of re-presentations? Mike Peyton in *New Scientist*, undated, here reproduced from Knorr-Cetina (1991) p. 5, one more link in the chain. Courtesy of Mike Peyton.

This cartoon does not pretend to be anything more than a caricature, but, as we have seen in the preceding pages, we have serious and detailed science studies documenting this chain-production of images on the basis of methodologies of quasi-ethnological laboratory studies,[76] of the iconography of scientific fields or of specific objects,[77] of cultural histories of drawing and printing,[78] and finally also as a side-product of historically accurate experimental replications.[79] Thus, images begat images.[80]

[76] See, e.g., Lynch & Edgerton (1988) on astronomical images and Amann & Knorr-Cetina (1988) on microbiology.

[77] See, e.g., Saunders (1995), Nickelsen (2000) and sec. 7.1 on botany, Hopwood (2006) on embryology, C. E. Jackson (1975, 1978, 1989) on birds and Dupré (2008) on telescopes.

[78] See, e.g., Johns (1998) and Kusukawa (2012) on the early-modern period and, e.g., these web-pages of giraffe images: www.girafamania.com.br/index/girafamania.htm and .../artistas/pintura-litografia.htm (both last accessed Sept. 17, 2013).

[79] See, e.g., Breidbach et al. (eds. 2010, 2011) and Heering et al. (eds. 2000).

[80] Daniela Bleichmar: "images produce more images," oral statement made during discussions at the 6th Spring School at Mao, Menorca 2011. During her talk there, as well as in Puig-Samper (2012) and Bleichmar (2012), the whole Spanish empire was interpreted "as a visual engine, spinning out images": no less than 12,000 images from expeditions into the Spanish empire outside Europe in Middle and South America between the late 1770s and the early 1800s.

8

Practical training in visual skills

Science and education were long sundered in the minds of their historians, in mutual ignorance or simply disinterest. Social historians of mathematics and physics, such as Gerd Schubring, Russell McCormmach, Kathryn Olesko and Josep Simon, argued for the crucial importance of mathematics and science teaching in understanding the emergence, stabilization and development of scientific paradigms. Andrew Warwick's case studies on late-Victorian and early Edwardian education in the Cambridge Mathematical Tripos and David Kaiser's papers on science pedagogy in gravitation theory and quantum electrodynamics renewed interest in the connection between research and pedagogy.[1] Feynman diagrams, but also very simple line drawings in early-modern astronomy texts, trained their readers in skills of mental visualization.[2] For the purposes of this book, what is most interesting is how scientists, engineers and physicians acquire their skill at creating and manipulating visual representations. Each discipline and each visual domain within these fields has its own very specific cognitive skills of pattern recognition and visual inference. Such skills have to be learned – by no means an easy or quick process. It might take years for a newcomer actively participating in the research practices to learn by example and see how to identify patterns or construct models. Although this learning is usually done in larger groups or teams, ultimately each individual has to work his or her way through a learning curve. Because different individuals differ slightly in training, background and personal goals, there will always be some difference in the skill ultimately obtained. This can yield a variation in local practices or a leap forward if one individual manages to combine two or more such skills in a new, fruitful way (see chaps. 3 and 4 for my view on how new visual cultures are formed). Any historiographic study of pedagogical practices will thus have to include biographical as well as prosopographic facets.

---

[1]See, e.g., Jungnickel & McCormmach (1986), Schubring (ed. 1991), Olesko (1991, 2006), Warwick (2003), Kaiser (ed. 2005, 2005a), and Mody & Kaiser (2008) or Simon (ed. 2012, 2013) for recent surveys.

[2]See Crowther & Barker (2013) on "training the intelligent eye" in early-modern astronomy.

8.1   Technical drawing in France, Germany and Britain

When the sociologist of technology Kathryn Henderson started her ethnomethodological field observations in various technical design centers in the 1990s, she
was overwhelmed by the ubiquity of a variety of visual materials and their cen-
trality in the everyday discourse and practice at these centers of inscription and
conscription. In her own words:

> In the world of engineers and designers, sketches and drawings are the
> basic components of communication, words are built around them. Vi-
> sual representations are so central that people assembled in meetings wait
> while individuals fetch drawings from their office or sketch facsimiles on
> whiteboards. Coordination and conflict take place over, on, and through
> drawings. Visual representations shape the structure of the work and de-
> termine who participates in that work and what its final products will be.[3]

Thus, it is not surprising that trainees in engineering, design or some other
technological expertise spend a considerable amount of time and energy on the
production of sketches, technical drawings and other forms of visual representation. This section focuses on the history of lessons in technical drawing,[4] to
impart the skill of producing and interpreting these drawings unambiguously.

   In the Middle Ages and the early-modern period, only very few select groups
of practitioners produced drawings that could reasonably be described as technical
drawings. Villard de Honnecourt's famous sketchbook from between 1220 and
1235 was a strange mixture of quick sketches of objects or buildings viewed, ideas
conceived while traveling, and notes for work in progress. The sketchbook was
most likely rather a show-and-tell source of illustrative drawings than a collection
of blueprints for manufacture.[5] The same is true of other early drawings on
technical subjects, such as gunner booklets in the tradition of Conrad Kyeser's
(1366–post-1405) treatise *Bellifortis* (1405) displaying several types of weaponry
and other instruments of war.[6] The reasons for the scarcity of any kind of technical drawings from these periods are threefold:

---

[3]Henderson (1999) p. 1; cf. here p. 69 on her notion of conscription devices.

[4]For historical surveys, see Ulrich (1958), Feldhaus (1959), Nedoluha (1960), Booker (1963),
Lipsmeier (1971), and Baynes & Pugh (1981) specifically on British examples.

[5]On the interpretation of this source and various other early machine drawings, see, e.g., the
contributions by Marcus Popplow, David McGee and Pamela Long in Lefèvre (ed. 2004).

[6]See Kyeser (1405/1967). For a thorough survey of other such *Büchsenmeister* booklets of the
15th and 16th centuries, see Leng (2002).

(a) A web of secrecy was spun around such knowledge or practical designs.

(b) Transmission was limited to personal transfer from master to pupil.

(c) Parchment and paper are brittle materials that do not preserve well unless special care is taken. Such an investment of time and effort is only made for relics considered of special importance. So countless such drawings and sketches were lost through deterioration or neglect. The few sources that have survived are either rare remnants in the taxonomy of codicology or demonstration drawings for special purposes, such as to impress a potential patron.[7] They often omit important details and not infrequently show imaginary constructions that would never work in practice.

An additional factor that inhibited the use of scaled technical drawings was that until the 19th century, pieces of work were unique in most areas of manufacture, each product slightly differing from the rest. Even screws and bolts were hand-made and pairs were custom-matched to fit properly after having been individually manufactured. The earliest exceptions to this general rule are found in architecture and shipbuilding. In the former, often whole series of cut stones had to be as identical as possible to speed up the process of building a wall or columns,[8] for instance. In shipbuilding apparently whole series of ships were built according to one set of templates indicating the outside curves of the hold and the exact dimensions of the internal ribbing.[9]

In machine design, fairly standardized technical drawings appeared only from 1770 on, thus clearly coinciding with the Industrial Revolution. Such companies included *Boulton & Watt* in Birmingham, founded in 1775 (producer of steam engines), *Bramah* (hydraulic presses) and *Maudslay* (machine tools and microme-ters for exact measurements).[10] Historians of technology and education agree that it was only in this late Enlightenment period that technical drawing finally became somewhat canonized, if not yet standardized. In this sense, technical drawings were indeed "a product of the industrial revolution, [... dependent] on the emergence of new forms of manufacture and organization" as well as on an in-

---

[7]For other late medieval and early-modern technical drawings, see Lefèvre (ed. 2004) and Lefèvre, Schöpflin & Renn (eds. 2003). Cf. also the database of machine drawings 1450–1650: http://dmd.mpiwg-berlin.mpg.de/home

[8]On this hotly disputed issue of early mass production in stone masonry, see Kimpel (2005).

[9]On early-modern shipbuilding, see, e.g., Booker (1963) pp. 68–71 and Alertz (1991). For British examples, see Baynes & Pugh (1981) chaps. 3–4 and here color pl. III.

[10]For superb sample technical drawings from these firms, see Baynes & Pugh (1981) chap. 4; cf. here sec. 4.1, p. 130 on Nasmyth, and sec. 5.2 on the indicator diagram, an invention by the *Boulton & Watt* company.

creasing social differentiation between design, construction and production.[11] In Britain, where the Industrial Revolution began, pronounced regional differences and competition as well as secrecy concerns impeded the emergence of a unified standard for drawings, however remarkable individual achievements by talented draftsmen may have been.

Major impulses came from revolutionary France,[12] a country with a strong tradition in systematic education for artisans, architects, mining experts and other technical specialists. Jean Baptiste Colbert (1619–83), minister under Louis XIV, had already tried to improve the standards of artisanal training by creating the first arts and crafts schools (*Écoles des Arts et Métiers*) in the 1660s. Foreign experts and artists were hired as teachers in Châlons-sur-Marne and Angers, for instance. The copperplate engraver and architect Charles Lebrun (1619–90) the later head of the royal manufactory of fine tapestries (*Gobelins*), founded an *Académie Royale de Peinture et Sculpture* in Paris in 1648; and in Austria, the engraver Jakob Mathias Schmutzer (1733–1816) likewise founded an engraver's academy in 1766. Under the *Ancien Régime*, the Parisian *École des Ponts et Chaussées* (1747), the *École des Mines* (1783) and the *École Royale du Génie* in Mézières (1748) were founded. A final breakthrough toward systematic teaching of technical drawing based on mathematics came with the founding of the *École Central des Traveaux Publics*, in the heat of the French Revolution 1794. One year later, it was renamed the *École Polytechnique*. Jacques-Élie Lamblardie (1747–97), the director of the *École des Ponts et Chaussées*, originally conceived this institution as a kind of preparatory school for learning the basic skills of civil engineering for two years prior to more specialized training elsewhere in bridge, fortification and street construction. When Gaspard Monge (1746–1818) was appointed professor of mathematics, physics and descriptive geometry there,[13] he pushed through his idea of unifying the basic education of all civil and military engineers to also include architecture, hydrology and other technical fields. On commission as *lieutenant-colonel du génie militaire* at the Mézière academy, Monge had worked as a draftsman (*dessinateur*). When Napoleon's forces

[11]Baynes & Pugh (1981) p. 11.

[12]On the history of French technical education, see, e.g., Nedoluha (1960) pp. 35ff., Artz (1966), Emmerson (1973) pp. 77ff., Ferguson (1992*b*) pp. 75ff., Sakarovitch (1995) and König (1997) pp. 200–17.

[13]Monge had been teaching at the Mézière school of engineers. He acquired reputation for finding a geometrical solution to a problem in fortification design, but his lectures there were not published and were considered a military secret. On Monge, see Brisson (1818), Dupin (1819), Taton (1951), Lipsmeier (1971) pp. 96ff., Booker (1961, 1963 chap. 9), Belhoste & Taton in Dhombres et al. (eds. 1992), and Grattan-Guinness (1990) pp. 115ff. and (2005).

marched into Italy and confiscated valuable paintings, such as by Titian, Monge was consulted about their restoration and exhibition at the *Louvre*. From 1798 to 1801, he was also a member of the Napoleonic expedition in Egypt. As head of the *Commission des Sciences et des Arts d'Égypte*, Monge oversaw the work conducted by the illustrators entrusted with documenting the French excavations and findings in Egypt. It was eventually published in the opulently illustrated multivolume *Description de l'Égypte* (1809–28), for which some 400 engravers produced 837 copperplates containing more than 3,000 images. Thus Monge was by no means only a geometer or mathematician. He was a multifaceted 'artist-scientist' who also delivered lectures on topics like woodcutting, stonecutting and sculpting, the art of engraving, shading, aerial perspective, and color perception when contrasts are juxtaposed. He thus personifies a strong and intense visuality.[14] This rich experience as a draftsman and applied geometer became the source of a superb course by Monge (taught for the first time in 1795) and a textbook on *Géometrie descriptive* (first published in the *Journal de l'École* in 1795 from a stenographed transcript of his lectures, later also as a book in 1798). Monge's idea was to base the teaching of drawing on rigorous geometry. He put great care into conceiving a step-by-step introduction to the basics of Euclidean, projective and descriptive geometry. The central task of all engineers, as Monge conceived it, was the graphic, two-dimensional rendering of a three-dimensional object.[15] That means mental exercises in projection techniques, including shadow casting and perspective. In order to learn the skills necessary for interpreting such 2D drawings correctly, and eventually to be able to draw the such themselves, the pupils needed more than lectures and demonstrations by their teacher. Monge realized that thorough practical exercises were indispensable. Descriptive geometry had to be instilled in the minds and hands of these *polytechniciens*, as this elite corps of engineers soon began to call themselves.[16]

Soon after its foundation in 1794, 45% of the *École Polytechnique*'s curriculum constituted drawing exercises, quite a heavy load. Descriptive geometry was split into basic lessons in *stéréotomie* (i.e., 2D rendering of 3D figures, both geometrically

[14]These last mentioned topics were included only in the later editions of Monge's textbook from 1827 onwards: see Monge (1798*b*) pp. vi, xii–xiii, 134f., 170–88. On the practice of stone-cutting in the Mézière curriculum and on connections to the Freemasons, see Lawrence (2003) pp. 1271ff.

[15]Monge (1798*b*) "la représentation graphique des objets," "augmentée de la théorie des ombres et de la perspective" (ibid., subtitle).

[16]On the history of the *École Polytechnique* and the corps spirit among its *élèves*, see, e.g., Fourcy (1828), Leverrier (1850/51), Bradley (1976), Shinn (1980), Daston (1986), Belhoste in Belhoste et al. (eds. 1994), and Grattan-Guiness (1990) chap. 3 and (2005).

on paper and physically working with real stone and woodworking samples), architecture and fortification. This was complemented by *l'art du dessin*, i.e., freehand drawing lessons on the human figure, landscapes, etc. In 1795, the *École* hired 40 draftsmen and model-makers. The steady rise in importance of chemistry, physics and other subjects caused the percentage of drawing lessons in the overall curriculum to drop to 11.5% in 1812 and to a mere 9.8% in 1894. But conceptually, these drawing lessons remained a central component of engineering instruction in France.[17]

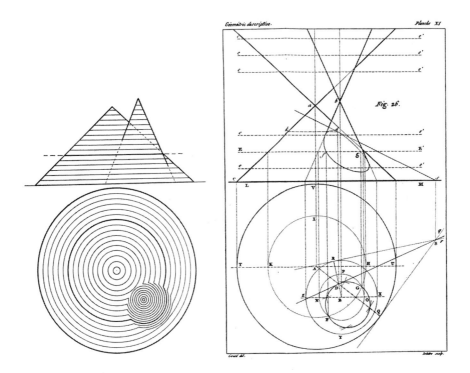

Fig. 81: Two intersecting cones in Monge's *Géométrie descriptive* (right). Drawing by Girard and engraving by Delettre, from Monge (1798) pl. XI, also reproduced in Booker (1963) pp. 102, 99.

Let us look at one example in Monge's *Géométrie descriptive*. The graphic representation of two circular intersecting cones is, according to the much older standard methods of orthogonal projection, from the side (*en face*) and from the top us-

[17] For the detailed curriculum of the *École polytechnique* and the above data, see its first institutional history, Fourcy (1828*b*) pp. 56–62, 376ff., its *Livre du centenaire* (Paris 1895), vol. 1, pp. 14, 30, 42, 86 and Lipsmeier (1971) pp. 101ff.

ing contour lines ('bird's eye view') in cartographic mode (both in fig. 81, left). Monge's construction according to descriptive geometry is much more involved (fig. 81, right). It also allows one to determine the curve formed by the intersection of the two cones, for instance.

Monge's textbook rapidly became standard in centralized France, with supplementary textbooks, for example by Hachette (1822), soon also covering technical drawing and other more specialized subjects. Jean Hachette (1765–1843) was Monge's tutor from 1794 to 1799 before becoming assistant professor at the *École polytechnique* (or 'X' as its *élèves* abbreviated their *alma mater*).[18] Monge's techniques also became known in America through a French civil engineer, Marc Isambard Brunel (1769–1849), who was trained in the French Navy in the late 1780s, migrated to the United States in 1793, where he became chief engineer of the City of New York before resettling in Great Britain in 1799. A former French officer and pupil of Monge, Claude Crozet (1790–1864), started lecturing in the USA in 1816 and also published the first American textbook on the subject in 1821. The US *Military Academy* at West Point, the *Virginia Military Institute*, the *Rensselaer Polytechnic Institute* and other engineering schools soon adopted large portions of the *École polytechnique's* curriculum.[19] A Spanish translation of Monge's textbook was promptly completed in 1803. The first full German translation of Monge's course, however, appeared only in 1900, but adaptations and abridged editions were published from 1828.[20] Institutionally, the Parisian *École polytechnique* as a role model quickly spread throughout Central Europe, once the efficiency and success of its integrative way of teaching had become clear. As a consequence, further polytechnic institutes were founded in Prague (1806), Vienna (1814), Berlin (1821 and 1831), Karlsruhe (1825), Munich (1827), Dresden and St Petersburg (both 1828), London (also in 1828, if the foundation of an engineering section at *King's College* is counted), Stuttgart (1829), Kassel (1830), Hannover (1831), Augsburg (1833), Liège/Lüttich (1835), Ghent (1835), Zurich (1855), Riga (1862), Aachen

---

[18]See Hachette (1822, 1829). Hachette also introduced the first course on machine theory and *science industrielle* at the *École polytechnique* in 1806; see Fourcy (1828*b*) p. 37 and sec. V, Grattan-Guinness (2005) pp. 235–7, 244. X's impact on civil engineering is also documented in Kurrer (2008) pp. 52ff., 291ff.

[19]See, e.g., Lawrence (2003) pp. 1270f. and 1278, note 8; Ferguson (1992*b*) pp. 77, 200, notes 17–18 and further references given there.

[20]See, e.g., Schreiber (1828/33). The projection techniques of the Karlsruhe architect Friedrich Weinbrenner (1766–1826) were also influential. Carl Friedrich Gauss reviewed the 3rd edition of Monge's textbook positively for the *Göttingische gelehrten Anzeigen* (July 31, 1813); quoted in Lipsmeier (1971) p. 104. On the international diffusion of Monge's descriptive geometry, see, e.g., Cunningham (1868), Booker (1961) pp. 26ff. and Booker (1963) pp. 130ff.

(1870), Lemberg (1871), Warsaw (1895), etc.[21]

Below these top-notch polytechnic institutions there existed in Germany artisanal academies (*Gewerbeakademien*) and schools of the applied arts (*Kunstgewerbeschulen*), a hierarchical system of secondary education (from the preparatory *Gymnasium*, qualifying for university study, to the *Realschule*, substituting mathematics and the applied sciences for instruction in Latin and Greek), down to the elementary *Volksschule*, which only offered a four-year course on very basic reading, writing and drawing skills. Finally, there were Sunday schools for supplementary instruction of young apprentices.[22] It was realized, though, that even a reduced teaching load quintessentially had to include basic drawing skills, even for lower-ranking apprentices in practical fields, such as woodworking or metalworking. They needed to understand the technical drawings used in the manufacturing processes at their future workplaces. From the middle of the 19th century onwards, drawing instruction in trade schools and artisanal academies were grouped into three sections: (i) elementary geometrical drawing (related to the basic theorems of geometry, as part of a mathematics class; (ii) descriptive geometry (a strongly deflated version of Monge's *Géométrie descriptive*); and (iii) applied graphic art (*Fachzeichnen*), sometimes still complemented by lessons in freehand drawing, even though that was sometimes criticized as valueless to the future profession.[23]

Further progress in the social differentiation of production processes raised the demand further for both active drawing skills and the passive ability to read technical drawings. During the period of promotorism (*Gründer-Zeit* 1871–73), this led to increased importance being attached to freehand and geometrical drawing and the graphic arts (*Fachzeichnen*) at German artisanal schools.[24] Old hierarchies, ranking freehand sketching uppermost as the crowning achievement of drawing instruction (cf. fig. 82), were thus tendentially inverted. Freehand drawing became the problematic part, supposedly distracting pupils from confining themselves to geometric rigor.

Although technical drawing was commonly rooted in both Euclidean geometry and perspectival theory, cultural factors within the different linguistic traditions led to remarkable differences in how practical geometry was taught and routinely

---

[21] See Nedoluha (1960) pp. 38f., Lipsmeier (1971) pp. 104ff., Emmerson (1973) pp. 87ff. and König (1997).

[22] For an excellent survey of this highly differentiated educational system and the place of technical and free drawing in each of these institutions, see Lipsmeier (1971).

[23] See Lipsmeier (1971) pp. 126–70 (with an emphasis on the province of Hessen c. 1850), and pp. 249ff. on the period 1874–1907.

[24] On the redoubling of all three curricular elements from 1864–89, see Lipsmeier (1971) p. 251.

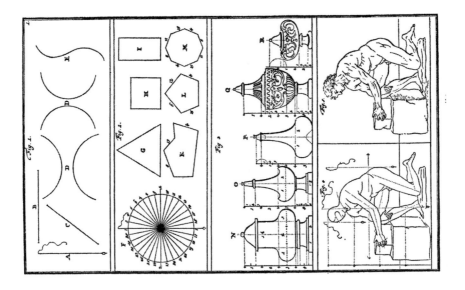

Fig. 82: An ideal typical sequencing of drawing lessons, starting with geometric lines and shapes (1 and 2), progressing on to simple descriptive drafting (3 and 4), to free figure drawing, as the final stage (5). From Preissler (1725*d*) vol. 1, pl.

implemented in the various nations. Their different emphases and blind spots have been neatly summarized in Alexander W. Cunningham's tabular survey (fig. 83), the first internationally comparative history of engineering drawing:

In France, Monge's pupil Victor Poncelet (1788–1867) and other *polytechniciens* and *normaliens* extended and deepened Monge's approach toward erecting a full theory of projective geometry and graphic statics. Projective geometry developed into a mathematical subdiscipline; later it merged into elasticity theory and developed further into separate subfields of mathematical physics and theoretical engineering.[25] In German-speaking countries, Ferdinand Redtenbacher (1809–63), an alumnus of the Vienna polytechnic and professor of mechanics and engineering at the Karlsruhe polytechnic since 1841, led the theoretization of engineering education. He was supported by his pupil Franz Reuleaux (1829–1905), who taught at the Zurich polytechnic from 1856 to 1864 and later became director of the Charlottenburg polytechnic in Berlin. Reuleaux replaced Monge's still rather descriptive geometry and kinematics with a *Theoretische Kinematik*.[26]

[25]See, e.g., Jean Dhombres in his introduction to Fourcy (1828*b*) pp. 62ff., Nedoluha (1960) pp. 48, 98; Timoshenko (1983), Scholz (1989) and Freguglia (1995).
[26]See Reuleaux (1875). Cf. in particular König (1999) pp. 16–27 on Redtenbacher and pp. 35–45 on Reuleaux.

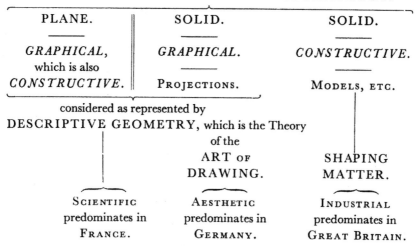

Fig. 83: Practical geometry as taught and practiced in France, Germany and Great Britain in the 19th century. From Cunningham (1868).

Thus, around 1865, a typical curriculum at polytechnic institutes only comprised between 15% and 30% drawing instruction. Twenty years earlier, the Augsburg polytechnic school required 10 hours each for courses in descriptive geometry, architectural and technical drawing, ornamental design and freehand sketching of organic figures, i.e., altogether 40 from a total of 55 hours per week during the first year of study. The latter two courses were continued in the second year, and architectural as well as technical drawing were even continued into the third year.[27] In the subsequent decades at the Berlin polytechnic, laboratory and drafting hours increased from c.35% of the total instruction time in 1881/82 to 45% in 1886/87 to 48% in 1888/89 and 1895/96 to well over 70% in 1898/99.[28]

Another tendency specific to Germany was an emphasis on drawing from models. Drawing lessons alone were evidently inadequate for generating good spatial intuition, so drawing from models became increasingly important, particularly as technical objects became more complex. The old remedy for this problem

---

[27] I.e., altogether roughly 50% of the overall curriculum. See Lipsmeier (1971) p. 108 on Augsburg and Emmerson (1973) pp. 154–7 on the *École Centrale* and the *Rensselaer Polytechnical Institute* (both 1850), MIT 1865, the University of Illinois 1867 and the *Pennsylvania Polytechnical Institute* 1862.

[28] Figures quoted in Hamilton (2001) p. 63 after Gispen (1989) p. 156.

was to resort to demonstration models from cabinet collections. They were occasionally displayed during the lecture on the tabletop.[29] In order to encourage the ability to translate 2D drawings into 3D structures and *vice versa*, teachers at German and Swiss technical colleges and polytechnics included drawing exercises explicitly requiring working from 3D models and sample objects. Drawing after models (*Zeichnen nach Modellen*) became another cornerstone of technical education in German-speaking countries, even though the original impulse to broaden visual education originally came from the painters and educators Ferdinand and Alexandre Dupuis in France.[30]

Compared with these heavily theory-ladened drawing practices in French- and German-speaking countries, those followed in the Anglo-Saxon world were much more pragmatic and production-oriented. In 1822, the Cambridge professor of chemistry and natural philosophy, William Farish (1759–1837), published his method of "isometrical perspective."[31] Accordingly, objectively similar lengths were supposed to be rendered on a drawing as lines of equal length (hence 'isometric' for 'same scale'). The complicated model machinery that Farish often used in his lectures had motivated him to develop this representation. This machinery had to be assembled beforehand and dismounted afterwards for storage. Farish wanted the drawings documenting this procedure for his assistants to be as clear as possible and free of perspectival distortion. As fig. 84 shows, isometric projection implies the choice of one and the same angle from which to view all three spatial axes. Usually $30°$ is chosen, since $\sin 30° = 1/2$, leading to an exact halving of lines or edges seen from this angle.

Isometric projection became better known from publications in the *British Cyclopedia* in 1835 and various textbooks on *Practical Geometry*, such as that from 1836 by Thomas Bradley, who taught at the Engineering Department of *King's College*, London, or J. F. Heather's *Elementary Treatise on Descriptive Geometry* (1851). This simple form of visual representation has been the favorite for the untrained eye ever since. It is used, for instance, in technical illustrations of maintenance and operation manuals. Shop mechanics and mechanical engineers usually prefer simple orthographic plots for production purposes, if necessary on several orthog-

[29] On this older tradition of model collections, see, e.g., Farish (1822) pp. 1f., Lipsmeier (1971) pp. 181ff.

[30] See Lipsmeier (1971) pp. 185ff. on the Dupuis brothers, and ibid., pp. 189ff. on their strong impact esp. in Germanic countries. On Austria, see esp. Nedoluha (1960) pp. 36–45.

[31] See Farish (1822). On Farish and isometry, see, e.g., Booker (1963) chap. 11, Purbrick (1998) and Lawrence (2003).

Fig. 84: A demonstration model, showing various types of power transmission, rendered in 30° isometric projection, with the light coming from the top left corner. From Farish (1822) pl. 2.

onal planes.[32] Monge's variant of descriptive geometry, in which two intersecting planes are just represented by their line of intersection, for instance, was considered "an abstract technique, difficult to grasp and of little value to the English in their efforts to establish a universal language of graphical communication across the syllabi of the engineering and architectural profession."[33]

In the 20th century, the advent of rationalization in production led to a further social differentiation of skills and professions. Technical drawings likewise developed further into some distinct types of inscription devices:[34]

---

[32]For a shopfloor example of the clash between these different representational styles, see Henderson (1999) pp. 37f. For further examples from the British context, cf. Emmerson (1973) chaps. 4, 7, Baynes & Pugh (1981), Purbrick (1998) and Booker (1963) chaps. 12–13; on the American scene, ibid., chap. 14 or Emmerson (1973) chap. 8.

[33]Lawrence (2003) p. 1273; see ibid., pp. 1274ff. on Peter Nicholson's "British system of projection," a parallel oblique projection, as a kind of "simple and visual" compromise solution between Monge's and Farish's versions.

[34]For the following taxonomy and good examples, see Ulrich (1958), Henderson (1999) p. 36 and König (1999) pp. 25f. For some of these entries, cf. also Baynes & Pugh (1981) pp. 12–21.

- the initial design drawings by hand or draft sketches;
- project outlines by the production office at the developmental stage;
- production blueprints with exact measurements of the component parts;
- detail views of complicated elements;
- specification diagrams with maximum tolerances, etc., for official certification;
- presentation and advertisement illustrations for potential clients, displaying the product, but not suitable for production purposes (to avoid stimulating unauthorized competition);
- maintenance schemes, specially made for users and service personnel, omitting many irrelevant details and concentrating on handling features;
- dimensional drawings and outlines of the basic idea for utility models and patent applications, not sufficient for actual production purposes either, but showing the principle of its operation;
- other types of complementary visualizations for the production line, such as block diagrams and flow charts, graphically summarizing the production process rather than the individual workpiece;
- and, finally, the whole range of 3D devices, including small-scale models, medium-size mock-ups and full-size prototypes.

The education and practice of engineers, designers and other technologists changed drastically with the invention of the electronic computer in the 1940s. The first graphical user interfaces emerged in the 1960s,[35] followed by computer drawing programs, such as Sketchpad, Paint, CorelDraw and Adobe Photoshop, and the boom in computer-aided design (CAD) programs and computerized production facilities such as computer numerically controlled (CNC) machines in the 1980s and 1990s. But in many fields, good technical drawing skills are still a cherished competency.[36]

> designers regard pencil and paper as more direct in getting a fleeting idea down as quickly as possible. [... Afterwards] they can then be entered into the computer, and electronic graphics capabilities can be used to

---

[35]The very first was invented in 1963 according to www.khulsey.com/history.html (accessed March 13, 2012).

[36]On the decline of formal instruction in drafting and the negative repercussions on an engineer's intuitive thinking, see Ferguson (1977, 1992). On the persisting relevance of drawing on paper, even in highly computerized contexts, see Henderson (1999), esp. chap. 8 on "mixed use practices, combining the electronic world and the paper world."

further refine them. Sketches can also be printed out again on paper when designers need to analyze the design as a whole. Again, at such an analysis phase, thinking is connected with the physicality of seeing and manually manipulating visual ideas. People employ such mixed practices to take advantage of the capabilities of the new tool while still maintaining important paper-world practices that are tied to thinking and analysis in visual terms.[37]

However, the heyday of highly sophisticated technical drawing by hand is over, and other skills now dominate engineering practice. It is still a visual culture, but very different from the one described in this section, which is centered on the late-18th and 19th centuries.

8.2   Slides, posters and plates in scientists' training

The idea to use optical projection to show images to broader audiences rather than just looking at precious individual copies of painted or printed images with just a few people at a time is quite old. The earliest implementation was called *laterna magica*. The earliest variants I am aware of are to be found in Giambattista Della Porta's *Magia Naturalis* of 1589 and in Athanasius Kircher's magisterial study on "the major art of light and shadows" from 1646.[38] Della Porta's *Natural Magic* still introduced the 'magic lantern' as a gadget mainly useful for irritating and confusing the uninitiated, whereas the Jesuit Kircher wrote a whole chapter on this device to lift its superstitious aura by carefully explaining its functioning to demystify it. Figure 85 is taken from Kircher's section "De Lucerna Magicæ seu Thaumaturgæ constructione." A large image of a skeleton (G) is projected onto a wall inside a dark chamber by a small device in which a fine pencil of light (here from an oil lamp I) is guided onto a small painted image on glass before enlargement by a simple lens (situated at H). In 1659, Christiaan Huygens (1629–95) introduced the important innovation of mounting a bi-convex lens at one of the two tube ends of the lantern, so that the projected image (which appears upside down) could be focused.

In the 17th and 18th centuries, so-called solar microscopes operated on a very similar principle to project enlarged images by daylight. Jean Paul Marat (1743–93) interestingly put this device to use to make hot fumes visible by their differing

---

[37] Henderson (1999) p. 206.

[38] See Della Porta (1589), translated into English in 1658, and Kircher (1646). On the early history of the 'magic lantern,' cf., e.g., Gosser (1981), Tom Gunning in Schwartz & Przyblyski (eds. 2004) p. 103 and Rossell (2008).

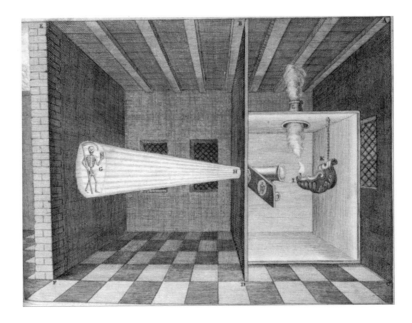

Fig. 85: Schema of a simple lantern-slide projection. From the second edition of Kircher (1646*b*) p. 768, which appeared in Amsterdam in 1671 and already includes Huygens's innovation. The image does not show that the projected image has to be inserted into the device upside down, and the slides are in front of – rather than behind – the lens tube, but these errors may have been a mistake by the unnamed engraver.

indices of refraction. He was persuaded to have thus demonstrated the existence of caloric.[39]

The major problem with these early devices was their dependence on strong natural sources of light, such as sunlight, only available on bright days and not in the evening or at night. With the invention of intense artificial light sources such as hydrogen/oxygen gas lamps and the electric arc, the 'magic lantern' lost its magic and became a common instruction aid in schools, churches and popular theaters, in scientific societies as well as in colleges and universities. These gadgets projected slides, consisting of a semi-transparent image, either a painting or a drawing and later mostly a photograph mounted on or between solid glass plates. The size of these slides ranged between c. 20×30 cm and just 2×3 cm, with 6×6 and 2.4×3.6 cm becoming two standard dimensions used until the late 20th century. The popularity of lantern slides declined with the development of motion pictures and the spread of radio technology, video and later television. Around

[39]See, e.g., Heering (2005) and Hentschel (2007) pp. 308ff.

the end of the 20th century, slide projectors went totally out of fashion, with the advent of digital images and beamer technology. Lantern slides had served as one of the most important and highly influential techniques of teaching and dissemination of science and technology for roughly 100 years.

In their heyday, slide projectors were omnipresent. The *Keystone View Company* of Pennsylvania, for instance, sold slides in bulk quantities. They could be purchased individually or in conveniently arranged larger sets of images, sometimes together with accompanying material that could be read out in the classroom or used for essay-writing assignments. Public providers such as the *Colonial Office Visual Instruction Committee* (COVIC) in Great Britain also resorted to this medium. Its aim was to "instruct the children of Great Britain about their Empire, and the children of the Empire about the 'Mother Country'." Slides were used in geography, biology, chemistry and physics, just to mention some subjects at school, but also for university education in anthropology, architecture, botany, zoology, mineralogy, geology and other divisions of natural history, medicine and various other research specialties such as spectroscopy.[40]

In all these areas, lantern slides were usually presented to pupils or students in darkened rooms so that their visual attention was totally absorbed by the displayed images carefully selected to maximize their effect. Repetition and slight variations in a theme lead to rapid recognition, and sooner or later the visual patterns are memorized, be they the appearance of a zebra or giraffe in zoology class, the intricate surfaces of various polished and etched solids in materials testing, or recurrent groups of spectral lines in spectroscopy. For the last example, I was able to document such educational techniques in greater detail and their impact on a larger group of students from *Wellesley College* (a women's college near Boston) between 1876 and 1916. From Sarah Frances Whiting's (1847–1927) lecture notes and instruction sheets for practice sessions, which were fortunately carefully preserved in the college archives, we learn that Whiting displayed important or typical spectra either as wall-hanging posters or as lantern slides and had her students practice redrawing those spectra until they memorized the characteristic line groups of key chemical elements and their exact locations in the spectrum. From various student notebooks and drawings, we can also infer how easily certain minor mistakes could creep into the students' drawings, despite the spectra having

---

[40]See Edyvean (1988), Hentschel (1999*b*), Abbott (2003), James R. Ryan in Schwartz et al. (eds. 2004) on the COVIC (quote from pp. 145f.), Bucchi in Pauwels (ed. 2005), Reiser (2010) on wall charts in these visual science domains; cf. http://legacy.mblwhoilibrary.org/leuckart/ on zoological wall charts by Rudolf Leuckart (1822–98). On lantern slides in materials testing, see here fig. .

Fig. 86: Main physics lecture hall at the Massachusetts Institute of Technology, Walker Building, Boston campus, 1890s, with several lantern slide projectors on the podium and at least four poster maps of the spectrum at the front and one big map to the right. Reproduced by permission of the MIT Museum, Cambridge, Mass., Department Photo Albums, scrapbook vol. 24, no. 4.

been displayed right in front of them.[41] This episode shows how important active (re)drawing is for the formation of a visual memory and Gestalt recognition skills. Wall-hanging posters and lantern slides – in our times, rather digital images and PowerPoint presentations – are thus necessary but do not suffice for training these skills. They have to be complemented by active learning stages. Students must not only look passively at images but also work actively with them. It is doubtful to me whether the demise of intense drawing lessons is fully compensated by our modern-day substitute technologies of digital drawing or Photoshop manipulation of digital images. Students who still used the traditional pen-and-ink drawing of the 19th and 20th centuries were probably much more intimately familiar with the patterns they drew than their modern counterparts ever will be.

[41] For details, see Hentschel (1999) and (2002) pp. 385–93.

8.3   X-ray atlases and training radiologists

When Conrad Wilhelm Röntgen (1845–1923) discovered x-rays in his laboratory in late 1895,[42] he had no idea what a huge impact this finding would have, not only on experimental physics, the field he taught at Würzburg University, but even more so on medicine, materials testing, astronomy and many other disciplines. On November 8, he had accidentally noticed that an electric discharge inside a partially evacuated Crookes tube, wrapped up inside blackened paper, nevertheless caused his barium platinocyanide screen to fluoresce visibly. He had been tracing cathode rays generated by such discharge processes.[43] Whatever it was that caused this fluorescence, it could not have been cathode rays, as the new rays had penetrated through the glass walls of the Crookes tube as well as the thick layers of paper, both of which were known to absorb cathode rays. Being a good and careful experimenter, Röntgen did not brush this observation aside, as several other experimenters who had noticed the phenomenon before him had done. Pursuing it further in a long series of experiments, he was thereby able to discover the new type of rays. He dubbed them "X-rays," but, particularly in the German-speaking world, they soon also came to be called *Röntgen-Strahlen* or roentgen rays. In the days following his initial observation, he first found out not only that the new rays could pass through thin paper and glass easily, but also that they were blocked by metal, bone and other dense matter. His hand left shadow-like traces on the screen when held between the tube and the screen. Another big leap forward was made when he found out that it was possible to record these traces. Thus far, they had only appeared faintly on a chemically sensitized screen. By means of photography, he could produce lasting traces as evidence, which helped enormously in achieving very rapid acceptance and replicability of his new findings. This medium also quickly transformed this branch of experimental physics into a new visual science culture – indeed into one of the most impressive visual cultures ever created. On December 22, 1895, Röntgen succeeded in photographing an exposure of his wife's hand, and a few days later he distributed printouts of this photographic negative depicting her marriage ring on a skeletal hand to about 70 fellow professionals in Germany, France, England and elsewhere. On New Year's Day, he wrote the first preliminary communication about his new findings. It was published in the proceedings of the Würzburg Society for Physical and

[42] On Röntgen's vita and on the discovery of x-rays, see, e.g., Glasser (1931) and Dommann (2007) chap. 1.

[43] On this research tradition that soon also led to the discovery of the electron and radioactivity, see, e.g., Eisenberg (1992) chaps. 1–3, Dahl (1997) and Müller (2004).

Medical Studies, and a series of others very soon followed, by himself and some of his colleagues in the field about successful replications. On January 5, 1896, the first newspaper article on these findings appeared in the Vienna daily *Die Presse* and was quickly picked up by a host of other sensationalist accounts. In a matter of a week, photographs of human hands semitransparent to x-rays and displaying all their fragile bones along with foreign objects such as metal jewelry or an intruding bullet[44] had become a standard ingredient of these early reports on roentgen rays and their discovery. Without such graphic illustrations, it is unlikely that the new technology for creating and recording the new rays would have had such a rapid acceptance and diffusion. Even the Kaiser wanted to be informed; Wilhelm II granted Röntgen a private audience in Berlin on January 12.[45] He was as intrigued as the artists who inferred a new feeling of space from radiographs. Among the general public, fears of becoming totally exposed to x-ray views were rampant in surrealistic sketches and sensationalist reports. The public reaction soon developed into what can appropriately be called a media hype. The crowning moment was the appearance of hypertrophic caricatures of ladies and gentlemen exposed down to their skeletons by x-ray photographers. To be able to peek beneath many layers of clothing, to actually look inside solid matter, was fascinating and at the same time scary for the public at large.[46] The reception among specialists from various disciplines was equally strong and enthusiastic. On the one hand, medical doctors quickly abandoned their usual reservations about technical apparatus that they had not been trained to handle. On the other hand, scientists in physics, chemistry and biology, besides engineers and technicians, were equally fascinated by the new possibilities and helped to improve the technology for generating and recording the rays. The quality of x-ray photographs quickly improved (cf. fig. 87 for an example from February 1896, taken only two months after the discovery of the new rays). High-quality images, in turn, helped to convert medical doctors entirely to the new technology. Within a few months of Röntgen's discovery, the first handbooks on x-rays and their applications appeared,

---

[44]On an episode from 1897 in which bullet fragments were detected in a boy's hand after the wound had healed, see Pasveer (1989) p. 368. Other examples are given, e.g., in Brecher & Brecher (1969) pp. 9ff., 56, pl. VI, Eisenberg (1992), Lemmerich (1995) pp. 111ff., Dommann (2001) and (2007) pp. 7ff., and Vera Dünkel in Bredekamp, Schneider & Dünkel (eds. 2008) pp. 145f.

[45]See, e.g., Schüttmann (1995) pp. 12ff. On the x-ray hype, cf. Dommann (2001), Keller (2004), Roth (2013) pp. 135ff., and Monique Sicard in Martinet (ed. 1994) pp. 28f., esp. on France.

[46]The cultural resonance with other developments in architecture, the arts and sciences, rendering solid structures increasingly transparent, is discussed by Asendorf (1989). Further examples are found in Henderson (1988) and Dommann (2007) pp. 19–21.

geared first and foremost to medical applications.[47] Another advantage of the new technology was often repeated: "the patient can now see with his own eyes the real condition of his abnormality before having a part of his body removed."[48] The new visual representations of the interior of the human body became an apparently direct mediator between doctors and their patients. The visual images seemed to speak for themselves. Complicated verbal medical explanations, riddled with physician's jargon, were anything but clear by comparison. That no-one at the time knew for sure what this new radiation, so-called x-rays for "unknown," actually was, bothered only a few people – most notably its discoverer – so long as the images they generated were good. Their ontological identification as a type of electromagnetic radiation only became unambiguous after x-ray interference was established in crystals in 1912.[49]

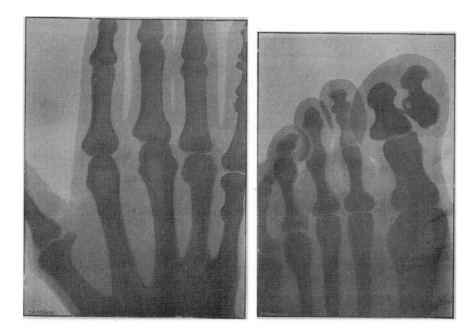

Fig. 87: Left: x-ray exposure revealing birdshot in a hand. Right: x-ray exposure of the toes of a foot, with a supernumerary phalanx. Both photographs from S. Rowland (Feb. 8, 1896) p. 364.

[47]Pasveer (1989) p. 361 quotes H. S. Ward: *Practical Radiography*, appearing in May 1896. Cf. also Kassabian (1901).

[48]Rowland (1896) p. 362; cf. also Eisenberg (1992) chap. 4, pp. 12–22.

[49]See Hentschel (2007) sec. 7.2.2 on the intense discussions and arguments about the various optional explanations raised at the time, sound waves or longitudinal ether vibrations among them.

The fascinatingly detailed sharpness of such images seemed to imply that x-ray images were unproblematic to use and easy from the start. But this assumption was mistaken. Few parts of the body are as easy to screen with x-rays as the palm of the hand. The torso presents a confusing overlapping of various layers of bones and inner organs, not to speak of the individual variations between different patients (see fig. 89). The fluctuations in quality of available photographic material at the time only complicated things, along with the uneven intensity and spectral distribution generated by the x-ray-emitting tubes and other technical limitations. The historian and sociologist of medicine Bernike Pasveer, the historian of technology Monika Dommann and the historian of art Vera Dünkel have all argued convincingly that the diagnostic message of x-ray images was by no means transparent or self-evident but had to be actively shaped by x-ray workers of various kinds (ranging from physicians to technicians to scientists). Pasveer, in particular, claims that 'skiagraphy' – the science of shadows, as radiology was also called in the early days – was constructively shaped by the following four routines:[50]

1. experimentation with the new technology, including ray-generating tubes, high-voltage generators, photographic materials and the specimens for examination;
2. direct comparison of x-ray images of corpses with radiographs of living patients, and likewise between images from healthy and ill patients;
3. 'translation' of diagnostic information acquired by other methods into the shadowy images of x-rays; and, finally,
4. comparison of new x-ray images with a rapidly accumulating inventory of stored standard images.

I would argue that these four routines are significant also from our comparative point of view, because we find the same routines at work in the early stages of several other visual cultures in science and technology. But staying with radiology, for the moment, it is significant that even insiders concede that "for the first five years, x-ray apparatus was more an interesting toy than a weapon of value in medicine [...] Looking back, it seems remarkable that any results could be obtained with such makeshift and unreliable apparatus – still more remarkable the

---

[50]On the following, see Pasveer (1989), quotes from pp. 361 and 377. On the social construction of medical images, cf. also Yoxen (1987), Lerner (1992), esp. p. 387 on the vanishing of the term 'skiagraphy' (which connotes haziness and uncertainty) in the 1920s, Cartwright (1995), Dommann (2007) pp. 14–23, 279ff., Dünkel in Bredekamp, Schneider & Dünkel (eds. 2008), p. 141 and Roth (2013) p. 154.

range of examinations attempted and their comparative success."[51] X-ray images were relatively unproblematic to use for the diagnosis of bone fractures and for the localization of foreign bodies (e.g., bullets, artificial teeth and needles); but attempted applications in internal medicine – for instance, in the diagnosis of tuberculosis or lung cancer – were far more complicated. When John Frederick Halls Dally (1877–?), assistant resident medical officer at the *Royal National Hospital for Consumption* in Ventmor, England, wrote a survey article on the use of roentgen rays in the diagnosis of pulmonary diseases in 1903, he did *not* claim that the new technology was a royal road to the art of diagnosis nor that it was infallible. He rather advised dual usage of old and new techniques side by side. "Without intuition or previous study the one is almost as incomprehensible as the other, but as we gaze the wealth of detail rises before our vision until finally we are able to interpret the meaning of streaks and shadows that to the untrained eye are meaningless."[52] Interestingly, he tried to convince his more traditionally minded colleagues by bringing diagnosis with x-rays into line with the older methods by means of a detailed analogy (see fig. 88).

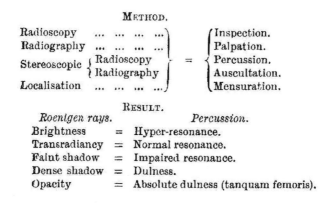

Fig. 88: Analogy between new and old diagnostic techniques. From Dally (1903) p. 1800.

This tabular comparison between old and new methods and results is symptomatic of a general tendency by new technologies to camouflage their novelty under an old cloak. When daguerreotype portraits were initially introduced, they were first framed like oil paintings. Similarly, cars first looked like motorized coaches. The

---

[51]Barclay (1949) p. 300. On the introdution of x-ray images into medicine, see Brecher & Brecher (1969) or Lerner (1992) for the USA and Canada; Lemmerich (1995) chaps. C and D on Germany.
[52]Dally (1903) p. 1806.

new diagnostic techniques were supposedly easier to swallow when an analogical relation was drawn between older techniques of exploring the interior of the human body. What makes this example so striking is the obvious disanalogy between visual and acoustical techniques in this comparison.

Initially, the handling and application of the new rays was neither controlled nor regulated in any way. As a result, various people who had more than fleeting contact with these exposures, especially doctors who regularly used the rays in their practices or patients who were repeatedly treated with them, developed clear side effects: initially a reddening of the skin and eyes; in worse cases, deeper inflammations, skin cancer, and in 1904 the first of several deaths.[53]

As the inherent dangers of this new technology began to come to light, there was a tendency to professionalize this new branch of medicine. It was increasingly practiced in hospitals and only very rarely outside of them. During the first five years, all kinds of people of very mixed backgrounds had been experimenting with x-rays and working with makeshift apparatus to generate and record them. In the early days, radiology was in fact "a piebald proceeding, [...] which fitted no one. The attendant wires and sparks suggested an electrician's work, surely; [...] but there were the plates, dark room and chemicals, considered usually the accessories of the photographer. Neither of these artisans, on the other hand, could be expected to intrude themselves so far into the realms of medicine as to offer a diagnostic verdict."[54] Consistent with radiology's heterogeneous roots in physics, photochemistry and photography, electrical engineering and medicine, the first societies for the new field, such as the British *Röntgen Society* (established in 1897) and the Dutch *Nederlandsche Vereeniging voor Eletrologie en Röntgenologie* (founded in 1901), usually admitted both medical and nonmedical members. After the first half-decade of such interdisciplinary work, some voices began to be heard demanding a stronger professionalization of radiology within medicine. By 1910, the growing availability of standardized apparatus helped to establish radiology as a standard routine of medical diagnosis. The skill and know-how of the pioneering experimenters had been black-boxed.[55] World War I

---

[53] According to Martinet (ed. 1994) p. 30, Clarence Dally, aged 39, was the first known fatal case after the amputation of the inflicted limb. Cf. also Brecher & Brecher (1969) chap. 7, Eisenberg (1992) chap. 11, Dommann (2007) chap. 7 and Metcalfe (1910) p. 433 on contemporary knowledge about the pathological action of x-rays on healthy skin.

[54] Brecher & Brecher (1969) p. 104. Swiss examples of this early hodge-podge of apparatus and skills are given in Dommann (2007) pp. 53–74.

[55] See, e.g., Brecher & Brecher (1969) chaps. 4, 9 and 15–17, or Dommann (2007) pp. 82ff. on the importance of stable, easy-to-use apparatus.

counteracted these tendencies, however, because as many hands as possible were needed to operate the radiological apparatus in field hospitals; and many volunteers (including the physicist Lise Meitner) were quickly trained to assist in the war effort. Consequently, the ultimate substitution of laymen for trained professionals only set in around 1920, when diplomas were issued and special examination procedures were instituted. Radiologists became broadly accepted and esteemed experts in the production and interpretation of x-ray images. Conversely, the more technically inclined parts of the complex task were increasingly left to so-called technical assistants, also called radiographers, and other technical service personnel of much lower status in the medical hierarchy at hospitals and university institutes.[56] In this differentiated and ranked system of medical and technical expertise, all these specialists insisted on demarcation from the layman: "There is a danger if radiograms fall into the hands of the lay public. They are misread. Only those with a medical education can comprehend the actual position of the parts shown in the radiogram and allow for the distortion produced by the method."[57]

Atlases of x-ray images taken from healthy models provided a kind of fixed point against which images of patients could be gauged.[58] Other atlases specialized in displaying exemplary diseases in order to train radiologists in recognizing the typical appearance and variations in visual indicators on radiographs.[59] Contrastive atlases showed plates of normal and pathological findings side by side.[60] Images taken from slightly different angles helped the reader to form a mental 3D image of, say, the patient's lung or abdomen, from the 2D shadow-photographs. Sometimes, control images were necessary to raise awareness about artifacts. Sources of such artifacts were overexposure or underexposure of the chemically sensitized emulsion or photographic plate, fluctuations in the so-called ray 'hardness,' i.e., the radiation characteristics generated by the x-ray tube (for which there were no

[56] For details on the various stages of this dual process of professionalization for radiologists and their technical assistants, see Pasveer (1989) pp. 363–7, Blum (1999), Dommann (2001) and (2003) chap. 3, and further sources cited there.

[57] Metcalfe (1910) p. 432. Cf. Moser (1939) and Dommann (2007) pp. 249ff. on Leonie Moser's learning the skills of x-raying as a medical assistant.

[58] See, e.g., atlases by Max Immelmann in Berlin 1900, Eiselsberg & von Ludloff in Berlin 1900, Rudolf Grashey in Munich 1905, 1917; *An Atlas of X-ray Interpretation: Bones* by J. F. Calder, 1987 or *The Chest X-Ray: A Systematic Teaching Atlas* by Matthias Hofer, 2007. The last two examples show that these atlases are still being produced today.

[59] See, e.g., the atlases with images from tuberculosis patients published by Otto Ziegler in Würzburg 1910, Hanns-Alexander & Arthur Beekmann in Leipzig 1927ff., Walter Brednow (1933), or atlases exhibiting typical war-related injuries (by Nicolai Gulecke in Berlin 1917).

[60] See, e.g., MacKendrick & Whittaker (1927). In Germany, *Archiv und Atlas der normalen und pathologischen Anatomie in typischen Röntgenbildern* was issued in more than 50 different editions.

reliable quantitative gauges during the first two decades of radiology), blurring caused by patient movement during exposure (which could take up to several minutes in the early days of this technique), and idiosyncratic variations in the shape and location of internal organs.[61] Therefore, "it is always necessary to control any one picture [...] by a second."[62] In all these different ways, image comparison (strategy no. 4 above) was – and still is – one of the most decisive steps in radiological diagnosis. A final strategy by which early radiologists learned to see with x-rays was retro-spectroscopy, i.e., careful checking of old plates against autopsy findings after patients, who in the past had been x-rayed and radiologically diagnosed, had died. The anatomical findings could then be compared against the faint shadow-traces on the plates and diagnostic clues could be obtained.[63]

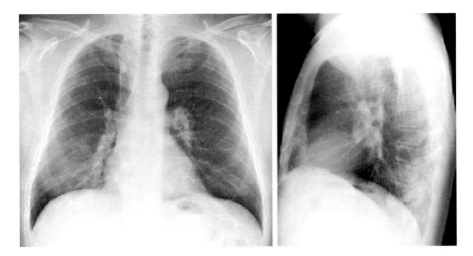

Fig. 89: Radiographs of human lungs with dorsal opacity, visible only in the side-view radiograph. From www.mevis-research.de/ (posted in March 2004). By courtesy of Prof. Hans-Holger Jend.

As the preceding examples show, the perspective of any single skiagram was one source of potential misdiagnosis. Not surprisingly, in 1896 the electrical engineer Elihu Thompson (1853–1937) suggested adapting stereoscopic techniques readily

[61]On these sources of error and the resulting importance of intense schooling in the visual analysis of radiographs, see, e.g., Grashey (1905) and Lerner (1992) pp. 386ff.

[62]Metcalfe (1910) p. 433. Cf. here chap. 7 on various kinds of interconnections between different scientific images.

[63]Sad and slightly cynical though this strategy is, because it comes too late for the patient involved, it was crucially important for the gradual build-up of radiological expertise; see Brecher & Brecher (1969) pp. 109ff.

available in normal photography (see here fig. 3 on p. 22) in order to "secure some indication in space of various imbedded solid objects."[64] The idea had already occurred to some radiologists to move both the x-ray tube and the photographic plate around the target sample while maintaining a constant distance between both. This blurs out the shadows of the other matter around the target as the shadow of the target intensifies. The results of these early efforts at body-section 'tomography' were not very convincing, though. Nevertheless, despite its "terrible technical quality," the technique added a new tool to the radiologist's toolbox for examining special cases. Tubercular cavities, for instance, were in fact indirectly attenuated in radiographs by the blurring of the overlying rib cage.[65] Another way to overcome the static and perspectival disadvantages of x-ray images was to dynamicize them. The easiest way to do so was to observe the rays using a fluorescent screen – the patient could move, or breath in and out, and the physician would be able to see the changing shadows on the screen. But none of this could be recorded for closer examination or analysis. The medium of cinematographic film certainly offered two new options: one could either film the fluorescent screen, or record the x-rays directly on film material specifically sensitized to those wavelengths.[66] This way, x-ray films of respiring or moving bodies were produced. However, x-ray atlases remained the most important medium for training radiologists. Aside from educational and research purposes, radiographs also fascinated experimenters and artists for their aesthetic qualities. Within a few months of Röntgen's discovery becoming generally known, the two Viennese experimental physicists Josef Maria Ludwig Eder (1855–1944) and Eduard Valenta (1857–1937) produced a set of 15 high-quality x-ray exposures that were printed as heliogravures at the Royal Imperial Institute for Instruction and Experimentation in Photography and Reproduction (*Lehr- und Versuchsanstalt für Photographie und Reproductions-Verfahren*, cf. here fig. 90 and Faber (2003): Note the artful layout of the x-rayed objects on the available surface and the delicate, subtle half-tones in the radiograph obtained).

[64]Thompson as quoted in Brecher & Brecher (1969) pp. 62f. Cf. Kassabian (1901) pp. 232–40, Eisenberg (1992) chap. 5 and Lerner (1992) p. 391.

[65]See Lerner (1992) pp. 391f. for this quote from an *American Roentgen Ray Society* conference in 1936. Cf. also Brecher & Brecher (1969) chap. 10, esp. p. 128. On the more sophisticated computed tomography (CT), see here p. 202.

[66]On early x-ray fluoroscopy and cinematography, see Janker (1931, 1949), James (1935), Mitchell (1935), Holmgren (1945), Curtis (2004), Dommann (2007) pp. 275f. and here sec. 7.4.

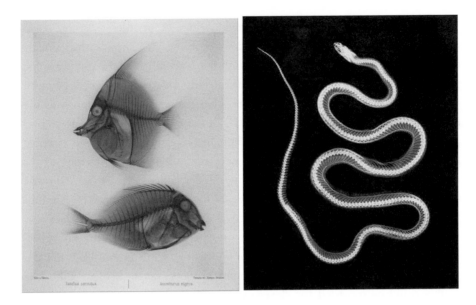

Fig. 90: Artful heliogravures of x-ray photographs. Left: Two flat fishes (moorish idol: *Zanclus cornutus*; and blue-lined surgeonfish: *Acanthurus nigros*). Right: An *Aesculap* adder. Both reproduced from Eder & Valenta (1896) pl. X and XV.

Later examples from various research fields in medicine, science and materials testing making use of x-rays confirm that radiographs have kept their strong aesthetic attraction. Even high-energy physicists, not particularly noted for refined taste, included two pretty radiographs of flowers and shells in a centennial issue on x-rays in the *Stanford Linear Accelerator Center* (SLAC) magazine.[67]

Altogether, x-ray images, atlases and films opened up a new avenue to seeing inside the human body. Skiagraphy, radiography or radiology, as this field was ultimately called, became a very important visual subculture of medicine. Its specialists go through extensive training in order to familiarize themselves thoroughly with the strange world of faint shadows. Through this training, radiologists acquire a particular skill in seeing *Gestalten* in images only exhibiting faint and strange shadows to the nonexpert. Just as in other visual science cultures, a particular aesthetics attended these virtuosic skills, together with an in-built fascination that still binds x-ray workers to their medium, even though the dangerous medical side effects of the new technology have forced them to work in close, lead-armored underground cabinets.

[67]See *Beam Line* 25, no. 2 (1995) pp. 6–7.

9

Mastery of pattern recognition

Philip Ball at the London office of *Nature* has claimed (1999) that pattern recognition is one of the most essential feats in which humans in general, and scientists in particular, excel. For a pattern to become recognizable, it first has to be viewed frequently enough. There must be sufficient chance to memorize and categorize it precisely, in order for it to be distinguishable from other similar patterns.

9.1 Visual inventories of possibilities

One way to achieve this is to draw various patterns alongside each other. Students or beginners can go through a visual inventory to study the features of each pattern at their own pace. Examples of various kinds of visual inventories abound in the sciences. One could start with the machine components (such as wheel and axle, screw, gear, pulley and wedge) into which Leonardo da Vinci mentally segregated complex machines.[1] More than two centuries later, Christopher Polhem (1661–1751) developed a "mechanical alphabet," i.e., a large set of some 80 wooden models displaying basic mechanisms, transmissions, gears, couplings, ratchets, and so forth. His idea was that an engineer with a concrete task at hand could systematically go through the repertory of miniaturized 3D models and search for the optimal solutions to his problem, thus combining letters of the 'mechanical alphabet' into complete words (machines) and sentences (sets of machines in factories). This 'mechanical alphabet' was actually used in teaching at the *Laboratorium mechanicum* – Sweden's first school of technology – and later also at the Institute of Technology, predecessor to Sweden's Royal Institute of Technology (*Kungliga Tekniska Högskolan*).[2] This creative way of using 3D models in teaching engineers was soon copied elsewhere and remained a standard part of the curriculum well into the 20th century, as is demonstrated by fig. 91.[3]

---

[1] Mainly in his *Codex Madrid*, which remained unpublished at the time; it anticipated much later systematizations, such as in Franz Reuleaux's *Kinematics of Machinery* (1875/76). On Leonardo's and Reuleaux's machine elements, see Moon (2007).

[2] On Polhem's mechanical alphabet, cf. Hallden (ed. 1963) pp. 36, 134f., 157f., Ferguson (1977) p. 835 and www.tekniskamuseet.se/1/318_en.html Thirty-two of these models are kept in the National Museum of Science and Technology (*Tekniska Museet*) in Stockholm; thirteen other models are on display in the Falun Mining Museum, both in Sweden.

[3] See Ferguson (1977) pp. 835f. and Hindle (1983) p. 135 for further French and US examples, i.e., Jakob Bigelow's 'Elements of machinery' shown at Harvard University, and similar tabular displays by Jean Hachette, Philippe Louis Lanz and Augustin de Betancourt, all pupils of Monge at the *École Polytechnique*.

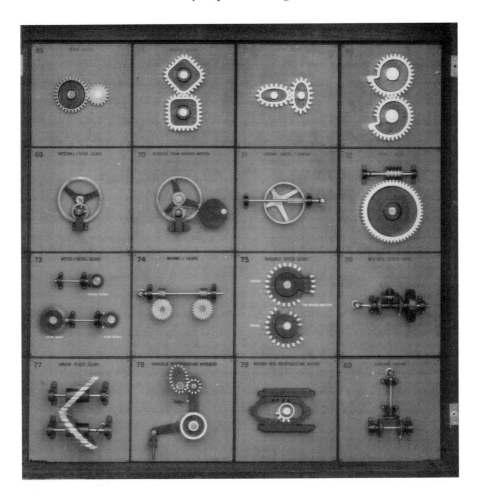

Fig. 91: Sixteen different types of gears. Display by W. M. Clark, 1929. This tray is one of altogether 10 such sets on display at the Newark Museum, New Jersey, USA. Reproduced by permission.

In order to emphasize the connection with visual thinking (to be covered more fully in chap. 10), I have selected three further examples: (i) the so-called Chladni figures as a late 18th-century visual inventory of modes of vibrations of surfaces; (ii) a survey of possible mechanisms for channel gating in late 20th-century protein biochemistry and (iii) a plate illustrating Linné's system of plant classification.

9.1.1   Chladni figures

Ernst Florens Friedrich Chladni (1756–1827) was the offspring of a long dynasty of Wittenberg theologians, academics and learned men.   After pursuing studies of jurisprudence and philosophy in Wittenberg and Leipzig, he obtained his law degree in 1782 from the University of Leipzig.   While researching acoustics, Chladni followed Robert Hooke's lead in examining the vibrations caused by a violin bow striking the edge of a glass plate sprinkled with flour.[4] Chladni sought to perfect this method and developed a technique to reveal the various modes of vibration of rigid surfaces.[5]

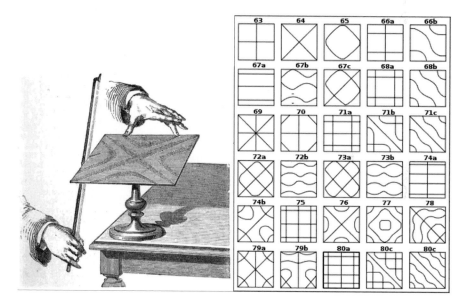

Fig. 92: Left: setting a plate into vibration by means of a violin bow. Right: the resulting Chladni figures classified by the number of nodes along the edges of the quadratic plate. From Stone (1879) p. 26, pl. 12 and Chladni (1802) pl. V, respectively.

Chladni's technique, first published in 1787, consisted in drawing a bow across the edge of a metal plate with a light dusting of sand on its upper surface (cf. the left part of fig. 92). The bowing continues until the plate reaches resonance.

---

[4]See Hooke's *Diary 1672–80*, ed. by Henry W. Robinson & Walter Adams (eds. 1935), p. 448, entry of July 8, 1680; on Hooke's pioneering contributions to various visual cultures, including microscopy and telescopy, see here pp. 30 and 159.

[5]On Chladni's research and results, see Ullmann (1996), Jackson (2006) chap. 2 and further sources listed there.

The vibration causes the sand to concentrate along the still nodal lines along the surface, thereby tracing their contours. By carefully choosing the points of maximum vibrations (where the bow is placed) and the nodal points (where no vibration can occur, either because of an inbuilt fixture or by a finger placed onto the surface), Chladni could generate a whole array of patterns, which he classified according to the number of nodal points along the edges (cf. the right part of fig. 92). Long before any theoretical calculation of these complex resonance patterns was possible, Chladni's visual technique thus allowed their experimental production and examination. Since Chladni did not have a tenured position, he earned his living from the sale of instruments, popular books and lectures delivered across Europe. Even the French emperor Napoléon Bonaparte was deeply impressed by the man who made sound visible.[6] The intricate patterns – soon called Chladni figures – became highly influential in practical acoustics and musical instrument-making and are still in use to examine resonance phenomena and vibrational patterns.

9.1.2   Channel gating in neurobiology

The historian of biosciences Maria Trumpler has shown how a combination of various techniques of interventions and forms of representation led to a convergence of images and models of the way nerve cell membranes operate. First, a host of possible mechanisms was identified, then simple model diagrams were created to depict these mechanisms as clearly as possible, and then methods to look for these functional processes were sought. The following survey of possible mechanisms for channel gating (fig. 93) comes from a textbook published in 1984.

This type of inventory of possible mechanisms is just one of many building blocks for mental modeling of processes in neurobiology. Trumpler showed how scientists combine various static 2D representations and models of varying degrees of complexity to form a "complex mental image which can incorporate all perspectives simultaneously, reflect differing time scales at will" and end up with a kind of mental "collage of various molecular models."[7]

---

[6]On this episode, see Jackson (2006) p. 34f., Stöckmann (2007) and Tkaczyk (2014). A retrospective tempera painting depicts Chladni, Le Cépède, Napoléon, Laplace and Bertholet; see www.mpiwg-berlin.mpg.de/de/forschung/projects/RGTkaczyk (accessed Dec. 8, 2013).

[7]Trumpler (1997) pp. 87ff. Cf. also Cambrosio et al. (2005) on Linus Pauling's equally multi-faceted visualization chains and narrative sequences of antibody formation; more generally, Hindle (1983), Gooding (2010) and here p. 331 on the complex interplay of 2D, 3D and 4D representations in science and technology.

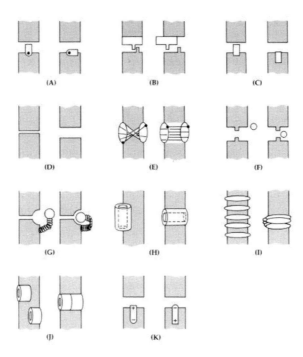

Fig. 93: Possible mechanisms for channel gating: rotating or sliding (A–C), shutting or twisting pores (D, E), blocking with separate particles (F, G), swinging in and out (H), assembling subunits (I, J) or blocking and releasing by electric charge (K). From Hille (1984) p. 335, fig. 2. © Sinauer Associates, Inc. Cf. Trumpler (1997) pp. 76ff. for the later specifications.

9.1.3   Linné's system of plant classification

Switching fields once more in order to underscore the ubiquity of the phe-nomenon, let us look at the 18th-century taxonomist Carl von Linné (Linneaus 1707–68), who introduced a new classificatory system into botany based on the number and form of stamens and pistils to determine how a plant was to be classed. A plate designed and drawn by Georg Ehret (1708–70) illustrates this sexual system of botanical classification. This first-rate botanical illustrator pub-lished it in a small booklet on the *Methodus plantarum sexualis in sistema naturae descripta* at Leyden in 1736, one year before Linnaeus himself released his much better known *Genera Plantarum* in print[8] (see here color pl. IV for Ehret's original

---

[8]On Linnaeus's classificatory system, see, e.g., Heller (1964), Stafleu (1971) and Müller-Wille (1999), also Hentschel (2007) pp. 65f. On Linnaeus's collaboration with Ehret and their joint work in George Clifford's (1685–1760) garden, see Calmann (1977) chap. 4, Charmantier (2011) and www.nhm.ac.uk/natureplus/community/library/blog/2011/04/01/item-of-the-month-no-8-

hand-colored drawing with pencil corrections that were incorporated into the final engraving). Linnaeus later supplied the readers of his *Philosophia Botanica* (1751) with a full-page plate of 62 different leaf shapes, apparently in the hope of having thereby exhausted the possible leaf forms in nature. The legend to his plate denotes terms specifying each of these shapes that botanists were henceforth able to use in clear and distinct formal descriptions of species. This became a ritualized part of Linnaeus's system of nature, requiring that every new species be formally described and classified accordingly.[9] As far as the colors of plants were concerned, the illustrator-brothers Franz Bauer (1758–1840) and Ferdinand Bauer (1760–1826) developed a numerical system by which to quickly denote 140 different hues (see here color pl. VI and VII). This allowed them to dispense with painting in the field. They just noted down the appropriate numbers from the color scale onto their black-and-white sketches and left the work of hand-coloring to a later stage back in their studio.[10]

Efforts to routinize the referencing of colors and to standardize their denominations abound in visual science cultures. We have already spoken of Lambert's color triangle (on pp. 121ff.) and the astronomer Charles Piazzi Smyth's plate of spectral colors (see p. 142). His father, Admiral William Henry Smyth (1788–1865) had imbued in him this refined sense of color from painters such as Giovanni Battista Lusieri (1755–1821). Smyth senior attempted in his *Sidereal Chromatics* (1864) a precise method to determine the colors of double stars. Expanding upon his *Bedford Cycle of Celestial Objects* (published in 1844), which had garnered a gold medal from the *Royal Astronomical Society* and a two-year presidency of this society, the *Sidereal Chromatics* provided both a theory for why double stars manifest colors and a method for determining their most exact recording. In detailed charts, Smyth senior compared more than 100 double stars with his own previously published observations and those of his fellow astronomer, Father Benedetto Sestini (1816–90). Smyth's famous color chart attempted to standardize this identification of double-star colors.[11]

---

april-2011–georg-ehrets-original-drawing-to-illustrate-linnaeus-sexual-system-of-plants.

[9] For a reprint and further commentary on Linnaeus' 1751 pl. I, see Nickelsen (2004a) pp. 178–81 and Charmantier (2011), who also shows that "Linné's thinking was profoundly visual and that he routinely used visual representational devices" (ibid., p. 365) such as sketches, analytic (structural) drawings, maps, tables and diagrams despite his reputation as a poor draughtsman.

[10] Cf. Hans-W. Lack (2000b) pp. 136 and 215, who found this long-lost color code in a Madrid archive, Lack & Ibáñez (1997), Nickelsen (2004a) pp. 204–7, Bleichmar (2012) pp. 97–100, Mabberley & San Pino (2012) and here chap. 11 on color taxonomies.

[11] On the earlier and further development of this branch of astronomical research, see, e.g., Malin & Murdin (1984), or Roth & Fosbury (2013) and further references given there.

9.2 The illusory pattern of Martian canals

Historians of astronomy have pointed out repeatedly that what motivated their actors to do their night-long surveys in the chilly domes of their observatories was an intrinsic fascination with their objects of study. The supposed 'canals' of Mars will serve here as an example of a search for patterns that went seriously awry. By studying such a story of scientific failure, I want to underline the difficulty and riskiness of this approach. Searches for patterns at the very limit of visibility may well hit upon something important, but they are equally likely to end up getting waylaid by illusory features that human observers project onto the phenomena. That is exactly what happened in this case. This episode captured the attention of the few professionals in planetary astronomy as well as the public at large.[12]

The planet Mars, the fourth closest to the Sun and relatively near to the Earth, has aroused the curiosity and awe of mankind throughout history. Its rather eccentric orbit could be reconstructed on the basis of Tycho Brahe's pre-telescopic quadrant measurements of its positions as a function of time and was the key to Kepler's planetary orbit theory. Any closer observation of the planet had to await the invention of better telescopes, since even Galileo's, Kepler's and Scheiner's early telescopes produced a too small field of view to be able to discriminate any details. Francesco Fontana (c. 1580–1656) was the first to claim to have seen a few darkish markings on its surface in 1636, and Christiaan Huygens (1629–95) made sketches of the most pronounced of them, nowadays known as Syrtis Major, in 1659. Assuming it was a permanent surface feature, Huygens could determine the rate of rotation of the planet and arrived at a surprisingly good estimate that was later also confirmed by Giovanni Domenico Cassini (1625–1712) in 1666. Cassini was the first to describe the white areas close to the planet's axis of rotation. They were termed 'polar caps' in open analogy to the Earth's ice shields at its northern and southern poles. That these white caps seemed to grow and recede periodically was interpreted as indicative of the existence of Martian seasons, again in analogy to terrestrial summer and winter periods. The darker areas were consequently interpreted as seas (*mares*) in analogy to dark areas on the Moon's surface and to terrestrial oceans. William F. Herschel (1738–1822) continued this trend when in the 1780s he interpreted other more rapidly changing surface features as indicators of "clouds and vapours" and consequently as proof of "a considerable but moderate atmosphere." Further observations left him in no doubt that there were living beings on Mars and that these inhabitants

---

[12]For good overall surveys, see Hetherington (1976), Sheehan (1996), Lane (2006) and Hoyt (1976) or Strauss (2001) with particular emphasis on Percival Lowell.

"probably enjoy a situation similar in many respects to our own."[13] I would like to point out that none of these observers mentioned so far were cranks, scientific outsiders or lunatics. They counted among the top astronomers of their day. These early episodes from the history of Martian exploration show the latent tendency of observers to form analogies between planetary and terrestrial features, i.e., to map relations between observed and well-studied phenomena from a familiar base domain (here, the Earth) onto a new and yet unknown target domain (here, Mars and its unexplored surface).[14] Also in analogy to terrestrial geography, the late 18th and early 19th centuries experienced a trend for mapping Martian topography. The German astronomers Wilhelm Beer (1797–1850) and Johann Heinrich von Mädler (1794–1874), who were famous for their map of the lunar surface, published in 1840 the first serious chart of Mars. Whenever Mars was in a favorable position and the seeing conditions were reasonably good, other observers followed suit in producing increasingly detailed emendations.[15] Whereas these maps grossly differed in depiction style and detail, many of the larger and more striking features, at least, stabilized. They were given names in a similar manner to the denotations of lunar craters and other surface features of our closest neighbor in space. As a representative of these Martian maps immediately before the new discoveries of 'canals,' I depict in fig. 94 the one by the British landscape painter and amateur astronomer Nathaniel Everett Green (1823–99), who taught at the *Royal Academy* in London and gave art lessons to Queen Victoria and other members of the royal family. Green observed the planet under good seeing conditions in Madeira with a 13-inch reflector during the summer of 1877.

Green's map, drawn in soft pencil, was carefully printed as a chromolithograph in sepia tones on three different plates. It is full of labels implying Martian/terrestrial analogies, such as 'Secchi Continent,' 'Dawes Ocean,' 'Fontana Land' or 'South Cap.' His use of certain drawing techniques is also reminiscent of terrestrial maps. Cross-hatching signified depth, with the darkest areas supposedly the deepest seas on the planet Mars. His charming naming system, intended to immortalize all previous observers of Mars, did not stick, but many of the features drawn on this map are retraceable on later maps. One feature, however, was added when the Italian astronomer Giovanni Virginio Schiaparelli (1835–1910),

---

[13]On these earliest observations, see Flammarion (1892), Hoyt (1976) chap. 1, quotes from p. 3, and further refs. to the primary literature cited there on p. 321 as well as in Sheehan (1996) chaps. 1–3.

[14]On analogies in science, see Hentschel (2010) and further literature cited there.

[15]On lunar cartography prior to 1877 see, e.g., Proctor (1888) and Sheehan (1996) chaps. 4–5. For a detailed explanation of criteria for 'good seeing' in astronomy, see Hoyt (1976) chap. 4.

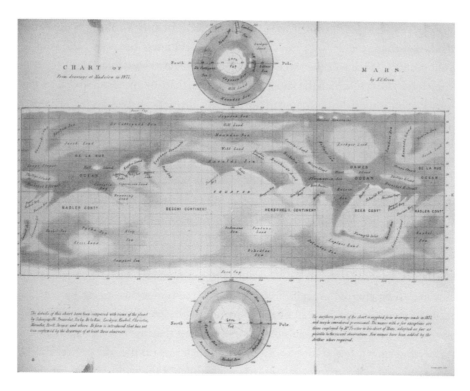

Fig. 94: A map of Mars, the central area in cylindrical projection and the two polar caps in orthographic projection. From Green (1880) pl., based on observations made in Madeira in August and September 1877. Only details confirmed by at least three observers were entered on this map.

who directed the Milan observatory from 1862 to 1900, published his Martian map of 1879 in the proceedings of the renowned *Accademia dei Lincei* in Rome.[16] Based on new observations made with his Merz refractor telescope during the 1877 opposition of Mars, Schiaparelli noted a couple of thin lines, apparently interconnecting the various oceans (*mares*), lakes (*lagos*) and huge rivers (*fluvius*) that were given such names as the Nile or Indus. In later maps, he incorporated a profusion of these thin lines, interconnected in a strange maze-like network mostly originating from one of the *mares*. Taking up a term already used in one of Father Secchi's Martian maps of 1863, the director of the observatory in Milan called these features '*canali*' (canals). Schiaparelli was keenly aware that interpreting these thin dark lines as 'canals' was a rather bold leap of the imagination for which no

[16]On the life and work of Schiaparelli, see MacPherson (1910), Knobel (1911), Sheehan (1997) and the documentation by Michele T. Mazzucato (dated Feb. 7, 2010) with further links and references.

proof was yet in sight. But it must have seemed obvious to him, on the other hand, that these canals were the communicating channels between various lakes and oceans.[17] Whatever they were, Schiaparelli was dead sure that they were real. In a letter to Green, which the latter quoted in an article for an astronomical review journal summoning others to join in the hunt for further canals, Schiaparelli wrote: "It is impossible to doubt their existence, as that of the Rhine on the surface of the earth."[18] The Italian astronomer was known for his meticulous care and superb drawing skills. In 1882, he did caution his readers that these features "may be designated as canals although we do not yet know what they are." Sometimes they appeared to be "vague and shadowy, other times clear and concise." In order to set these new features apart from subjective hallucinations or rapid atmospheric turbulence effects, Schiaparelli insisted, though, that while their "aspect and degree of visibility are not always the same," their topographic arrangement on the Martian surface "appears to be invariable and permanent." Hence, he reasoned, they had to be charted as well.[19] Maps such as the one shown in fig. 95 have to be understood as composites of many nights of observations. Schiaparelli and others who observed the canals always emphasized that they never saw them all at once but only a few at a time. Their exact positions on the Martian disk were then measured micrometrically. Schiaparelli estimated the width of the canals to be from 30 to 120 km, an enormous size compared with man-made canals. Their lengths were equally surprising. In some cases, they amounted to no less than 80° of the Martian surface (roughly 5,000 km).

During the next opposition of Mars in January and February 1882, Schiaparelli confirmed another observation that he had already made in 1879: At least 20 of these canals appeared as closely parallel, double lines. He emphasized that he had made sure that this phenomenon, which he called 'gemination,' was not an optical illusion. The phenomenon remained stable even when telescopes and eyepieces were exchanged. Schiaparelli declared: "I am absolutely sure of what I have observed."[20] Continued observation during further oppositions of Mars in 1882 and subsequent years suggested to him that this strange gemination

---

[17] See Schiaparelli (1877/78), e.g., on p. 337: "punto importante nella topografia di questa regione per le multiple communicazione, che il lago della Fenice, ha coi mari e con canali circostanti."

[18] Schiaparelli in a letter to Green, dated Oct. 27, 1879, quoted in Green (1879) p. 252, who was unable to see any canals from his home in St John's Wood, London.

[19] All previous quotes from Garret P. Service's English translation of Schiaparelli (1882) pp. 220f.

[20] Schiaparelli (1882a) p. 221, English transl. by Service. Cf. MacPherson (1910) p. 471 and Sheehan (1997) p. 13. This new feature of gemination, which was observed by few other observers, Sheehan explains as a result of a worsening pathological condition of Schiaparelli's eyeball.

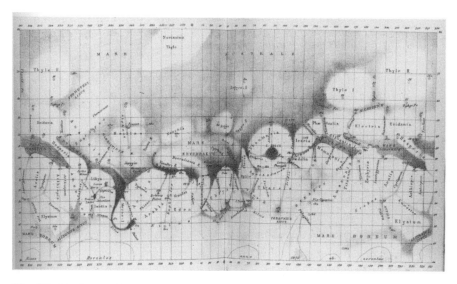

Fig. 95: Martian map by G. V. Schiaparelli, based on observations during the Martian opposition of 1879. From Schiaparelli (1880/81) pl. II, lithographed by Bruno e Salomons in Rome.

seemed to be periodic and seasonal. It correlated with the Martian spring of the northern hemisphere, which Schiaparelli interpreted as a striking color change of the reddish-ocre portions of the planet that he and many other observers before had identified as 'continents.' It thus seemed as if the two ice caps of Mars partially melted and flowed through the maze of channels into the various oceans situated mostly around its equator. As on Earth, that was the warmest region of the planet because solar rays hit the area nearly orthogonally throughout most of the year. He interpreted the color changes on the Martian 'continents' and the coincident appearance and disappearance of canals to be dual indicators of changes in climate and vegetation in the Martian ecosphere.[21] By continuing this maze of hypothetical analogies constructed between the known facts about the terrestrial atmosphere and Martian conditions, it seemed quite natural also to assume a closed water cycle with evaporation in warmer regions and condensation as clouds and precipitation in the polar regions. In a series of articles for the popular journal *Natura ed Arte*, Schiaparelli spun this web of speculations further about *la vita sul pianeta Marte*.

---

[21] See, e.g., Schiaparelli's letter to E. B. Knobel in 1882, quoted in Knobel (1911) p. 284: "Il est indubitable que les taches de Mars sont sujettes à des variations considérables de couleur et d'intensité; les lignes obscures ne sont pas toujours visibles, et ne sont pas toujours bien nettes. De là en partie les différences entre les observateurs dont on pourra peut-être donner l'explication plus tard, lorsque notre connaissance des phénomènes sera plus complète."

The interest it aroused among the general public escalated around 1900 into a veritable hype about possible life forms on Mars.[22]

The refractor telescope that Schiaparelli used in Milan for all of his early Martian observations had an objective lens aperture of only 8.75 in. (22 cm). That was much smaller than the largest telescopes of his time, whose apertures could already measure up to 36 in. (91 cm) in 1888.[23] The planetary details purportedly observed were right at the theoretical diffraction limit. Why did astronomers at other reputed sites such as Greenwich in London or at the much better equipped *Lick Observatory* on Mt Hamilton, then boasting the world's largest refractor, not discover these *canali* long before Schiaparelli in Milan? And why was he the only one to see them for nine years? This was surprising, but not unheard of. For it was well known among astronomers that telescope size and quality were but two of the factors determining the overall quality of images. Astronomical seeing, i.e., the local atmospheric conditions, and especially the homogeneity and turbulence-free state of the air above the observatory, were at least as important. For that reason, astronomers left the proximities of big cities and moved to isolated highland locations or, better still, to islands with high volcanic peaks.

Perhaps Schiaparelli had just simply been especially lucky in chancing upon a day or two with particularly good seeing? As he himself admitted and explained to many colleagues and laymen trying also to observe these strange *canali*, one had to be patient and await the brief moments of exceptionally good seeing, when minor air turbulence suddenly abated, ceasing to distort the view, and briefly lifting the maya veil that usually occluded the phenomena. This same reason was put forward for all failed attempts to photograph the canals. The long exposure times necessary to photograph such a relatively faint and small object as Mars did not catch those fleeting good moments that the human eye and memory could detect. They disappeared in the camera's integrated image from a longer exposure period. The trend of small to medium-size telescopes yielding the best views of the supposed 'canals' continued in the coming decades, and the conventional correlation between telescope size with image quality seemed definitely broken. When astronomers failed to see Schiaparelli's channels using their powerful refractors, they even got the recommendation to perhaps try again with a smaller telescope – sometimes this switch even worked!

---

[22]For a commented reprint of his essays about life on Mars, see Schiaparelli (1893–1909). Cf. also Hoyt (1976) chap. 12 and Manara & Chlistovsky (2001) pp. 5–7.

[23]On Schiaparelli's telescopes in Milan, see, e.g., Schiaparelli (1893–1909) pp. 11f.

Eventually, a few other astronomers convinced themselves that they were also seeing what Schiaparelli had described in print. In 1886, Joseph Perrotin (1845–1904) and Louis Thollon (1829–87) at the Nice Observatory announced that they, too, had seen the canals through their telescope, and similar claims by various other minor figures soon confirmed them. Not all of these 'confirmations' were worth very much. François Joseph Charles Terby's (1847–1911) observations made with an 8-inch refractor in Louvain, Belgium, illustrate this. His explicit goal from the start was: "verification of the admirable maps of M. Schiaparelli, maps which are yet, even today, it is necessary to confess with regret, the object of an entirely unjustifiable suspicion."[24] He emphasized how difficult it was to see narrow canals with an 8-inch refractor under unfavorable weather conditions:

> We took inspiration from the principle announced by some great observers: 'Often,' they say, 'one can see well what one especially seeks.' [...] Likewise, we have sought the canals of Mars in the regions where we knew that M. Schiaparelli has proved them to exist; we have taken care to calculate in advance the approximate longitude of the central meridian for each observation, and, map in hand, we have patiently and obstinately pursued these very difficult details. [...] the observer [...] would be discouraged very quickly if he was not sustained in advance with an unbreakable faith in the truth of the Milan results; this faith alone, in effect, could inspire him with the perseverance and, I may say, the obstinacy necessary.

The most important and influential support for Schiaparelli's claims came from Flagstaff, Arizona, however. Percival Lowell (1855–1916) had funded a new observatory there, which began operations in June of 1894 and was equipped with an 18-inch Brashear refractor.[25] The Flagstaff instrumentation was thus quite similar to the new 49-centimeter refractor in Milan with lenses from the Merz company (the successors to Fraunhofer), and the telescope mechanics and support from Repsold Bros. in Hamburg. The image of Mars could be enlarged 500×.

Lowell and his small team of assistant observers immediately turned their attention to observations of Mars and other closer planets, such as Venus and Mercury, spurred on by the wish perhaps to be the first to prove the existence of extraterrestrial life. Lowell was in close contact with Schiaparelli[26] and was one of his earliest converts about the 'discovery' of canals on Mars. His agenda

---

[24]Terby (1892) p. 479; the following quote comes from the same page.

[25]On Lowell, who had been trained by Benjamin Peirce at *Harvard University*, and the history of his observatory, see Strauss (2001), Hoyt (1976) chaps. 2–3 and 5 and further sources there.

[26]On their exchange of letters, see Manara (2005) and Manara & Chlistovsky (2001).

was to provide watertight proof and confirmation of Schiaparelli's claims, which were still very controversial within the small community of planetary astronomers throughout the world. During the 1894 opposition, Lowell and his first assistant Andrew Ellicott Douglass (1867–1962) made 917 drawings of the planet's disk and 57 sketches of the southern polar cap. In later observing seasons, this data continued to pile up into thousands of drawings and sketches, to form the basis of a detailed 'aerography' of the Martian surface. A special approach was developed in Flagstaff for these graphic recordings. Lowell insisted on not mixing the records of different observers – knowing how much such drawings depend on the idiosyncratic style of each draftsman. He had each observer keep his own recording book into which detailed drawings of the planet and its features were made over and over again, almost every night when seeing conditions were good enough to obtain any useful result. Substantial changes in such series of observations were then interpreted as evidence of temporal changes on the planet's surface, especially if different observers independently arrived at the same conclusion, each on the basis of their own series of drawings.[27] These graphic data were supplanted by other physical measurements, such as spectroscopic examinations or polarimetric measurements of the light emitted from Mars (both in search of oxygen and water).[28] The controversial canals were seen, at least by Lowell and Douglass, from the very start of their observation program. But another assistant, William Henry Pickering (1858–1938), was not able to see the *canali*. This misfortune was explained away by the assumption that Pickering simply did not have the "acute eye" needed to see such intricate features.[29] Lowell also continued to develop the vague hypotheses by Schiaparelli into a full-blown theory of the Martian atmosphere, which became the subject of several of his books to appear in 1895, 1906 and 1908 and attracted a wide readership inside and beyond astronomer circles. The *Lowell Observatory Bulletin* and *Annals* also carried regular updates on global topographic maps of Mars and interesting surface details seen at Flagstaff. Martian globes (fig. 96), made the same way as lunar or terrestrial globes,[30] brought the Arizonan insights onto the desks of countless laymen.

[27] One example of the cinematography of apparent changes on the Martian surface over time as recorded by one observer is Pickering (1892) pl. XXX.

[28] See Hoyt (1976) chap. 9, and Campbell (1894), according to whom spectroscopic measurements yielded no discernible difference in the spectra of the Moon and Mars and thus no evidence of a hydrous Martian atmosphere.

[29] On this episode, see Hoyt (1976) p. 64.

[30] On the history and distribution of Martian globes, see Blunck (1995/96); on lunar globes, cf. Blunck (1999); on the production of terrestrial and stellar globes, Mokre (2008).

Fig. 96: Three globes from the Lowell Observatory, Flagstaff, with Martian canals and polar ice caps, from 1898 (right), 1903 (left) and 1905. Like Martian maps, these globes (with diameters of c. 15 cm each) were accrued composites of many individual observations. Photograph by Joe Orman, with permission. See joeorman.shutterace.com/Observatories/observatories_lowell_0749.jpg (accessed Jan. 2014).

Lowell, more outspoken than Schiaparelli, interpreted his findings on canals as clear evidence of intelligent design. As his key argument, Lowell repeatedly used "their singular arrangement that is most suggestively impressive. They have every appearance of having been laid out on a definite and highly economic plan. [...] they are almost without exception nearly geodetically straight, supernaturally so, and this in spite of their leading in every possible direction [and] of apparently uniform width throughout their length."[31] How tempting it was to read artificially made creations into the structures apparently observed on Mars becomes even more obvious in a plate produced by one of Lowell's close associates, Edward S. Morse (1838–1925), for a popular book.

As shown in fig. 97, Morse compared aerial views of man-made surface structures, such as railroad tracks, roads or irrigation canals, with the patterns sketched by Schiaparelli and by Lowell and his associates. To the layman reading these kinds of publications, which still continue to appear to this day, the similarity

---

[31] Lowell (1893) p. 649. More of the kind can be found in Lowell's three books on Mars.

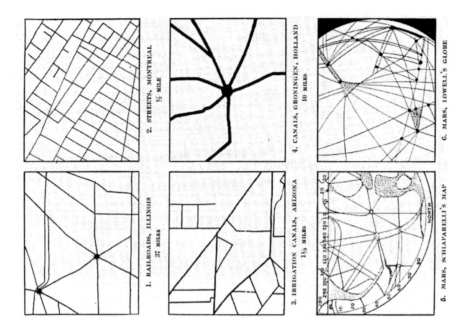

Fig. 97: Comparison of terrestrial surface features with Martian canals. From Morse (1906) pl. VI.

of these patterns was sufficient proof that intelligent life existed on Mars. This plate and its context thus represent another problem with the search for patterns in science. There is no guarantee that two patterns sharing surface similarities – such as Martian and Dutch canals – were created by the same kinds of processes; pattern similarity does not guarantee causal relatedness.

That this methodological problem lay at the heart of this episode only became clear when the resistance to Schiaparelli's and Lowell's claims finally mounted beyond a simple reiteration of the inability to replicate the observations in Milan and Flagstaff elsewhere. Many people pegged high hopes on scientific photography. But for an awfully long time the very small and diffuse images obtained by Carl Otto Lampland (1873–1951) and a few others elsewhere dashed these hopes.[32] One could count oneself lucky if a few of the large surface features, such as the white polar caps or some of the large *mares*, were discernible at all on these tiny photographs or their coarse-grained enlargements. But the dispute about Martian channels could not be decided by them. It also did not help either way when

[32]On this episode and a few facsimiles of the photographs obtained, see, e.g., Hoyt (1976) chap. 11.

Lowell and his associates started to claim in 1896 that features resembling *canali* were visible not only on Mars but on various other closer planets as well, most prominently on Venus and Mercury.[33] Did this mean that these other planets also had Earth-like atmospheres, canals and all the rest? Or did it mean that the same observational mistake or interpretational error that had perhaps crept into the Lowellian observations of Mars was now also showing up in their observations of other planets? Both options were aired at the time, but it was a clear impasse. In 1903, the British assistant astronomer Edward Walter Maunder (1851–1928) at the *Royal Observatory* in Greenwich resolved to examine experimentally a question that thus far had been neglected in the heated debates pro and contra Martian channels. On the assumption that all Martian channel claims were mistaken, what else was it that these experienced observers had seen? It was clear to all participants in the debate that Schiaparelli as well as Lowell were personalities of integrity, quite in contrast to the flood of journalistic expositors of their findings who just rode on the wave of popular fascination with life on Mars to create sensationalist accounts (sometimes replete with fantasy images) of how the Martian surface would look from up close.[34] Both Schiaparelli and Lowell had built up reputations on careful and painstaking observational work and both had nothing to gain but much to lose by any intentional fabrication of false claims, which could thus safely be ruled out. What else could it have been that these observers had mistakenly taken as evidence for Martian channels?

Maunder's answer to this intriguing question was as simple as it is surprising:[35] Schiaparelli, Lowell and company had fallen into a trap set by our sensory physiology. Our vision has an inbuilt tendency to connect isolated dots into lines and to interpret subtle boundaries of color or brightness as lines. Maunder had already suspected as much in a critical article for the generalist journal *Knowledge* in 1894. Between July 1902 and May 1903, he followed up his own earlier claim with a series of experiments. A class of boys aged between 12 and 14 from the *Royal Hospital School* at Cambridge was shown sketches of certain regions of the Martian surface without any canals drawn in, but otherwise as faithful to the available various topographic maps of Mars as possible. All the boys were totally ignorant of what they were being shown and were given no hint about what they were supposed to see. They were just asked to draw "all that they could see and be sure

---

[33] See Hoyt (1976) chap. 8, again with a few facsimiles of Lowell's drawings.

[34] For a few examples of such images, taken from Flammarion's *Les Terres du Ciel*, first published in 1884, see Canadelli (2009) pp. 458–9.

[35] On the following, see Maunder (1894) and Evans & Maunder (1903), quotes on pp. 488 & 492.

of, each for himself, without noting what his neighbors were drawing."[36]

A considerable number of these boys, especially those sitting a medium distance away from the shown sketches, tended to include lines in their sketches that varied in length from 3 to 6 inches. These lines were very similar to the Martian canals drawn by observers in Milan and Flagstaff. When irregular dots were introduced into black-and-white diagrams of Martian surface features by Schiaparelli himself, the boys also tended to connect those dots into straight, or nearly straight, lines, "joining up two or more black dots and prolonged to the 'nearest' sea." Maunder thus felt compelled to conclude

> that markings having all the characteristics of the canals of Mars can be seen by perfectly unbiased and keen-sighted observers upon objects where no marking of such a character actually exists. They are in a sense truly 'seen,' not imagined, because they are the natural rendering by the eye of real markings of a different character. [... Thus] the canals of Mars may in some cases be, as Mr. Green suggested, the boundaries of tones or shadings, but [...] in the majority of cases they are simply the integration by the eye of minute details too small to be separately and distinctly defined.[37]

Note how the Cambridge astronomer carefully avoided imputing personal fault. He tactfully underlined that the human eye could not but interpret certain signs in certain ways. As can be imagined, Lowell and his coworkers were neither happy nor satisfied with this experimental demonstration. Who would want his observational capabilities compared against a bunch of Cambridge school boys?

The debate continued on for quite a while. Simon Newcomb (1835–1909) at the US Marine Observatory in Washington, DC, tried to explain the canals as an effect of optical diffraction. That would explain why the strange canali were seen best in telescopes with smaller apertures (leading to the strongest unwanted effects). Lowell and his coworkers retorted that nothing had been proven yet; but the tide had turned. The Gestalt fixation on canals had somehow been broken. Several of Lowell's assistants left Flagstaff and/or changed fields, and eventually the adherents of Martian canals died out. Since then, the communis opinio is that Maunder's sensorial physiological explanation for the strange findings is basically correct. Finer drawings by Eugenios M. Antoniadi (1870–1944), obtained at the Meudon 33-inch refractor, further clarified the situation by breaking the hard lines of Schiaparelli's drawings down into finer details (cf. fig. 98).

Even the diehards had to give in when Mariner missions organized by NASA since the 1960s began to map Mars photographically by satellite. In 1975, two

---

[36]Cf. ibid., pl. 18–19 for facsimiles of the images shown to the boys and some of their drawings.
[37]Evans & Maunder (1903) pp. 497 and 499.

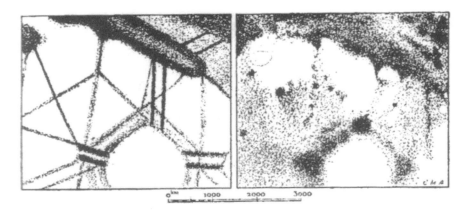

Fig. 98: Comparison of lithographed drawings of the Martian region of Elysium, drawn by G. V. Schiaparelli between 1877 and 1890 with later drawings by Eugenio Antoniadi, based on observations between 1909 and 1926 at the much more powerful Paris Meudon 33-inch telescope.

Viking missions even explored the red planet physically with a landing unit and transmitted their images back to Earth. Accordingly, the Martian surface is an extremely arid, rocky and dusty, wildly rugged landscape. Its atmosphere is thin, cold and bone-dry, exposed to extreme daily temperature variations from −83°C to −33°C and devoid of any surface water. Periodic storms only blow dust across the Martian surface,[38] quite different from what Schiaparelli and Lowell had imagined.

The moral of this little tale is that pattern recognition is not an easy business. It can easily go astray, especially if the all-too-human observers are swayed too much by suggestive analogies to well-known terrestrial phenomena. The foregoing is also a lesson about how finicky human vision is. It is optimized to resolve potential threats quickly and thus evolutionarily preprogrammed to over-interpret rather than under-interpret sensory perception. That is why humans are still quicker than modern computer software programs in detecting visual patterns. But this speed comes at a price. Our vision occasionally plays tricks on us. The fact that Schiaparelli suffered from red–green color blindness must have further distorted his acute perception of color boundaries and edges, besides the human eye's construals of figment lines.[39] Patterns clearly perceived by an observing subject or by sophisticated blur masking and deconvolution algorithms in modern image analysis software are not necessarily present in the observed object.[40]

---

[38] On the history and results of Mars exploration missions, see Morton (2002) and Moore (2006).

[39] For more on these physiological effects of sensory perception and on their replication in modern high-speed digital image definition software programs, see Dobbins & Sheehan (2004).

[40] Another case in point is the controversial resolution of the structure of nebulae, initially only observed and drawn by pioneers such as Lord Rosse, George Bond and William Tempel, a story

9.3   Electron microscopy

Another field in which the skill of pattern recognition is crucial is electron microscopy (EM). One could argue that the same holds true for conventional light microscopy, since even under a normal microscope seemingly well-known everyday objects like a hair or a needle look quite different from what we would expect. But the observer's dependence on the preparation method and on photographic recording increased dramatically in EM, to the point that it has even been called a veritable image science.[41] Because in EM thin objects are examined structurally by a beam of electrons artificially accelerated and focused by means of magnetic lenses, no object is 'seen' directly. Electron microscopes rather record a specimen's interaction (in the form of absorption and diffraction) with the electron beam. In transmission electron microscopy (TEM), this electron beam actually passes through the specimen, and the specimen's electron shadow is then viewed on a fluorescent screen or recorded on photographic film. This resembles x-ray photographs, but with cathode rays replacing Röntgen's x-rays. In scanning electron microscopy (SEM), on the other hand, the reflection and back-scattering of a tiny electron beam is recorded and an 'image' of the specimen is reconstructed on the basis of this interaction. Both techniques only work when the full beam path in which the sample is placed has been fully evacuated. For both types of electron microscopes, the specimen has to be thinly coated to avoid its immediate destruction and evaporation upon bombardment by electrons accelerated to energies of 50–200 keV. But usually the specimen is destroyed or severely altered during the act of observation anyway, and the recorded photograph is the only remaining trace of it. All kinds of image analysis and contrast enhancement techniques are needed for its further analysis. The point of the exercise is to increase image resolution, which is always limited to the order of magnitude of the wavelengths by which one 'observes' the specimen. Whereas light microscopy is limited to 3,000 Å (or 300 nm), electrons of momentum p have a wavelength equivalent to $\lambda = h/p$ of order of magnitude 1 nm or less. Thus, in principle, a theoretically 1,000-fold increase in resolution is attainable.[42] This gain in spatial resolution is obtained at the high price of much more complicated and convoluted access to the

beautifully told by Omar W. Nasim (2008)–(2012).

[41] See Breidbach (2005) on EM as *Bildwissenschaft*. Cf. Rasmussen (1997), Gabor (1948, 1957), Hawkes (1985) and Qing (1995) on the history of EM.

[42] On the de Broglie wavelengths of matter, see Kragh (1999) pp. 164f. and further literature cited there. One of the earliest articles using wave–particle duality to estimate the systematic errors of electron microscopes caused by diffraction is Ardenne (1938).

specimens: They are no longer visually examined but are indirectly reconstructed from data on transmission or reflection and scattering.

The first steps toward what later became prototypes of electron microscopes in the late 1920s were taken by Hans Busch (1884–1973) and Ernst Ruska (1906–88) in the context of experiments to improve cathode-ray oscilloscopes.[43] At the high-tension laboratories of the polytechnical *Technische Hochschule Berlin*, Ruska constructed the first electromagnetic analogue of an optical collector lens, for which he earned his *Diplom* in electrotechnology in 1930. Together with his supervisor Max Knoll (1897–1969), Ruska then also succeeded in designing the electromagnetic analogue of an optical two-lens combination in optical telescopes and microscopes.[44] As fig. 99 shows, the first images obtained were just trying to retrieve known structures with the new image-forming technology in order to test its strengths and weaknesses.

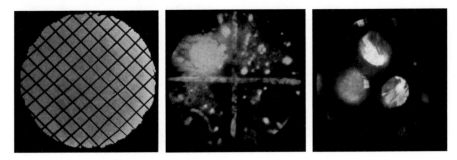

Fig. 99: Three early electron microphotographs from the early 1930s: the first EM image of a fine wire net (magnification factor 12×) and the first EM images of a cathode surface (80×) and of a gold foil behind a supporting honeycomb plate (11×), as the very first TEM image of a thin foil. From Knoll & Ruska, c. 1931 (left), and Brüche, c. 1932; all reproduced in Brüche (1957) pp. 601–2.

In December 1933, Ruska had already reached magnifications of 8,000× and even 12,000× and thus exceeded the maximum of what light microscopes were attaining. Ruska was then working on television apparatus, but in January 1937, the *Siemens AG* founded a new electron microscopy laboratory in Berlin-Spandau,

---

[43]On the pioneering work by Busch, Knoll and Ruska, see Ruska (1979, 1984) and Qing (1995) chaps. 2–3 and the appendix, pp. 143–7.

[44]In an early talk about his work, Ruska still spoke of his gadget as analogous to the Kepler telescope. The term electron microscope was first used by him in a publication submitted in Sep. 1931 and appearing in early 1932. That was when the link to the Davisson–Germer experiments on electron waves was first made by Fritz Houtermans, Dennis Gabor and Manfred von Ardenne: see Qing (1995).

where Ruska and his collaborator Bodo von Borries (1905–56) built the first prototypes of electron microscopes (still in 1937). Only one year later, the first commercially available electron microscopes were sold to laboratories in science and industry.[45] From that time on, several other companies and scientific instrument manufacturers also started to develop EM, most notably the research group around Ernst Brüche at the *Allgemeine Elektrizitäts-Gesellschaft* (AEG) in Berlin;[46] Manfred von Ardenne (1907–97) in a semi-private laboratory in Berlin-Lichterfelde, where he broke a world record in obtaining a resolution of 30 Å in 1940 (surpassed by others in 1946); *Carl Zeiss* in Jena and Oberkochen; and the *Radio Corporation of America* (RCA) in New Jersey. RCA produced a "universal" electron microscope (called EMU) that became quite popular because it was much easier to handle and more robust than all previous models.[47]

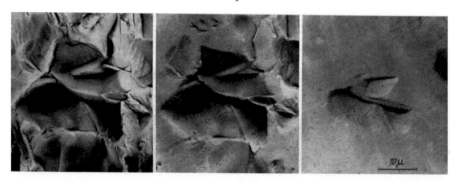

Fig. 100: Electron microphotographs of a glass surface at three different stages of polishing. A clear inspection of details much smaller than 1 μm is possible (cf. the scale of 10 μm at the bottom of the right image). These samples were obtained by H. Poppa (undated) by coating the glass at three stages throughout the polishing. From Brüche (1957) p. 606.

Until 1944, Siemens sold approximately 30 high-end electron microscopes working at electron energies of 75 keV and yielding enlargements of up to 30,000×, thus allowing a point resolution of up to 13 nm (see fig. 100). Roughly 200 scientific papers were written based on the use of this new research technology. Soon

[45]See, e.g., Grünewald (1976) with a critical commentary by Ruska a year later, Ruska (1984), Niedrig (1987), Qing (1995) and further sources quoted there.

[46]See, e.g., Brüche (1943, 1957), Grünewald (1976) and Qing (1995) chap. 4 and the appendix, pp. 141–3.

[47]See Ardenne (1940), Reisner (1989), Rasmussen (1997) chap. 1 and Rasmussen & Hawkes in Bud & Warner (eds. 1998) pp. 382–5.

after World War II had ended, Siemens restarted production of their electron microscopes; their postwar model UEM 100 was replaced by the even more powerful Elmiskop series in 1954, which sold more than 1,000 copies in the following decade.[48] Particularly microbiologists and material scientists happily seized the new opportunities for structure examination offered up by this new research technology, financially well supported by a flurry of grants from government and grant agencies in the late 1940s and 1950s.[49] During the 1960s and 1970s, scientific instrument makers in the Far East, like *Hitachi* and JEOL, gained dominance in the market, but the recent inventions of aberration-corrected electron optics, which drastically improved image quality and resolution once more (up to 0.08 nm), renewed interest among European instrument developers and companies in this technique (cf. fig. 101 below and figs. 119f. on pp. 370f. for examples).[50]

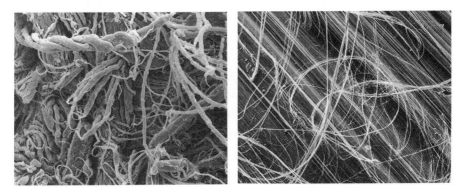

Fig. 101: Two modern high-resolution EM images for comparison against fig. 100. Left: suede, the inner side of fine fibrous leather, enlarged 225×; right: chrysotile, the mineral form of asbestos, enlarged 3,650×. From Breger (1995) pp. 36f. © Dee Breger, Micrographic Arts.

Despite the drastically different images generated by light microscopy and EM, the striking feature in the history of the development of this new technique is that researchers and technicians alike strove to make EM images as similar as possible to light microscopy images. EM images were construed and seen as extensions of traditional imagery – just a bit more refined and thus offering additional informa-

---

[48]For these data, see Urban (2007) p. 40. According to Brüche (1957) p. 610, of the 1,300 electron microscopes in operation worldwide in 1955, only c. 60 (or 5%) were located in Germany, whence they had originated two decades before.

[49]On aspects of funding, see Rasmussen (1997), esp. in biology.

[50]Ibid., p. 41 and references to primary literature given there; for samples of modern high-quality EM images see, e.g., Breger (1995), www.micrographicarts.com or Newbury & Williams (2000).

tion on structure.[51] Thus, for the characterization of materials, TEM has rapidly become one of the most crucial research technologies for the metallurgist, ceramics specialist or crystallographer. It is used to examine, for instance, crystal defects and second-phase precipitates, including dislocations, interphase irregularities and grain boundaries, now operating in the 1 Å range.[52] For biological applications, it took a bit longer because the dehydration in the strong vacuum necessary for EM and the beam's strong interaction with organic matter changed and often destroyed organic specimens rapidly: chemically unaltered samples lose one-third of their dry weight in the first few minutes of viewing.[53] SEM, introduced commercially only in 1965, has been appreciated for its broad range of magnification factors and the high plasticity of its images. They have a three-dimensional quality, strong contrast and vast focal depth, often 50–100% of the width of the image field.[54]

Because of the huge gap in magnification between light microscopes and electron microscopes, these attempts at a quasi-continuous extension of vision *qua* analogy into the submicroscopic realm of structures well within the orders of magnitude of macromolecules were quite risky and, in fact, often failed. In order to arrive at secure results, EM images have to be used in conjunction with other analytical techniques such as x-ray diffraction, electron and x-ray spectrometry. To bridge the several orders of magnitude between EM images and conventional light microscopy images, a combination of visual strategies was deployed:

- The same specimen was examined by EM under low magnification and by light microscopy to produce a kind of visual translation table for structures as they were supposed to look by the new technique.[55] Since EM images were vastly better in image sharpness and definition, this was also excellent advertisement of the superiority of the new technique.

- Where the same magnification could not be obtained for the two different techniques, one at least tried to get as close to the optical range as possible

---

[51]On this mode of regarding EM as a mere extension of the image worlds of the light microscope: Breidbach (2005) p. 164, providing examples from early neuroanatomy; Rasmussen (1997) chap. 6.

[52]More on materials science contexts in Newbury & Williams (2000) and Cahn (2001) pp. 92ff., 110ff.

[53]On these negative repercussions of EM observation on samples, and on the necessary preparatory techniques to circumvent them, see Grubb & Keller (1972), Johnson & Cantino (1988) and Rasmussen (1993) p. 232.

[54]Ibid., pp. 324–6, 332–8 and 341 on their current prices.

[55]For an example of such a translation plate from Burton et al. 1942, see Rasmussen (1997) pl. 6.5. For a link to studies by Hacking and Franklin, who also pointed out this same strategy of calibration employed in other experimental fields, see ibid., pp. 13, 253.

so that a certain overlap between the old and the new could raise the impression that well-known features were recognizable, simply seen under much higher resolution, but still sharing key features or markers.[56]

- Maximization of focus, minimization of background debris and a choice of samples with the least obvious preparatory damage were further precautions against artifacts.[57]

- Another strategy was to vary the preparation techniques as much as possible for one class of specimens (for instance, rapid cryogenic freeze-drying or drying at room temperature, smearing or ultrafine cutting, metal coating or the production of replicas by coating with a thin removable film) and then to search for patterns recurrent in the EM images of all the different specimens. These were interpreted as intrinsic to the sample, the variations were suspected to be artifacts of the preparation.[58]

To give just one example of a grossly misinterpreted EM image: A physician and infertility researcher in New York City, Frances Seymour, had EM images of human sperm taken at RCA and rushed to print them in the *Journal of the American Medical Association* before having gone through the micrograph approval process that RCA had wisely instituted to prevent such rash conclusions. Her "lousy" photographs became the object of derision for her fantastic, totally unfounded interpretation of features such as "suction mouths in the head, a body, and a tail."[59]

At any rate, despite occasional over-interpretation, EM rapidly became a well-established visual culture. It was an image-centered science in the sense of being even totally dependent on photographic images as the basis of all further processes of inference. The obtained microphotographs were the veritable 'data' with which the scientists and their technical assistants worked. The coated samples are usually damaged or even completely destroyed during the process of 'observation.' Discrimination between fact and artifact as well as pattern recognition are crucial skills that every beginner has to master to be able to interpret EM photographs reliably. In order to make it easier to 'see' the specimens three-dimensionally, modern electron microscopes enable a kind of stereo-seeing mode. The samples

---

[56]Rasmussen (1997) pp. 235ff. speaks of 'landmarks' in conscious analogy to cartographic maps of unfamiliar terrain.

[57]For examples, see Rasmussen (1993) and (1996) p. 339.

[58]See again Rasmussen (1997) pp. 56f. for examples of this strategy in research practice.

[59]Quotes from a disgusted entry in T. F. Anderson's notebook, dated June 26, 1941, and quoted with further commentary on this notorious episode in Rasmussen (1996) p. 337.

are actually just turned by 5–10° between two exposures, thereby creating a pair of stereoscopically analyzable images with vastly improved illusory depth. Further enhancement tricks include injecting metallic vapor deposits on the specimen from the side, which produces sharper contour lines and a kind of shadow-casting. Optimal focusing was made much easier with the introduction of so-called 'wobblers' that periodically vary the angle under which the specimen target is hit to artificially increase the aperture of the illumination to help find the best focal plane. In order to gauge the size of structures, sometimes minute artificial polystyrene balls of known size are sprinkled onto the sample prior to exposure. Good size estimates can then be made from the length of their 'shadows' in the EM image.[60]

Despite all these improvements over the years, there is still a lot of implicit knowledge behind the skills of preparing the samples, and coaxing out contrast and subtle features in the photochemical manipulations of EM photographs in the darkroom. More up-to-date manipulations with the CCD camera are made by a digital image enhancement program in an intricate interplay with image simulation techniques. An interesting example of this is the work by Aaron Klug (*1926) at *Birkbeck College* in London (and later in Cambridge, England), who ran up against the limitations of spatial resolution in the 1960s while examining viruses. He started a series of experiments to understand how certain spiral 3D structures would be mapped onto the 2D image-plane of the electron micrograph. Combining x-ray diffraction and electron microscopy, Klug devised a new imaging technology that he called 'crystallographic electron microscopy.'[61]

Over a span of 40 years, EM remained an arcane technique relying on an "operator-intensive instrument" for which many weeks of training are necessary "just to run (certainly not master)."[62] The handling and operation has not altered substantially. It still consists of large, water-cooled electromagnetic lenses that still have to be protected against acoustic and mechanical vibrations, and the output still is either an analogue fluorescent screen and/or photographic film. Especially the complex art of specimen preparation is still very finicky, despite all the advances made in the use of ultramicrotomes since 1950.[63] The mind-boggling

---

[60]Such improvements are described and illustrated in Brüche (1957), Gabor (1957), Gordon, Bender & Herman (1970) and Breger (1995).

[61]See Gooding (2010) p. 24; cf. also Rasmussen (1997) chap. 5 on virus research using EM.

[62]Both quotes are from Newbury & Williams (2000) pp. 340f. Cf. also Lynch (1985) and Rasmussen (1997) pp. 146f. on EM training. For a web-based EM resource for research and education see http://em-outreach.ucsd.edu/.

[63]On these continuities, see Rasmussen & Hawkes in Bud & Warner (eds. 1998). Cf. Reimer

beauty of EM photographs was – and still is – an obvious factor explaining the strong fascination with this 'art' of micro-observation for practitioners and laymen alike (cf. here sec. 12.2 on recent beauty contests between EM images). EM images still figure prominently in the popular media, with electron microscopes titulated as "super-microscopes," "super-eyes" or "new magic eyes." They have a naïve reliance on the immediate, unbroken interpretability of these images as "molecular landscapes, mountains, cliffs, furrows, and strange desert formations."[64] Fundamental differences in the way images are generated are completely ignored. But, as we have already seen, this tendency to domesticate, to assimilate old standards of seeing, was also at play in the scientific actors' own approaches to these images.

9.4   Interobserver and intraobserver variability in CT scans

Given modern medicine's extreme dependence on imaging technologies´– such as x-ray radiology, NMR, PET and CT scans – and given the extent to which correct interpretation of an image depends on a radiologist's skill, one might well ask: How reliable are diagnoses made on the basis of medical images? This question does come up in medical practice. Physicians and patients alike are interested in assessing the quality and reliability of these visual techniques and the people performing them. In the following, I will focus on one diagnostic test for lung cancer by means of computed tomography (CT) scans, conducted in 2003 by a Houston-based team of oncologists and radiologists in the Department of Diagnostic Radiology and Biostatistics of the University of Texas, headed by Jeremy J. Erasmus (*1958).[65]

The procedure by which they checked for interobserver and intraobserver variability in the measurement of lung cancer was as follows. Images of 'non-small' cell carcinomas of a study group of 33 patients with malignant lung tumors[66] were presented to five radiologists with thoracic fellowship training and more than four years of post-training experience. They accessed these images in the form

---

(1968) on the earlier state of the art, and Bracegirdle (1978) chap. 4, Niedrig (1987) p. 55 and Rasmussen (1993) pp. 232f. on the historical development of ultramicrotomes. Cf. also Baird & Shew (2004) p. 153 on black-boxing and automated adjustment in EM.

[64] Quotes from popular articles on EM in Rasmussen (1997) pp. 224f. The popularization of EM in this vein already begins with an article on 'Alice in Electronland' by Ladislaus Marton (1943).

[65] On the following, see Erasmus et al. (2003) and Peitgen et al. (2011). Similar findings on the interobserver variability of mammograph screening by US radiologists were published by Beam et al. (2002).

[66] All patients had cancerous regions measuring at least 1.5 cm. In this sample, the average sizes varied between 1.8 and 8 cm, with a 4.1 cm average. For smaller samples, the error rates are even higher than what is reported below.

of film prints based on CT scans of critical layers of the patients' lungs. The measurements were repeated after 5–7 days, and their tumor size estimates were statistically analyzed. When the five radiologists were given the second set of images, they were led to believe that they were taken from the same patients at a later date. They naturally expected a certain amount of growth of the carcinoma (if untreated) or recession (if treated), but no direct comparison with the older images was allowed. In fact, they were given the exact same images again, in order to test the variability of their size estimates. *A priori* one would expect that – statistically, at least – the same size estimates should result. *De facto* the two measurements differed by an increase of more than 20% in every third instance. Thus the carcinoma was classifiable as progressive. In every seventh case, the difference was so strongly negative (by more than 30%) that the carcinoma would have been classifiable as recessive in partial response[67] to a successful medical therapy. Despite eliminating all other factors that might lead to interpretation differences in the CT scans, there was a shockingly high variability in the interpretation of one and the same image, seen by the same radiologist under identical conditions and with the same luminescent screen. The surprising conclusion by Erasmus and his collaborators was:

> Measurements of lung tumor size on CT scans are often inconsistent and can lead to an incorrect interpretation of tumor response. Consistency can be improved if the same reader performs serial measurements for any one patient.[68]

All findings discussed so far concern so-called intraobserver variability, i.e., the consistency with which one observer judges the size of certain malignant tumors. As concerns interobserver variability, the findings were equally disconcerting: Depending on the overall shape and structure of the tumor, interobserver variability ranged between 140% for estimates of the maximum diameter and overall size of irregular, spiculated tumors (such as in the right part of fig. 102A and B in the center of the dark right half) and only 7% for the maximum diameter and 14% in the overall size of lobular, well-marginated tumors (left part of fig. 102).

This finding runs counter to the general assumption that such measurements of CT images yield comparable and reproducible values independently of the radiologist involved. What are the reasons for such a strong variability? First of all, there may have been a certain expectation for temporal change. But since this might have gone in either direction (growth or shrinkage), the statistical

---

[67] Both according to the international norm RECIST specified in 1999.
[68] Erasmus et al. (2003) p. 2574.

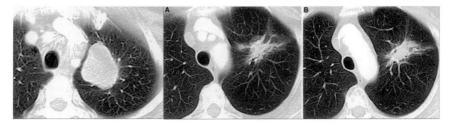

Fig. 102: Two different forms of lung cancer as seen on CT scans. Left: well-defined and clearly delineated nearly spherically shaped. A, B: irregular, spiculated tumors. From Erasmus et al. (2003) pp. 2575f. Reprinted with permission. © 2003 American Society of Clinical Oncology.

importance of this subjective factor is unclear. The influence of objective factors related to the CT imaging technology and its inbuilt limitations is much more certain. CT and NMR scans are digitally produced images with a fixed unit length for the smallest image element. Instead of an infinite number of image points, one only has a finite spatial grid of volume elements (also called voxels). Each voxel can only be given one gray tone or color value, and the overall quality of the CT or NMR scan is limited by the total number of voxels available. It currently lies in the range of 512 × 512 voxel per layer for CT scans, corresponding to the image resolution of a 0.25 megapixel camera and thus less than even a simple handheld apparatus with 640 × 640 voxels (or 0.3 megapixels) and far less than high-quality digital cameras with more than 10 megapixels. This means that unlike digital photographs, the effect of pixeling is still considerable. For the marginal voxels of, say, 1 mm size, there is uncertainty about whether the voxel should be counted as part of the tumor, or as part of its environment. The imaging specialist Heinz-Otto Peitgen (*1945), who became world famous for his studies on the visualization and aesthetics of fractals in the mid-1980s and has been working since then in medical image analysis, has called this error source the 'partial volume uncertainty.' A simple image of a square shown unpixelated and pixelated (at the top and bottom of fig. 103) demonstrates how apparently small uncertainties can accumulate into considerable uncertainties as to the total volume of a given figure once pixeling and blurring are superimposed.

As a trained mathematician, Peitgen was able to argue that uncertainties that may easily amount to tumor volume uncertainties of 60% or more are intricately bound to the imaging technology at hand. Not even a skilled and experienced modern radiologist could overcome it by subjective interpretation. In his opinion, computer-aided tumor volumetry can reduce the uncertainties in the volume estimates by factor 5. In order to develop such complex computer support

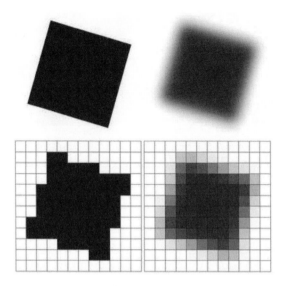

Fig. 103: Effects of digital pixeling (from top to bottom) and blurring (from left to right) on volume estimates. From Peitgen et al. (2011) p. 266. © Nova Acta Leopoldina, by permission.

systems for the measurement and interpretation of digital images, Peitgen founded the interdisciplinary *Center for Medical Image Computing, MeVis Research GmbH* in Bremen in 1995. In 2009, *MeVis Research* became an institute of the *Fraunhofer-Gesellschaft*, today called *Fraunhofer MEVIS*. Furthermore, Peitgen and his co-workers founded the *MeVis Medical Solutions AG* in 1997 as a commercial enterprise for developing software for cancer diagnosis and surgical support.[69] The still considerable resistance by the medical community to such computer-aided systems is intimately linked to issues of professional pride. Highly skilled radiologists understandably insist on the ultimate superiority of their trained eye, their favored 'instrument' for the interpretation of the rising flood of digitally produced data.

[69]For further examples of the potential of such computer-aided support, see Peitgen et al. (2011).

10

Visual thinking in scientific and technological practice

Throughout this book we have encountered several examples of 'visual thinking' in science and technology. To Gottlob Frege, the early Ludwig Wittgenstein or the logical empiricists, it would have been anathema to speak about something like 'visual thinking' because for these philosophers 'thinking' was intimately linked with words bound together in sentences. Only words could possibly have any 'truth value' and 'meaning,' whereas images are at best fitting or adequate, useful or misleading, but never 'true' or 'untrue' in the full-blown sense. Against this preponderance of language within philosophy of science since its 'linguistic turn,' the art psychologist, pedagogue and media scientist Rudolf Arnheim (1904–2007) managed to fight for his concept of 'visual thinking,' first presented in full in his book of that title from 1969.[1] As lecturer on art psychology at the *New School for Social Research* in New York (until 1966), at the *Carpenter Center for the Visual Arts* at Harvard University (1968–74) and at the University of Michigan in Ann Arbor (1974–84), he exerted much influence on psychology and cultural theory worldwide.

In his book, as well as in many articles, Arnheim combatted the prevalent "myth of the priority of language" and defended his notions of an "intelligence of seeing" and of "perception as knowledge."[2] Borrowing from the insights of the Berlin school of *Gestalt* psychologists, in which he had received his own academic training, Arnheim insisted on the difference between 'seeing of' and 'seeing as.' The latter goes far beyond the mere physiological act of sensory perception. It includes an active interpretation of what is seen as a specific *Gestalt*, defined as an antithesis to a mere sum or bundle of constituents, i.e., as "an articulated whole, a system, within which the constituent parts are in dynamic interrelation with each other and with the whole, an integrated totality within which each part and subpart has the place, role, and function required for it by the nature of the whole."[3] 'Seeing something as a *Gestalt*' for Arnheim

---

[1] Arnheim, born in Berlin, had studied within the Berlin school of *Gestalt* psychology, writing his PhD thesis in 1928, thereafter working as editor of the weekly cultural journal *Weltbühne*. In 1933, he emigrated, first to Rome, where he worked on an encyclopedia of cinematography, then to London in 1939, and finally to the USA in 1940.

[2] See Arnheim (1969*c*), quotes from pp. 10 and 24.

[3] Wertheimer (1985) p. 23.

was thus not passive reception of a light ray on the human retina, but active appropriation and interpretation of this sensory impulse by the human brain and intellect. He rather compared attentive observing and 'seeing as' with the active, groping exploration by humans of their environment, an active search for patterns already familiar to them, and careful comparison of all new observations against this huge reservoir of known *Gestalten*. Thereby, Arnheim could easily explain striking features of human visuality, such as the surprising speed with which we recognize objects or with which we spot irregularities in a heap of data, as long as the data are presented to us in visualized form. We are much slower at spotting these same irregularities in a table or in other nonvisual representations. Abstraction to him was successful typologization of frequently appearing *Gestalten* around certain clusters of phenomenologically similar forms; and verbal concepts were rather the end point than the initiation of these efforts at structuring our environment.[4] True to the tradition of *Gestalt* psychology, which since its inception had been actively interested in the actual reasoning processes of mathematicians and scientists, Arnheim also devoted some of his attention to the use of images and mental models in scientific and technological reasoning processes. Examples that he picked included Cusanus on infinity, Leonardo on anatomy, Kepler on ellipses and Gauss on non-Euclidean geometry. Finally – and this is very relevant to our layered approach to visual cultures – Arnheim also realized the high relevance of visual thinking to education and practical training. His examples for the "pedagogical gaze" encompass the pedagogical classics by Comenius and Pestalozzi, experimental demonstrations in physics courses and close analysis of problems in science teaching by means of visual aids.[5]

Since Arnheim, many psychologists and cognitive scientists, artists and art historians, sociologists, and theoreticians of culture have asked themselves exactly how this 'visual thinking' works and in what sense it operates differently from normal thinking. Historically minded engineers, such as Eugene S. Ferguson or Walter Visconti, the architect Tom F. Peters, as well as historians of technology, such as Brooke Hindle, have also provided plenty of case studies on visual thinking in the engineering realm.[6] Out of the vast array of excellent examples from this

---

[4] On Arnheim's ideas about abstraction, degrees of *Prägnanz* (singularity or simplicity and stability) and concepts as climaxes (*Begriffe als Höhepunkte*), see Arnheim (1969c) chap. 10, esp. pp. 175ff., taking as nice examples acute and obtuse triangles.

[5] See Arnheim (1969c) chaps. 15f. on scientific and educational uses of images and models; cf. also Crowther & Barker (2013) on informational vs. didactic functions of early-modern astronomical diagrams.

[6] See Ferguson (1992), Vincenti (1993), Hindle (1983, 1984), Peters (1998) or Hamilton (2001).

terrain, we will later pick bridge construction (in sec. 10.3). But beforehand, let us first address some more general points and then look at various instructive examples from the geosciences.

The American psychologist Howard Gardner has argued that special mental faculties are indeed involved in the processing of visual sensory impressions and images, and that there is indeed something like 'spatial intelligence' as distinct from verbal intelligence.[7] To Robert Root-Bernstein, this faculty rather amounts to a third "kinaesthetic" base-skill aside from visual and verbal skills, especially if it is applied to active forms of motion such as dancing or subtle eye-to-hand coordination, for instance in experimentation, remote-control guidance or minimally invasive surgery.[8] Microscopic observation provides excellent examples of skilled hand-to-eye coordination.[9] At any rate, spatial intuition most definitely includes:

- recognizing the same object when viewed from various angles;
- drawing an iconic representation of some viewed object;
- imagining an object known from one angle as it would look if regarded from another *vista*;
- conjuring up mental imagery, as well as mentally playing with it or transforming it, etc.

The British empiricist philosopher David Hume (1711–76) already noted in 1748 that "to join incongruous shapes and appearances costs the imagination no more trouble than to conceive the most natural and familiar objects." In the mid-19th century, the physiologist, physicist and philosopher Hermann von Helmholtz (1821–94) stated that "memory images of purely sensory impressions [...] may be used as elements of thought combinations without it being necessary, or even possible, to describe these in words." In his *Analysis of Sensations* (1886), his Austrian colleague Ernst Mach (1838–1916) pointed out that pairs of irregular shapes are much more readily recognized to be identical in form if they are presented in the same orientation.[10] This last finding implies that if they are presented in different orientations, the observer's mind first has to perform some (mental) transformation work to compare these shapes. Other experimental

[7]See Gardner (1985), esp. chaps. 8–9 on spatial and bodily-kinaesthetic intelligence, and ibid., pp. 178f. on Piaget's developmental studies concerning spatial understanding and mental imagery.

[8]See, e.g., Root-Bernstein (2001) p. 301 or (1999) chap. 4 on 'imaging,' chap. 9 on 'body thinking,' and chap. 11 on 'dimensional thinking.'

[9]See, e,g., Keller (1996) and Carusi (2011) pp. 323ff.

[10]For the references to these primary works and further literature on mental images and their transformation, see Shepard & Cooper (1986) pp. 1–8.

psychologists and cognitive scientists in the second half of the 20th century have confirmed that special spatial thought processes are indeed involved, for instance, in mentally rotating a complicated form or imagining how it would look from the other side. Thus, for instance, to recognize the set of cubes in fig. 104 as identical (in sets a and b), and topologically different to the others (in set c) takes each observer time, and, as has been confirmed in experiments, proportionately more time the more the two angles differ from each other (see the linear increase in time with angle of rotation in the right half of fig. 104). It is thus indeed as if the observer slowly turns the object in his or her mind until the two images either match (a and b) or don't match (c). Likewise, it takes time to recognize objects when portions are not visible, when the images are degraded, etc.[11]

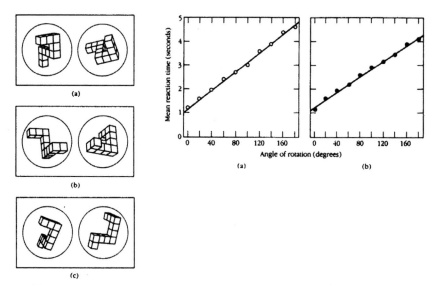

Fig. 104: Experiments on mentally rotating complicated objects. Left: a complicated array of cubes, shown at different angles. Test subjects were asked whether the two images are congruent. Right: the times taken between posing the question and getting the right answer are plotted against the angles of rotation of the image pairs. From Metzler & Shepard (1974) pp. 159 and 161. Reproduced with the kind permission of Prof. Roger N. Shepard.

---

[11]On these fascinating experiments and their controversial interpretation, see, e.g., Metzler & Shepard (1974), Shepard & Cooper (1986) part I, Pylyshyn (1973, 1981), Kosslyn (1990, 1994) and Miller (1996) pp. 266ff.

Follow-up work with the aid of computer-controlled corneal reflectance eye-tracking systems led to the claim that human information processing during these mental exercises proceeds in three successive stages: (i) a search phase, in which the eye seeks out particularly characteristic features of the two forms to be compared; (ii) the actual mental transformation stage, in which the mental rotation is performed together with sequences of comparisons to check congruence; and (iii) a confirmation stage, in which all other features of the figures are verified for congruency. The times for stage (ii) again rose linearly with the angular difference between the two views. Children and the elderly needed longer than regular adults. A similar proportionality between time and angle was found with blind test subjects presented with two objects to determine by touch whether the two three-dimensional shapes were identical.[12]

Further follow-up work shows that these psychological experiments on human visuality can be generalized to more complex levels of *Gestalt* recognition and perception. Consider, for instance, how the topology of a non-rigid manifold maintains its structure during mental transformations, such as the three-holed torus in fig. 105. With difficult tasks like this, we are already close to everyday practices like undoing a knot and topological arguments in geometrodynamics (cf. sec. 3.4).

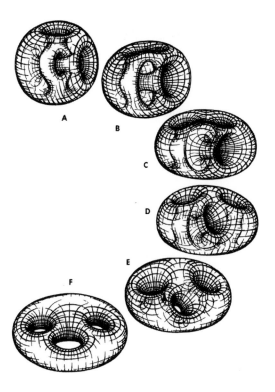

Fig. 105: Topological deformations of a two-dimensional manifold of genus 3, which turns out to be topologically equivalent to a three-holed torus. From Shepard & Cooper (1986) p. 260. © 1982 Massachusetts Institute of Technology, by permission of The MIT Press.

[12]On this follow-up work, see Shepard & Cooper (1986) pp. 171–86.

This indicates that the mental rotation under examination is not a specifically visual process but rather a mental operation. Metzler and Shepard see their example (fig. 104) as representative of visual thinking in Arnheim's sense. They argue that this kind of mental rotation plays a key role not only in how we go about furnishing our apartments but likewise in geometric problem-solving, engineering design or stereochemistry, to name just three examples from the vast scope of this book.

Is there really a higher visual logic in image processing – a *Bildlogik*? Can one argue with images, and, if so: What are the guidelines of this discourse (or rather 'viscourse'? – see more on this concept on pp. 262ff.)? Are there equivalents to the Aristotelian syllogisms in which our visual arguments can be classified?

I think it is fair to say that answers to all of these questions are far less straightforward than one might perhaps wish. Generally speaking, we realize that any such 'logic' of visual representations is tightly bound to the medium and communicative context of the given representation. In other words, movies and television have a different 'logic' than copper plates in an early-modern book or photomechanical prints in a modern science journal. Each of these media has its own cultural settings and constraints.[13] Another point that can be stated in a general form is that visual logic is not predicative like simple sentences, in which a predicate is ascribed to an object. An image might display an object as having some predicate, but in the image a lot more goes on at once than in a sentence. Several sentences follow each other linearly. (The human capacity to process several sentences uttered at once is fairly limited!) thus channeling arguments and discourses into a linear mold, hyperboles, allusions to already uttered things and other recursive features of language notwithstanding. By contrast, each image opens up a large associative field of meaning relations, so the linkages between different images are much less linearly organized and much more open to interpretation than are typical sentences.[14] So the real question for us is less *whether* images have a logic of their own than *how* they convey their meaning. We shouldn't only ask what images (or models) do to us, but also *how* they achieve the means and ends ascribed to them.[15]

One purpose built into images in science and technology is a condensed survey of information: one graph or diagram can summarize hundreds or thousands of measurements. It can also reveal tendencies or patterns otherwise hidden in

[13] A good survey of the various theoretical efforts to come to grips with visual thinking is provided in the introduction to Hessler & Mersch (2009), esp. p. 14 on cultural "dispositive."

[14] On this fundamental difference between images and text, see especially Boehm (1994, 2007).

[15] This is the meaning of Mitchell's famous question: "What do pictures really want?" See Mitchell (1996), who deliberately researches the "lives and loves of images."

the masses of data, another in-built purpose of many scientific inscriptions. A third purpose is to assist in scientific or technological problem-solving. *Gestalt* psychologists and cognitive scientists from all over the world have worked on the various stages of this very complex process and have identified the following components:[16]

- recognizing central features of the problem (extracting its core);
- stripping away irrelevant features and simplifying, if necessary;
- modularizing complex problems (splitting into various smaller tasks);
- removing mental blocks and functional fixation through a creative, open-minded search for all alternatives and options ('productive thinking');
- finding adequate problem representations (modeling or abstracting);
- restructuring of problems (reformulating or translating into another framework in which handling the problem becomes somehow easier);
- recognizing similarities to other problem situations already encountered;
- choosing appropriate strategies to solve the problem from a reservoir of heuristics; and
- transferring solutions from other areas of knowledge or experience.

Visual representations can be helpful in many of these steps: to depict an object or a process forces the observer to simplify and to concentrate on important features; it quasi-automatically suggests visual analogies to other areas of knowledge, and it often also helps to find the right framing or modeling.

10.1   Gooding on Faraday and fossils

Several researchers in the cognitive sciences, art history and media psychology have taken up Arnheim's claims on visual thinking. Since our focus is on the natural and engineering sciences, I choose here the work of David Gooding (1947–2009). This professor of the history and philosophy of science at the Science Studies Centre in the Department of Psychology at the University of Bath took up Arnheim's interdisciplinary research agenda in a most fruitful way. His strongly comparative approach – like Svetlana Alpers's (see sec. 1.5) – will be another master narrative for us in what follows.

---

[16]For concise summaries, see, e.g., Ohlsson (1984), Wertheimer (1982) and (1985) pp. 23ff., and Sternberg & Davidson (eds. 1995).

In various case studies, ranging from Leonardo da Vinci's attempts to capture turbulence in running water, to Michael Faraday's efforts to come to grips with electromagnetic induction effects, to x-ray diffraction or the modern understanding of magnetic anomalies in mid-ocean deep-sea ridges, Gooding always tried to understand "how visual thinking works."[17] Here I pick out his nice example of palaeontological reconstructions of certain arthropods from the geological stratum called the Burgess Shale. These long since extinct species with no modern analogues lived in a Middle Cambrian faunal complex and their traces were found in British Columbia, Canada.[18] The few samples preserved as fossils in sedimentary rock of the Canadian Rockies had been flattened considerably, however, before petrification had set in. Sometimes only parts were preserved, and one always only sees certain segments in one orientation or another. Thus the difficult task for the researchers was to reconstruct the complete shape of the living organism from these skimpy remnants. The chain of representation in which this reconstruction was achieved is nicely illustrated by the sequence of three images taken from the research literature shown in fig. 106.

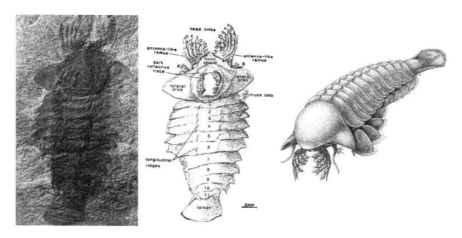

Fig. 106: Three steps in reconstructing *Sanctacaris uncata*. Left: photograph of a prepared petrification as found in British Columbia. Center: labeled drawing based on this photograph. Right: Marianne Collins's perspectival reconstruction based on several fossils and drawings. From Briggs & Collins (1988), ill. 3, fig. 1B on pl. 71, pp. 781f., fig. 6 on p. 792. © John Wiley & Sons.

---

[17] Gooding (2004*a*) p. 278. Cf. also Gooding (2004*b*, 2006), and Gooding (1990) for examples from his Faraday studies.

[18] See, e.g., Briggs & Morris (1983) and Briggs & Collins (1988) for two palaeontological primary texts, Rudwick (1972/85) and Gould (1989) for semipopular historical expositions, and Gooding (2004*a*) pp. 281ff. and (2004*b*) pp. 561ff. for a historico-cognitive analysis.

In a very simplified scheme, the succession of visual representations that were used would read as follows:[19]

(1)   'raw' data (here a fossil imprint)
       → prepared fossil imprint
       → unretouched photographic image(s)
       → retouched photographic image(s)
       → *camera lucida* image (drawn from these photographs)
       → diagrammatic representation of photographed fossils (labeled drawings)

While the sixfold sequence (1) as well as the threefold sequence in fig. 106 both seem to suggest a fairly linear process of 'translation' and interpretation, factual research practice is much more complicated, with many switches between 2D drawings and 3D reconstructions following each other in cyclic iteration until a satisfactory representation is obtained. Some *camera lucida* drawings lead to new ideas on how to retouch the photograph, some drawn diagrams will then also show inferred features actually missing from the original fossils because they had broken off, and initially each new fossil find will slightly alter the reconstructions. But the most important interplay that Gooding noticed was the "dialectical play between 2D-images of patterns, 2D-images of sections of 3D-structures, 2D-images of 3D-structures and actual 3D-structures."[20] "Passing countless hours rotating the damned thing in my mind" was essential for getting a grip on the 3D form of the organism from its various two-dimensional segments or impressions. In other words, the indispensable skill necessary to do good work in this area of paleontology is to have "a knack for making three-dimensional structures from two-dimensional components and inversely, for depicting solid objects in plane view."[21] Put schematically, we get:

(2)   diagrammatic representation, construed as a section
       → mental rotation
       → mental image of rotated section
       → mental comparisons
       → attempted match between drawing(s) and photograph(s)

---

[19] According to Gooding (2004*b*) p. 562, here already slightly amended. Cf. also Briggs & Williams (1981) on the restoration practices for flattened fossils, and Gould (1989) pp. 17f., 85ff. on the necessity to complement photographs by detailed drawings of the specimens.

[20] Gooding (2004*b*) p. 563.

[21] Ibid., pp. 563 and 567, quoting Harry Whittington in Gould (1989) pp. 92 and 101, who comments further on Whittington's special skill at 3D visualization. For structural parallels to protein biochemistry, see Trumpler (1997), esp. pp. 88f. on their dissatisfaction with images on paper and their "desire for a three-dimensional image moving over time."

In a further step of abstraction, Gooding condensed his insight into a kind of flow diagram (see fig. 107) showing the various types of visual inferences. First is the move from an observed but still unresolved phenomenon A to a first depiction as a first tentative pattern B. Then follows the inference of a structure C and/or of a process D (like the above flattening of a 3D form into a nearly 2D petrification), with iterative cycles of modified B, C and D. As he emphasizes, sometimes these visual transformations act in the direction of a complexity reduction, especially so at the beginning, but later these transformations might just as well go into the other direction of adding further detail. Thus simplification and complexification are dialectically intertwined and reiterated until a satisfactory scientific understanding is reached. This interplay is actually another deeper reason for the frequently found successions of various forms and techniques of representation in science and technology (as already discussed in sec. 7.2). Scientists vary their images in both directions, 'bottom up' from objects or data to interpretation and 'top-down' from identified patterns or hypothetical models. Features derived from these models are checked against observable patterns of their object. As is also demonstrated in Gooding's example in fig. 106, this nonlinear, recursive cycle of visualizations in various degrees of complexity does not forbid words and other symbolic signifiers. Notice the many written labels in the middle of the figure, which are absent from both the photograph of the fossil and the finished 3D reconstruction, although they could just as well have been mounted there, too. Because science aims at integrating various layers of knowledge, this image–text interaction is just as much a part of visual thinking as the dynamic vacillation between the number of dimensions considered or in the degree of simplification and complexity sought.

Each visual culture in science or technology has its own, very specific, rules of transformation guiding these inferences. So undoubtedly, specific domain knowledge and specific imaging techniques will be used. Thus every historical case study will be singular in these specificities. However, beneath the surface, Gooding's pattern–structure–process scheme (for which he introduced the acronym PSP) is present at the deeper level of visual inferences to be found in science and technology across the board. Gooding's claim that these kinds of visual reasoning structures meandering back and forth between simple abstractions (B), more complex 3D representations (C) and hypothetical inferences to processes (D) are clearly general features in a broad range of scientific fields, reflecting deep, cognitively based continuities. Superb examples from the realm of engineering have been provided for the inventors Alexander Graham Bell, Thomas Edison,

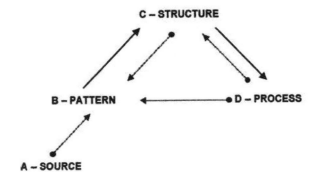

Fig. 107: Visual inferences from unresolved phenomenon A, to a first depiction as a tentative pattern B, the further inference to a structure C or to a (four-dimensional) process D, and back, with iterative cycles ACD and/or ADC. From Gooding (2004*a*) p. 286; cf. Gooding (2010) pp. 26f. With permission of Maney Publishing: www.maneyonline.com/isr

James Lovelock, Samuel Morse and Charles Steinmetz.[22] We will notice them in our two next examples from crystallography around 1800 and bridge construction around 1900.

10.2   Crystallographic puzzles: space models and x-ray diffraction

That many minerals found in nature actually occur in particular, regular shapes must have been known to humans from the earliest times. Some of the hardest of these, such as the quite commonly occurring quartz crystal, might even have been used as tools for drilling holes or scraping animal hides. Efforts to classify the various crystal forms and theories about how these crystals might have formed came much later, though.[23] Careful observers in the 17th and 18th centuries came up with two basic hypotheses:

(a) a concretion of round balls (Kepler 1611, Hooke 1665, Huygens 1678);

(b) a concatenation of small building blocks that supposedly already have the germ shape that the larger crystals would grow into by spatial aggregation (Gassendi 1678, Guglielmini 1688, 1705; Haüy 1784, 1801).

The earliest of these texts, a study on the six-pointed snowflake by the astronomer Johannes Kepler,[24] already makes the perceptive observation that, de-

---

[22]See Hindle (1983) pp. 85–126 and Root-Bernstein (1999) pp. 50ff.

[23]For broad surveys of the development of crystallography, see Burke (1966), Laudan (1987), Lima-de-Faria (1990) and Kubbinga (2012).

[24]On Kepler's *Strena Seu de Nive Sexangula* and its interesting mathematical assumption about the densest packing whose strict proof could only be found by Thomas C. Hales in 1998, see

pending on the piling of successive layers, two different structures would result that later came to be called cubic and hexagonal. These early modern observers may well have inferred this insight from piles of canonballs, for instance. The perceptive eye can thus easily spot these two basic structures. Robert Hooke took up these considerations in his *Micrographia* (1665), as did Nicolaus Steno in *De solido intra solidum naturaliter contento dissertationis prodromus* (1669). Hooke drew various more complicated crystals and simplified piles of balls in arrangements that mimic the basic shapes and facet angles of such crystals (cf. here the left part of fig. 108). Steno also started to speculate about their processes of growth. The step from perfectly spherical to elliptical elements was then taken by Christaan Huygens in 1690 in order to explain the feature of double refraction in Islandic spar (cf. the right part of fig. 108).

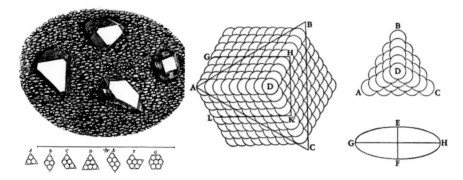

Fig. 108: Left: Hooke (1665) correlates crystal forms with regular arrangements of spherical balls. Right: Huygens (1690/1912) chap. 5 on the biaxial structure of calcite crystals formed by ellipsoids.

The basic problem with this kind of ball-pile approach was that no such miniature spherical balls could be found in crystals. Microscopic studies by Leeuwenhoek, Freind, Bourguet and others rather revealed that even the most minute components of crystals observable in the best microscopes of the time were sharp-edged and not at all roundish. Some crystals, such as Icelandic calcite, could easily be split into ever smaller pieces with a sharp knife, but all of them seemed to have the same shape as the large crystal. This observation, repeated countless times and becoming a staple of lectures in natural history, strongly supported the alternative conceptualization of crystals as an aggregation of units

http://en.wikipedia.org/wiki/Kepler_conjecture and further references given there.

of the same shape as the large crystal. During the 18th century, instrument makers such as Arnould Carangeot (1742–1806) constructed the contact goniometer to allow quite precise determinations of the angles between the facets of crystals large enough for such measurements. They required perfect alignment of the crystal surfaces with the adjustable legs of the instrument (cf. fig. 109).

Using these contact goniometers, the mineralogist Jean-Baptiste Louis Romé de l'Isle (1736–90) spelled out a first phenomenological regularity in his *Essai de Cristallographie* in 1772:[25] the angles between neighboring surfaces of similar orientation are constant in all crystals of the same mineral species, whereas the lengths of the sides might vary substantially. This finding became the "first law" of the new discipline of crystallography when these observations were systematized further. Romé de l'Isle also realized that truncations of corners or edges could transform the principal forms of a crystal into many variants. He illustrated his finding in impressive plates demonstrating the large variety of side lengths and truncations retaining constant main angles across a sampling of crystals out of the same mineral. He even had 3D models of such crystals made in order to make this finding better 'graspable' to nonmathematical minds.[26] The project of transferring the interplanar angles of natural crystals to a set of terracotta models even promoted the invention of the contact goniometer. Subscribers to the second edition of Romé de l'Isle's *Crystallographie* (published in four volumes in 1783) were offered a complete set, altogether no less than 448 ceramic models, as a supplement to the dozens of plates. Quite a few of these terracotta, or later bronze, models are preserved in natural history collections and geological or mineralogical institutes all over the world (e.g., in Paris, Vienna, Haarlem, Utrecht, the *British Museum* in London, etc.).[27]

Romé de l'Isle was not the first to use the term 'crystallography.' It had been introduced by Moritz Anton Cappeller (1685–1769) in 1723, but a research tradition – and later a separate discipline of crystallography – branched off from mineralogy only at the end of the 18th century. One of the outstanding crystallographers working along these lines was René-Just Haüy (1743–1822), son of a simple linen weaver.[28] Haüy had initially studied theology and worked as a priest

---

[25]See Romé de L'Isle (1772, 1794), Buée (1804), Wiederkehr (1977) and Wilson (1994) pp. 50–3.

[26]On the production of 448 different types of such models by Romé de l'Isle's assistants Claude Lhermina, Arnould Carangeot and the artist Swebach-Desfontaines, see, e.g., Touret (2004).

[27]On the importance of these crystal models for the formation of crystallography, see Touret (2004) pp. 46ff. For online samples, see www.mineralogy.eu/models/01_model.html and en.wikipedia.org/wiki/Crystal_model (both last accessed Sept. 26, 2012).

[28]On Haüy, see, e.g., Cuvier (1823/27), Kunz (1918), Weber (1922), Wiederkehr (1977/78)

Fig. 109: Left: a contact goniometer designed by Arnould Carangeot to measure crystal angles. This particular instrument was used by Haüy during his tenure at the Muséum d'Histoire Naturelle in Paris. Photograph from Touret (2004) p. 45. Courtesy of the Royal Academy, Amsterdam. Right: a portrait of Haüy measuring the angles of a calcite crystal with his contact goniometer. Engraving by Johann Anton Riedel (1733–1816) after a drawing by Félix Massard (1776–?), first published in 1812.

and teacher since 1770 at the *Collège Cardinal Lemoine*. His superb work in crystal-lography earned him in 1802 the successorship to Louis Jean-Marie Daubenton (1716–99) and his former assistant Déodat Gratet de Dolomieu (1750–1801) as professor of mineralogy at the *Muséum national d'Histoire naturelle* in Paris, where he taught until his death. Since 1809 Haüy also held the chair for mineralogy at the *Sorbonne* in Paris. Haüy started his conceptualizations from the elementary observation that the mechanical cleavage of larger crystals of minerals of one kind always yields smaller elements of the same basic shape, regardless of the initial outline and proportions of the large crystal. Calcium carbonate crystals, for instance, which occur in nature in at least three different shapes (pentagonal, hexagonal and rhomboid), ultimately always yield rhomboid fragments.

and Wilson (1994) pp. 53–6; cf. Lacrois (1945) with excerpts from Haüy's correspondence, and Bilodreau-Guinamard (1997) on Haüy's unpublished manuscripts. On the further development of crystallography since Haüy, see Wiederkehr (1978), Maitte (2001), Kubbinga (2012) and further literature cited there.

The Swedish chemist Torbern Bergman (1735–84) had made a similar observation in 1773,[29] but Haüy generalized his finding well beyond singular establishments of fact. He inferred that the rhomboid shape was what he then called the *forme primitive* of this class of minerals.[30] His task was to deduce all the other shapes that crystals of this mineral might assume out of this elementary unit. This deduction – or, as we should rather say, this inference – included a hefty amount of visual thinking. First of all, macroscopic crystals had to be conceived as composed of submicroscopically small polyhedrons, which he dubbed *molécules intégrantes*, i.e., microcrystals containing all the chemical components of the crystal in constant proportions reflecting their overall composition.[31] The primary form (*forme primitive*) of each crystal was simply the densest packaging of these *molécules intégrantes* piled on top of each other, layer by layer. Deviant secondary forms (*formes sécondaires*) of such crystals were obtainable by algorithmic omission of so-called *molécules soutractives*, i.e., groups of polyhedrons omitted in the next layer at the edges or corners (cf. figs. 110 and 111 for a few examples). This kind of visual thinking in terms of 3D building-blocks provided Haüy with a relatively simple and straightforward explanation for the mysterious polymorphism of many minerals occurring in more than one crystal form in nature (e.g., aragonite).

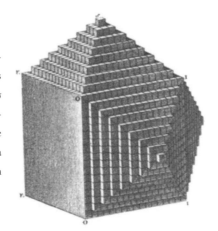

Fig. 110: Haüy's virtual construction of deviant forms of crystals. Haüy mentally piles a layer decremented by one row of *molécules intégrantes* on top of a cubic block. After repeating this observation and letting the size of the *molécules intégrantes* diminish below a sensory threshold, he obtains a crystal with truncated edges. From Haüy (1793) plate.

[29] See Hoykaas (1955) pp. 319ff.

[30] See Haüy (1782*a*) on calcite and (1782*b*) on garnet, Haüy (1802) on other calcareous minerals, and Buée (1804) and Hoykaas (1955) pp. 320ff. on the similarities and differences between Romé de l'Isle's and Haüy's systems.

[31] On the following, see, e.g. Haüy (1815), Weber (1922) and Wiederkehr (1977/78).

Haüy allowed only three basic *formes primitives* in the core of his mental crystal constructs. For the mineral olivine (or *péridot* in French, cf. here fig. 112), he distinguished the simple tetragonal *péridot primitive* (later renamed cubic), *péridot continue* (later redesignated (ortho)rhombic), and *péridot quadrangle* (today called hexagonal). But how would it be possible to construe other crystal forms, for example, the rhomboid dodecahedron, from such basic constituents as plain squares of the cubic variant? Figures 110-111 demonstrate how this could be done, by selectively building up layer upon layer and tapering off certain sectors along the edges or at the corners ("figures de décroissemens sur les bords ou sur les angles"). In principle, arbitrary diminutions (*décroissemens*) or truncations (*troncatures*) are possible; but in order to reduce arbitrariness and complexity, Haüy assumed that only the simplest of these mathematical possibilities actually occur in nature. *A priori* simple numerical relations were preferred to more complicated ones, thus (1 : 2)1/2 for cubes, (2 : 1)1/2 for rhomboid dodecahedrons, and (1 : 8)1/2 for the regular tetrahedron. That means, in the taper of crystal surfaces from one layer to the next, only one or two rows or corner units are omitted, not five, ten or even more. He also assumed that these *décroissemens* would have to be applied more or less symmetrically at the borders and corners of the crystal layers.

> "This idea is so satisfying and so true, that in general nature tends to uniformity and simplicity. The ratios between the dimensions of the limiting forms possess this last property in a remarkable way."[32]

Haüy's 'law of symmetry,' as the last assumption was also called, became one of the earliest usages of the principle of symmetry in the natural sciences.[33] On the basis of this quasi-Platonic assumption, Haüy believed it possible to 'construct' all known crystals. And conversely, given a particular crystal, Haüy could put forward an educated guess as to what the primitive form of the *molécule integrante* would be, and what *décroissemens* were at play in this sample during its growing stage hundreds of thousands of years ago.

Plates such as that in fig. 111, with their subtle perspectival shortenings of nontrivial crystal shapes, were actually quite demanding for engravers and lithographers to execute. Therefore Haüy chose to depict his 3D crystals in parallel projection rather than in linear perspective. He had some assistance from mathematicians for the plates in volume 5 of his *Traité de Minéralogie* of

---

[32] Haüy (1815); cf. also Wiederkehr (1977/78) p. 293 and Blondel-Mégrelis (1981) pp. 292ff.

[33] On the fascinating history of ideas concerning this principle, which had cropped up in antiquity under a different meaning but was reimported into physics from crystallography relatively late, in the 19th and early 20th centuries, see Hon & Goldstine (2008).

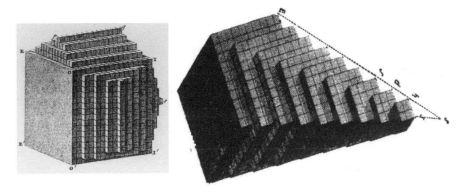

Fig. 111: Left: Haüy's 1801 piling of cubic *molécules integrantes* to form the pentagonal dodecahedron of pyrite. Notice the decrescence of the layers in the proportion 2:1, yielding an interfacial angle at pq of 127°, closely corresponding to the 128° in the empirical crystal. Right: Haüy's 1804 construction of the scalenohedron.

1801. One was a colleague of Haüy in the *Académie des Sciences*, Gaspar Monge, professor of mathematics at the *École Polytechnique*, and the foremost exponent of descriptive geometry.[34] Aside from Monge, several students at the *École des Mines*, the chemist Vauquelin and the librarian Clouet also helped in the production of Haüy's plates. The students Trémery and Champeaux also helped calculate the horizontal projection of 3D crystals onto 2D surfaces. Despite all these efforts, a few ambiguities remained unresolvable in this 2D parallel projection.

Like his predecessor Romé de l'Isle, Haüy also decided to work with 3D models in teaching his students and laymen the subtleties of crystal growth by the selective addition and subtraction of layers. Unlike Romé, Haüy chose to work with wooden models, which were crafted under his close supervision.[35] Figure 112 shows a few samples preserved together with Haüy's original labeling.

Haüy's research program amounted to a methodological revolution in crystallography. He was able to distinguish many kinds of minerals and crystals that hitherto had been incorrectly intermingled. For instance, he could distinguish no fewer than 16 different mineral species in what had previously been subsumed under the mineral collectors term 'schorl,' a similar differentiation of 6 different species became possible in the mineral class of zeolites, and zircon was differentiated into 5 species. Haüy himself self-confidently announced: "Nowhere else

---

[34]On Monge and descriptive geometry, see secs. 5.1 and 8.1. On crystal depiction, cf. Mss. 1400 in the Parisian *Muséum National d'Histoire Naturelle* and Bilodreau-Guinamard (1997) pp. 344f.

[35]On these models and their production, see Saeijs (2004) and Touret (2004) pp. 51ff.

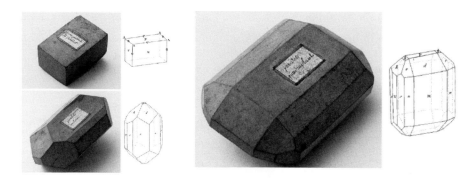

Fig. 112: Pear-wood models of olivine in its *forme primitive* and two varieties, designed and labeled by Haüy c.1800 (5–8 cm in size). From the collection acquired by M. van Marum for Teyler's Museum in Haarlem. From Touret (2004) p. 55. Courtesy of the Royal Academy, Amsterdam. The adjacent drawings are from Hauy's *Traité de Minéralogie* (1801), vol. 5, p. LX, figs. 198, 201, 202.

does the old-fashioned natural history contain so many errors."[36]

In fact, Haüy's writings can be used as a convenient marker for the differentiation between mineralogy and crystallography. The former, a much older field of natural history, was the mother discipline from which crystallography sprouted just around then. The term 'crystallography' was first used in 1723 by Moritz Anton Cappeller (1685–1769), but for a long time the field was still considered part of mineralogy. The actual branching-off occurred under the intense mathematization characteristic of the early 19th century. The formalized treatment of spatial symmetries pioneered by Haüy continued to inspire his pupils and followers. Whereas the first edition of Haüy's first textbook in 1801 was still entitled *Traité de Minéralogie*, a separate *Traité de Cristallographie* appeared in 1822, indicating that the latter field had completely emancipated itself. In 1815, the crystallographer Christian Samuel Weiss (1780–1856) renamed Haüy's three types of basic structure – *péridot primitive*, *péridot continue* and *péridot quadrangle* – into cubic (*kubisch*, ortho)rhombic and hexagonal; and the mineralogist Friedrich Mohs (1822) added the still missing two classes of monoclinic and triclinic. In the following decades, other crystallographers and applied mathematicians, such as Johann Friedrich Christian Hessel (1796–1872), Ludwig August Seeber (1793–1855), Auguste Bravais (1811–63), Arthur Schoenflies (1853–1928), Yevgraf Fedorov (1853–1919) and Pierre Weiss (1865–1940), fed the increasingly mathematical field with further insights into regular patterns in space (space groups) and their classification – a

[36] Haüy (1801) vol. III, p. 67.

new branch of applied mathematics had been born.[37] Once these regular crystallographic patterns had become firmly established, deviations from the ideal spatial forms shifted to the forefront of research: the study of line defects, dislocations and the temporal spread and growth of such spatial inhomogeneities originated in the 1930s and today marks an important strand of materials science, full of visual thinking.[38]

A superb example of a visual analogy taken from an early 20th-century episode in crystallography is that provided by Max von Laue (1879–1960) between x-ray diffraction and optical interference. Laue had just finished a review article for the *Enzyklopädie der mathematischen Wissenschaften* on wave optics. Thus he was primed to performing an analogical transfer of interference at a crossed grating, which had been treated in some paragraphs of his *Magister* thesis. He simply applied this basic idea to a process of interference of much shorter wavelengths (i.e., x-rays) passing through the regular grid of a regularly formed crystal. For the optical case, he had just shown that crossed gratings would not yield the normal bright and dark fringes. Instead of the normal stripes of light, only tiny points of light would show as the points where various semicircular interference fronts generated by each of the two crossed gratings intersect. Now, he expected a similar phenomenon to occur when a parallel beam of roentgen rays were led through a monocrystal with its various layers occurring at regular distances. The observations then conducted in Munich did, in fact, yield such regular patterns of points that somehow depended on the symmetries and orientation of the crystal through which they were guided (see fig. 113). Even though it still took quite some effort to translate the point patterns into crystal symmetries, this finding was a brilliant confirmation of the wave-like nature of x-rays, which was then still being heavily disputed.[39]

Von Laue's crystallographic method was soon refined and successfully applied to all kinds of materials, not only by teams in Munich, but particularly by William Henry Bragg (1862–1942) and his son William Lawrence Bragg (1890–1971).[40] In Germany, the application of x-ray diffraction to minerals was slow to catch

[37] On this process, see, e.g., Wiederkehr (1978), Maitte (2001), Hon & Goldstine (2008) and Kubbinga (2012).

[38] See Smith in Wechsler (ed. 1988), Newbury & Williams (2000) and Cahn (2001) pp. 91ff., 110ff. for both experimental and theoretical examples.

[39] See Kubbinga (2012) p. 24 and Eckert (2012). On the arguments pro and contra the wave-like nature of x-rays, cf. Wheaton (1983) and Hentschel (2007) sec. 7.2.7.

[40] On the two Braggs and their experimental method, see, e.g., W.L. Bragg (1913, 1914, 1933ff., 1945), Jenkin (2011) and further sources mentioned there.

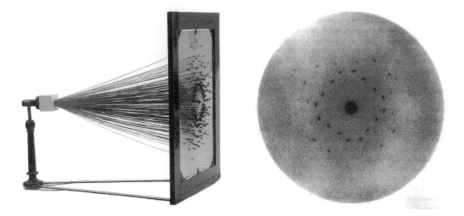

Fig. 113: Left: model demonstrating the generation of a Laue diagram from a crystal (symbolized by the white cube) through which an x-ray bundle is sent from the left. Repeated interference of these rays inside the layered crystal leads to a few x-ray bundles being reinforced, which produce an intricate pattern of radiated points on a photographic plate held perpendicular to the ray behind the crystal. From Weber-Unger (ed. 2009), p. 30. © Wissenschaftliches Kabinett Simon Weber-Unger. Right: one of the first Laue diagrams of a copper vitriol crystal. From Bragg (1945) fig. 4.

on, because crystallography belonged to the empire of mineralogists. They were institutionally far away from physicists, who were the guardians of the new research technology. That is why Cambridge, Manchester and London became international centers for roentgen-crystallographic techniques. Pioneering work on the structure of DNA (made to crystallize by strong cooling) was performed by Rosalind Franklin (1920–58) and Maurice Wilkins (1916–2004) in the 1950s, first at *King's College* and then at *Birkbeck College*. The American physical chemist Linus Pauling, whom we have already met as an important protagonist of visual thinking in chemistry (see p. 220), was also working along this line, experimentally examining various complex organic molecules. Pauling was the first to publish his suspicion that DNA has a screw-like or helical structure. Watson and Crick were the first to come up with the suggestion of a *double* helix, based on one of Franklin's unpublished x-ray-diffraction photographs of DNA in its B form, which Wilkins had secretly shown to Watson and Crick on January 30, 1953, without Franklin's prior consent.[41]

---

[41] On this infamous episode, which also shows how crucial it has become in modern science to have prompt access to highest-quality visual representations, see, e.g., Watson (1964), Olby (1974, 1994) and Keller (1996) 107ff.

This versatile technique of von Laue diagrams had the serious disadvantage that it only worked for relatively large monocrystals. Many other materials only have minute monocrystals or more irregularly grown polycrystals. For these, another method developed in Göttingen in 1915 by Pieter Debye (1884–1966) and Paul Scherrer (1890–1969), was more useful. The Debye–Scherrer technique uses layers of finely pulverized crystals oriented at random. Thus, instead of yielding more or less sharp points – as in the von Laue diagram – the monochromatic x-ray beam yields circularly symmetric rings whose intensity and radius provide the clue to the substance's microstructure.

With Kathleen Lonsdale (1903–70), who joined the crystallography research team headed by William Henry Bragg at the *Royal Institution* in 1924 and became the first tenured professor of chemistry and head of the Department of Crystallography at *University College London*, these sophisticated techniques of crystal analysis found their way back into the visual culture of design. Inspired by one of Lonsdale's lectures, Dr. Helen Megaw (1907–2002), assistant director of research at the *Cavendish Laboratory* in Cambridge, wrote an article on the beauty of crystals repeating Lonsdale's suggestion that crystal structures might be of utility to designers. In preparation for the 1951 Festival of Britain, a few interested x-ray crystallographers worked together with 26 leading British manufacturers in the so-called "Festival Pattern Group" – a unique project involving scientists, designers and manufacturers. Diagrams of atomic structures and x-ray diffraction patterns thus inspired various patterns on curtains, wallpaper, carpets, lace, dress fabrics, ties, plates and ashtrays, which were displayed at the Regatta Restaurant during the festival. This, in turn, inspired many other designers worldwide.[42] Advanced science had found its way into the visuality of modern design of the 1950s.

10.3 Suspension bridge construction

Unlike conventional bridges, which are supported by pillars or arches from below, suspension bridges span broader spaces, such as wide rivers or channels, without relying on support from underneath. First and foremost, reliable tests were required for mechanical durability and elasticity of new construction materials, such as wrought iron or steel cabling, in order to assure that they could carry the weight of such wide spans. But totally new techniques for calculating the static and dynamic stability of such bridges were required as well. Their design implies a

---

[42]For further historical details and examples, see Thomas (1951), Lapage (1961) pp. 59f., and the online exhibition www.wellcomecollection.org/whats-on/exhibitions/from-atoms-to-patterns.aspx (last accessed July 22, 2012).

high degree of visual thinking, and so it is no accident that among the pioneers of these designs we find evident representatives of visual cultures. Interestingly, the first suspension bridges, namely simple rope bridges, are found in jungle cultures of China, Tibet, Bhutan, India, as well as South America.[43] European-style suspension bridges have sturdy steel cables or wrought iron chains suspended between high towers at both ends of the bridge, plus vertical suspender cables, also called hangers, carrying the weight of the traffic deck below. Since all loads on the bridge are transformed into tension in the main suspension cables, these must be anchored in firm ground beyond the two ends of the bridge. The main cables continue beyond the abutment pillars to deck-level supports, and further to solid anchors in the ground. Initially, the side railing served as a stiffening truss.

The first European design for a suspension bridge is found in a book from 1595 on *machinae novae* by the Croatian polymath Fausto Veranzio (1551–1617). It includes drawings for a timber-and-rope suspension bridge, as well as for a hybrid suspension and cable-stayed bridge using iron chains. The first such suspension bridge with iron chain links had to wait until the early 19th century, though. US Justice James Finley (1756–1828) actually built one spanning just 70 feet over Jacobs Creek in Westmoreland County, Pennsylvania, in 1801.[44] The elegance and simplicity of suspension bridges soon fascinated architects, engineers and painters. The Scottish landscape artist Alexander Nasmyth, whom we have already met (in sec 4.1, p. 130) as the father of the engineer James Nasmyth, was toying with "bow and string" bridge constructions already in 1794.[45] It took another three decades, though, for the new technique to advance enough for the first larger suspension bridges to be attempted. Gathering experience with the new material in more traditional bridge support structures, the Scottish engineer, architect and stonemason Thomas Telford (1757–1834) eventually mustered the courage to attempt a suspension bridge for the Menai Strait across the Stanley Sands between Anglesey and the Welsh mainland, measuring a 580-foot (c. 180 m) free span. Metallurgical techniques for making sturdy iron cables were not yet developed enough for such lengths. So Telford chose individually linked 9.5-foot (2.9 m) iron-bar chains rather than cables (cf. fig. 114 bottom right for a close-up of the iron eyebar connectors).

---

[43]See Navier (1823) pt. I, Drewry (1832) 1ff. or Peters (1987) pp. 13–33 on non-European precursors and on the first documentable European prototypes.

[44]See Finch (1941) p. 74, Peters (1987), Kranakis (1997) chaps. 1–3 and sources cited there. Cf. www.si.edu/mci/english/research/past_projects/iron_wire_bridge.html (accessed July 21, 2011).

[45]See Nasmyth (1883) pp. 45–7, including reproductions of Nasmyth's designs.

Fig. 114: The first large suspension bridge across the Menai Strait, designed by Thomas Telford and built in the years 1819–26. The ironwork for the bridge was cast at the Hazledine Foundry in Coleham, Shrewsbury. Top: Thomas Telford's design for the Clifton suspension bridge. From the *Bristol Mercury*, February 1830. Bottom left: detail from a lantern slide, preserved at Shrewsbury Museums Service (SHYMS: P/2005/1121). Image sy9964. Bottom right: the eyebar connectors between the c. 3-m-long iron bars, from www.flickr.com/photos/roj/5542447647/sizes/l/in/photostream/ On early indications of problems with this design in stormy weather, see, e.g., Provis (1839) and Billington (1977).

Subsequent generations of specialist structural engineers and architects optimized suspension-bridge designs. John August Roebling (1806–69) became most famous. Born in Mühlhausen, Germany, he migrated to the United States of America in 1831 after studying at the Berlin *Bauakademie*. Settled in Pennsylvania, he soon started to contract railway-route surveying jobs and work on smaller construction commissions. In his private workshop he continued to improve on techniques for making and handling rope strand cables, invented in 1934 by the German senior mining engineer Julius Albert (1787–1846). Roebling had learned of them during his training in Germany. Because such cables were composed of several strands of helically twisted wire, should one strand break, it would not immediately have catastrophic consequences – a precondition for the use of longer cables as supports for suspension bridges. Initially, wrought iron was used for each strand. Later, that was supplanted by steel fibers. Roebling received contracts for various bridges from the mid-1840s onward. They bridged the Allegheny River near Chicago, the Monongahela River near Pittsburgh, the Niagara River on the border between the USA and Canada, and the Kentucky River near Cincinnati. With a 322 m span, his bridge across the Ohio River in Cincinnati (finished in 1867 and later called the Roebling Suspension Bridge) was the largest of its kind in the world until he broke his own record with the Brooklyn Bridge in New York (see fig. 115). It connects Brooklyn and Manhattan by a span of 486 m and was completed posthumously in 1883. In addition to vertical suspenders, it uses auxiliary inclined stays ('storm cables') to improve its rigidity. It was stiffened even further when a second-level track was installed, making the bridge able to withstand heavy traffic and storms.[46] One generation later, the Swiss-born civil engineer Othmar H. Ammann (1879–1965), a graduate of the Zurich polytechnic who had migrated to the USA in 1904, constructed the George Washington Bridge across the Hudson River. It had no stiffening girders despite its huge span of 1,066 m. Its weight-bearing cables of 90 cm diameter each consisted of 26,108 parallel steel-wire strands of 5 mm diameter.[47]

Materials testing procedures also improved and became increasingly institutionalized alongside these advances in design and planning. Dozens of materials testing stations were founded in Britain, Germany, Austria, Switzerland, France and the USA toward the turn of the 19th century and into the 20th century.[48]

---

[46]On Roebling, see Finch (1941) pp. 77f., Steinman (1950), Billington (1977) pp. 1655ff. and (1983) chap. 5, Wittfoht (1983), Hindle (1984), and Kahlow (2006) with listings of archival material.

[47]On Ammann, see Stüssi (1969, 1974) and Billington (1977) pp. 1657ff., (1979) and (1983) chap. 8.

[48]On the institutionalization of materials testing and materials science, see Hentschel (2011a).

Fig. 115: The Brooklyn Bridge, designed by John A. Roebling and completed in 1883. Albumen photographic print, taken by George P. Hall & Son in 1896. From the Library of Congress Prints and Photographs Division Washington, D.C., DIGITAL ID: (intermediary roll film) pan 6a19666.

Stronger materials made it feasible, in principle, to reduce the diameter of cables and the size of cable-supporting towers. The new art of structural engineering with reinforced concrete created increasingly elegant, slender structures that were much admired. A few structural engineers were skeptical about the tendency to downscale the stiffening trusses that Roebling and other bridge builders of the 19th century had deemed absolutely necessary. Some had reported resonance problems for suspension bridges, caused by certain forms of periodic excitation, whether set off by marching soldiers or gusts of wind. But these warning signs were marginalized until catastrophe struck. On November 7, 1940 strong side winds caused the Tacoma Narrows Bridge to collapse from the aeroelastic flutter of its suspended surface (see fig. 116). When opened on July 1, 1940, just a few months before, the Tacoma bridge with a span of 853 m was the third largest suspension bridge in the world. But its extremely slim traffic deck lacked the necessary torsional rigidity and was essentially a thin tar-covered ribbon of steel with a strong tendency to vibrate and resonate. The bridge earned the nickname "galloping Gertie" as a result and became quite a tourist attraction for daredevil drivers and pedestrians. The crucial moments of its disintegration were recorded on film, making this catastrophe world-famous. It triggered intense debates and revisions in the obligatory structural reinforcements, which had been trimmed down too much in the evolutionary history of bridge design.[49]

In the latter part of the 19th and the early 20th centuries, far larger spans were achieved, most famously perhaps the 2.7 km-long Golden Gate Bridge (finished

[49]See Farquharson (1941), Finch (1941), Billington (1977ff.) and Ferguson (1992*b*) pp. 186f., 213, with further sources. For online access to the film showing the collapse of the bridge, see www.youtube.com/watch?v=j-zczJXSxnw or www.youtube.com/watch?v=xox9BVSu7Ok (black-and-white sound version, both accessed July 22, 2011).

Fig. 116: Tacoma Narrows Bridge shortly before its collapse on November 7, 1940. From Ferguson (1992*b*) p. 187. Cf. https://archive.org/details/SF121 for a film of the resonance collapse

in 1937) linking San Francisco with the Bay Area. Its free span measures 1.28 km at 67 m above ground.[50] Denmark's Great Belt Bridge (*Storebæltsbroen*, opened in 1998) has a 1,624 m span. The longest suspension bridge in the world since 1998 is the Akashi-Kaikyo Bridge in Japan, also known as the Pearl Bridge, with a central span of 1,991 m.[51]

Throughout the history of suspension bridges, a compromise between visual elegance and technical soundness had to be found. This anticipates a point to be made in chap. 12 on the intricate interplay between aesthetic judgment and rational argument. The choices by structural engineers and architects were determined by the extent of their knowledge about static stability and (aero)dynamically induced oscillations as much as by their aesthetic judgments. The beauty of flat catenaries and the gracefulness of slim suspensions competed against the clumsy necessity for vertical stabilizing trusses for flexible decks. In most cases, acute visual thinking and structural intuition prevented the worst.[52] But sometimes the balance tipped

[50]On the history of the planning and construction of this gateway to California, see Strauss et al. (1938ff.) and Van der Zee (1986).

[51]Good surveys of the new art of structural engineering are provided by Billington (1983) and Kurrer (2008). Photographs of the most recent suspension bridges can be found on the Internet.

[52]See, e.g., Rothmann or Blair Birdsall in the discussion of Billington (1978) pp. 246 and 1031. Cf. also Eduardo Torroja in Billington (1979) p. 672: "the design process was neither purely rational nor purely visual but rather both together."

too far in the direction of formal elegance. Occasional reality checks were evidently necessary. A sophisticated mathematical theory of suspension bridges (without stabilizers) had been developed as early as 1823 by Claude-Louis-Marie-Henri Navier (1785–1836), instructor at the Parisian *École des Ponts et Chaussées*. But, as Navier's own practical failure shows with his ill-fated chain suspension bridge across the Seine in 1826, highflying theory does not always guarantee success.[53] Leon S. Moisseiff's application of W. J. M. Rankine's and Wilhelm Ritter's elastic deflection theory did not prevent the Tacoma disaster in 1940, either. This theory predicted that longer bridges would permit more flexibility.[54] Aerodynamic stability under unusual conditions, such as strong winds, could not be calculated reliably until well into the 20th century, so the only remedy for civil engineers and architects aside from empirical formulas and rules of thumb, was to build models of their bridges and subject them to testing in wind tunnel experiments.[55] The expert Daniel Dicker conducted a critical inquiry into the evolution of suspension bridges. This professor of structural engineering at the *State University of New York* in Stony Brook and recipient of the 1967 Norman Medal for his analysis of the Tacoma Narrows Bridge failure reached the conclusion that, until 1940, suspension-bridge design had "been based more on feel than on fact."[56] And even in our modern-day bridge constructions, things can still go very wrong, notwithstanding rigorous testing by computer simulations and sophisticated static and dynamic computations. The Millenium Bridge in London, built in 2000 to connect the north and south banks of the River Thames, had to be closed down again after just two days of limited access. It was found that large numbers of pedestrians could set it swaying, which necessitated major structural revamping that lasted almost two years.[57]

[53] See Navier (1823*b*), esp. pp. 249–317 on the controversy after the *Pont des Invalides* disaster. Cf. Picon (1988), Kranakis (1997) pp. 97–204, Cannone & Friedlander (2003) and Kurrer (2008) on the historical development of such structure theories.

[54] See Moisseiff (1933), Stuessi (1974) pp. 36–44 and Kranakis (1997) pp. 161f.

[55] On this "infant stage" of aerodynamic calculability, which was ended by Steinman (1943/45), see, e.g., George Schoepfer in the discussion on Billington (1978) p. 378. Steinman (1946) p. 20 states that "the aerodynamic aspect was forgotten by a generation that preached the virtues of flexibility without recalling its hazards." One rule of thumb cited by Steinman is that the truss or girder depth should not be less than 1/100th of the bridge's span; more such empirical rules are found in Drewry (1832) pp. 159ff. and Kranakis (1997) pp. 42ff.

[56] See Anon. (1969) p. 61. It provides Dicker's simplified criterion for stability evaluation, another rule of thumb for the critical wind speed as a function of cable tension, span and air density. Cf. also Ammann (1966) for the mix of scientific, technical and aesthetic criteria going into the design of the Verrazano Narrows Bridge spanning 4,260 ft. It was one of the first for which extensive computer calculations for stress and deformation factors were carried out.

[57] See http://en.wikipedia.org/wiki/Millennium_Bridge_%28London%29 (accessed Sep. 9, 2011).

11

Recurrent color taxonomies

Most of the visual representations discussed so far exclude one important dimension: color. The primary reason why color is frequently neglected in the context of scientific and technical illustrations is that until recently any color illustration or reproduction was exorbitantly expensive and therefore relatively rare. Hand-coloring of drawings or printed illustrations occurred in fields where the specific color of a depicted object was quintessential, such as urinoscopy (see below) or dermatology, botany (cf. here secs. 6.1 and 7.1) or mineralogy (discussed in sec. 11.2). In cartography, it became standard to emphasize the borders between different countries or regions in differently colored tints. After chromolithography was developed in the mid-19th century, a few other fields very occasionally also made use of this option. But until cheaper photomechanical printing procedures were devised in the 20th century, color illustrations remained a relatively rare occurrence in medicine, science and technology. This changed drastically with the spread of desktop publishing and affordable color printers in the late 20th century. Nowadays color is virtually omnipresent, even in scientific publications, compared with foregoing centuries.

Color played a much more important role in scientific observation and experimentation beyond the world of printed sources, however. One intriguing early example is the medical branch of urinoscopy.[1] In the Middle Ages and early-modern period, it was a widely practiced 'art' to infer illness from the color, consistency and possibly admixtures of a patient's urine. A broad array of manuscripts and early printed works indicate that physicians used a differentiated set of terms to gauge color as their main indicator. The historians of medicine Karl Sudhoff and Charles Singer have established the existence of an iconographic tradition of drawings and early printed plates (many of them carefully hand-colored).[2] Over a dozen, often up to 20, urine flasks labeled with carefully chosen color terms are arranged in a circle, sometimes grouped into four sets corresponding to the four basic humoral characters. A superb hand-colored illustration can be found

---

[1] On the historical roots of this field, which today is called urinalysis, see Armstrong (2007) and the following note.

[2] See the commentary by Sudhoff & Singer to Ketham (1491*b*) pp. 90–2, 100 for examples, including two early 15th-century manuscripts, and Ketham (1491).

in one copy of Ulrich Pinder's treatise from 1506, *Epiphanie medicorum* (see here color pl. II). On the title verso of Johannes de Ketham's *Fasciculus Medicinae* of 1491, the urine flasks are not colored in but are carefully labeled with color terms indicating a refined sense of color.   For instance, Ketham distinguishes between "dull citrus fruit" and "not so dull citrus fruit," between "dull gold" and "pure molten gold," between western and eastern saffron, etc.[3] As this example also shows, early color terms qualify the base colors by adjectival attributes derived from direct comparison with familiar objects. The limits of this comparative color labeling become clear with descriptions such as "greenish color of urine, such as the color of an animal's liver," "dark blue ... like black wine," or "green color ... like a green plant stem."[4]

Various aids were developed in order to help observers find the right color designations for their verbal descriptions of subtle nuances. In the 18th century, color dictionaries began to appear listing the few dozen basic terms (e.g., red or blue) along with several hundreds of their nuances (e.g., carmine red or Prussian blue). The most elaborate of these dictionaries was the *Farbenlexicon* (1780) by Christian Friedrich Prange (1756–1836), an art educator based in Halle. He lectured in the philosophical faculty at the local university and became head of its drawing school in 1822. Prange's lexicon listed no fewer than 4,608 hues together with recipes for mixing them and provided samples on 48 plates, each of them with $6 \times 16 = 96$ fields colored in by hand (see here top right of color pl. VIII, for a sample).[5] These color matrices were horizontally sorted by basic color, starting with dark blue, shading over into lighter blues and violets, then into red, yellow, green and brown tones. Each of these basic tones grow paler along the vertical axis. Since each color field was numbered, each color nuance could be uniquely identified by plate and field number. In the application part of his dictionary, Prange implemented this easy-to-use numerical taxonomy to describe the colors of flower and insect specimens, simply citing the plate and field numbers from his plate volume for each color area.[6] This simple and unambiguous color-coding was also used in exchanges between natural historians and their illustrators

[3]Ketham (1491*b*) p. 2 title verso.

[4]Ibid.

[5]See Prange (1780), esp. pp. 477–572 for the extensive tables of color terms, his separate plate volume with 48 color plates, and pp. 393ff. on "the application and usage of colors to imitate nature" (*Anwendung und Gebrauch der Farben zur Nachahmung der Natur*).

[6]For instance, the stag beetle and rhinoceros beetle were described by Prange (1782) p. 411 as colored in the dark violet tones on plate XXXIX, nos. 49–50, and 33–4, respectively; and the base color of the buttercup (*Anemona pulsatilla*) was identified on plate VII, no. 82, with the shading using nos. 1ff. and highlights in nos. 84ff. (ibid., p. 393).

to determine the exact tones and shadings of plates for hand-coloring by the 'illuminators.'[7]

Artists and artisans were the main audience for such works. Wallpaper and textile manufacturers, for instance, were the fields of practice in which the push for standardizing color designations was strongest. Various plates with normed color samples documenting production standards are preserved in the Parisian *Musée des Arts et Métiers*, and the director of the Gobelin Manufactory, Michel-Eugène Chevreul (1786–1889), was one of the most influential analysts of colors and their contrasts.[8] The classificatory 18th century spawned many efforts to develop rational color taxonomies and color systems.[9] Around 1800, intense debates about whether one could develop these systems out of a dual base (such as Goethe's color system grounded in a basic conflict between light and dark), a triple base (such as Thomas Young's red, green and violet or Hermann von Helmholtz's blue, green and red), a quadruple base (such as in several systems for painters) or Newton's famous system of five primary and two secondary colors (red, yellow, green, blue and violet vs. orange and indigo). As I have shown elsewhere, Newton's color taxonomy with its inbuilt acoustic analogy had a lasting impact well into the 19th century on visuality within spectroscopy and beyond.[10]

During the 19th century, several new printing techniques came into play that made it possible to avoid the labor- and cost-intensive hand-coloring of plates, which had hitherto been the only way to make imprints in color. Chromolithography, in particular, became popular in fields in which color information was absolutely necessary. For the printing of his highly complex spectrum maps, the Astronomer Royal of Scotland, Charles Piazzi Smyth (see here p. 142), relied on the reputed establishment of W. & A. K. Johnston in Edinburgh, "geographers to the Queen" since 1837. Specialized in topographic and geographic maps, these printers had all the skills necessary to translate his artistic drawings into the print medium, either as photolithographs or multi-stone chromolithographs – a process that could take up to one year, with several exchanges of proofs between the finicky astronomer and the increasingly stressed printers. His color plate, printed

---

[7]For examples from botanical expeditions, see Bleichmar (2012) pp. 97–100 or Lack & Ibáñez (1997), Lack (2000*b*) pp. 137 and 215 (cf. here color pl. VI); on color-designation systems in dentology, designed to match shades of crowns, see Elkins (2007) pp. 126–30.

[8]On Chevreul, see, e.g., Chevreul (1839), Stromer (ed. 1998) chap. 17 and Robin Rehm (in prep.).

[9]See, e.g., Schäffer (1769), Mayer (1775), Lambert (1772) and here sec. 4.1, furthermore Prange (1782) pp. ix–xix, xxix–xxxi or Stromer (ed. 1998), Kuehni (2008) and Spillmann (2010) for further literature.

[10]See, e.g., Goethe (1791/92, 1810ff.), Young (1807*b*) p. 345, Helmholtz (1852, 1867) and Hentschel (2002*a*) chap. 1, (2006*c*) and (2014*b*) on Newton's color system and its impact.

in Edinburgh, indicates the fine taxonomy of color terms with which Smyth described his spectra, a man as garrulous as his drawing pen was prolific.[11] The German physical chemist Wilhelm Ostwald (1853–1932) spent the last 20 years of his life fighting for his own chemically inspired color taxonomy. By 1944, his *Farbenfibel* had been reprinted 15 times, and various other publications by him on color harmonies and taxonomies were discussed controversially among scientists, artists, art historians and the broader public. Ostwald also interacted with the textile industry and with kindergarten teachers pushing for more drawing instruction for young children. Thus he exemplifies many facets of typical protagonists of visual science cultures.[12] In 19th-century biology, the high-quality color lithographs made by Ernst Giltsch for the marine zoologist Erich Haeckel, who was himself a gifted amateur artist, are another case in point.[13]

In the 20th century, one of the most influential color ordering systems was created by the artist and professor at the *Massachusetts Normal Art School*, Albert Henry Munsell (1858–1918). His system using decimal notation instead of color names dates back to 1898. Three independent dimensions of the color space are distinguished and represented cylindrically in Munsell's 1905 *Color Notation* and in his *Atlas of the Munsell Color System* of 1915: (i) five principal hues are distinguished (red, yellow, green, blue and purple), with five intermediate hues halfway between them and further nuances indicated by degrees within a horizontal circle; (ii) chroma represents the "purity" or saturation of the color and is measured radially outward from the neutral (gray) vertical axis; (iii) value (also called lightness) is indicated along the vertical axis of the cylinder from 0 (black) to 10 (white). In 1929, Munsell's color atlas was published as *The Munsell Book of Color* by the Munsell Color Company (founded in 1918). After the *Optical Society of America* completed a careful psychophysical recalibration in the 1940s, its use was recommended by *American Standards* for the classification of colored surfaces. Munsell's color taxonomy is still fairly widespread, especially in the USA and in Japan, in fields such as architecture and engineering and for the classification of consumer goods, such as for matching beer color in breweries. Botanists classify plants taxonomically and identify diseases leading to tissue discoloration according to Munsell's plant tissue charts. Munsell's system is particularly popular in the heavy machine industry (especially in ship building), in cosmetics and in forensic pathology as well as for

---

[11]See here p. 142, esp. Warner (1983), Brück (1988), or Hentschel (2008*a*).

[12]On Ostwald's *Farbenfibel*, see, e.g., http://home.arcor.de/unesma/hinweis.htm and www.colorsystem.com/?page_id=862& lang=en.

[13]See here p. 228 and esp. Breidbach (2006) or Richards (2008); on color in scientific illustrations more generally, cf. Lapage (1961) pp. 75f.

geological and archeological applications. Soil or earth, in which archeological objects are embedded, is classified by color, after water has been squirted on a sample for comparison against Munsell's color chart. The anthropologist Charles Goodwin, who studied the situated practices of students learning how to handle this color taxonomy in the early 1990s, has pointed out:

> Finding the correct category is not an automatic or even an easy task (Goodwin 1993). The very way in which the Munsell chart provides a context-free reference standard creates problems of its own. The color patches on the chart are glossy, while the dirt never is, so that the chart color and the sample color never look exactly the same. Moreover, the colors being evaluated frequently fall between the discrete categories provided by the Munsell chart.[14]

To us, living in the 21st century, with its high-speed computers endowed with powerful graphics cards, desktop publishing programs and cheap color printers, it is hard to imagine how valuable colored prints used to be in earlier centuries. Yet this issue of color taxonomy is still with us. This becomes apparent the moment you enter Photoshop or any other more sophisticated graphics program and are confronted with having to learn how to handle the 'translation' between various up-to-date color coding schemes (RBG vs. CMN, DIN vs. CIE vs. OSA vs. NCS, etc.).[15]

The issue of subjectivity versus objectivity was also superimposed on the color discourse. To the one camp, "colour [...] is a construction of the brain. There are no colours in the outside world."[16] To the other camp, colors are one of the most striking and immediately given features on Earth, and the frequent neglect of color phenomena within science is rather a fundamental defect of this Newtonian science (in Goethe's artistic eyes, for instance). The geographical explorations of faraway or hard-to-reach regions, for which the 18th century is famed, brought naturalists and adventurers into contact with new, fascinating and often very colorful phenomena. In this chapter, we will look at the efforts to rationalize the impression of color intensity in the deep blue sky observable high up in the Alps (sec. 11.1) and in minerals and gemstones within the context of 18th and early 19th-century mineralogy (sec 11.2).

---

[14]See Goodwin (1994) pp. 608–9. Cf. Nickerson (1976), Kuehni (2002, 2008), Landa & Fairchild (2005), Stromer (ed. 1998) pp. 102–5 on the fundamental ideas and on the historical development of Munsell's system and http://en.wikipedia.org/wiki/Munsell_color_system plus links there for color reproductions of Munsell's charts, atlases and further references.

[15]On the preceding, see, e.g., Stromer (ed. 1998) chaps. 39–58.

[16]Zeki (1999) p. 83.

11.1   Gauging the blue of the sky: cyanometry

Horace Bénédict de Saussure (1740–99) was a Genevan naturalist who became famous for his scientific Alpine observations.[17] From his base in Geneva, he regularly organized expeditions into the regions around Chamonix and the Mont Blanc massif. He published a detailed description of these journeys and observations in his four-volume oeuvre *Voyages dans les Alpes* (1779–96). Unlike other explorers and mountain climbers, his main motivation was not the sheer sport of being the first to accomplish a given feat. It was scientific interest in recording all kinds of phenomena and their alteration with altitude. He and his accompanying assistants and mountain guides carried a great assortment of fragile instruments up high mountains, including precision clocks, thermometers, barometers and custom-designed hygrometers, to carefully record local temperature, pressure, air humidity and other parameters. A drop in the barometer reading allowed de Saussure to gauge the altitudes they reached. Taken together, the many instrument readings offered him a full picture of the local climate as a function of altitude, time of day and weather. That allowed him to test various meteorological theories proposed by Jean-André Deluc (1727–1817).[18] It was on these mountain excursions that de Saussure noted the striking intensity of the sky's color at higher altitudes. On the highest mountain tops that he could reach, it took on a deep dark blue, closer to black than would normally be imagined or depicted in paintings. This apparently constant phenomenon, never before analyzed, revealed surprising variation with height. De Saussure, as a good and keen observer, sought to make this impression more systematic and precise. To determine the blueness of the sky quantitatively, he created an instrument that he called a cyanometer.[19]

De Saussure brought along a kind of proto-cyanometer on his first ascent of Mont Blanc in August 1787, just one year after this highest mountain peak in Europe had ever been reached by a human. It consisted of 16 sheets of paper pre-colored in various tints of blue, ranging from the deepest royal blue available, to the palest light blue renderable with Prussian blue. He graded these nuances, starting with the number 1 for the darkest tint, up to 16 for the lightest. Identical sets of these 16 hues, made from these painted sheets, were left with his assistant Évèque, who stayed at the base camp in Chamonix, and with his colleague Senebier, who remained further down by Lake Geneva, for later comparison of the

---

[17] On de Saussure's life and work, his visuality and his context, see, e.g., Sigrist (ed. 2001).

[18] On Deluc's and de Saussure's theories on the generation and distribution of heat in the upper atmosphere, see, e.g., Hentschel (2007) sec. 4.4.7, pp. 330–8 and further sources cited there.

[19] On the following, see de Saussure (1779–96) vol. 4, §§2083–6 and (1787, 1788, 1790, 1792).

different simultaneous readings. On the day of de Saussure's and his son's ascent, they each looked toward zenith at midday from where they were and gauged the sky's blueness against the 16 hues available on the scale. De Saussure's 'reading' fell between the first and second grades, his son's, who had stopped along the way, fell between the fifth and sixth and Senebier's in Geneva was closest to the seventh. This confirmed de Saussure's suspicion that the depth of the blueness correlated with the height of the observer above sea level (all other parameters such as weather, cloud density, etc., being equal). But the remaining uncertainties in the observer's estimates also told him that a more refined set of blue shades was needed for a definitive comparison and gauging of the blue sky. For the *Mémoires de l'Académie des Sciences de Turin* for the years 1788–89, de Saussure increased the number of blue nuances to 40 (cf. fig. 117).

By 1790, the year in which this memoir finally appeared, de Saussure had increased the number of blue gradations to 52. In various Italian, Swiss and German journals, de Saussure described this refined scale and how to make it. He also set forth his comparative measurements taken with it at various hours of the day, and in various directions from the horizon to the zenith.[20] Interestingly, he did not prescribe the number of gradations of blue. He left open the size of his circular array of tapered blue rectangles and hence the total number of gradations. That made it adaptable to the sophistication with which other observers wished to discriminate neighboring shades of blue. Its degree of fineness was adjustable to the schooling of the observer's eye in distinguishing color nuances.[21]

[20]See, e.g., de Saussure (1779–96) §§2083–6 for the data, and Prévost (1806) for an attempt at their quantitative analysis in terms of the height distribution of light-absorbing vapors.

[21]For details on the manufacture of these cyanometers, see Breidbach & Karliczek (2011). They confine themselves to the 52 color intervals found in the most sophisticated cyanometer among the holdings of the Genevan *Musée d'Histoire des Sciences*.

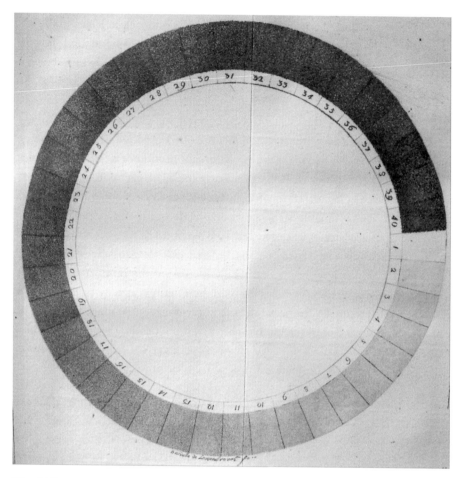

Fig. 117: Horace Bénédict de Saussure's cyanomètre with 40 gradations of blue, c. 1788. Stipple engraving by Bardello de Lewenbrouck, printed monochromatically in a bluish tint as plate X in de Saussure (1790*a*). For a color reproduction of Saussure's most refined cyanometer with 52 gradations, preserved in the Genevan Musée d'Histoire des Sciences, see here color pl. IX.

The historian and philosopher of biology Olaf Breidbach in Jena and his collaborator André Karliczek recently replicated de Saussure's cyanometer following his instructions. Discrepancies in the readings off their replica by different observers amount to no more than 2–3 grades in the scale of 52 shades of blue. So intersubjective agreement, if not perfectly achievable, is still pretty good, given the very subtle character of the quality to be observed and gauged. One has to take certain precautions, though: not to conduct the observations at close quarters; to avoid direct exposure to sunlight during observation; and not to choose a part of the sky too close to the solar disk. Breidbach and his collaborators also argued that de Saussure's cyanometer design concurs with his theory about how the sky's intensity increases with altitude. Briefly put: the denser the haze, the fainter the blue. Their replication is impressive, but I am skeptical as far as this hypothetical link to his theory is concerned, because in that case de Saussure would rather have created his various shades of blue by increasing the number of layers of white washes over a deep blue base to map the filtering capacity of increasingly dense light-absorbing air. In principle, this inverse way of producing a range of blue tints would have been readily available, given the painting techniques in use. The procedure de Saussure actually chose to produce his graduated scale was successive glazings over of dilute color.

The cyanometer never did become an extremely popular device and has remained more or less unknown among laymen. In the late 18th and early 19th centuries, there was a modest wave of interest in this kind of color gauging of the blue of the sky, however. One finds in the published literature observational cyanometer reports by the Swiss physicist Pierre Prévost (1806), the Scottish experimentalist David Brewster (1830) and the German explorer of the Himalayas Hermann Schlagintweit (1852), and there are even some in Alexander von Humboldt's *Kosmos* (1858). The scale is portrayed in contemporary reference works, such as Gehler's *Physikalisches Wörterbuch* and the *Edinburgh Encyclopædia* (1830). Several observers conceded fairly limited accuracy in the readings obtainable with this gauge. It was riddled with other problems of its own, such as rapid fading of the disk of comparison tints upon regular exposure to light. One user, a forestry secretary from Coburg, complained that it remained very difficult to compare the dry, pale and "dead" colors of the cyanometer with the clear, transparent and "vivid" blue of the sky. This user actually preferred to use a glass-tube cyanometer filled with water in which a blue pigment was dissolved in various degrees of concentration. Others suggested using a wedge-shaped hollow glass prism filled with a blue fluid in order to construct a more objective gauge of successive dilutions of blue.[22] The cyanometer even caused some ripples beyond science, with no lesser a personage than Privy Councillor Johann Wolfgang von Goethe (1749–1832) recommending it to the painter Johann Heinrich Meyer (1760–1832), for whom he had two especially made. The explorer Captain John Ross (1777–1856) carried a cyanometer made by de Saussure's instrument maker T. M. Paul in Geneva on his Antarctic expedition, but apparently did not make much use of it. Alexander von Humboldt (1769–1859) did record cyanometer readings during his South American tours.[23] The instrument is still in use as a quick, easy-to-use and handy means of gauging the color intensity of the sky. As with other visual instruments, it demands considerable experience and practice to work with it properly. Therefore, a much reduced scaling of blue is used in pedagogical contexts (see fig. 118).[24]

Fig. 118: Geneva schoolchildren gauging the sky with a simplified cyanometer as part of the European Discoveries project. Courtesy of David Jasmin, head of the affiliated La main à la pâte Foundation (see www.fondation Lamap.org/fr/europe-decouvertes, last accessed Jan. 2014).

[22]See Gehler (1787ff.), suppl. vol. 5 (1795), pp. 491–3, 538–41, Göbel (1818) with quotes from p. 242 and Breidbach & Karliczek (2011) pp. 24–7 for further references.

[23]See Werner (2006) sec. 2; on Humboldt's visual thinking in other areas of research, esp. on spatial distribution and isoline mapping, cf. Camerini in Mazzolini (ed. 1993), Godlewska (1999, 2001) and Rupke (2001).

[24]Cf. http://myfrencheasel.blogspot.com/2008/07/how-to-build-cyanometer.html for step-by-step instructions on how to make your own simple cyanometer.

11.2   Mineralogical color codes

Gemstones and minerals have fascinated humans since the earliest times. Count-less color terms actually derive from the vernacular names of minerals from which natural pigments were produced (by pulverization). At the transition from the late-medieval to the early-modern period, Paracelsus's signature theory established the color of minerals as the most important property for their classification.[25] Theophrast von Hohenheim (1493–1541), or Paracelsus as he preferred to call himself, had spun a dense web of speculative interconnections between the micro-cosm and the macrocosm. In Paracelsus's view of nature at least, the reddish color of the planet Mars or the red appearance of certain minerals, such as hematite, were intimately linked to the redness of blood, from which the ferrous mineral hematite even derived its name. Astrological predictions of bloodshed during periods in which Mars was observable were as 'logical' in this world-view, as was administering powders of the reddish mineral to 'cure' blood-related illnesses.

> We humans on Earth perceive everything inside the mountains from out-ward signs and similarities; likewise all the particularities of the herbs and everything inside rocks. Nought which is in the ocean deep, in the firma-ment on high, cannot be discerned by man [...] All this comes from the *key signature*.
>
> He who does not become an astronomer by *key signatures* is no *astronomer*; he who does not philosophize by *key signatures* is no *philosopher*; he who does not cure by *key signatures* is no physician.[26]

A straightforward consequence of this obsession with 'signatures' was an overem-phasis on a mineral's color as a presumably infallible indicator of its potent ingre-dient and medical effect on the human body. A concentrated dose of hematite powder was supposed to heal an anemic condition, whereas blood-letting would cure a flushed feverish complexion, which the doctor inferred as an excess of red-ness in the blood. Even though chemical composition, mechanical, structural and crystallographic properties gained in importance in mineralogy of the 17th and 18th centuries, color remained central. Haüy's predecessor Louis J. M. Daubenton

---

[25]On the following, see, e.g., Hiller (1942, 1952), or Schott in Hentschel (ed. 2010).

[26]"Wir Menschen auf Erden erfahren alles, was in den Bergen ist, durch äussere Zeichen und Gleichnisse, ebenso alle Eigenschaften der Kräuter und alles, was in den Steinen ist. Nichts ist in der Tiefe des Meeres, in der Höhe des Firmaments, das der Mensch nicht erkennen kann [...] Das alles kommt durch sein *signatum signum*. / Wer nicht aus dem *signato signo* Astronomus wird, der ist kein *Astronomus*; wer nicht aus dem *signato signo* philosophirt, kein *philosophus*; wer nicht aus dem *signato signo* arzneiet, der ist kein Arzt." (emphasis in original; transl. by A. M. Hentschel).

(1716–1800), for instance, divided minerals into seven classes, exactly corresponding to the seven color zones of the Newtonian spectrum of light, namely, violet,
indigo, blue, green, yellow, orange and red. Daubenton also described a procedure
to determine the exact color of a crystal by illuminating it with solar light cast
inside a dark chamber. A colorless, clear quartz crystal positioned next to it is
illuminated by light spectrally decomposed through a prism. The precise angular
position of the prism at which the clear quartz crystal and the crystal under examination exhibit the same color hue is noted and taken as the numerical equivalent
of the crystal's color.[27] At the start of his career, Daubenton was an anatomical
illustrator of Comte de Buffon's *Histoire Naturelle et Particulière*. For more than 50
years, he worked as curator of the natural history collections of the royal cabinets
in Paris. Buffon's zoology attracted his interest to paleontology (i.e., the study
of petrified traces of plants and animals), and it was along this trajectory that
Daubenton began to study mineralogical specimens. It was incidentally under
Daubenton's influence that René-Just Haüy (see p. 334) was also motivated away
from botany toward mineralogy.

The geologist and mineralogist Abraham Gottlob Werner (1749–1817) also
defended color as the primary feature for identification. This phenomenological
marker, he reasoned, was easily and directly discernible, whereas chemical composition and the microscopic crystal structure, both indispensable in classification,
could only be inferred upon conducting complicated angular measurements or
chemical experiments. Werner was the son of an ironworks foreman in Upper
Lusatia. He studied from 1769 at the *Bergakademie* in Freiberg in Saxony, which
had been founded two years previously and rapidly became the foremost training
school for mining officials throughout the German lands. After advanced studies at the University of Leipzig, Werner became mining inspector and professor
of mineralogy at the Freiberg mining academy, where he taught from 1775 to
1817.[28] Many of his hundreds of pupils later became professors at universities
throughout Europe, so Werner's mineralogical school dominated the geosciences
of that period.[29] In an influential publication from 1774, mining inspector Werner
established a differentiated terminology to describe the external characteristics of
minerals, including color, outward form, luster and surface structure. The hardness of minerals was later added by Werner's pupil and successor in Freiberg,

---

[27] On Daubenton's *Tableaux méthodique des minéraux* from 1784, see Laudan (1987).

[28] On Werner and the "school of Freiberg," see, e.g., Groth (1925) pp. 149ff., Laudan (1987)
chaps. 5 and 8 and Wilson (1994) pp. 99–100.

[29] For good surveys on the history mineralogy in this period, see Laudan (1987) and Fritscher &
Henderson (eds. 1998).

Friedrich Mohs (1773–1839), who developed a quantified scale to gauge this attribute. Color, still Criterion Number One in Werner's set of characteristics, became a highly differentiated field, with an extremely detailed set of terms offered by Werner to describe the minute differences in shade and hue. They were readily translated into English by Patrick Syme (1774–1845), a Scottish flower-painter active in the "Wernerian and Horticultural Society of Edinburgh."[30] In fact, Syme extended Werner's suite from 79 to 108 "standard colors." These could be modified in intensity. By adding such qualifiers as "pale, deep, dark, bright, and dull" or tinting them with admixtures of gray, black or brown, the total number of different hues was multiplied to over 30,000.[31] As an example, here are a few of his terms in the gray, yellow and red ranges as translated by Syme:

"greyish white, ash grey, smoke grey, French pearl grey, yellowish grey, blueish grey, greenish grey, blackish grey, greyish black [...]

sulfur yellow, primrose yellow, wax yellow, lemon yellow, gamboge yellow, King's yellow, saffron yellow, gallstone yellow, honey yellow, straw yellow, wine yellow, Sienna yellow, ochre yellow, cream yellow [...]

tile red, hyacinth red, scarlet red, vermillion red, aurora red, arterial blood red, flesh red, rose red, peach blossom red, carmine red, lake red, crimson red, cochineal red, veinous blood red, brownish purple red.

As a scholar in Freiberg in 1767, Werner started to expand the already large collection of minerals and fossils, the so-called *Stuffencabinett* of 1765, into a huge repository of more than 10,000 pieces. It was displayed in the *Stuffensaal*, opened in 1768, containing several glass cabinets, and was continually expanded into museum proportions. The Freiberg collection also grew as a result of the imposition in 1765 of an obligatory deposit of samples of any minerals offered for sale by traders in the area (*Niederlage verkäuflicher Minerale*). The *Bergakademie* had a right of first purchase to all products out the local mines, which was essential to its acquisition of very rare, unusual finds. The rapid growth of the Freiberg collection led to further subdivisions:

---

[30] Syme's English translation: *Werner's Nomenclature of Colours: With Additions, arranged so as to render it highly useful to the arts and sciences. Annexed to which are examples selected from well-known objects in the animal, vegetable, and mineral kingdoms*, published in Edinburgh in 1814, is available via Google Books.

[31] See Werner (1774/1814*b*) pp. 5–7, and, on the following, ibid., pp. 19ff.

- a systematic oryctognostic collection (c. 8,000 pieces), ordered according to Werner's system;
- a collection featuring external attributes in characteristic specimens;
- a mineral identification practice collection (specifically for students);
- a gemstone collection;
- a specialized collection of pseudomorphs;
- a geognostic collection (based on current hypotheses of mineral genesis);
- a geographic collection (reflecting the known regional distributions);
- a suit collection (representing custom combinations of different minerals);
- and a collection of fossils.

This copious collection of minerals and fossils was used not only for research and public presentation, but especially for teaching at the academy. The Freiberg students thus received hands-on training with real samples in the laboratories for physical chemistry – the first student laboratories of this kind in the German provinces). Apart from real minerals and crystals, the students also got to handle 3D models serving to illuminate issues such as crystalline structure, formation and later also chemical composition. Ample use of topographic, geographic and geological maps, diagrams, and drawings helped school and enhance their visual thinking skills. Freiberg students were also required to draw minerals and crystals themselves. Here we have practically the full set of indicators of a visual science culture.

12

Aesthetic fascination as a visual culture's binding glue

The words "beautiful," "pretty" or "pleasing" in praise of objects or processes are quite endemic to science and technology. Aesthetic admiration and the pleasure of looking at scientific things has been fairly widespread throughout the ages and across the disciplines; but it is not so easy to find such utterances expressed openly in print. Scientists – and technologists likewise – are all too often reserved about spelling out their delight. It is most easily found in the "peripheral written record of letters of tenure evaluation, eulogies or award nominations. There, where the rhetorical setting seems to demand it, the scientist relaxes. And praises the beautiful molecule."[1] That the chemist and Nobel laureate Roald Hoffmann (* 1937) bothered to collect evidence of 'molecular beauty' in the *Journal of Aesthetics and Art Criticism* is a rare, open documentation confronting this aesthetic judgment. Occasionally, remarks about 'beauty' are tucked away as irrelevant sidelines, as ancillary feelings, as slightly embarrassing subjective deviations from the 'purely objective' main line of research. Sometimes they are mistaken as evidence for the 'truth.' Here I take them very seriously, but as indicators of a drive that creates a deep, emotional attachment to particularly visual cultures of science and technology. As already explained in chap. 2, to me these indicators of aesthetic appreciation are an important layer of visual cultures and should not be dismissed off-hand. They are more than merely subjective overtones of scientific practice – they are an integral part of it, with important repercussions for its cognitive content. On this dual emotional/cognitive aspect of aesthetic judgment, here once again our initial crown witness:

> We *feel* that these molecules are beautiful, that they express essences. We feel it emotionally, let no one doubt that. But the main predisposition that allows the emotion, here psychological satisfaction, to act, is one of knowing, of seeing relationships.[2]

Among these cognitive aspects of images depicting molecules are, of course, re-lations of symmetry and asymmetry, and degrees of simplicity and complexity. In addition, molecular images encourage mental manipulation and thought experi-ments about relative stability or molecular dynamics. They are fun to play with,

---

[1] Quote from chemist Roald Hoffmann (1990) p. 191.
[2] Hoffmann (1990) p. 202 (emphasis orig.), with a lot of "beautiful" examples.

or – in the case of 3D models – to look at from many different angles. Nor are chemists alone in this. The same is true of many other fields.

Javier DeFelipe has recently shown that several early neuroanatomists profited substantially from their artistic talent and aesthetic pleasure in illustration work. They include the studies of the intricacies of the nervous system by Golgi, Fusari and Retzius. Santiago Ramón y Cajal (1852–1934), for instance, admitted in his autobiography that "like the entomologist hunting for brightly coloured butterflies, my attention was drawn to the flower garden of the grey matter." The pyramidal cells in the brain region appeared to him like "butterflies of the soul"; "the garden of neurology offers the investigator captivating spectacles and incomparable artistic emotions. In it, my aesthetic instincts were at last fully satisfied."[3] Indeed, his drawings of neurons and their intricate network patterns of synaptic interconnections somewhat resemble later works by Wols or other artists of the *art informel* movement from the 1950s.[4]

The Cambridge-based historian of biology Anne Secord has shown how much botany also profited by a widespread sense of pleasure in looking at finely executed drawings and paintings of plants. In particular, she documented how early 19th-century botanists consciously utilized this aesthetic appreciation to mobilize new recruits and to train neophytes in the art of recognition and classification. Like Roald Hoffmann, Secord also emphasizes that the close study of hand-drawn images or printed plates provides "both sensory enjoyment *and* rational pleasure."[5] It is an effective stimulus in the early stages of learning; but it is just as much so for expert botanists.[6] Martin Kemp concurs, even claiming that "no modern science has a more visual history than botany and no scientific illustrations have been more widely admired for what we may call aesthetic pleasure."[7] I am not so sure about the superlative, though. Other fields of natural history, e.g., zoology or geography, and later many natural sciences have a similar record of adoration

---

[3] Excerpts from Ramón y Cajal's *Recuerdos de mi vida – Historia de mi labor científica* of 1917 in English translation and commented upon in Javier DeFelipe's blog at Oxford University Press (2013).

[4] For further examples of such interconnections between science and art, see DeFelipe (2013), Blossfeld (1936), Schmidt & Schenk (eds. 1960), Borsig & Portmann (1961) and Kepes (1963, 1965).

[5] See Secord (2002) pp. 28ff., quote from p. 29 (emphasis added). Cf. ibid., p. 49 for quotes from c. 1840 about reducing the "dryness" of classificatory botany, and about "how easily and pleasantly [it is learned] when understood by means of our drawing."

[6] Woodcuts by Leonhart Fuchs and his artesanal team (cf. here sec. 6.1) were admired by both groups alike. Repeated reprintings of famous botanical works, such as Fuchs (1542) or Besler (1713), are a case in point for this dual usage. See also Lack (2001) as one example of the equally popular genre of historical selections of well-crafted and 'beautiful' botanical plates.

[7] Kemp (1996) p. 197; cf. Nickelson (2006) for other functions of botanical illustrations.

for beautiful specimens.[8] The following section documents the strong aesthetic
drive behind the motivation to assemble mineralogical cabinets and collections.

12.1   Mineralogical cabinets and collectors

Collecting earths, rocks and minerals is as old as mankind. Initially, samples of flint,
for instance, were apparently only collected for utilitarian purposes. Knocking
two flints together yields sparks that help ignite a fire; the sharp-edged fractures
of flint were useful for making spearheads and other implements for hunting
and combat. Very early in the history of mankind, these utilitarian motifs were
evidently supplanted by some kind of aesthetic appreciation – shiny shells and
minerals like onyx were collected for adornment. Rare and colorful stones became raw materials for jewelry. Other minerals were used pharmaceutically in
pulverized form or in ritual or medical practices, for example, as amulets.[9] With
the intensification of mining in early-modern times, mining towns and companies
also started to amass mineral collections, essentially as a teaching repertory of
ores to show newcomers what "good" and "bad" samples look like and which
ones are worth extracting. Native silver from the St Georg mine at Schneeberg
in Saxony counts among the earliest mineral specimens from European collections.[10] Spectacularly large or intricate silver, gold and copper nuggets (all three
of these metals occur naturally in elementary form) ended up in court collections
across Europe. They were initially stored in a 'curiosity cabinet' (*Wunderkammer*)
together with other interesting objects retrieved from the empire or donated by
foreign visitors.[11] The aesthetics of these royal, ducal or papal collections used to
be a hodge-podge of remarkable and precious objects. They were stored in often
overfilled, elaborately ornate vitrines or wooden cabinets without any discernible
order. The pharmacy of Louis XIII, for instance, exhibited Chinese pottery,
corals from the Mediterranean, intricate clocks and inlaid woodwork side by side
with rare minerals and precious gemstones in a highly individualized, handcrafted

---

[8]For explicit admissions of aesthetic criteria in astronomy and astrophysics, see Garfinkel,
Livingston & Lynch (1981) and Lynch & Edgerton (1988); on MRI, see Burri (2008) pp. 174–8; on
materials science and engineering, Cahn (2001) pp. 92ff., 110f.

[9]For an impressive overview of mineral collecting through the ages, see Wilson (1994).

[10]We know that this sample, depicted in Wilson (1994) p. 16, had been collected in 1477; it is
now exhibited in the *Staatliche Museum für Mineralogie und Geologie* in Dresden, Saxony.

[11]On this tradition of *Kunst- und Wunderkammern* and on important examples from Naples to
Halle, see, e.g., Schlosser (1908), Giuseppe Olmi in Mazzolini (ed. 1993), Findlen (1994), Wilson
(1994) pp. 18–42 and Müller-Bahlke (1998); Bann (1995) on this scopic "regime of curiosity."

wooden cabinet resembling a shrine decorated with royal insignia.[12] Unique, in fact often idiosyncratic or even strange, objects were on display, 'curious' pieces – hence the term 'curiosity cabinet,' by which these 'wonder chambers' were also known. The only system to them was their lack of one, as none can reasonably subsume coins, anatomical models, seal rings, jewels and petrifications. Fine mineral specimens were often decontextualized and mounted on richly ornate gilt stands or artificial sculptures. The only umbrella (if there was one) was the claim to grant universality that their magnificent collector sought in their nimbus. These polished *preciosa* glistened in shrines designed to make the viewer shrink duly in admiration of these symbols of wealth and omniscience.[13]

In 1671, Louis XIV issued an edict regulating the royal collections at the *Jardin du Roi*, which housed the botanical and zoological collections, and the royal pharmacy, where medicinal plants and mineral specimens were kept. Upon his death in 1715, these collections became the *Cabinet du Roi*, which was later reorganized by Bernard Jussieu (1699–1777). When the world-famous naturalist and classifier Georges Buffon (1707–78) became its curator in 1739, he decided to inaugurate it as one of the first public museums in Paris, five years before the *Louvre* with the royal art collection. Royal collections and derivative museums were thus an important institutional context in which natural history could – and did – thrive. The curators of these collections – among whom we find the foremost natural historians of the age – incrementally changed the aesthetics of display from a tohu wa-bohu of curiosities arranged for impressive effect to a rigorous array of labeled objects of well-defined character. Categorized into orders, classes, species, kinds, etc., the objects were arranged according to the various systems of natural classification under discussion during the 18th century. Systematized collections in natural history became increasingly specialized and presented an ever-growing number of natural kinds inside long rows of plain vitrines, carefully labeled and grouped according to the natural classification preferred by the curator.[14] Not only the size of the collections rose, but also the level of precision with which each sample was catalogued, described (see here sec. 11.2 on color terminology)

---

[12]For a reconstruction of the "pharmacy" of Louis XIII, still on display in the basement of the *Musée d'Histoire Naturelle* in Paris, see Wilson (1994) p. 61.

[13]On the rhetoric of 'curiosity cabinet' displays and their similarity to shrines, see Bann (1995). For nice examples, see the 'shrine of rarities' ("Raritätenschrank") of Georg Hinz (c.1666) in Bredekamp et al. (eds. 2000) p. 136 or here color plate X for a French example.

[14]On this transformation from flabbergasting wonder chambers (16th–17th c.), to transparent naturalia cabinets (18th c.), to methodically inventorized special collections (19th–20th c.), see, e.g., Olmi in Mazzolini (ed. 1993), Bann (1995) and te Heesen in Dürbeck et al. (eds. 2001).

and classified. By 1789, the French king's mineral cabinet, for instance, had grown to c. 2,600 pieces, but more than 50 private collections already far exceeded that number, ranging from 4,000 to over 40,000 specimens.[15] Abraham Gottlob Werner's own collection in Freiberg totaled over 10,000 samples, subdivided into 8,043 in the systematic collection, 309 in his collection of distinguished features, 1,398 of gem species, 1,045 in his petrological collection, and 168 in a set of large and particularly aesthetic samples chosen for display.[16]

Private collectors during the Renaissance, such as Agricola, Gesner, Johannes Kentmann (1518–74) and Ulisse Aldrovandi (1522–1605), initiated this move away from an idiosyncratic display of all kinds of treasures toward a much more matter-of-fact store in mineral cabinets with labeled drawers or plain glazing.[17] The *trompe l'oeil* painting of the mineral cabinet of Alexandre Isidore Leroy de Barde (1777–1828), shown here as color pl. X, marks the transition phase from the scopic "regime of curiosity" to the "regime of display" (Bann 1995): while the collection is already purged of polished gems, petrified wood, corals, shells or other non-mineralogical treasures, Barde's presentation still follows the old ideal of arrangement by size and aesthetic criteria for maximum effect, not by group, chemical composition or any of the other serious criteria of mineralogical classification.[18] All the minerals are still openly displayed in an individualized wooden casement, without any glass or other form of protection against dust, and not yet hidden away in uniform drawers. The identity of painter and collector reminds us that for ages the mineral kingdom has been one of the well-known sources of inspiration for artists, and that the cultural split that later drew these two worlds further and further apart had not yet occurred.

By 1799, 75 different public mineral museums and permanent exhibits existed in Europe, from Almadén to Zurich and Stockholm, and from Dublin to St Petersburg.[19] In the United States, the private collection of James Smithson (1765–1829)

[15] For a tabular survey of 55 collections formed before 1799 and ranging between some 4,000 and 100,000 specimens, with their prices (where available), see Wilson (1994) pp. 46f.; cf. also Werner (1778) for advice to the collector on how to build up, classify and present such collections.

[16] See Wilson (1994) pp. 99f., www.gupf.tu-freiberg.de/freiberg/fg bilder/min sammlung.html and www.universitaetssammlungen.de/person/95 (both accessed Sept. 4, 2013) on the current state of the collections, into which Werner's collection has been integrated, expanded to more than 80,000 crystals, etc.

[17] Kentmann's collection has not survived, but we have his drawing of simple wooden cabinets with a set of 26 numbered drawers that housed his collection, which became a model for countless later natural history cabinets: see Wilson (1994) p. 25.

[18] On the development of mineralogy as a science and its systems of classification, see, e.g., Groth (1925), Laudan (1987) and further literature mentioned here in sec. 11.2.

[19] See Wilson (1994) pp. 152f. for a complete list with founding years, and pp. 157–99 for a census

became the base-stock of the *Smithsonian Institution*, founded in Washington, DC on the basis of his testamentary will to create "an establishment for the increase and diffusion of knowledge among men," supported by the hefty sum of over $550,000 shipped in 105 bags of 1,000 gold sovereigns to the Philadelphia mint. Even though his large mineral collection, also bequeathed to the new institution, was later destroyed in a devastating fire, the mineral collection of the *Smithsonian Institution* became one of the world's largest, now comprising approximately 350,000 mineral specimens and 10,000 gems; a considerable portion is also listed or even digitized online.[20] Apart from printed and often heavily illustrated inventories of these collections, there exist hundreds of mineral guide books to aid the collector in identifying his or her own samples gathered on excursions or bought at one of the annual mineral fairs conducted worldwide from Munich in Germany to Minas Gerais in Brazil. Posters on these trade events and popular books on minerals abound, with high-quality images of extraordinary samples epitomizing their beauty not just in my opinion but also for any avid collector.

A similar resonance between art and science exists in the neighboring field of geology. The *Art Institute of Chicago, Department of Museum Education*, offers an exhibition on "Art and Geology," and in the *24th International World Congress in the History of Science, Technology and Medicine*, a day-long session on geology in art and literature featured topics such as "The art of geological mapping" and the work of John Ruskin (1819–1900), "whose career in the mid-Victorian period united the fields of literature, art and geology."[21] Coffee-table books entitled *The Art of Geology* entice onlookers (more than readers) to flip through their fine selections of breathtaking color images of volcanic eruptions, sedimentary structures and fault lines, many of them surprisingly close to celebrated works of *art informel*. The deeper connection between them is that abstract artists of the *informel* – like geologists – ultimately study (and adore) structures.[22]

---

of 1,200 mineral collectors 1530–1799.

[20] See Wilson (1994) pp. 82–3 and http://mineralsciences.si.edu/collections.htm (accessed Sept. 1, 2013) for data about c. 90% of their specimen, 10% also available as digital images.

[21] See www.artic.edu/aic/collections/exhibitions/American/resource/1241, geologyinart.blogspot.de/ and www.ichstm2013.com/programme/guide/s/S112.html (all accessed Dec. 8, 2013).

[22] Samples for this kind of publication include Moores (ed. 1988) and van Diver (1988); the hidden connection to abstract art is made explicit in Schmidt & Schenk (eds. 1960).

12.2 Beauty contests for technical drawings or electron-microscope images

Throughout this monograph, we have repeatedly encountered expressions of 'beauty' in connection with images deriving from scientific or technical contexts. David Kirkaldy's masterful technical drawings of engines for the London Great Exhibition of 1851 became prominent exhibits and were later published as steel engravings. For the Paris World Fair of 1855, some of his highly illusionistic drawings were even chosen as a presentation gift to Napoleon III; and the London *Royal Academy of Arts* exhibited selections of his drawings of Napier's transatlantic iron steamship *Persia* in 1854 and 1861. "Masterworks" by him and dozens of other superb technical illustrators were collected in an anthology entitled *The Art of the Engineer*, which also documents the multifarious use of such drawings far beyond the technical realms of construction and documentation.[23]

This practice of singling out outstanding images as aesthetically pleasing is by no means limited to highly competitive international fairs and high-profile achievements, but can be found in various other contexts, technical and scientific alike. When C. T. R. Wilson succeeded in developing his "expansion apparatus for making visible the tracks of ionizing particles in gases" (soon informally called Wilson's cloud chamber), the photographs obtainable by this new research technology were welcomed as a new gateway into the subatomic world, as a 'direct' and allegedly unmediated form of permanent recording "free from distortions" – this is notwithstanding their having been heavily reworked and retouched to remove scratches, etc. On top of it all, Wilson's photographs were frequently called "beautiful."[24] Later offsprings from this "image tradition" in what later became elementary particle physics are still found far beyond the confines of nuclear reactors and particle accelerators.[25] They have transgressed the invisible but otherwise highly effective borderlines to the world of fine arts and are now a fairly popular source of inspiration for artists such as Jeannie Therrian-Gottschalk. Visual and cultural transfers in the other direction also occur.[26]

The topoi of 'beauty' and 'art' in science and natural history have already received considerable attention in the literature on the histories of several sciences and the fine arts.[27] Whereas some of the alleged parallels – between, say, Ein-

---

[23]See Robertson (2013) p. 128 and Baynes & Pugh (1981) pp. 170f.; cf. also Purbrick (1998).

[24]On the history of the cloud chamber, see Galison & Assmus (1989); cf. Chaloner (1997) on the uncontroversial persuasive power and undisputed evidentiary status of Wilson's photographs.

[25]This 'image' tradition is profiled against a 'logic' tradition of electronic counters: Galison (1997).

[26]On these transfers, see Elkins (2007) pp. 15–18, (2008) pp. 156ff. and sources quoted there.

[27]See, e.g., Henderson (1983), Ellenius (ed. 1985), Lapage (1961), Wechsler (ed. 1988), Dance (1989), Miller (1996), Kemp (2000) and Krohn (ed. 2006).

stein's and Picasso's destruction of Euclidean space, claimed by both Henderson (1983) and Miller (1996) – remain fairly superficial, others are indicative of intense interaction and actual resonances between members of thought collectives considered to be very remote from each other, both cognitively and culturally. In the scientific estate of one of the pioneers of electron microscopy, Ernst Brüche (1900–85), I found a photograph album of electron-microscope images compiled by his co-workers. Some of the images were commented upon with captions such as 'the Picasso of electron microscopy'; others strongly remind me of action paintings by Jackson Pollock or abstract sculptures by Max Bill.[28] During the 1960s and 1970s, when these electron-microscope images were made, a whole string of books were published exposing striking parallels between contemporary art and natural patterns.[29] The audience for these publications and exhibitions was by no means limited to "physically not particularly enlightened, but aesthetically easily pleasable people."[30] The intense fascination with these new image worlds, first arising in academic circles, spread beyond this initial milieu. The public at large was equally gripped by these striking parallels and surprisingly intricate patterns. The whole lifework of personalities such as György Kepes (1906–2001), professor of visual design at *MIT* in Cambridge, Massachusetts, consisted in the aesthetic and intellectual digestion of these surprisingly deep interrelations between fields apparently far removed from each other.[31]

Often it is not so easy to say how clear these parallels were to the actors at the time they made or constructed their visual images and how these consciously perceived parallels – if they existed – then guided their selection, framing and composition. Assumptions of a linear and stepwise transition from scientific motif to its artistic re-rendering seem naïve, but we do not have good studies on the kinds of associations, resonances and subconscious processes that might be going on during such transformations. New perspectives on these processes, conducted as soon after the actual imaging practices and as close to the actors as possible, might be gained from these considerations in future studies. Such issues are by no means only relevant to historical case-studies and long-since-dead authors. They are still acute in present-day science and technology. In 2004, the

---

[28] See Brüche's estate kept in the archive of the Mannheim *Landesmuseum für Technik und Arbeit*, Nachlaß Brüche, box 8.

[29] Cf., e.g., books like Schmidt & Schenk (eds. 1960/67) on *Kunst und Naturform*, with a foreword by the biologist and natural philosopher Adolf Portmann (1897–1982).

[30] A quote from a letter by A. Thiessen to H. Bethge, quoted in a paper by Falk Müller in Hessler (ed. 2006), note 28.

[31] See Kepes (1944, 1963, 1965); cf. Schmidt & Schenk (1960) and Borsig & Portmann (1961).

former director of Lamont/Columbia's SEM and x-ray microanalysis facility, Dee Breger, founded *Micrographic Arts* to offer customized photomicrographic imagery commercially. A glimpse into the Internet, e.g., under Google Image search terms such as 'electron microscopy + art,' 'micro beauties' or 'micro monsters' also yields interesting hits. A gallery of electron-microscopically generated images crops up that were retouched by Adobe Photoshop or another image-processing program. Under this search, I found the prizewinning shot shown here in fig. 119, representing a whole genre in the gray zone between science and art.

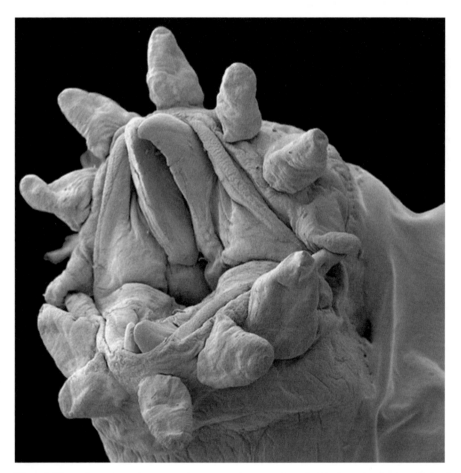

Fig. 119: Scanning electron microscope image of a hydrothermal worm obtained by Philippe Crassous, submitted to the 2011 FEI Owner Image Contest, and published by the newspaper *The Telegraph* in 2011. Available online (accessed Aug. 28, 2013) at www.telegraph.co.uk/science/picture- galleries/8677290/Scanning-electron-microscope-images-reveal-hidden-horror-and-beauty.html. Image courtesy of FEI. Cf. also fig. 101 on p. 313.

Such scientific 'beauty contests' are not limited to (electron-) microscopy, of course. Since 2000, *Science* magazine and the US *National Science Foundation* (NSF) have been jointly sponsoring an annual "International Science and Engineering Visualization Challenge." Images of scientific or technological objects deemed 'most beautiful' are awarded in five different categories: photography, illustrations, informational posters and graphics, interactive games, and videos. Two mathematicians received the prize in 2006 for their computer-generated images of complex geometrical surfaces, but all sorts of microscopic images were also conspicuous in this competition (see fig. 120 for a composite).[32]

Fig. 120: Composite of various prizewinning images for the 2006 Science & Engineering Visualization Challenge, organized by the NSF with the comment "Some of science's most powerful statements are not made in words." From www.nsf.gov/news/special_reports/scivis/challenge.jps (accessed July 15, 2011, no longer online). By permission of the NSF.

Following a similar vein, scientists and photographers at the *Rochester Institute of Technology* have repeatedly organized exhibitions on *Images from Science*. Scientific photography from all areas of research and development could be submitted, which were then selected according to criteria of aesthetics and visual impact as well as skill required to obtain the image. The jury consisted of a mixed assortment of scientists, photographers, photo-journalists, educators and practicing artists. Several dozens of photographs that underwent the concerted scrutiny of these experts were shown in a traveling exhibition to various campuses and published in a colorfully illustrated catalogue.[33]

When the Federation of German Research Institutes, the *Helmholtz-Gesellschaft*, recently organized a similar travelling exhibition featuring current research per-

[32] On the 2006 prize, see, e.g., Nesbit and Bradford (2006) and Rocke (2010) p. xi. More generally, cf. http://www.nsf.gov/news/special_reports/scivis/.

[33] See Davidhazy & Peres (eds. 2002, 2008).

formed at its 17 prestigious institutions with altogether more than 30,000 employees, these exhibits were also dominated by high-class photographs ranging
from satellite images to the submicroscopic dimensions of nanoscale tubes.[34]
Condensed texts about what is shown on these images (and occasionally also
how they were taken) were distributed in slim supplementary leaflets. However,
the gripping large-scale photographs with minimalist textual commentary made
the greatest impact in this exhibition as it traveled across the Federal Republic
of Germany in 2009–12. The title of this impressive exhibition *Wunderkammer
Wissenschaft* ("Science: Chamber of Wonders") made it crystal-clear that in both
cases, the intended audience was viewers, not readers – their target was the gaze,
not the brain of the public.

The beauty of many scientific images acts remarkably like a binding glue across
the board on various social levels. First of all, it binds together all experts in a
visual domain in their viscourse on how to improve their own images. It serves as
a mutual motivator in arduous pursuit of the cause. Sometimes, "really good" or
"beautiful" images also help to convince a scientific community of the existence
of an unexpected phenomenon or a strange effect still under heavy dispute.
Thus "seeing is believing," as various examples from materials science in a recent
textbook also confirm: "an image which is sharp, coherent, orderly, fine textured
and generally aesthetically pleasing is more likely to be true than one which is
coarse, disorderly and indistinct."[35] Finally, some images, such as those chosen
for the above-mentioned exhibitions, also fascinate a much larger group of people
outside the specialty. It has been observed that some scientific communities,
such as modern astronomers and astrophysicists or even high-energy-particle
physicists and mineralogists, produce two kinds of images: on the one hand,
'pretty pictures' for conference posters, advertisement brochures, annual reports
and other publications aimed at a broader audience; on the other hand, 'scientific'
images for use only among specialists, perhaps occasionally for publication in
specialized research journals, but certainly not for wider distribution.[36] The
tricky point here is that both types of images are supposed to represent the same
objects – be it the Crab Nebula or a biochemical virus – although, in fact, pretty

---

[34] Helmholtz-Gesellschaft (ed. 2009–12) and www.wunderkammer-wissenschaft.eu/ausstellung
/aktuelles/ or www.helmholtz.de/aktuelles/veranstaltungen/wunderkammer_wissenschaft/ (both
last accessed July 20, 2012).

[35] Cahn (2001) pp. 91f., there supported by one of the earliest optical micrographs of a dislocation
in a silicon crystal, "decorated" with copper, which makes the dislocation "visible."

[36] On this distinction, see, e.g., Elkins (1995) p. 557 and A. Müller et al. in Liebsch & Mößner
(eds. 2012); cf. Beaulieu (2002) and Racine et al. (2005) on fMRI.

pictures tend to deviate much more from the object depicted than the scientific images, for instance through intense use of strongly chromatic false colors or other serious manipulation of the content. The *Hubble Space Telescope* images distributed by NASA for public relations purposes were, in fact, heavily attacked by the experts. The German astronomer and astrophysicist Rudolf Kippenhahn (∗ 1926) criticized them as serious astronomical images overly reworked (falsified?) for popular consumption.[37] Likewise, brain images with high-activity color zones (see here color pl. XVI) are often overinterpreted as alleged proof of a 'neuro-realism' or 'neuro-essentialism' when such fMRI images are taken out of their research context and acquire wider symbolic meaning in the public arena. Experts take these images as but one representation of their data. That is convenient because it allows, for example, rapid verification of the spatial resolution and quality or specificity of a brain response. The public tends to place too great a reliance on these suggestive images as icons of 'neuro-essentialism' and is thereby seduced into grossly overinterpreting them.[38]

[37] Kippenhahn (2006); cf. Villiard & Levay (2002), Ullrich (2003*b*), Elkins (2007) pp. 10-15, Sabine Müller in Groß & Westermann (eds. 2007) pp. 93–112, 189–218, Kessler (2012) and Roth & Fosbury (2013).

[38] On this danger, see Beaulieu (2002) and Racine et al. (2005) on fMRI in the public eye.

13

Issues of visual perception

The springboard of this chapter is the following striking observation: Surprisingly many of those actively engaged in optical, microscopic or spectroscopic research during the 19th and 20th centuries shared a common interest in physiological issues – think of Thomas Young, Fraunhofer, Brewster, and Zöllner on color perception, or Young, Matthiessen, Abbe, Janssen and Listing on the physiology of the eye. What we obviously have here is a resonance between various visual science cultures: 'mapping' in spectroscopy and geography, 'sensitizing' in spectroscopy, photography and physiology. The link between various scientific visual domains and sensory physiology since 1800 might be part of a broader cultural trend, since it also crops up in other fields; but here I will just pick out two examples: solar astrophysics and scientific photography.[1]

13.1 Jules Janssen: black drops and solar granulation

We have already met the French astronomer Jules Janssen (in sec. 7.4) while we were discussing the insatiable urge by scientists and engineers to obtain 'moving images' of rapidly changing objects or of processes either too fast or too slow to be observed visually. Janssen had invented the 'photographic revolver' to record the exact timing of the transit of Venus across the solar disk in 1874. The Harvard historian of science Jimena Canales has shown how this episode – usually simply recounted as part of the prehistory of cinematography – is also multiply linked to the history of sensory physiology.[2] First, exactly determining a time is intrinsically bound with the following problem: Each observer will have a different reaction time between seeing something happen and actually pressing a button, reading off a time, or executing some other kind of registration procedure. Systematic experiments in controlled settings had revealed that, to a good approximation, such readings by different observers can be somewhat coordinated if a 'personal equation' is attached to each of them. This means that a certain delay time (in some cases, also anticipation time) has to be subtracted or added to his or her respective readings to arrive at an intersubjectively valid reading.[3] Second, the

---

[1] For other examples and the broader context, see part II of Dürbeck et al. (eds. 2001).
[2] See Canales (2002). Cf. also Sicard (1998), Schaefer (2001) and further references there.
[3] On this personal equation issue, see Schaffer (1988) and critically Hentschel (1998b).

exact determination of, for instance, Venus's apparent contact with the Sun yields another very specific problem of human vision. It is related to the borders between zones of very different light intensity. Various expeditions to observe the passage of Venus prior to 1874 had already shown a strange 'black drop' effect. When Venus comes close to the solar rim, the circular image of the planet, showing as a dark circle against the very bright background of the solar surface, somehow deforms into a droplet shape, with a neck developing towards the rim long before the actual contact (cf. fig. 121, left). This phenomenon, also called *goutte* or *ligament* in the French literature, was extremely disturbing because it made it practically impossible to determine the exact time of contact, i.e., the time when the tangents of Venus's outer rim and the Sun's rim coincide. Physiological laboratory experiments on observing dark spots against a white background moving toward white or black border zones confirmed this effect. Various possible explanations for the black-drop effect were suggested at the time, ranging from an optical illusion to the refraction of light in Venus's atmosphere or diffraction around Venus. Today we know that the main cause is ordinary smearing of images mainly due to multiple small-angle scattering of light by temperature variations in our Earth's atmosphere and to diffraction within the telescope (see the right half of fig. 121 for a digital simulation).[4]

These discrepancies were highly vexing. Despite all the precautions taken and the ambition since 1769 to get very precise measurements, astronomers still got "strange divergences" between the times given by various observers, even when they were standing side by side. The unavoidable conclusion was a fundamental "defectiveness of observations" with "only discordant results."[5] As Canales has convincingly shown, Janssen's photographic revolver must thus be seen as an effort to overcome this impasse and to achieve intersubjectivity by a mechanically automated and standardized procedure in which the human observer no longer intervenes. According to Janssen and his French astronomer-colleague Hervé Faye (1814–1902), photography was the only way out of this embarrassment: "The sole method offering complete guarantees is photographic observation [...]. This kind of observation suppresses the observer and with him the anxiety, fatigue, bedazzlement, haste, the errors of our senses, in a word, the always suspicious intervention of our nervous system."[6] However, as Canales also shows, this was by no means as straightforward as Janssen may have believed at the time. First, the

---

[4]Cf. Schaefer (2001) and Pasachoff et al. (2004) for further commentary.
[5]See Schaefer (2001), Pasachoff et al. (2004), Canales (2002) pp. 592ff. and cited sources.
[6]Faye (1870) p. 543. Cf. also Sicard (1998) pp. 47, 61 and Daston & Galison (1992).

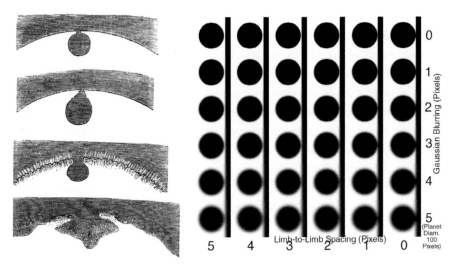

Fig. 121: Left: The 'black-drop' phenomenon during the transit of Venus as sketched in 1769. Right: Modern digital recreation by J. Westfall of the black-drop effect of an unsmeared image in row 0, with increasing Gaussian blurring downward to typical daytime seeing conditions in row 5. From Forbes (1874) p. 49 and Schaefer (2001) p. 330. Courtesy of Prof. Bradley E. Schaefer.

daguerreotype records of these observations likewise had to be examined visually (with the aid of a microscope, of course). It turned out that the same black-drop effect, so disturbing in direct observation of the passage, also showed up in these later microscopic examinations, simply because there the astronomers were again looking at border zones between brightness and darkness. Second, photographing introduced other sources of error, such as solarization effects from overexposure and unequal expansion of the photographic emulsion from thermal heating during exposure. Thus, contrary to popular folklore, Janssen's photographic revolver did not immediately disperse the mist around the exact timing of the passage of Venus. It is a first example, though, of the occasionally close connection between visual observation, issues of intersubjective registration and inscription of these observations, and sensory physiology.

Janssen continued to work in areas where exactly this interplay was central. He may have been more interested in physiological issues than other researchers, since his doctoral thesis was a study on the absorption of radiant heat in the eye[7]

[7]See Janssen (1860). On Janssen's vita, see, e.g., Bigourdan (1908) and Levy (1973). On his photographic revolver, see here p. 247. On his links to various other visual cultures, most notably architecture, see here p. 147. On his later work in spectroscopy, see Hentschel (2002a) pp. 102ff. and further biographical sources mentioned there.

and he also cooperated with a physician in designing an ophthalmoscope. The 1874 expedition to Japan to observe the transit of Venus then turned his attention to photographing the Sun and smaller structures on its surface, such as sunspots and so-called solar granulation.

This term, granulation, is a metaphor that signals its origin in a mental process of comparing fundamentally new phenomena never seen or described before to known features of our world. In fact, when solar granulation was discovered around 1860, the earliest observers compared it to things as different as interlacing leaves, a willow-leaf pattern, rice grains, pores and bubbles in boiling liquid (cf. fig. 122).[8] These published illustrations, in turn, strongly influenced the perception of later observers (such as Warren De la Rue or John Herschel) about what they expected to see through their telescopes. This was particularly tricky, since the surface structures in question, today called solar granulation, with an average diameter of roughly 1 arcsec, were at that time right at the resolving limit of telescopes equipped with a quite modest aperture and a resolving power of about 100.[9] Photographic recording of granulation was still not possible, and visual observations of the surface of the Sun were chronically hampered by turbulence in the Earth's atmosphere ('daytime seeing' caused by the solar heat).

The early quarrels about how to interpret solar granulation show how differently one and the same observation can be interpreted and rendered by different observers. Janssen was very much aware of this effect, which we today would reinterpret as evidence of *Gestalt* psychology. 'Seeing of' is not yet 'seeing as,' and the latter *Gestaltsehen* is culturally and stylistically preconditioned, to some extent beyond the control of the observing individual. On the occasion of a conference in Toulouse in 1888, Janssen confronted this issue head-on and discussed how various astronomers since early-modern times had portrayed another phenomenon on the solar surface: sunspots. Starting with drawings by Fabricius (1611), then moving on to drawings by Galileo (1612) and Scheiner (1626), William Herschel (1801) and his son John Herschel (1837), and finally comparing them with contemporary sunspot drawings by Tacchini and Langley, Janssen concluded:

[8]On Nasmyth 1860, Nasmyth & Herschel 1861, E. J. Stone & Nasmyth 1864, and Huggins 1865/66, see Bartholomew (1976) and various other primary sources cited there. Cf. also Hentschel in Hentschel & Wittmann (eds. 2000) pp. 23–5.

[9]According to Bartholomew (1976) pp. 266, 282, the persistence of Nasmyth's misleading willow-leaf metaphor was due to John Herschel's enduring support of it.

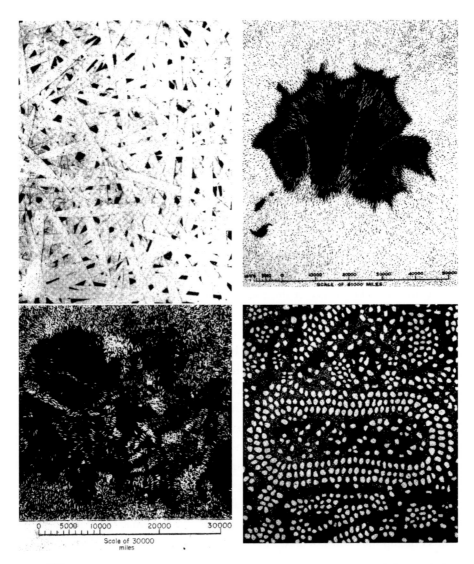

Fig. 122: Four different visual conceptualizations of solar granulation. Top left: interlacing leaf pattern (Nasmyth 1860); top right: willow-leaf pattern (Nasmyth & Herschel 1861); bottom left: rice grains, drawn by E. J. Stone & Nasmyth 1864; bottom right: pores (Huggins 1865/66). Reproduced with commentary in Bartholomew (1976) pl. I–IV and in Hentschel & Wittmann (eds. 2000) p. 24.

This short examination is sufficient for showing to us the disaccord that
exists even among the best observers when observing solar phenomena. It
convincingly demonstrates that the true method for observing them is to
obtain, firstly, images drawn by the sun itself.[10]

Given this odd phenomenon of solar granulation, it is not surprising that Janssen
would not be content with a mere subjective recording of what he personally
might have seen under the best observing conditions. As an outspoken advocate
of photography as an astronomical research technology, Janssen wanted nothing
short of a photograph of the phenomenon. To him, photographs were the true
retina of the scientist,[11] more reliable and less riddled with subjective interpretation, providing lasting images produced by the fleeting phenomena themselves.[12]
But getting a photograph of the solar granulation was very difficult, simply because the Sun is such an extremely hot body (c. 6000°C). It shines so brightly
that normal photographic plates immediately 'solarized,' i.e., they showed signs
of severe overexposure. Furthermore, the human eye (and the human interpreter
behind it) is somehow able to average out the rapidly changing disturbing effects
of thermal convection currents that lead to irregular fluctuations of good and bad
seeing across the solar surface. The trained eye is able to optimize the image
it registers, while the photographic plate, set to a short exposure time to avoid
overexposure and solarization, simply yielded the momentary view with all its in-
built seeing defects. Longer exposures through thicker filters led to a summation
of all the local fluctuations and thus to an even worse image. Another source
of error was solar irradiation, as observed, for instance, in photographs of solar
protuberances during a total eclipse. The bright protuberances seemed to creep
into the overlapping lunar disk as it obscured the solar surface. Consequently, no-
one had been able to produce a photograph of the solar granulation even though
this surface feature had been known for quite a while already and photographic
techniques had been around since 1839. Because of the abnormal light intensity,

[10]Janssen (1887*b*) p. 39 and (1888*c*) p. 33, Engl. transl. by Canales (2002) p. 605. On early-
modern sunspot representations, see Biagioli (2006), Bredekamp (2007) and van Helden & Reeves
(eds. 2010). Cf. here p. 40 for a similar point of ambiguity with respect to sunspots as interpreted
by earlier Chinese stargazers and Saturn's rings.

[11]See Janssen (1883*a*) p. 23 and (1883*b*) p. 128 (according to Gunthert (2000) note 23, not already
in 1874 as is often claimed!): "Aussi n'hésité-je pas à dire que la plaque photographique sera bientôt
la véritable rétine du savant."

[12]Janssen (1883*b*) p. 122: "La photographie y substitue l'image matérialisée du phénomène
lui-même" and Janssen (1877*c*) p. 1251 on photographs as providing "vrais rapports d'intensité
lumineuse des diverses parties de l'objet qui lui donne naissance." This position clearly resonates
with the ideology of 'mechanical objectivity' as defined by Daston & Galison (1992, 2007).

even a normal photograph of the Sun had been achieved by Hippolyte Fizeau (1819–96) and Léo Foucault (1819–68) only in 1845, i.e., no earlier than six years after the invention of the daguerreotype.[13]

The reasons for all these failures gave Janssen clear clues about how to proceed when he began his studies on solar photography in 1874:[14]

(i) He radically shortened the exposure times. With the help of highly qualified mechanics and instrument-makers, he obtained fast shutter systems allowing exposure times of only 1/2,000 s, with a self-estimated precision of only ± 1/10,000 s in 1877, and only 1/100,000 s by 1883.[15]

(ii) He examined the image of the Sun in various spectral regions and found out that particularly well-defined images resulted in the violet range of the spectrum near spectral region G. For most of his subsequent solar photographs, he filtered the sunlight so that only this violet range came through, which greatly improved the overall definition of his images and essentially circumvented chromatic aberration.

(iii) He carefully experimented with the available photographic emulsions, finally settling on a collodion iodine-bromuriate with iodine dominating. Its peak sensitivity is just in the violet range of the spectrum region chosen for his solar photographs. The chemical developing procedures were likewise carefully optimized.[16]

(iv) He selected a high-quality achromatic objective of 13.5 cm diameter.[17]

(v) Corresponding to this large lens, he also enlarged the size of his photographic plates, from initially 12 to 15, then to 20 and finally to 30 cm width. For special cases, he even experimented with 50 and 70 cm plates.

The resulting images of Janssen's 'photoheliograph' were extremely clear (*netteté extraordinaire*), were sufficiently large to avoid disturbing effects like solar irradiation, and even allowed a further 10× magnification for close-up examinations

[13] On the history of solar photography, see Janssen (1883*b*) and (1896) pp. 91–3, Lankford (1984), Rothermel (1993) and Pang (1994/95, 1997*b*).

[14] On the following, see Janssen (1877*a–c*, 1878, 1883*b*, 1896, 1903). On Janssen's other work and vita, see pp. 247 and 78 here.

[15] Janssen (1877*c*) p. 1252, (1883*b*) p. 127 and (1896).

[16] For photochemical details, see Janssen (1877*c*) p. 1252, (1878) p. 695 and (1896) pp. 94–7.

[17] According to Janssen (1896) pp. 93 and 101, his camera was built by M. Prazmowski, "le servant et habile constructeur"; ibid., pp. 98f. for further details on Janssen's photoheliograph.

and printing as Albertypes (1877) and heliogravures (1896).[18]

When Janssen finally succeeded in obtaining photographic records of solar granulation in 1877, he had just moved into the newly founded Meudon Astrophysical Observatory as its first director. With its superior equipment, Janssen continued to improve his photographs and finally published enlarged sections of his exquisite plates in the first volume of the *Annales de l'Observatoire d'Astronomie physique de Paris* in 1896 (cf. fig. 123 for two samples).

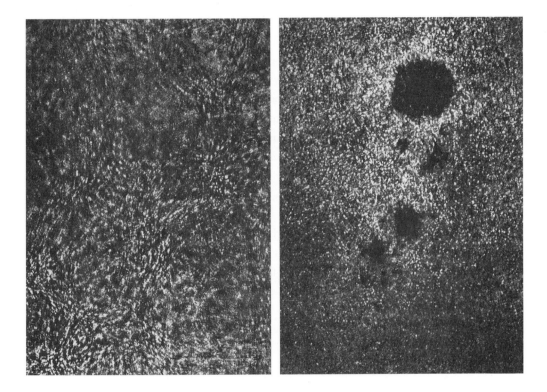

Fig. 123: Two samples of Jules Janssen's photographs of solar granulation. Left: the central region, showing zones of good and bad definition, taken in 1878. Right: granulation around a group of sunspots, taken in 1894. From Janssen (1896) heliogravure pl. III and IV (printed by Fillon & Meuse in the large plate size of 22 cm × 17 cm from a section enlarged tenfold.

[18] Quote from Janssen (1896) p. 95 and (1878) p. 700 verso, where he emphasized that his *épreuves photoglyptiques* had been "obtenu sans aucune intervention de la main humaine." For reproductions of two prints of 1877 photographs on albumized paper, see Sicard (1998) p. 59. For later samples, see fig. 123 here.

Janssen argued that his images proved that all previous debates were moot about whether the solar granulation really was like willow leaves or rice grains, or whatever else. The photographs showed the granulation to be of varying sizes, between 1/10 of a arc-second and 3–4 arcsec.[19] They also varied in form, from small circular cells to slightly larger, somewhat polygonal shapes. These granules seemed to be present all over the solar surface, but were quite sharply circumscribed in some zones of his photographs and diffuse or nearly vanishing in other zones. Their rapid changes in shape and position in his series of photographs taken with his photographic revolver proved what a mutable feature of the solar surface they were. The only explanation Janssen could think of was that the granules were situated at slightly different depths of the photospheric layer, in which they were exposed to strong convection currents induced by the very high temperatures. Close comparisons of various different plates with a magnifying glass led Janssen to postulate the existence of a *réseau photosphérique*, an irregular network of figures, generally rounded or polygonal, but sometimes also more rectilinear in shape, with typical intervals and sizes of roughly one arc-minute. He interpreted their rapidly changing shape and definition as indicative of "violent motions that upset the granular elements," somewhat analogous to "atmospheric clouds."[20]

This interpretation was not far off the mark. What Janssen could not have known when he published these findings and their interpretation in 1877 was that convective zones tend to develop these cell-like patterns if they are situated between a hotter lower layer and a colder upper layer. In 1900, Henri Bénard (1874–1939) experimentally created such conditions in his laboratory and observed what are now called Bénard cells (fig. 124). They are the perfect and profound visual analogy to what Janssen had photographed on the solar surface.[21] We have thus seen how a scientist like Jules Janssen was repeatedly confronted by issues of sense physiology throughout his long career as one of France's outstanding astronomers and astrophysicists. Physiological concerns played an integral role in Janssen's visuality. In the next section, we will see that this is not an isolated finding.

---

[19]These numbers were quoted in Janssen (1878) p. 697. In Janssen (1896) p. 105, he settled on typically 1–2 arcsec, with the smallest granules down to 1/2 and 1/4 arcsec.

[20]See Janssen (1877*b*) p. 775, (1877*c*) p. 1254, (1883*b*) p. 123 and (1896). According to Sicard (1998) p. 55, Janssen later developed a strong interest in volcanoes in search of analogies to the phenomena observed on the solar surface.

[21]See Bénard (1900). Cf. also Aubin (2008) and here p. 48 on early cinematographic recordings of this phenomenon.

Fig. 124: The formation of Bénard cells in a heated liquid. Left: two convective cells, heated from below, with two downstreams next to each other. Center: small section (c. 7 cm), photograph taken from above the liquid. Right: temporal shifts of Bénard cells. Although the overall network pattern is stable, each cell shape and position fluctuates. From Bénard (1900) pp. 1262, 1270.

13.2   Recording the invisible

The cases of x-rays and solar granulation discussed in secs. 8.3 and 13.1 already point to the issue of photography as a crucial technique by which to register processes normally invisible to the human eye on a photographic plate, a film strip or a CCD for later careful inspection, possibly also enlargement, contrast enhancement or artificial sharpening of focus. Through such photographic or digital recording, the invisible is transformed into something visible, a process thoroughly fascinating not only for past generations of scientists and technologists but also for specialists and the general public today.

When Ottomar Volkmer (1839–1901), president of the Viennese society *Photographische Gesellschaft* and since 1892 director of the imperial state press, *kaiserlich-königliche Hof- & Staatsdruckerei*, published his booklet on "the photographic recording of the invisible" in 1894, x-rays had not yet been discovered. So his examples were largely selected from the equally fascinating field of high-speed photography: ballistic projectiles. Interest in this highly specialized area of applied research had risen during the Franco-German war of 1870/71, when both sides accused each other of using bullets designed to expand on impact, thus limiting penetration and increasing the diameter and severity of the resulting injury. Such expanding bullets, also known as dumdum bullets, were banned by international convention but, in fact, soldiers directly hit by enemy projectiles often exhibited those telltale crater-like wounds.[22] Could it be that the supersonic speed with which projectiles from modern guns were ejected carried along a cone of compressed air that upon impact could create such gaping cavities? How could this be verified?

---

[22] On this issue, see, e.g., Cohen & Seeger (eds. 1970).

The flight of a projectile at high speeds was way beyond what the human eye is able to follow. In France, Austria-Hungary and Great Britain, research projects were started on developing objective recordings of such supersonic projectiles in flight. In Britain, the inventor Thomas Skaife (1806–76) assisted a team at the Royal Arsenal, Woolwich in South London with his home-made camera. In Germany, the photographer Ottomar Anschütz (1846–1907) experimented in Buckau near Magdeburg on military proving grounds and on the *Krupp* testing site near Meppen. And in the Austrian Empire, Ernst Mach (1838–1916) and his son Ludwig (at that time working in Prague), as well as Peter Salcher (1848–1928) and co-workers at the Austrian Marine Academy in Fiume, were all working toward the same goal. They utilized a technique invented by August Toepler (1836–1912) in 1864, so-called 'schlieren photography.'[23] The *Schlierenmethode*, as it is called in German, makes use of the fact that compression of air (or indeed of any transparent medium) changes its refractive index. Variations in the refractive index n caused by these motion-induced density gradients in a fluid refract light entering obliquely in the form of a strongly collimated light beam. This refraction leads to a slight spatial variation in the intensity of the light viewed from the side. If (and only if) one succeeds in photographing the projectile just at the moment it passes a zone prepared with such collimated sidelight, the variation in density around the projectile can be recorded in the manner of a shadowgraph system. The efforts by all these experimenters were thus channeled into designing a camera and/or experimental setup allowing such instantaneous registration. The mechanical shutters available at the time were not able to capture minuscule time intervals of the order of $1/500,000$ s (or even shorter). Mach and his co-workers succeeded after implementing the idea that the metallic projectile be made to close an electric circuit upon entry into the prepared zone.[24] This set off an electric spark for a split second generated by a charged Leyden jar. The flash of light exposed the image on the photographic plate (see fig. 125). To avoid stray light, these sophisticated experiments had to be conducted in total darkness.

Capturing shock waves around a projectile at supersonic speeds on a photographic plate and thus effectively freezing time was an engineering feat in itself. Yet Ernst Mach took additional pleasure in the success of these experiments. As a rigorous phenomenalist, he restricted his ontology to directly observable things or processes. Like Bishop Berkeley before him, Mach also restricted reality to

---

[23]See, e.g., Volkmer (1894) pp. 3–12, Krehl & Engemann in Hoffmann & Berz (eds. 2001) and http://en.wikipedia.org/wiki/Schlieren_photography (last accessed Sept. 9, 2013).

[24]On Mach's and Salcher's experimental setup, see, e.g., Volkmer (1894), Merzkirch & Seeger in Cohen & Seeger (eds. 1970), Hoffmann & Berz (eds. 2001) pp. 23ff. and Pohl (2002/03).

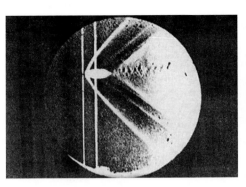

Fig. 125: Two schlieren photographs taken by Ernst Mach and Peter Salcher of a projectile coming from the right. We clearly see several V-shaped cones of compressed air (shock waves) around the projectile and the turbulent zone directly behind the projectile. The two straight lines near the left margin are metallic wires connected to the flash. The left photograph was taken in March 1887 and is reproduced here from Mach & Salcher (1887). The right exposure was taken several years later. From the Deutsches Museum, Munich, Mach papers, CD52642.

what can be sensed. Everything else was "metaphysical" to him – and to him that meant that it had to be ignored. The ability to photograph the V-shaped cones of compressed air in front of a projectile as well as the turbulence behind it had special significance. Both of these features – long hypothetically suspected to exist by obvious analogy with water waves around a moving boat – finally became 'real' processes of the world to him. As I have argued elsewhere, much of Mach's experimental work can be understood in its entirety as following this research agenda: to make invisible things visible.[25] Mach became famous for his experiments in sensory physiology and for his critical historical studies, which also inspired the young Albert Einstein. As evidence of his acute skill at prejudice-free observation, we close this chapter on issues of visual perception with an image that Mach drew to represent what we actually see if we look out on the world with just one eye open (see fig. 126) – namely, something quite different from the standard plane representation we have become accustomed to.

Mach was obsessed with obtaining high-quality visual representations of the fleeting phenomena observed or examined in his laboratory; and he had a strong interest in physiological aspects of vision (*inter alia*, he discovered the so-called

[25]See Hentschel (1987).

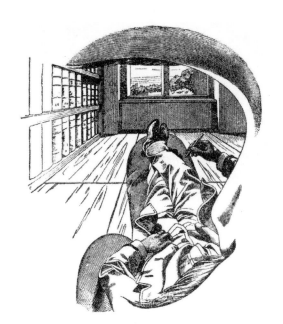

Fig. 126: The world as seen with only one eye open. From the chapter on "anti-metaphysical preliminaries" in Ernst Mach's classic *Die Analyse der Empfindungen und das Verhält-nis des Physischen zum Psychischen* (1886/1900), fig. 1.

Mach bands appearing under certain conditions in zones of strong contrast). Through popular lectures, often supported by self-designed 3D models (such as his wave-chain model), wall-hanging posters and carefully crafted plates in his heavily illustrated semipopular books, he also transported this visual research impetus into the realm of teaching and popularization. His visualization of supersonic shock waves thus also followed the same research imperative of making the invisible visible. Mach also repeatedly exercised visual thinking, most famously perhaps in his rejection of Newtonian absolute space on the basis of Newton's celebrated thought experiment with a rotating bucket. Mach imagined the walls of the rotating bucket growing thicker and thicker, then made the mental transition to an infinitely thick shell of mass around the bucket's center. These heavy masses then became equivalent to the sum of all fixed stars surrounding each point of the universe. Mach thus inferred that those heavy masses constituted the only legitimate reference point in the universe. Mach also predicted that inertial drag effects should show up inside a massive rotating shell.[26] Whatever could not be made visible ultimately did not exist for the phenomenalist Mach. Thus he rejected Newton's absolute space as a chimera; if someone talked of atoms in his presence, he would retort: "Habn's aans g'sehn?" ("Have you ever seen one?").[27]

---

[26] On Newton's and Mach's famous thought experiment that later inspired Einstein to develop his general principle of equivalence, see Hentschel (2014*a*) and further refs. given there.

[27] On Mach's phenomenalism and anti-atomism, see Blackmore (1972) and Hentschel (1987).

14

Visuality through and through

With the example of Ernst Mach in chap. 13, we have come full circle in our grand tour through many dozens of different visual domains in science, technology and medicine. As just shown, Mach's oeuvre epitomizes all layers of visual cultures as spelled out systematically in chap. 2 of this comparative study, especially good drawing skills, obsession with visual resources, visual thinking and a spread of all of the above into his teaching and into his private life. Mach's life and oeuvre also exemplifies the 19th-century mania for making everything visible,[1] a strand which also perfectly fits for Kekulé, Maxwell, Plateau and countless other figures discussed previously.

Mach also fits our prosopographic grid of typical representatives of visual science cultures (as developed in secs. 4.3 and 4.4). Growing up on a farm in Moravia, he cultivated many visual hobbies (from butterfly collecting to photography to designing little engines), but did not do well at school, and consequently took an apprenticeship in carpentry before going back to high school and later studying mathematics and physics at Vienna University. Thus he has the artisanal background – so atypical of scientists in the socially stratified 19th century and yet so characteristic of those individuals who so often became pioneers of visual cultures in science, technology and medicine. Throughout our grand tour, we have encountered many other representatives of this *a priori* unlikely, yet highly characteristic, combination of unusual background, skills and visual inclinations. Avid Victorian photographers (remember Brewster's, Herschel's and Talbot's albums), art-loving 20th-century American spectroscopists (remember W. W. Morgan's avocation as "picture lady," voluntarily explaining select pieces of high-brow fine art to school children; cf. here p. 147) and highly skilled 17th-century draftsmen (such as Robert Hooke, Christopher Wren and Christiaan Huygens) all became pioneers of new visual domains, be they spectroscopy and scientific photography, microscopy, planetary astronomy or magic lantern projection (cf. p. 278). Many of them – like Ernst Mach – did not confine themselves to 2D representations, but strove to get a fuller, spatial impression, thus either becoming pioneers in stereo-photography (such as Brewster, cf. here pp. 144ff.) or designing freely rotatable 3D models of their objects (for instance, Huygens with respect to Saturn's rings; cf. p. 40).

---

[1] See Kelley et al. (eds. 2009) p. 25 on the *Sichtbarkeitsmanie* of the 19th century.

Nowadays, one would perhaps rather strive for naturalistic cinematography or a good computer simulation, but the essential goal – to get the best possible visual insight into the inner workings of the object or process at hand – has remained as vivid and insatiable as in former times. These findings, obtained here by a comparative glance at visual cultures *tout court*, apply across the ages from at least the early-modern period onward until the late 20th century. Nevertheless, there are certain times and contexts that definitely nourished such visual inclinations (e.g., the early-modern and Victorian periods in Great Britain, the late *ancien régime* in France, and the television-dominated era of the USA during the 1960s and 1970s), whereas other periods and contexts were not 'iconophile' (see here p. 113 for a few examples of such 'iconophobe' environments). Consequently, visual cultures in science, technology and medicine are not distributed equally over time and space, but rather lumped in certain clusters of resonance. A few important locations for such resonances between different visual domains covered here include:

- Florence around 1400, with a broad interest in the practical applications of mathematics and especially of geometry, not only among number-crunching calculators (*Rechenmeister*), but even among artists and architects, goldsmiths and stonemasons. In this setting, linear perspective thrived as a new world vision (*Weltanschauung*), with mathematics as the new *lingua franca*;

- Nuremberg around 1500: Albrecht Dürer as transmitter of perspectivalism to the North and as a bridging figure between mathematics, arts and artisans;

- the Netherlands of the 16th and 17th centuries, with the obsessions of naturalistic description and a strong 'mapping impulse' (Alpers);

- Glasgow around 1770, with James Watt and Joseph Black collaborating, which led straight to the indicator diagram exploited by Watt's steam engine company throughout the world; via Carnot and Clapeyron, who learnt about this during their engineer-sojourns in Russia – this knowledge fed back into so-called Carnot-cycles as the core of the new field of thermodynamics;

- Paris around 1800, with the newly founded *École polytechnique*, where Gaspard Monge developed the new descriptive geometry, Lazare and Sadi Carnot the new mathematical theory of machines, and a plethora of highly skilled instrument-makers providing the material resources on the basis of which this culture flourished;

- Paris 1890–1910, with astronomer Janssen, physiologist Marey, physico-chemist Perrin, cell biologist Comandon, and yet others, all fascinated with chronophotography, moving images and early cinematography;

- the USA in the 20th century, with large suspension bridges built across rivers or straits (several famous ones just in the New York City metropolitan area).

The reason for this frequent clustering of visual domains in time and space is a structural feature in the genesis of visual science and technology cultures, for the first time clearly spelled out in this book. I have systematically elaborated and exemplarily illustrated this feature in chaps. 4 and 5. New visual cultures are frequently formed by creative individuals already well embedded in one or even several visual cultures, such as technical drawing (a subject taught intensely at 19th century polytechnics but not at universities). They dare to try out these skills in domains where they had hitherto been ignored and then prove to be highly useful in generating new knowledge or connections. The systematic pattern of dozens of examples unfolded in the foregoing chapters is thus a transfer from one already highly developed and established visual domain to another, new domain. For instance, spectroscopic maps received their impulse via a transfer from cartographic mapping, 19th-century technical drawing and innovative graphic displays such as Charles Minard's *carte figurative* of Napoleon's invasion of Russia from Gaspard Monge's projective geometry and his intense visual training for all students at the *École Polytechnique*. Late 20th-century medical imaging technology profited from digital imaging specialists with backgrounds in crystallography (Lauterbur) or astronomy (Bracewell). Surprisingly, many of these transfers are not between neighboring fields, but between two domains rather far afield, which makes them hard to trace and easy to overlook – correspondingly, few of these transfers have been covered in the existing literature so far – a vast territory of hidden but pertinent links still lies ahead of us.

Such resonances and transfers exist not only between different visual cultures, but also between different visual representations within each of these cultures. In chap. 7, we dealt with near-synchronic concatenations of representations and with diachronic chains of different images, following each other in an unending quest for better quality and closer proximity to the object or process under scientific examination. We met the obsession of physicians, scientists and technicians to get "the best" image possible, and also noticed that often this insatiable demand was the trigger for the development of better recording and/or printing techniques. We noticed how many of our pioneers of visual science cultures either themselves became experts in these printing techniques or stood in close contact with such specialists (see various examples in chaps. 4 and 6). Aesthetic fascination proved to be the binding glue of these loosely formed communities (chap. 12).

Throughout this monograph, "visual cultures" in medicine, science, and technology have been taken to be multilayered. Their material traces – most obvious to the historian who starts out to study them – encompass drawings and paintings, wax models and moulages, architectural or molecular 3D models, photographs, digital images stored in computer memories and a myriad of other visual aids, microscopes and telescopes, pantographs and stereoscopes, among many more visualization gadgets. Which of the above are part of the package characterizing a given visual culture or a more specialized visual domain is dependent on the local and disciplinary context. Undoubtedly, though, each of these visual domains creates a specific visuality that has to be learned and mastered by all newcomers to this field, be they the pioneers who first build it up or the students who decide to study it. Time and again, we have encountered the institutionalized forms in which such inculcation into a visual culture or domain is achieved: for instance, drawing classes for microscopists and technicians, laboratory courses for chemists and hands-on sessions on model-building for architects. Frequent repetition and intense study of wall-hanging charts, illustrated books, plates, albums, 3D models, films, computer simulations, etc., all have a lasting effect: they generate superb skills in pattern recognition and bind practitioners in visual cultures emotively strongly to their objects, which they perceive to be gripping, fascinating, and beautiful. Material, educational, disciplinary, cognitive and emotional strands are all integral parts of this complex package. The schooled visuality of members of such visual cultures is no straitjacket, but rather an aid in seeing and discerning what others would not even notice nor recognize as relevant. This ability of seeing as – i.e., seeing a slight darkening of a sector in an x-ray image or an MRI scan as an indicator of a lesion or as a meaningless artifact – is what distinguishes a trained radiologist from a layman. Like all scientific skills, this ability of *Gestalt* recognition does not come from nowhere, neither does our historiographic understanding and reconstruction of it. Re-learning or historically accurate replication are two strategies practiced by some of my colleagues.[2] Intense study of *all* the pertinent sources – not just influential publications, but also the material and cognitive relics of former scientific practices (including pedagogy and popularization); not only the professional activities, but also the private avocations of our actors – this is the pathway that the rest of us will have to go in our quest for a fuller picture.

---

[2] Especially in the program that originated at the University of Oldenburg under Falk Riess in the 1980s, now still practiced at the Universities of Flensburg and Jena; cf., e.g., Heering et al. (eds. 2000), Staubermann (2007) and Breidbach et al. (eds. 2010).

LIST OF ABBREVIATIONS

AIP American Institute of Physics, College Park, Maryland
ASCE American Society of Civil Engineers, New York
ATP adenosine triphosphate (a coenzyme)
BJHS *British Journal for the History of Science*, Cambridge
CAD computer-aided design programs/detection
CAT computed axial tomography
CCD charge-coupled device
CNC computer-numerically controlled machines
CRAS *Comptes Rendus hebdomadaires de l'Académie des Sciences*, Paris
CT see CAT
DNA deoxyribonucleic acid (of genes)
DSI Database of Scientific Illustrators, University of Stuttgart
EM electron microscopy
EMI Electric and Musical Industries, Hayes and London
FMRI functional magnetic resonance imaging
FONAR field focusing nuclear magnetic resonance imaging
HSPS *Historical Studies in the Physical (and Biological) Sciences*, Berkeley
IWF Institut für den Wissenschaftlichen Film, Göttingen
JHA *Journal for the History of Astronomy*, Cambridge
JWCI *Journal of the Warburg and Courtauld Institutes*, London
MIT Massachusetts Institute of Technology, Cambridge, Massachusetts
MNRAS *Monthly Notices of the Royal Astronomical Society*, London
MRI magnetic resonance imaging
MRT magnetic resonance tomography
NASA National Aeronautics and Space Administration, Washington, DC
NMR nuclear magnetic resonance
NSF National Science Foundation, Arlington, Virginia
PET positron emission tomography
Phil. Mag. *The London, Edinburgh and Dublin Philosophical Magazine*
PSP pattern–structure–process scheme (Gooding)
PTRSL *Philosophical Transactions of the Royal Society of London*
RAS Royal Astronomical Society, London
SEM scanning electron microscopy
SHPS *Studies in the History and Philosophy of Science*
STM scanning tunneling microscopy
TEM transmission electron microscopy

Recommended pathways into the secondary literature

This book ends with a relatively long list of select primary and secondary literature. Given the fact that in early 2014 the international WorldCat database lists more than 125,000 separate publications under the keywords 'visual culture' in English alone,[3] this bibliography of circa 2,000 entries can by no means be complete. It is necessarily limited to what has been used and consulted by me while writing this book. Nevertheless, this literature cited is perhaps still too much for a beginner or student seeking quick orientation in this topic. Further guidance might well be imperative, since – as I explained in the introduction – unfortunately many pertinent secondary texts turn out to be rather disappointing. Therefore I have decided to precede my detailed bibliography by a few short personal recommendations on introductory texts that I found helpful and inspiring for readers interested in the broad array of visual representations used in various sciences and technologies, their changes over the course of history, as well as their technical production and preconditions, their communicative functions and cultural resonance.

Normally, the first inclination would be to search for compilations of carefully selected classics in the field, but the two standard readers on the market, Mirzoeff (ed. 1998) and Evans & Hall (eds. 1999) are perhaps suitable for students of cultural and media studies with strong 'postmodern' inclinations, but fairly disappointing to historians of science, technology or medicine.[4] For these latter, I rather recommend *The Nineteenth-Century Visual Culture Reader*.[5] If I had to choose only one introductory text, I would opt for Luc Pauwels's *Visual Cultures of Science: Rethinking Representational Practices in Knowledge Building and Science Communication* (2005). As a second choice, I would add the volume edited by Horst Bredekamp, Birgit Schneider & Vera Dünkel (2008): *Das Technische Bild. Kompendium zu einer Stilgeschichte wissenschaftlicher Bilder*, with the obvious limitation that it is published in German with no English translation available. While the first of these two introductory anthologies is structured by areas of application and discipline used in visual culture studies, the second is rather structured by methodology and topic in a very original, inimitable way. On visual research methods, the most comprehensive introduction was published by Gillian Rose in 2001 (but get the

---

[3]See www.worldcat.org/search?qt=worldcat_org_all& q=visual+culture ; the number refers to my search on April 15, 2011 yielding 42,742 entries; by Sept. 11, 2013, this total had increased to 83,020, and by Feb. 16, 2014 to 126,049.

[4]The second of these two readers, embellished with a Mapplethorpe photograph on its cover, is quite explicitly postmodern; cf. Evans & Hall (eds. 1999) p. iv.

[5]See Vanessa Schwartz & Jeannene Przyblyski (eds. 2004), therein esp. papers 1, 3, 13, 15, 18.

expanded 3rd edition of 2012); The *Sage Handbook of Visual Research Methods*, edited by Eric Margolis and Luc Pauwels in 2011, strongly focusses on applications in the social sciences.

For surveys on the changes in the uses and preferred types of images in different historical contexts, I refer readers to Brian Ford's *Images of Science* (1992) or Harry Robin's *The Scientific Image* (1992) as two very condensed versions covering the whole history "from Plato to NATO," so to speak, or, as Robin put it in his subtitle, *from Cave to Computer*, in a single volume of some 200 pages. Alternatively, one can turn to the five-volume series *Album of Science* under the editorship of I. B. Cohen. Volume 1 covers visual representations in antiquity and the Middle Ages (with a strong inclination toward the latter, induced by this volume's editor John E. Murdoch and appeared in 1984. Volume 2 covers the Renaissance and early-modern period up to 1800 (edited by I. B. Cohen himself). Volume 3 treats the 19th century (edited by L. Pearce Williams). Volumes 4 and 5 cover the 20th century, split into the physical sciences and the biological sciences (edited by Owen Gingerich and Merriley Borell, respectively). Noteworthy selections of impressive technical illustrations are offered by Baynes & Pugh (1981) and VEB (1989).

Among the wealth of other analyses of scientific illustrations in the 20th century, the volume by Martin Kemp (2000) on images from *Nature* deserves special mention, since he tackles a broad array of images taken from that leading science journal and analyzes them from the perspective of an art historian, specializing on scientific images from Leonardo da Vinci to the millennium simulation of cosmological evolution. Needless to say, various other publications by Kemp – like those of his art-historian colleagues Ernst Gombrich, Horst Bredekamp and Samuel Edgerton – are also highly commendable.

Some of the readers of this book might be interested in more detailed analyses of images in specific disciplines or research areas. I could not possibly cover all specialties, of course, as there are endlessly many of them. I would like to pick out a few outstanding studies here as masterful in breadth and depth. The first of these truly exemplary, paradigmatic studies is Ann Shelby Blum's *Picturing Nature. American Nineteenth-Century Zoological Illustration* (1993). It is a model of its kind for her coverage of issues of technical production as well as representational adequacy and expression, the interplay of artisan-illustrators and naturalists or scientists, and printing, as well as issues of reception. Other excellent studies on this interplay include Lapage (1961), Knight (1977) and Kusukawa (2012). For botany, we have richly illustrated anthologies of historical illustrations, such as Blunt (1950) and Lack (2001), detailed compilations of illustrated books, such as Claus Nissen

(1950, 1951/66, 1969/78) and Magee(2009), and the excellent analyses of copy relations between botanical illustrations by Kärin Nickelsen (2000–2006), to be used side by side. Kevles (1997) provides a good survey on medical imaging in the 20th century, whereas Cartwright (1995) exemplifies the feminist approach, nowhere more important than in this particular field. For spectroscopy, I have tried to analyze all pertinent sources in Hentschel (2002*a*); for optical microscopy, I recommend Wilson (1995) and Schickore (2007); for 20th-century variants of it, see, e.g., Rasmussen (1997), Baird & Shew (2004), Hennig (2005), Granek & Hon (2008), Mody (2006) and Mody & Lynch (2010); for holography, see Johnston (2005–2006); and for astronomy, see Bajac (2010). In 1999, Axel Wittmann and I coedited an anthology of specialized papers, including a broad historiographic survey that will be useful to those seeking further advice on the literature of this blatantly visual science culture. Popular images of astronomy have been analyzed by Susanne Utzt (2004).

As far as photography is concerned – both as a representational technique and as an art form in its own right – the secondary literature is so huge and dispersed that a selection is really difficult and remains somewhat arbitrary. With respect to scientific photography, I find the following books most noteworthy and useful: Darius (1984) and Gaede (ed. 1999) as samplers of scientific photographs; Thomas (1997) and Keller, Faber & Gröning (2009) for historical analyses of scientific photography; and Schaaf (1992) as an outstanding example of studies close to primary source materials, in this case the papers, letters and early photographic experiments of Fox Talbot and John Herschel, two pioneers of photographic processes in the mid 19th century.

The complex terrain of color schemes and encodings is explored in Sherman (1981), Stromer (ed. 1998) and Spillmann (2010). On the long and arduous training of visuality, see Ferguson (1992) for contexts of technical education, Dommann (2003) on radiology, and furthermore Elkins (2000) on "How to use your eyes" in 32 other visual domains ranging from x-ray images to perspective picture, and from moth wings to fingerprints.

More specialized texts are summarized in a helpful way in three essay reviews: two by Nikolow & Bluma (2002) and by Cornelius Borck (2009) on visualization and the public, the other by myself, focusing on texts dealing with 'image practice' in science and technology (Hentschel 2011*a*).

I hope that these brief remarks are helpful as initial guidance through this dense forest of secondary literature on visual culture.

BIBLIOGRAPHY

Abbott, Alison: Visual zoology. Historic zoological wall-charts in Pavia, *Nature* 421 (2003): 580.

Ackermann, James S. (1978) Leonardo's eye, *JWCI* 41: 108–46 & pl.

Adelmann, Ralf, Jan Frercks, Martina Hessler and Jochen Hennig (2009) *Datenbilder. Zur digitalen Bildpraxis in den Naturwissenschaften*, Bielefeld: Transcript.

Alberti, Leon Battista (1440) *De pictura*. (*a*) mss. 1435; (*b*) 1st ed. Basel; (*c*) in German transl. by Hubert Janitschek in *Kleinere kunsttheoretische Schriften*, Vienna: Braumüller, 1877; (*d*) *On Painting and On Sculpture*, ed. & transl. by Celic Grayson, London 1972; (*e*) *On Painting*, ed. & transl. by C. Grayson & Martin Kemp, London 1991.

Alertz, Ulrich (1991) *Vom Schiffbauhandwerk zur Schiffbautechnik: Die Entwicklung neuer Entwurfs- und Konstruktionsmethoden im italienischen Galeerenbau (1400–1700)*, Hamburg: Kovač.

Alpers, Svetlana (1976/77) Describe or narrate? A problem in realist representation, *New Literary History* 8: 15–41.

— (1978) Seeing as knowing: A Dutch connection, *Humanities in Society* 1: 147–73.

— (1983) *The Art of Describing: Dutch Art in the Seventeenth Century*, Chicago: University of Chicago Press.

— (1996) Visual culture questionnaire, *October* 77: 26.

Amann, Klaus & Karin Knorr-Cetina (1988) The fixation of (visual) evidence, *Human Studies* 11: 133–69.

Ammann, Othmar H. (1966) Verrazano-Narrows bridge: conception of design and construction procedure, *Journal of the Construction Division. Proceedings of the ASCE* 92: 5–21.

Ancient Sunspot Records Research Group (1977) A re-compilation of our country's records of sunspots through the ages, *Chinese Astronomy* 1: 347–59.

Andersen, Kirsti (2007) *The Geometry of an Art. The History of the Mathematical Theory of Perspective from Alberti to Monge*, Berlin: Springer.

Andrew, Edward Raymond (1955) *Nuclear Magnetic Resonance*, Cambridge: Cambridge University Press.

— (1980) N.m.r. imaging of intact biological systems, *PTRSL* B289: 471–81.

— (1984) A historical review of NMR and its clinical applications, *British Medical Bulletin* 40, no. 2: 115–19.

Anon. (1882) Alexander Nasmyth, *Art Journal* 34: 208–9 & pl.

— (1883) Review of Nasmyth (1883), *Quarterly Review* 155: 389–419.

— (1890) The mechanical inventions of James Nasmyth, *The Engineer* 16 & 23 May: 406–7, 426.

— (1911) Microkinematography, *Nature* 88, no. 2198: 213–15.

— (1969) Stiffening systems vs. aerodynamics, *Civil Engineering – ASCE* 39,6: 61.

— (1981) First clinical trials of diagnostic NMR, *Radiology/Nuclear Medicine Magazine* June: 8–12.

— (1993) *The Molecular Art of Irving Geis*, Cambridge, MA: MIT.

Anschütz, Richard (1909) Life and chemical work of Archibald Scott Couper, *Proceedings of the Royal Society of Edinburgh* 29: 193–273.

— (1929) *August Kekulé*, vol. 1: *Leben und Wirken* and vol. 2: *Abhandlungen, Berichte, Kritiken, Artikel, Reden*, Berlin: Verlag Chemie.

Arago, Dominique-François-Jean (1854ff.) *Oeuvres Compètes*, Paris: Gidot.

Araya, Agustin A. (2003) The hidden side of visualization, *Techné* 7, no. 2: 27–93.

Archer, Mildred (1962) *Natural History Drawings in the India Office Library*, London: HMSO.

Ardenne, Manfred von (1938) Die Grenzen für das Auflösungsvermögen des Elektronenmikroskops, *Zeitschrift für Physik* 108: 338–52.

— (1940) *Elektronen-Übermikroskopie: Physik, Technik, Ergebnisse*, Berlin: Springer.

Armstrong, J. A. (2007) Urinalysis in Western culture: a brief history, *Kidney International* 71: 384–7.

Arnheim, Rudolf (1969) *Visual Thinking*, (*a*) Berkeley: University of California Press; (*b*) in German transl.: *Anschauliches Denken*, Cologne: DuMont, 1972, (*c*) rev. ed. with new foreword, ibid. 1996.

Arnold, J.T., S.S. Dharmatti & M.E. Packard (1951) Chemical effects on nuclear induction signals from organic compounds, *Journal of Chemical Physics* 19: 507.

Artz, Frederick B. (1966) *The Development of Technical Education in France 1500–1800*, Cambridge, MA: MIT Press.

Asendorf, Christoph (1989) *Ströme und Strahlen: Das langsame Verschwinden der Materie um 1900*, Giessen: Anabas (Werkbund-Archiv no. 18).

Aubin, David (2008) 'The memory of life itself': Bénard cells and the cinematography of self-organization, *SHPS* A39: 359–69.

Bach, Carl & Richard Baumann (1914) *Festigkeitseigenschaften und Gefügebilder der Konstruktionsmaterialien*, Berlin: Springer, 1st ed.; 2nd exp. ed. 1921.

Baigrie, Brian Scott (1996) *Picturing Knowledge. Historical and Philosophical Problems Concerning the Use of Art in Science*, Toronto: University of Toronto Press.

Baird, Davis (1989) Instruments on the cusp of science and technology: the indicator diagram, *Knowledge and Society* 8: 107–22.

Baird, D. & A. Shew (2004) Probing the history of scanning tunneling microscopy, in: D. Baird, J. Schummer & A. Nordmann (eds.) *Discovering the Nanoscale*, Amsterdam: 145–56.

Bajac, Quentin (2010) *Dans le champ des étoiles–les photographies et le ciel*, Paris: RMN.

Bal, Mieke (2003) Visual essentialism and the object of visual culture, *Journal of Visual Culture* 2,1: 5–32; Critical commentaries on this paper, ibid., 2,2: 229–60, and a reply by M. Bal, ibid., 260–8.

Balázs, Béla (1924) *Der sichtbare Mensch oder die Kultur des Film*, Vienna: Deutsch-Österreichischer Verlag; reprint Frankfurt: Suhrkamp, 2001.

Baldasso, Renzo (2006) The role of visual representation in the scientific revolution: a historiographic inquiry, *Centaurus* 48,2: 69–88.

Ball, Philip (1999) *The Self-Made Tapestry. Pattern Formation in Nature*, Oxford: Oxford University Press.

Bann, Stephen (1995) Shrines, curiosities and the rhetoric of display, in Lynn Cooke & Peter Wollen (eds.) *Visual Display. Culture beyond Appraisal*, Seattle: Bay Press, 14–29.

Banta, H. David (1984) Embracing or rejecting innovations: clinical diffusion of health care technology, in: *The Machine at the Bedside*, Cambridge: Cambridge University: 64–94.

Barclay, A. E. (1949) The old order changes, *British Journal of Radiology* 22: 300–6.

Barley, Stephen R. (1986) Technology as an occasion for structuring evidence from observation of CT scanners and the social order of radiology departments, *Administrative Science Quarterly* 31: 78–108.

Barringer, David (2006) Raining on evolution's parade, *I.D. Magazine* 53, no. 2 March/April: 58–65.

Bartholomew, C. F. (1976) The discovery of solar granulation, *Quarterly Journal of the Royal Astronomical Society* 17: 263–89 & pl. I–IV.

Bauer, Malcolm I. & P. N. Johnson-Laird (1993) How diagrams can improve reasoning, *Psychological Science* 4,6: 372–8.

Baumann, Richard (1920) Metallographie, in: Otto Lueger (ed.) *Lexikon der gesamten Technik und ihrer Hilfswissenschaften*, Stuttgart/Leipzig, suppl. 1: 449–54.

Baumann-Schleihauf, Susanne (2001) Zum 500. Geburtstag: Kräuterbücher und die Fuchsie erinnern an Leonhart Fuchs, *Pharmazeutische Zeitung Online* 6, http://www.pharmazeutische-zeitung.de/index.php?id=titel_06_2001.

Baume Pluvinel, A. de la (1908) Jules César Janssen, *Astrophysical Journal* 28: 89–99.

Baxandall, Michael (1972) *Painting and Experience in Fifteenth-Century Italy: A Primer in the Social History of Pictorial Style*, (*a*) Oxford: Clarendon Press; (*b*) German transl. *Die Wirklichkeit der Bilder. Malerei und Erfahrung im Italien des 15. Jahrhunderts*, Frankfurt: Athenäum, 1987.

Baynes, Ken & Francis Pugh (1981) *The Art of the Engineer*, Surrey: Lutterworth and New York: Overlook.

Beale, Lionel Smith (1854) *The Microscope, and Its Application to Clinical Medicine*, London: Highley; 4th ed. under the new title: *How to Work With the Microscope*, London: Harrison 1868.

Beam, C. M., P. M. Layde & D. S. Sullivan (2002) Variability in the interpretation of screening mammograms by U.S. radiologists, *Archives of Internal Medicine* 156: 209–13.

Beaulieu, Anne (2002) Images are not the (only) truth: brain mapping, visual knowledge, and iconoclasm, *Science, Technology and Human Values* 27: 53–86.

Beck, Friedrich (1989) Max von Laue, in: Klaus Bethge & Horst Klein (eds.) *Physiker und Astronomen in Frankfurt*, Frankfurt am Main: Alfred Metzner, 24–37.

Beham, Hans Sebald (1528) *Dises buchlein zeyget an und lernet ein maß oder proporcion der ross: nutzlich iungen gesellen, malern und goldschmiden.* Nürnberg: Peypus.

Belhoste, Bruno, Amy Dahan Dalmedico & Antoine Picon (eds. 1994) *La formation polytechnicienne, 1794–1994*, Paris: Dunod.

Bell, Trudy (2007) Roger Hayward: Forgotten artist of optics, *Sky & Telescope* 114,3: 30–7.

Bénard, Henri (1900) Les tourbillons cellulaires dans une nappe liquide, *Revue Générale des Sciences* 11: 1261–71, 1309–28.

Benfey, Theodor (ed. 1966) *Kekulé Centennial*, Washington: American Chemical Society.

Beniger, James R. & Dorothy L. Robyn (1978) Quantitative graphics in statistics: a brief history, *The American Statistician* 32,1: 1–11.

Berger, Peter (1959) Johann Heinrich Lamberts Bedeutung in der Naturwissenschaft des 19. Jahrhunderts, *Centaurus* 6: 157–254.

Besler, Basilius (1713) *Hortus Eystettensis*, Eichstätt; reprint Munich: Kölbl, 1964.

Beyen, H. G. (1939) Die antike Zentralperspektive, *Jahrbuch des Deutschen Archäologischen Instituts* 54: 47–72.

Biagioli, Mario (2006) *Galileo's Instruments of Credit: Telescopes, Images, Secrecy*, Chicago: University of Chicago Press.

Bičák, Jiři (1978/2009) The art of science: interview with Professor John Archibald Wheeler conducted in 1978, published in English in 2009: *General Relativity and Gravitation* 41: 679–89.

Bigg, Charlotte (2008) Evident atoms: visuality in Jean Perrin's Brownian motion research, *SHPS* A 39, no. 3: 312–22.

— (2011) A visual history of Jean Perrin's Brownian motion curves, in: Lorraine Daston & Elizabeth Lunbeck (eds.) *Histories of Scientific Observation*, Chicago: University of Chicago Press: 156–79.

Bigg, Charlotte & Jochen Hennig (eds. 2009) *Atombilder. Ikonografien des Atoms in Wissenschaft und Öffentlichkeit des 20. Jahrhunderts*, Göttingen: Wallstein.

Bigourdan, G. (1908) J. Janssen, *Bulletin Astronomique* 25: 49–58.

Billington, David P. (1977) History and aesthetics in suspension bridges, *Journal of the Structural Division* (ASCE) 103: 1665–72 and the discussion, ibid. 104 (1978): 246–9, 378–80, 619, 732f., 1027–35, 1174–6; 105 (1979): 671–84.

— (1983) *The Tower and the Bridge: The New Art of Structural Engineering*, New York: Basic Books.

Bilodreau-Guinamard, Bénédicte (1997) Les manuscrits de René-Just Haüy conservés à la bibliothèque centrale du Muséum national d'histoire naturelle de Paris, *Revue d'Histoire des Sciences* 50 (1997): 335–53.

Binnig, Gerd & Heinrich Rohrer (1985) The scanning tunneling microscope, *Scientific American* 253: 50–6.

— (1987) Scanning tunneling microscopy, *Reviews of Modern Physics* 59: 615–25.

— Christoph E. Gerber & Edmund Weibel (1983) 7 × 7 reconstruction on Si (111) resolved in real space, *Physical Review Letters* 50, no. 2: 120–3.

Biringuccio, Vannoccio (1540): *Pirotechnia*, (*a*) Venice; (*b*) Engl. transl. by Mrs. Gnudi & C. S. Smith, Cambridge, MA: MIT Press, 1966.

Blackmore, John T. (1972) *Ernst Mach: His Work, Life and Influence*, Berkeley: University of California Press.

Blakemore, Michael & J. B. Harley (1980) Concepts in the history of cartography, *Cartographica* 17,4: 1–120.

Blanford, H. F. (1878) Janssen's new method of solar photography, *The British Journal of Photography* 25: 543–5.

Bleichmar, Daniela (2012) *Visible Empire: Botanical Expeditions and Visual Culture in the Hispanic Enlightenment*, Chicago: University of Chicago Press.

Bloch, Felix & W. W. Hansen & M. E. Packard (1946) The nuclear induction experiment, *Physical Review* 70: 474–85.

Blondel-Mégrelis, Marika (1981) Le modèle et la théorie, analyse d'un exemple – la crystallographie de Haüy, *Revue Philosophique de la France et de l'Étranger* 171: 283–96.

Bloom, Terrie F. (1978) Borrowed perceptions: Harriot's map of the moon, *Journal for the History of Astronomy* 9: 117–22.

Blossfeld, Karl (1936) *Urformen der Kunst*, Berlin: Wasmuth, 1936a; in Engl. transl.: *Art Forms in Nature*, Schirmer/Mosel, 1994.

Blum, Ann Shelby (1993) *Picturing Nature. American Nineteenth-Century Zoological Illustration*, Princeton, NJ: Princeton University Press.

Blum, Martina (1999) Mediating technology: the feminization of radiographers, *Working Paper. Munich Center for the History of Science and Technology*.

Blume, Stuart S. (1992) *Insight and Industry. On the Dynamics of Technological Change in Medicine*, Cambridge, MA: MIT Press.

Blunck, Jürgen (1995/96) Die Geschichte der Globen des Mars und seiner Monde, *Der Globusfreund* 43/44: 257–64.

— (1999) Deutsche Pionierarbeiten der Monddarstellung mit erhabenem Relief, *Der Globusfreund* 47/48: 293–303.

Blunt, Wilfrid (1950) *The Art of Botanical Illustration. An Illustrated History*, (*a*) London: Collins, 1st ed.; (*b*) 3rd ed. 1955 (reprint 1970); (*c*) exp. 4th ed., New York: Dover 1994.

— (1979) *The Illustrated Herbal*, London: Thames & Hudson/Metropolitan Museum of Art.

Bock, Hieronymous (1539) *Das Kreütter Buch, Darinn Underscheidt, Namen vnnd Würckung der Kreutter, Stauden, Hecken vnnd Beumen, sampt jhren Früchten, so inn Deutschen Landen wachsen*, (*a*) Straßburg, 1st ed.; (*b*) 2nd illustrated ed. 1546; (*c*) Latin transl. 1552;

Bodmer, George R.(1997) Technical illustration of Huxley, in: *Thomas Henry Huxley's Place in Science*, Athens: University of Georgia Press 1997: 277–95.

Boehm, Gottfried (1994) *Was ist ein Bild?* Munich: Fink.

— (1995) *Beschreibungskunst – Kunstbeschreibung: Ekphrasis von der Antike bis zur*

Gegenwart, Munich: Fink.

— (2007) *Wie Bilder Sinn erzeugen*, Berlin: Berlin University Press.

Bogen, Steffen & Felix Thürmann (2003) Jenseits der Opposition von Text und Bild. Überlegungen zu einer Theorie des Diagramms und des Diagrammatischen. In: Alexander Patschovsky (ed.) *Die Bildwelt der Diagramme Joachims von Fiore*, Ostfildern: 1–22.

Bonhoff, Ulrike Maria (1993) *Das Diagramm. Kunsthistorische Betrachtung über seine vielfältige Verwendung von der Antike bis zur Neuzeit*, PhD, University of Münster.

Booker, Peter Jeffrey (1961) Gaspar Monge (1746–1818) and his effect on engineering drawing and technical education, *Newcomen Society for the Study of the History of Engineering and Technology* 34: 15-34.

— (1963) *A History of Engineering Drawing*, London: Chatto & Windus.

Boon, Timothy (2008) *Films of Fact. A History of Science in Documentary Films and Television*, London & New York: Wallflower.

Bopp, Karl (ed. 1915) Johann Lamberts Monatsbuch, *Abhandlungen der königlich-bayerischen Akademie der Wissenschaften* 27, no. 6.

Borck, Cornelius (2009) Bild der Wissenschaft. Neuere Sammelbände zum Thema Visualisierung und Öffentlichkeit, *NTM* 17, no. 3: 317–27.

Borrell, Merriley (1986) Extending the senses: the graphic method, *Medical Heritage* 2: 114–21.

— (1989) *Album of Science. The Biological Sciences in the Twentieth Century*, New York: Scribner's.

Borsig, Tet Arnold von & Adolf Portmann (1961) *Verborgene Kunstformen*, Berlin: Herbig.

Bottomley, Paul A., Howard R. Hart, William A. Edelstein et al. (1983) NMR imaging/spectroscopy system to study both anatomy and metabolism, *Lancet* 2: 273–4.

Bracegirdle, Brian (1978) *A History of Microtechnique: The Evolution of the Microtome and the Development of Tissue Preparations*, (*a*) 1st. ed. Ithaca: Cornell University Press; (*b*) 2nd ed. London: Science Heritage, 1987.

Bracewell, Ronald H. (1956) Strip integration in radio astronomy, *Australian Journal of Physics* 9: 198–217.

Bracewell, R. H. & A. C. Riddle (1967) Inversion of fan beam scans in radio astronomy, *Astrophysical Journal* 150: 427–34.

Bradley, Margaret (1976) A scientific institution for a new society: the École polytechnique 1794–1830, *History of Education* 5: 11–24.

— (1981) Franco-Russian engineering links: the careers of Lamé and Clapeyron, 1820–1830, *Annals of Science* 38: 291–312.

Bragg, William Lawrence (1913) The diffraction of short electromagnetic waves by a crystal, *Proceedings of the Cambridge Philosophical Society* 17: 43–57 & pl. II.

— (1914) The analysis of crystals by the x-ray spectrometer, *Proceedings of the Royal Society London* A 89: 468–89.

— (1933ff.) *The Crystalline State: A General Survey*, London: Bell, 1933 (vol. 1) 1937 (vol. 4); 2nd exp. ed. 1965.

— (1945) *The History of X-Ray Analysis*, London: Longmans, Green & Co.

Bragg, Sir William Henry & Prof. W. L. Bragg (eds.) 1928/1930: *Stereoscopic Photographs of Crystal Models*, London: Adam Hilger Ltd., (a) 1st series 1928; (b) 2nd ser. 1930 (see also review by Herlinger 1928).

Brain, Robert Michael (1996) *The Graphic Method. Inscription, Visualization, and Measurement in Nineteenth-Century Science and Culture*, PhD University of California.

Brain, R. M. & Norton Wise (1994) Muscles and engines: Indicator diagrams and Helmholtz's graphical methods, in: Lorenz Krüger (ed.) *Universalgenie Helmholtz: Rückblick nach 100 Jahren*, Berlin: Akademie-Verlag: 124–45.

Braitenberg, Valentin (1984) Tentakeln des Geistes, *Kursbuch* 78, Dec.: 35–45.

Brand, Stewart, Kevin Kelly & Jay Kinney (1985) Digital retouching, *Whole Earth Review* 44: 42–9.

Brandes, Dietmar (1987) *Paradiesgarten der Botanik: Alte Herbarien*, Braunschweig: Institut für Pflanzenbiologie.

Brandstetter, Thomas (2011) Time machines – model experiments in geology, *Centaurus* 53: 135–45.

Braun, Ludwig (1898) Ueber den Werth des Kinematographen für die Erkennisse der Herzmechanik, *Verhandlungen der Gesellschaft Deutscher Naturforscher und Ärzte* 69: 185–8.

Braun, Marta (1992) *Picturing Time. The Work of Etienne-Jules Marey (1830–1904)*, Chicago: University of Chicago Press.

Braunmühl, Anton von (1891) *Christoph Scheiner als Mathematiker, Physiker und Astronom*, Bamberg: Buchner.

Brecher, Ruth & Edward M. Brecher (1969) *The Rays: History of Radiology in the United States and Canada*, Baltimore: Williams & Wilkins.

Bredekamp, Horst (1997) Das Bild als Leitbild. Gedanken zur Überwindung des Anikonismus, in: Ute Hoffmann, Bernward Joerges & Ingrid Severin (eds.) *LogIcons: Bilder zwischen Theorie und Anschauung*, Berlin: Sigma: 225–43.

— (1999) *Thomas Hobbes visuelle Strategien. Der Leviathan: Urbild des modernen Staates. Werkillustrationen und Portraits*, Berlin: Akademie-Verlag.

— (2003) A neglected tradition? Art history as Bildwissenschaft, *Critical Inquiry* 29: 418–28.

— (2004) *Die Fenster der Monade. Gottfried Wilhelm Leibniz' Theater der Natur und Kunst*, Berlin: Akademie-Verlag.

— (2005*a*) *Darwins Korallen. Die frühen Evolutionsdiagramme und die Tradition der Naturgeschichte*, Berlin: Wagenbach.

— (2005*b*) Denkende Hände – Überlegungen zur Bildkunst der Naturwissenschaften, in: Monika Lössl & Jürgen Mittelstraß (eds.) *Von der Wahrnehmung zur Erkenntnis*, Berlin: Springer: 109–32.

— (2005*c*) Die zeichnende Denkkraft. Überlegungen zur Bildkunst der Naturwissenschaften, *Interaktionen* 14: 155–71.

— (2007) *Galilei der Künstler. Der Mond, die Sonne, die Hand*, Berlin: Akademie-Verlag.

— (2011) Über die Unabschließbarkeit der künstlerischen Medien, in: *Wissenschaft und Kunst. HRK Jahresversammlung 2011*: 19–59.

Bredekamp, H., Jochen Brüning & Cornelia Weber (eds. 2000) *Theatrum Naturae et Artis. Theater der Natur und Kunst*, Berlin: Henschel (3 vols.).

Bredekamp, Horst & Franziska Brons: Fotografie als Medium der Wissenschaft. Kunstgeschichte, Biologie und das Elend der Illustration, in: Christa Maar & Hubert Burda (eds. 2004) *Iconic turn. Die neue Macht der Bilder*, Cologne: Dumont: 365–81.

Bredekamp, Horst & Pablo Schneider (eds. 2006) *Visuelle Argumentationen: Die Mysterien der Repräsentation und die Berechenbarkeit der Welt*, Munich: Fink.

Bredekamp, Horst, Gabriele Werner & Angela Fischel (eds. 2003*a*) *Bildwelten des Wissens. Kunst-historisches Jahrbuch für Bildkritik* 1, no. 1.

— (2003*b*) Bildwelten des Wissens, ibid.: 9–20.

Bredekamp, Horst, Birgit Schneider & Vera Dünkel (eds. 2008) *Das Technische Bild. Kompendium zu einer Stilgeschichte wissenschaftlicher Bilder*, Berlin: Akademie-Verlag.

Breger, Dee (1995) *Journeys in Microspace: The Art of the Scanning Electron Microscope*, New York: Columbia University Press.

Breidbach, Olaf (2002) Representation of the microcosm: the chain of objectivity in 19th c. scientific microphotography, *Journal of the History of Biology* 35: 221–50.

— (2005) 'Schattenbilder.' Zur elektronenmikroskopischen Photographie in den Biowissenschaften, *Berichte zur Wissenschaftsgeschichte* 28: 160–71.

— (2006) *Visions of Nature: The Art and Science of Ernst Haeckel*, Munich: Prestel.

Breidbach, O., Peter Heering, Matthias Müller & Heiko Weber (eds. 2010) *Experimentelle Wissenschaftsgeschichte*, Munich: Fink.

Breidbach, O. & André Karliczek (2011) Himmelsblau – das Cyanometer des Horace-Bénédict de Saussure (1740–1799), *Sudhoffs Archiv* 95: 1–29.

Breidbach, O., Kerrin Klinger & Matthias Müller (2013) *Camera Obscura: Die Dunkelkammer in ihrer historischen Entwicklung*, Stuttgart: Steiner.

Brewster, David (1830) Cyanometer, *Edinburgh Encyclopaedia* 7: 593.

— (1856) *The Stereoscope: Its History, Theory and Construction*, London: Murray.

Bridson, Gavin & Geoffrey Wakeman (1984) *Printmaking & Picture Printing. A Bibliographical Guide to Artistic & Industrial Techniques in Britain 1750–1900*, Oxford: Plough Press.

Bridson, G. & Donald W. Wendel (1986) *Printmaking in the Service of Botany*, Hunt Institute for Botanical Research.

Bridson, G. & James J. White (1990) *Plant, Animal and Anatomical Illustration in Art & Science*, St Paul: Bibl.

Briggs, Derek E. G. & D. Collins (1988) A Middle Cambrian chelicerate from Mount Stephen, British Columbia, *Palaeontology* 31: 779–98.

Briggs, D. E. G. & Simon Conway Morris (1983) New Burgess shale fossil sites reveal middle Cambrian faunal complex, *Science* 222: 163–7.

Briggs, D. E. G. & S. Henry Williams (1981) The restoration of flattened fossils, *Lethaia* 14: 157–64.

Brinkhus, Gerd (2001) *Leonhart Fuchs (1501–1566), Mediziner und Botaniker*, Tübingen: Stadtmuseum Tübingen.

Brisson, Barnabé (1818) *Notice historique sur Gaspard Monge*, Paris: Plancher.

Brodie, Benjamin C. (1866) The calculus of chemical operations, *PTRSL* 156: 781–859.

— (1867) On the mode of representation afforded by the chemical calculus as contrasted with the atomic theory, *Chemical News* 15: 295–305.

Brown, Alexander C.: see Crum Brown

Brownson, C. D. (1981) Euclid's optics and its compatibility with linear perspective, *Archive for History of Exact Sciences* 24,3 (1981): 165–94.

Brüche, Ernst (1943) Zum Entstehen des Elektronenmikroskops, *Physikalische Zeitschrift* 44: 176–80.

— (1957) 25 Jahre Elektronenmikroskop, *Die Naturwissenschaften* 44: 601–10.

Brück, Hermann Alexander & Mary Teresa Brück (1988) *The Peripathetic Astronomer. The Life of Charles Piazzi Smyth*, Bristol: Adam Hilger.

Brück, Mary Teresa (1988) The Piazzi Smyth collection of sketches, photographs, and manuscripts at the Royal Observatory, Edinburgh, *Vistas in Astronomy* 32: 371–408.

Brunfels, Otto (1530–6) (*a*) *Herbarum vivae icones*, ... (*b*) German transl. of the first part of 1530: *Contrafayt Kreuterbuch*, Straßburg: Han Schotten zum Thyergarten, 1532 (*c*) Ander Teyl des Teutschen Contrafayten Kreuterbuchs, ibid. 1537; (*d*) reprint of (*a, b*) Grünwald near Munich: Kölbl, 1964.

Brusius, Miriam, Katrine Dean & Chitta Ramalingam (eds. 2013) *William Henry Fox Talbot Beyond Photography*, New Haven & London: Yale University Press.

Bryant, John & Chris Sangwin (2008) *How Round is Your Circle? Where Engineering and Mathematics Meet*, Princeton: Princeton University Press.

Bryson, Norman, Michael Ann Holly & Keith Moxon (eds. 1994) *Visual Culture. Images and Interpretations*, Hanover & London. Wesleyan University Press.

Buchwald, Jed Z. (1989) *The Rise of the Wave Theory of Light. Optical Theory and Experiment in the Early Nineteenth Century*, University of Chicago Press.

Bud, Robert & Deborah Warner (eds. 1998) *Instruments of Science. An Historical Encyclopedia*, London: Routledge.

Buée, Abbé (1804) A letter to Mr. ∗ ∗ ∗, on M. Romé de l'Isle's and the Abbé Haüy's theories of crystallography, *Phil. Mag.* 19: 159–72.

Burke, J. G. (1966) *Origins of the Science of Crystals*, Berkeley: University of California Press.

Burke, Peter (1972) *Culture and Society in Renaissance Italy*, London: Batsford.

Burri, Regula Valérie (2000) MRI in der Schweiz. Soziotechnische, institutionelle und medizinische Aspekte der Technikdiffusion eines bildgebenden Verfahrens, *Preprints zur Kulturgeschichte der Technik*, no. 10 (also online).

— (2008) *Doing Images: Zur Praxis Medizinischer Bilder*, Bielefeld: Transcript (Technik – Körper – Gesellschaft, vol. 2).

Burri, R. V. & Joseph Dumit (2007) Social studies of scientific imaging and visualization, in: Edward Hackett et al. (eds.) *The Handbook of Science and Technology Studies*, Cambridge, MA: MIT Press, new ed.: 297–317.

Byard, Margaret M. (1988) A new heaven: Galileo and the artists, *History Today* 38: 30–8.

Cahn, Robert (2001) *The Coming of Materials Science*, Oxford: Pergamon Press.

Callendar, H. L. (1911) The caloric theory of heat and Carnot's principle, *Proceedings of the London Physical Society* 23: 153–89.

Calmann, Gerta (1977) *Ehret, Flower Painter Extraordinary. An Illustrated Biography*, Oxford: Phaidon.

Cambrosio, Alberto, Daniel Jacobi & Peter Keating (1993) Ehrlich's "beautiful pictures" and the controversial beginnings of immunological imagery, *Isis* 84: 662–99.

— (2005) Arguing with images: Pauling's theory of antibody formation, *Representations* 89: 94–130.

Campbell, W. W. (1894) The Spectrum of Mars, *Astronomy & Astrophysics* 13: 752–60.

Canadelli, Elena (2009) "Some curious drawings" – Mars through Giovanni Schiaparelli's eyes: between science and fiction, *Nuncius* 24,2: 439–64.

Canales, Jimena (2002) Photographic Venus: the 'cinematography turn' and its alternatives in late 19th century science, *Isis* 93: 585–613.

— (2006) Movement before cinematography: the high-speed qualities of sentiment, *Journal of Visual Culture* 5: 275–94.

— (2011) Desired machines: cinema and the world in its own image, *Science in Context* 24, no. 3: 329–59.

Cannone, Marco & Susan Friedlander (2003) Navier: Blow-up and collapse, *Notices of the American Mathematical Society*, January: 7–13.

Cardwell, D. S. L. (1989) *From Watt to Clausius. The Rise of Thermodynamics in the Early Industrial Age*, Ames (orig. ed. 1971).

Carlson, B. & M. E. Gorman (1990) Understanding invention as a cognitive process: The case of Thomas Edison and early motion pictures, 1888–1891, *Social Studies of Science* 20: 387–430.

Carnot, Sadi (1824) *Réflexions sur la puissance motrice du feu et sur les machines propres à développer cette puissance* (*a*) Paris: Bachelier; (*b*) reprint in *Annales de l'École Normale Supérieure* 1872; (*c*) German transl. by W. Ostwald: *Betrachtungen über die bewegende Kraft des Feuers und die zur Entwickelung dieser Kraft geeigneten Maschinen*, Leipzig: Engelmann, 1909; (*d*) Engl. transl. in: Eric Mendoza (ed.) *Reflections on the Motive Power of Fire by Sadi Carnot and other Papers … by É. Clapeyron and R. Clausius*, New York: Dover, 1960 (reprint 2005); (*e*) critical ed. with surviving manuscripts, ed. by Robert Fox, Manchester: Manchester University Press, 1986.

Carter, B. A. R. (1970) Perspective, *Oxford Companion to Western Art*, ed. by H. Osborne, Oxford: Oxford University Press: 840–61.

Cartwright, Lisa (1995) *Screening the Body. Tracing Medicine's Visual Culture*, Minneapolis & London: University of Minnesota Press.

Carusi, Annamaria (2008) Scientific visualizations and aesthetic grounds for trust, *Ethics and Information Technology* 10: 245–54.

— (2011) Computational biology and the limits of shared vision, *Perspectives on Science* 19: 300–36.

— (2012) Making the visual visible in philosophy of science, *Spontaneous Generations* 6,1, available online at http://spontaneousgenerations.library.utoronto.ca/index.php/SpontaneousGenerations/article/view/16141.

Cassirer, Ernst (1955) *The Philosophy of Symbolic Forms*, New Haven: Yale University Press (Engl. transl. & intro. by Charles W. Hendel, German orig. 1923–25).

Cat, Jordi (2001) On understanding: Maxwell on the methods of illustration and scientific metaphor, *Studies in History and Philosophy of Modern Physics* 32,3: 395–441.

— (2013) *Maxwell & Sutton: A Binocular Look at the Maxwell–Sutton Collaboration in Color Photography*, London: Routledge.

— (2014) *Master and Designer of Fields: James Clerk Maxwell and Constructive, Connective and Concrete Natural Philosophy*, Oxford: Oxford University Press, 2014.

Cavalcanti, Daniel D., William Feindel, James T. Goodrich, T. Forcht Dagi, Charles J. Prestigiacomo & Mark C. Preul (2009) Anatomy, technology, art, and culture: toward a realistic perspective of the brain, *Neurosurgical Focus* 27,3: 1–22.

Chadarevian, Soraya de & Nick Hopwood (eds. 2004) *Models: The Third Dimension of Science*, Stanford University Press.

Chakravartty, Anjan (2010) Informational vs. functional theories of scientific representation, *Synthese* 172: 197–213.

Chalmers, A. F. (1986) The heuristic role of Maxwell's model of electromagnetic phenomena, *SHPS* 17: 415–27.

Chaloner, Clinton (1997) The most wonderful experiment in the world: a history of the cloud chamber, *British Journal for the History of Science* 30: 357–74.

Chansigaud, Valérie (2007) *Histoire de l'Ornithologie*, Paris: Delachaux & Nestle.

— (2009) *Histoire de l'Illustration Naturaliste*, Paris: Delachaux & Nestle.

Chapman, Allan (1997) James Nasmyth, astronomer of fire, *1997 Yearbook of Astronomy*: 143–67.

Charmantier, Isabelle (2011) Carl Linnaeus and the visual representation of nature, *Historical Studies in the Natural Sciences* 41,4: 365–404.

Chastel, André (ed. 1961) *Leonardo da Vinci on Art and Artists*, New York: Orion;

reprint New York: Dover 2002.

Chen, Xiang (2000): *Instrumental Traditions and Theories of Light. The Uses of Instruments in the Optical Revolution*, Dordrecht: Kluwer.

Cherry, Deborah (2005) *Art: History: Visual: Culture*, Oxford: Blackwell; also published as special issue of *Art History* 27, no. 4.

Chevreul, Michel Eugène (1839) De la loi du contraste simultané des couleurs et de l'assortiment des objets colorés, Paris; English transl. by Charles Martel as: *The Principles of Harmony and Contrast of Colours*, London: Longman, Brown, Green and Longmans, 1854.

Chladni, Ernst (1787) *Entdeckungen über die Theorie des Klanges*, Leipzig: Weidmanns Erben & Reich.

— (1802) *Die Akustik*, Leipzig: Breitkopf & Härtel.

Choulant, Ludwig (1920) *History and Bibliography of Anatomic Illustration*, University of Chicago Press (orig. German 1852 rev. by transl. Mortimer Frank et al.).

Christensen, Lars Lindberg, Robert Fosbury & Robert Hurt (eds. 2009) *Verborgenes Universum*, Weinheim: Wiley VCH: 125–9.

Clapeyron, Émile (1834) Puissance motrice de la chaleur, (*a*) *Journal de l'École Royale Polytechnique*, issue 24, vol. XIV: 153–90; (*b*) German transl. by K. Schreiber: *Annalen der Physik* (3) 59 (1843): 446ff., 566ff.; (*c*) annot. reprint: *Abhandlung über die bewegende Kraft der Wärme*, Leipzig: Akad. Verlagsgesellschaft, 1926; (*d*) Engl. transl. in: Eric Mendoza (ed.) *Reflections on the Motive Power of Fire by Sadi Carnot and other Papers ... by É. Clapeyron and R. Clausius*, New York: Dover, 1960/2005.

— (1844) Rapport sur un mémoire de M. Clapeyron, relatif au règlement des tiroirs dans les machines locomotives, et à l'emploie de la détente, *CRAS* 18: 275–82.

Cleveland, William S. (1984) Graphs in scientific publications, *American Statistician* 38: 261–9.

Cohen, H. Floris (2007) Reconceptualizing the scientific revolution, *European Review* 15, no. 4: 491–502.

— (2011) *How Modern Science Came Into the World*, Amsterdam: Amsterdam University Press.

Cohen, I. Bernard (1980) *Album of Science. From Leonardo to Lavoisier, 1450–1800*, New York: Scribner's.

Cohen, Robert S. & Raymond J. Seeger (eds. 1970) *Ernst Mach: Physicist and Philosopher*, Dordrecht, Reidel (Boston Studies in Philosophy of Science, no. 6).

Comandon, Jean (1909) Cinématographie à l'ultra-microscope, de microbes vivants et des particules mobiles, *CRAS* 149: 938–41.

— (1929) Le micro-cinématographie, *Protoplasma* 6: 627–33.

Comandon, J., C. Levaditi & S. Mutermilch (1913) Étude de la vie et de la croissance des cellules *in vitro* à l'aide de l'enregistrement cinématographique, *Comptes Rendus hebdomadaires des Séances et Mémoires de la Société de Biologie* (65) 1, no. 74: 464–7; 75: 457.

Consentius, P. (1872) Das atempo-Zeichnen an der Königlichen Gewerbe-Akademie zu Berlin, *Monatsblätter für Zeichenkunst und Zeichenunterricht* 8: 66–70.

Copeland, R. (1901) Obituary of Charles Piazzi Smyth, *MNRAS* 49: 189–96.

Corey, Robert B. & Linus Pauling (1953) Molecular models of amino acids, peptides, and proteins, *Review of Scientific Instruments* 24: 621–7.

Cormack, Allan M. (1979) Early two-dimensional reconstruction and recent topics stemming from it, Nobel lecture Dec. 8; online at: www.nobelprize.ort/nobel-prizes/medicine/laureates/1979/Cormack_lecture (accessed on 5 Mar. 2012).

— (1983) Computed tomography: some history and recent development, *Proceedings of the Symposia in Applied Mathematics* 27: 35–42.

— (1992) 75 years of Radon transform, *Journal of Computer-assisted Tomography* 16, no. 5: 673.

Couper, Archibald Scott (1858): Note sur une nouvelle théorie chimique, (*a*) *CRAS* 46: 1157–60; (*b*) On a new chemical theory, *Phil. Mag.* (4) 16: 104–16.

Crary, Jonathan (1988) Techniques of the observer, *October* 45: 3–35.

— (1995) *Techniques of the Observer. On Vision and Modernity in the Nineteenth Century*, Cambridge, MA: MIT Press.

Croft, William J. (2006) *Under the Microscope: A Brief History of Microscopy*, Singapore: World Scientific.

Crosby, Ranice W. (1991) *Max Brödel: The Man Who Put Art into Medicine*, Berlin: Springer.

Crosland, Maurice (1962) *Historical Studies in the Language of Chemistry*, London: Heinemann.

Crowther, Kathleen M. & Peter Barker (2013) Training the intelligent eye: understanding illustrations in early modern astronomy texts, *Isis* 104: 429–70.

Crum Brown, Alexander (1861) *On the Theory of Chemical Combination. A Thesis*, (*a*) Edinburgh: MD thesis; printed privately in 1879; (*b*) also available online.

— (1864) On the theory of isomeric compounds, *Transactions of the Royal Society of Edinburgh* 23: 707–19.

— (1867) Remarks on Sir Benjamin Brodie's system of chemical notation, *Phil. Mag.* (4) 34: 129–36 (= criticism of Brodie (1866, 1867)).

Cunningham, Alexander W. (1868) *Notes on the History, Methods and Technological Importance of Descriptive Geometry, Compiled with Reference to Technical Education in France, Germany & Great Britain*, Edinburgh: Edmonston & Douglas.

Curtis, Scott (2004) Still/moving: digital imaging and medical hermeneutics, in: Lauren Rabinowitz & Abraham Geil (eds.) *Memory Bytes. History, Technology and Digital Culture*, Durham & London: Duke University Press: 218–54.

— (2005) Die kinematographische Methode. Das 'bewegte Bild' und die Brownsche Bewegung, *Montage/av* 14,2: 23–43.

— (2009) Between observation and spectatorship: medicine, movies, and mass culture in Imperial Germany, in: Klaus Kreimeier & Annemone Ligensa (eds.) *Film 1900: Technology, Perception, Culture*, Eastleigh and Bloomington: John Libbey/Indiana University Press: 87–98.

Cuvier, Baron (1823/27) Éloge historique de M. René-Just Haüy, lu le 2 Juin 1823, *Recueil des Éloges Historiques lus dans les Séances publiques de l'Institut Royal de France*, Paris, 3 (1827): 123–75.

Dahl, Per F. (1997) *Flash of the Cathode Rays. A History of J. J. Thomson's Electron*, Bristol: Institute of Physics.

Dalby, W. E. (1903) The Education of engineers in America, Germany and Switzerland, *Engineering* 75: 600-603.

Dally, J. F. H. (1903) On the use of the Roentgen rays in the diagnosis of pulmonary tuberculosis, *The Lancet*, 27 June: 1800–6.

Damadian, Raymond V. (1971) Tumor detection by nuclear magnetic resonance, *Science* 171: 1151–3.

— (1980) Field focusing n.m.r. (FONAR) and the formation of chemical images in man, *PTRSL* B 289: 489–500.

Damadian, R. V., M. Goldsmith & L. Minkoff (1977) NMR in cancer: FONAR image of the live human body, *Physiological Chemistry and Physics* 9: 97–108.

Damadian, R. V., L. Minkoff, M. Goldsmith & J. Koutcher (1978) Field-focusing nuclear magnetic resonance (FONAR), *Die Naturwissenschaften* 65: 250–2.

Damadian, R. V., M. Stanford & J. Koutcher (1976) Field focus nuclear magnetic resonance (FONAR); visualization of a tumor in a live animal, *Science* 194: 1430–2.

Dance, S. Peter (1989) *The Art of Natural History*, London: Bracken Books.

Darius, Jon (1984) *Beyond Vision*, Oxford: Oxford University Press.

Daston, Lorraine (1986) Physicalist tradition in early nineteenth-century French geometry, *SHPS* 17: 269–95.

— (2011) *Observation as a Way of Life: Time, Attention, Allegory*, Uppsala: Wikströms.

Daston, L. & Peter Galison (1992) The image of objectivity, *Representations* 40: 81–128.

— (2007) *Objectivity*, New York: Zone Books (see Hentschel (2008*b*) for a review).

Davidhazy, Andrew (ed. 2008) *Images from Science 2: An Exhibition of Scientific Photography*, Rochester, NY: RIT Cary Graphic Arts Press.

Davidhazy, A. & Michael Peres (eds. 2002) *Images from Science: An Exhibition of Scientific Photography*, Rochester, NY: RIT Cary Graphic Arts Press.

Davies, Jim, Nancy J. Nersessian & Ashok K. Goel (2005) Visual models in analogical problem solving, *Foundations of Science* 10: 133–52.

Daxecker, Franz (1992) Christoph Scheiner's eye studies, *Documenta Ophthalmologica* 81: 27–35.

— (1994) Further studies by Christoph Scheiner concerning the optics of the eye, *Documenta Ophthalmologica* 86: 153–61.

— (1996) Das Hauptwerk des Astronomen P. Christoph Scheiner SJ 'rosa ursina sive sol' – eine Zusammenfassung, *Berichte des naturwissenschaftlich-medizinischen Vereins in Innsbruck*, suppl. 13, 1–82.

— (2006) *Der Physiker und Astronom Christoph Scheiner*, Innsbruck: Wagner.

DeFelipe, Javier (2013) *Cajal's Butterflies of the Sourl: Science and Art*, Oxford: Oxford University Press and posted in their blog on Nov. 9, 2013 at http://blog.oup.com/2013/11/cajal-butterflies-of-the-soul-cerebral-cortex/.

Della Porta, Giambattista (1589) *Magiae naturalis sive de miraculis rerum naturalium*, (*a*) Naples 1589; (*b*) *Natural Magic*, London: T. Young & S. Speed, 1658.

Dennis, Michael Aaron (1989) Graphic understanding, *Science in Context* 3: 309–64.

Deregowski, Jan B. (1972) Pictorial perception and culture, *Scientific American* 127, Nov.: 82–8.

Derksen, A. (2005) Linear perspective as a realist constraint, *Journal of Philosophy* 102: 235–58.

Dhombres, Jean et al. (eds. 1992) *L'École Normale de l'An III. Tome I: Leçons de mathématique, edition annotée des cours de Laplace, Lagrange, et Monge*, Paris: Dunod.

Dickerson, Richard E. (1969) *The Structure and Action of Proteins*, New York: Harper & Row.

— (1983) *Hemoglobin: Structure, Function, Evolution and Pathology*, Menlo Park, CA:

Benjamin Cummings.

— (1997) Irving Geis, molecular artist, 1908–1997, *Protein Science* 6, no. 11: 2483–4.

Dickinson, H. W. & Rhys Jenkins (1927) *James Watt and the Steam Engine*, Oxford; reprint London: Encore Ed., 1989.

Dikovitskaya, Margaret (2006) *Visual Culture: The Study of the Visual after the Cultural Turn*, Cambridge, MA: MIT Press.

Dobbin, Leonard (1934) The Couper quest, *Journal of Chemical Education* 11: 331–8.

Dobbins, Thomas A. & William Sheehan (2004) The canals of Mars revisited, *Sky & Telescope* 107: 114–17.

Doherty, Meghan (2012*a*) Discovering the 'true' form: Hooke's *Micrographia* and the visual vocabulary of engraved portraits, *Notes and Records of the Royal Society* 2012: 1–24.

— (2012*b*) Creating standards of accuracy: Faithorne's *The Art of Graving* and the Royal Society, in: Rima D. Apple (ed.) *Science in Print: Essays on the History of Science and the Culture of Print*, Madison, WI: University of Wisconsin Press, 2012: 15–36.

Doi, Kunio (2006) Diagnostic imaging over the last 50 years: Research and development in medical imaging science and technology, *Physics in Medicine and Biology* 51: R5–R27.

Dommann, Monika (2001) Durchleuchtete Körper – die materielle Kultur der Radiographie 1896–1930, *Fotogeschichte* 21: 41–58.

— (2003) *Durchsicht, Einsicht, Vorsicht: Eine Geschichte der Röntgenstrahlen, 1896–1963* Zurich: Chronos.

Dössel, Olaf (2003) Geschichte der bildgebenden Verfahren in der Medizin, in: Ewald Konecny, Volker Roelcke & Burghard Weiss (eds.) *Medizintechnik im 20. Jahrhundert*, Berlin: IDE-Verlag: 59–92.

Douard, John W. (1995) E. J. Marey's visual rhetoric and the graphic decomposition of the body, *SHPS* 26, no. 2 175–204.

Douglas, Bedi & Bruce Joyce (1994) *Seeing the Unseen: Dr. Harold E. Edgerton and the Wonders of Strobe Alley Collins*, Cambridge, MA: MIT Press.

Dragga, Sam (1992) Evaluating pictorial illustrations, *Technical Communication Quarterly* 1.2: 47–62.

Drewry, Charles Stewart (1832) *A Memoir of Suspension Bridges*, London: Longmans.

Dupin, Charles (1819) *Essai historique sur les travaux scientifiques de Gaspard Monge*, Paris: Bachelier.

Dupré, Sven (2008) Newton's telescope in print: the role of images in the reception of Newton's instrument, *Perspectives on Science* 16: 328–59.

— (2010) Art history, history of science, and visual experience, *Isis* 101: 618–22.

Dürbeck, Gabriele et al. (eds. 2001) *Wahrnehmung der Natur – Natur der Wahrnehmung. Studien zur Geschichte visueller Kultur um 1800*, Dresden: Verlag der Kunst.

Dürer, Albrecht (1525) *Unterweysung der Messung mit dem Zirckel und Richtscheyt in Linien, Ebenen, und gantzen Corporen*, Nürnberg; reprint Nördlingen: Uhl, 1983.

— (1532) *Vier Bücher von menschlicher Proportion*, Nürnberg; French transl. in 1557; reprint Nördlingen: Uhl, 1980.

Dussauge, Isabelle (2008) *Technomedical Visions: Magnetic Resonance Imaging in 1980s Sweden*, Stockholm: Dissertation, KTH Royal Institute of Technology.

Eckert, Michael (2012) Disputed discovery: the beginnings of x-ray diffraction in crystals in 1912, *Zeitschrift für Kristallographie* 227: 27–35.

Eco, Umberto (1990) Die Karte des Reiches im Maßstab 1:1, in: U. Eco: *Platon im Striptease-Lokal*, ed. and transl. by Burkhart Kroeber, Munich: Hanser: 85–97.

Eden, Murray (1984) The engineering–industrial accord: inventing the technology of health care, in: *The Machine at the Bedside*, Cambridge University Press: 49–64.

Eder, Josef Maria & Eduard Valenta (1896) *Versuche über Photographie mittelst der Röntgen'schen Strahlen*, Vienna: Kaiserlich-Königliche Lehr- & Versuchsanstalt für Photographie und Reproductions-Verfahren.

Edgerton, Harold E. (2000) *Exploring the Art and Science of Stopping Time*, Cambridge, MA: MIT Press.

Edgerton, Samuel Y. (1975) *Renaissance Rediscovery of Linear Perspective*, New York: Harper & Row.

— (1984) Galileo, Florentine disegno and the "Strange spottednesse" of the moon, *Art Journal* 44, no. 3: 225–32.

— (1991) *The Heritage of Giotto's Geometry: Art and Science on the Eve of Scientific Revolution*, Ithaca: Cornell University Press.

Edvyean, Robert G. J. (1988) Henry Clifton Sorby (1826–1908): Studies in marine biology – the algal lantern slides, *Archives of Natural History* 15: 35–44.

Edvyean R. G. J. & Chris Hammond (1997) The metallurgical work of Henry Clifton Sorby and an annotated catalogue of his extant metallurgical samples, *Historical Metallurgy* 31, no. 2: 54–85.

Ehret, Georg Dionysius (1736) *Methodus plantarum sexualis in sistemate naturae descripta*, (*a*) Leiden; (*b*) reprint Stockholm: Hagelin Rare Books, 2000.

Eisenberg, Ronald L. (1992) *Radiology: An Illustrated History*, St Louis, MO: Mosby.

Elgin, Catherine Z. (2004) True enough, *Philosophical Issues* 14: 113–31.

Elkins, James (1987) Piero della Francesca and the Renaissance proof of linear perspective, *Art Bulletin* 69: 220–30.

— (1988) Did Leonardo develop a theory of curvilinear perspective?, *JWCI* 51: 190–6.

— (1995) Art history and images that are not art, *Art Bulletin* 77: 554–71.

— (1999) *The Domain of Images*, Ithaca: Cornell University Press.

— (2000) *How to Use Your Eyes*, London: Routledge.

— (2002) Preface to the book A Skeptical Introduction to Visual Culture, *Journal of Visual Culture* 1: 93–9.

— (2003) *Visual Studies: A Skeptical Introduction*, New York & London: Routledge.

— (2007) *Visual Practices Across the University*, Munich: Fink.

— (2008) *Six Stories from the End of Representation. Images in Painting, Photography, Astronomy, Microscopy, Particle Physics, and Quantum Mechanics, 1980–2000*, Chicago: University of Chicago Press.

Ellenius, Allan (ed., 1985) *The Natural Sciences and the Arts. Aspects of Interaction from the Renaissance to the 20th Century*, Uppsala: Acta Universitatis Upsaliensis.

Emmerson, George (1973) *Engineering Education: A Social History*, New York: David & Charles.

Entwisle, A. R. (1963) An account of the exhibits relating to Henry Clifton Sorby, shown at the Sorby centenary in Sheffield 1963, *Metallography*: 313–26 & plates.

Erasmus, Jeremy J., Gregory W. Gladish, Lyle Broemeling, Bradley S. Sabloff, Mylene T. Truong, Roy S. Herbst & Reginald F. Munden (2003) Interobserver and intraobserver variability in measurements of non-small cell carcinoma lung lesions: implications for assessment of tumor response, *Journal of Clinical Oncology* 21: 2574–82.

Ertl, Thomas (ed. 2010) *10 Jahre VIS, 3 Jahre VISUS*, University of Stuttgart.

Euclid: *Thirteen Books of Euclid's Elements*, ed. and transl. by Thomas Heath, (*a*) 1st ed. Cambridge: Cambridge University Press, 1908; (*b*) reprint New York: Dover 1956; (*c*) ed. of the Greek original by J. L. Heiberg as *Elementa* in it Euclidis opera omnia, Leipzig: Teubner, vols. 1–5, 1883 ff.

— Optics, (*a*) transl. by Henry Edwin Burton, *Journal of the Optical Society of America* 38 (1945): 357–72; (*b*) *Euclidis Optica* ed. by J. L. Heiberg in *Euclidis opera omnia*, Leipzig: Teubner vol. 7 (1895): 1–121; (*c*) annot. French transl. by Paul ver

Eecke, Paris & Bruges, 1959 (see Kheirandish (ed. 1999)).

Evans, David S. (1989) The life of Charles Piazzi Smyth, *Sky and Telescope* 98: 158–60.

Evans, D. S. et al. (1988) *Herschel at the Cape: Diaries and Correspondence of Sir John Herschel 1834–38*, Austin: University of Texas Press.

Evans, J. E. & E. W. Maunder (1903) Experiments as to the actuality of the 'canals' observed on Mars, *MNRAS* 63: 488f.

Evans, Jessica & Stuart Hall (eds. 1999) *Visual Culture: The Reader*, London: SAGE.

Eytelwein, Johann Albert (1798) Von dem Nutzen eines Wasserstandsskale, nebst Anweisungen zur Verfertigung derselben, *Sammlung nützlicher Aufsätze und Nachrichten* 2,1: 25–8 & pl. II.

Faber, Monika (2003) Josef Maria Eder und die wissenschaftliche Fotografie 1855–1918, in: Monika Faber & Klaus Albrecht Schröder (eds.) *Das Auge und der Apparat. Eine Geschichte der Fotografie aus den Sammlungen der Albertina*, Paris: Seuil and Vienna: Albertina: 142–69.

Farish, William (1822) On isometrical perspective, *Cambridge Philosophical Transactions* 1: 1–19 & 3 pl.

Farquharson, F. B. (1941) Aerodynamical stability of suspension bridges with special reference to the Tacoma Narrows Bridge, *University of Washington, Engineering Experiment Station Bulletin* 116, no. 1.

Faye, Hervé (1870) Sur l'observation photographique des passages de Vénus, *CRAS* 70: 541–8.

Fehrenbach, Frank (1997) *Licht und Wasser. Zur Dynamik naturphilosophischer Leitbilder im Werk Leonardo da Vincis*, Tübingen: Wasmuth.

— (2002) *Leonardo da Vinci. Natur im Übergang. Beiträge zu Wissenschaft, Kunst und Technik*, Munich: Fink.

Feldhaus, Franz Maria (1959) *Geschichte des technischen Zeichnens*, Wilhelmshaven: Kuhlmann.

Ferguson, Eugene S. (1977) The mind's eye: nonverbal thought in technology, *Science* 197: 827–36; also reprinted in *Leonardo* 11 (1978): 131–9.

— (1992) *Engineering and the Mind's Eye*, (*a*) Cambridge, MA: MIT Press; (*b*) German transl.: *Das Auge des Ingenieurs*, Basel: Birkhäuser, 1999.

Field, J. V. (1987) Linear perspective and the projective geometry of Girard Desargues, *Nuncius. Annali di Storia della Scienza* 2, no. 2: 3–40.

— (1988) Perspective and the mathematicians: Alberti to Desargues, in: *Mathematics from Manuscript to Print 1300–1600*, New York: 236–63.

— (1997) *The Invention of Infinity: Mathematics and Art in the Renaissance*, Oxford: Oxford University Press.

Fieschi, Caroline (2000) L'illustration photographique des thèses de science en France (1880–1909), *Bibliothèque de l'École des Chartes* 158: 223–45.

Files, Craig (1996) Goodman's rejection of resemblance, *British Journal of Aesthetics* 36, no. 4: 398–412.

Finch, James Kip (1941) Wind failures of suspension bridges, or evolution and decline of the stiffening truss, *Engineering News Record* 126: 7–79.

Findlen, Paula (1994) *Possessing Nature. Museums, Collecting, and Scientific Culture in Early Modern Italy*, Berkeley: University of California Press.

Fink, Daniel A. (1971) Vermeer's use of the camera obscura, *Art Bulletin* 53: 493–505.

Fiorentini, Erna (2005) Instrument des Urteils. Zeichnen mit der camera lucida als Komposit, *Max-Planck-Institut für Wissenschaftsgeschichte, Berlin, Preprints* no. 295.

FitzGerald, George Francis (1902) Foundations of physical theory: function of models, in G. F. FitzGerald *Scientific Writings*, London: Longmans & Green: 163–9.

Flammarion, Camille (1875) Le passage de Venus – résultats des expéditions français, *La Nature* 3: 356–8.

— (1892) *La Planète Mars et ses conditions d'habitabilité*, Paris: Gauthier-Villars, (*a*) 1st ed.; (*b*) 2nd ed. 1909.

Forbes, George (1874) *The Transit of Venus*, London: MacMillan.

Forbes, James D. (1849) Hints towards a classification of colours, *Phil. Mag.* (3) 34: 161–78.

Ford, Brian (1992) *Images of Science: A History of Scientific Illustration*, London: British Library.

Ford, Kenneth W. (1993) Interview with Dr. John Wheeler at Jadwin Hall, Princeton University, December 6, 1993, American Institute of Physics, Maryland.

Fortner, Hans (1933) Die Punktweg-Methode: Ein Verfahren zur quantitativen Auswertung von Mikro-Kinematogrammen, *Zeitschrift für wissenschaftliche Mikroskopie und mikroskopische Technik* 30: 1–62.

Foster, Hal (1988) *Vision and Visuality*, New York: The New Press.

Fourcy, Ambroise L. (1828) *Histoire de l'École Polytechnique*, (*a*) Paris; (*b*) reprint, with introduction and annotation by Jean G. Dhombres, Paris: Belin, 1987.

Fourney, Johann Heinrich (1780) Éloge de M. Lambert, *Nouveaux Mémoires de l'Académie royale des Sciences et Belles-Lettres. Année 1778*, Berlin: 72–90.

Fox, Daniel M. (1988) *Photographing Medicine. Images and Power in Britain and America since 1840*, New York: Greenwood Press.

Francoeur, Eric (1997) The forgotten tool: the design and use of molecular models, *Social Studies of Science* 27: 7–40.

— (2000) Beyond dematerialization and inscription. Does the materiality of molecular models really matter? *Hyle* 6: 63–84 (also available online).

Francoeur, E. & Jerome Segal (2004) From model kits to interactive computer graphics, in: De Chadarevian & Hopwood (eds. 2004), 402–32.

Frank, Gustav (1898) Wünsch, Christian Ernst, *Allgemeine Deutsche Biographie* 44: 317–20.

Frank, Gustav (2005) Musil contra Balázs – Ansichten einer 'visuellen Kultur' um 1925, *Musil-Forum* 28: 2003/04 (publ. 2005): 125–52.

— (2006) Textparadigma kontra visueller Imperativ: 20 Jahre *Visual Culture Studies* als Herausforderung der Literaturwissenschaft. Ein Forschungsbericht, *Internationales Archiv für Sozialgeschichte der Literatur* 31: 26–89.

Fraunhofer, Joseph (1815) Bestimmung des Brechungs- und Farbenzerstreuungs-Vermögens verschiedener Glasarten, in Bezug auf die Vervollkommnung achromatischer Fernröhre, (*a*) *Denkschriften der königlichen Akademie der Wissenschaften, München* 5 (publ. 1817): 193–226 (and 3 pl.); (*b*) reprint in Fraunhofer (1888): 1–32; (*c*) English transl.: 'On the refractive and dispersive power of different species of glass in reference to the improvement of achromatic telescopes, with an account of the lines or streaks which cross the spectrum,' *Edinburgh Philosophical Journal* 9 (1823): 288–99 & pl. VII.

— (1888) *Gesammelte Schriften*, ed. by E. Lommel, Munich: Akademie, in Kommission Franz.

Freguglia, P. (1995) De la perspective à la géométrie projective, in: Radelet-de Grave & Benvenuto (eds. 1995), 89–102.

Fried, Johannes & Stolleis, Michael (eds. 2009) *Wissenskulturen. Über die Erzeugung und Weitergabe von Wissen*, Frankfurt am Main: Campus.

Friedhoff, Richard M. & William Benzon (1989) *Visualization, the Second Computer Revolution*, New York: Abrams.

Fritscher, Bernhard & Fergus Henderson (eds. 1998) *Toward a History of Mineralogy, Petrology and Geochemistry*, Munich: IGN (Algorismus 23).

Fuchs, Leonhart (1542) *De historia stirpium commentarii impensis et virgillis elaborati...*, (*a*) Basileae [Basel], Officina Isingrimiana [Isengrim]; (*b*) reprint Munich: Kölbl

1981; (*c*) in Engl. transl. *The Great Herbal of Leonhart Fuchs: De historia stirpium comentarii insignes*, Stanford: Stanford University Press, 1999 (2 vols., ed. by Frederick Meyer, Emily Trueblood & John Heller).

— (1543) *New Kreüterbuch* (*a*) Basel: Isengrim; (*b*) facsimilated reprint of Fuchs's handcolored personal copy, Cologne: Taschen 2001; (*c*) Engl. intro. by Klaus Dobat & Werner Dressendorfer (eds.) *The New Herbal of 1543*, Cologne: Taschen, 2001; (*d*) in Dutch transl.: *Den nieuwen Herbarius, dat is, dboeck van-den cruyden*, 1549.

— (1545) *Läblische Abbildung und Contrafahtung aller Kreüter, so der hochglert herr Leonhart Fuchs der arzney Doctor inn dem ersten theyl seins neüwen Kreüterbuchs hat begriffen in ein kleinere form auff das aller artlichest gezogen damit sie füglich vonn allen mögen hin unnd wider zur noturfft getragen unnd gefurt ward*, (*a*) Basel, Isingrin; (*b*) reprint Grünwald near Munich: Kölbl 1980.

— (2001) *Die Kräuterbuch-Handschrift des Leonhart Fuchs*, ed. by Brigitte Baumann, Helmut Baumann & Susanne Baumann-Schleihauf, Stuttgart: Ulmer.

Füsslin, Georg (1993) *Optisches Spielzeug oder: Wie die Bilder laufen lernten*, Stuttgart: Füsslin.

Gabillard, Robert (1951) Mesure du temps de relaxation T_2 en présence d'une inhomogénéité du champ magnétique supérieure à la largeur de raie, *CRAS* 232: 1551–3; in Engl.: A steady-state transition technique in nuclear resonance, *Physical Review* 85 (1952): 694f.

Gabor, Dennis (1948) *The Electron Microscope: Its Development, Present Performance and Future Possibilities*, Brooklyn, NY: Chemical Publishing Co.

— (1957) Die Entwicklungsgeschichte des Elektronenmikroskops, *Elektrotechnische Zeitschrift* 78: 522–30.

Gaede, Peter-Matthias (ed. 1999) *Die unendliche Reise. Wissenschafts-Fotografie entdeckt neue Welten*, Hamburg: Geo.

Galilei, Galileo (1610*a*) *Sidereus Nuncius*; (*b*) reprint with commentary in *Opere* vol. 3 (1890); (*c*) transl. with commentary by Albert van Helden: *Sidereus Nuncius of the Sidereal Messenger*, Chicago: University of Chicago Press, 1996; (*d*) in German transl. with commentary by Hans Blumenberg: Frankfurt: Suhrkamp, 2002.

— (1613) *Istoria e dimostrazioni intorno alle macchie solari*, Rome: Mascardi; reprint Bruxelles 1967.

Galison, Peter (1979) Minkowski's space-time: from visual thinking to the absolute world, *HSPS* 10: 85–121.

— (1997) *Image and Logic. A Material Culture of Microphysics*, Chicago: University of

Chicago Press.

— (2000) The suppressed drawing: Paul Dirac's hidden geometry, *Representations* 72: 145–66.

Galison, P. & Alexi Assmus (1989) Artificial clouds, real particles, in: David Gooding, Trevor Pinch & Simon Schaffer (eds.) *The Uses of Experiment*, Cambridge: Cambridge University Press, 1989: 225–74.

Galison, P., Caroline A. Jones & Amy Slaton (eds. 1998) *Picturing Science, Producing Art*, New York: Routledge.

Galison, P. & David Stump (1996) *The Disunity of Science. Boundaries, Contexts and Power*, Stanford: Stanford University Press.

Galton, Francis (1874) *English Men of Science: Their Nature and Nurture*, London: MacMillan.

Garcia-Düttmann, Alexander (2002) The ABC of visual culture, or a new decadence of illiteracy, *Journal of Visual Culture* 1, no. 1: 101–3.

Gardner, Howard (1985) *Frames of Mind. The Theory of Multiple Intelligence*, New York: Basic Books.

Garfinkel, Harold, Eric Livingston & Michael Lynch (1981) The work of a discovering science construed with materials from the optically discovered pulsar, *Philosophy of the Social Sciences* 11,2: 131–58.

Garrison, R. F. (1995) William Wilson Morgan (1906–1994) *Publications of the Astronomical Society of the Pacific* 107: 507–12.

Gascoigne, Bamber (1986) *How to Identify Prints. A Complete Guide to Manual and Mechanical Processes from Woodcut to Ink-Jet*, New York: Thames & Hudson.

Gattegno, Caleb (1969) *Towards a Visual Culture: Educating through Television*, New York: Outerbridge & Dienstfrey; German transl.: *Das Fernsehen, eine Herausforderung für Bildung und Erziehung*, Hannover: Schroedel, 1975.

Gavroglu, Kostas & Ana Simões (2011) *Neither Physics nor Chemistry: A History of Quantum Chemistry* Cambridge, MA: MIT Press.

Gaycken, Oliver (2002) A drama unites them in a fight to the death: Some remarks on the flourishing of a cinema of scientific vernacularization in France 1909–1914, *Historical Journal of Film, Radio and Television* 22: 353–74.

Geertz, Clifford (1973) *The Interpretation of Culture*, New York: Basic Books.

— (1983) *Local Knowledge. Further Essays in Interpretative Anthropology*, New York: Basic Books; German transl.: *Dichte Beschreibung: Beiträge zum Verstehen kultureller Systeme*, Frankfurt: Suhrkamp, 2002.

Gehler, Johann Samuel Traugott (1787ff.) *Physikalisches Wörterbuch* ..., (*a*) Leipzig:

Schwickert, 1st ed.; (*b*) 2nd ed. 1825ff.

Geison, Gerald L. (1981) Scientific change, emerging specialties, and research schools, *History of Science* 10: 20–40.

Geison, G. L. & Frederic L. Holmes (eds. 1993) Research schools: historical reappraisals, *Osiris* 8.

Geissler, Julius (1889) Winke für Autoren, die graphische Wiedergabe von Illustrationen zu wissenschaftlichen Arbeiten betr., *Virchows Archiv* 115: 557–60.

Geus, Armin: Christian Koeck (1758–1818), der Illustrator Samuel Thomas Sömmerrings, in: Gunter Mann & Franz Dumont (eds.) *Samuel Thomas Sömmerring und die Gelehrten der Goethezeit. Sömmerring-Forschungen*, Stuttgart & New York 1985: 263–78.

Giere, Ronald N. (2010) *Scientific Perspectivism*, University of Chicago Press.

Gilbert, Peter (1972) Iterative methods for the three-dimensional reconstruction of an object from projections, *Journal of Theoretical Biology* 36, no. 1: 105–17.

Gill, Arthur T. (1969) Early stereoscopes, *The Photographic Journal* 109: 546–99, 606–14, 641–51.

Gillis, Jean Baptiste (1966) Leben und Werk von Kekulé in Gent, in: *Kekulé und seine Benzolformel*, Weinheim: VCH: 33–54.

Gingerich, Owen (1989) *Album of Science. The Physical Sciences in the Twentieth Century*, New York: Scribner's.

Gispen, Kees (1989) *New Profession, Old Order: Engineers and German Society 1815–1914*, Cambridge: Cambridge University Press.

Glasser, Otto (1931) *Wilhelm Conrad Röntgen und die Geschichte der Röntgenstrahlen*, Berlin: Springer (Röntgenkunde in Einzeldarstellungen, no. 3); Engl. version: Springfield, Ill.: Thomas, 1945.

Gloor, Baldur (1958) *Die künstlerischen Mitarbeiter an den naturwissenschaftlichen und medizinischen Werken Albrecht von Hallers*, Bern: Berner Beiträge zur Geschichte der Medizin und der Naturwissenschaften, No. 15.

Göbel, Forstsekr. (1818) Ueber das Kyanometer, *Journal für Chemie und Physik* 24: 238–52.

Godlewska, Anne Marie Claire (1999) The powerful mapping metaphor, in *Geography Unbound. French Geographic Science from Cassini to Humboldt*, Chicago: University of Chicago Press: 129–48.

— (2001) Von der Vision der Aufklärung zur modernen Wissenschaft – Humboldts visuelles Denken, in: O. Ette & W. L. Bernecker (eds.) *Ansichten Amerikas*, Frankfurt am Main: 157–94.

Goethe, Johann Wolfgang von (1791/92*a*) *Beiträge zur Optik*, 2 vols.; (*b*) annot. reprint in *Beiträge zur Optik und Anfänge der Farbenlehre, 1790–1808*, ed. by Ruprecht Matthaei, Weimar: Böhlau, 1951 (Die Schriften zur Naturwissenschaft, part 1, text vol. 3).

— (1810) *Zur Farbenlehre*, (*a*) Tübingen: Cotta, 1810 (2 vols.); (*b*) in *Die Schriften zur Naturwissenschaft*, ser. I, vols. 4–7. Weimar: Böhlaus Nachfolger, 1957–87 (see also Zehe (1992)).

Golan, Tal (1998) The authority of shadows: the legal embrace of the x-ray, *Historical Reflections* 24: 437–58.

Goldsmith, Evelyn (1984) *Research into Illustration: An Approach and a Review*, Cambridge: Cambridge University Press.

Golinski, Jan (1998) *Making Natural Knowledge. Constructivism and the History of Science*, Cambridge: Cambridge University Press.

Gombrich, Ernst (1960) *Art and Illusion. A Study in the Psychology of Pictorial Representation*, Princeton: Princeton University Press.

— (1982) *The Image and the Eye: Further Studies in the Psychology of Pictorial Representation*, Ithaca: Cornell University Press; 2nd ed. London: Phaidon, 1984.

Gooding, David (1990) *Experiment and the Making of Meaning*, Dordrecht: Kluwer.

— (2004*a*) Envisioning explanations, *Interdisciplinary Science Reviews* 29: 278–94.

— (2004*b*) Cognition, construction and culture: visual theories in the sciences, *Journal of Cognition and Culture* 4: 551–93.

— (2004*c*) Seeing the forest for the trees – visualization, cognition and scientific inference, in: *Scientific and Technological Thinking*, Mahwah, NJ: Erlbaum: 173–217.

— (2006) From phenomenology to visual field theory: Faraday's reasoning, *Perspectives on Science* 14: 40–65.

— (2010) Visualizing scientific inference, *Topics in Cognitive Science* 2: 15–35.

Goodman, Nelson (1969) *Languages of Art. An Approach to a Theory of Symbols*, London: Oxford University Press, 1st ed.; 2nd ed.: Indianapolis: Hackett, 1976.

Goodsell, David S. (2003*a*) Illustrating molecules, in: E. Hodges (ed.) *The Guild Handbook of Scientific Illustration*, Hoboken, NJ, Wiley: 267–70.

— (2003*b*) Looking at molecules – an essay on art and science, *ChemBioChem* 4: 1293–8.

Goodwin, Charles (1997) The blackness of black: color categories as situated practice, in: Lauren B. Resnick et al. (eds.) *Discourse, Tools, and Reasoning: Essays on Situated Cognition*, Berlin: Springer, 111–40.

— (1994) Professional vision, *American Anthropologist* 96: 606–23.

Gordon, R., R. Bender & G. T. Herman (1970) Algebraic reconstruction techniques (ART) for three-dimensional electron microscopy and x-ray photography, *Journal for Theoretical Biology* 29: 471–82.

Gördüren, Petra & Dirk Luckow (eds. 2010): *Dopplereffekt. Bilder in Kunst und Wissenschaft*, Cologne & Kiel: Dumont & Kunsthalle Kiel.

Gosser, H. Mark (1981) Kircher and the laterna magica – a reexamination, *Society of Motion Picture and Television Engineers Journal* 90: 972–8.

Gotz, Karl Otto & Karin Gotz (1979) Personality characteristics of successful artists, *Perceptual and Motor Skills* 49: 919–24.

Gould, Stephen Jay (1989) *Wonderful Life. The Burgess Shale and the Nature of History*, New York: Norton.

— (2000) Abscheulich! Atrocious! *Natural History* 109, no. 2: 42–9.

Gradmann, Christoph (2009) *Laboratory Disease: Robert Koch's Medical Bacteriology*, Baltimore: Johns Hopkins University Press.

Grafton, Anthony (2000) *Leon Battista Alberti: Master Builder of the Italian Renaissance*, New York: Bill & Wang.

Granek, Galina & Giora Hon (2008) Search for asses, finding a kingdom: the story of the invention of the scanning tunnelling microscope (STM), *Annals of Science* 65: 101–25.

Grashey, Rudolf (1905) Fehlerquellen und diagnostische Schwierigkeiten beim Röntgenverfahren, *Münchener Medizinische Wochenschrift* 52: 807–10.

Graßhoff, Gerd, Hans-Christoph Liess & Kärin Nickelsen (2001) *COMPAGO: Der systematische Bildvergleich*, Bern: Bern Studies in the History and Philosophy of Science.

Grattan-Guinness, Ivor (1990) *Convolutions in French Mathematics, 1800–1840. From the Calculus and Mechanics to Mathematical Analysis and Mathematical Physics*, Basel: Birkhäuser.

— (2005) The *École Polytechnique*, 1794–1850: differences over educational purpose and teaching practice, *The American Mathematical Monthly* 112,3: 233–50.

Green, Nathaniel Everett (1879) Mars and the Schiaparelli canals, *Observatory* 3: 252.

— (1880) Observations of Mars, at Madeira in Aug. and Sep. 1877, *Memoirs of the Royal Astronomical Society* 40: 123–40 & pl.

Groß, Dominik & Stefanie Westermann (eds. 2007) *Vom Bild zur Erkenntnis – Visualisierungskonzepte in den Wissenschaften*, Kassel: Kassel University Press (Studien des Aachener Kompetenzzentrums für Wissenschaftsgeschichte, no. 1).

Groth, Paul Heinrich (1925) *Entwicklungsgeschichte der mineralogischen Wissenschaften*, Berlin: Springer.

Grubb, D. & A. Keller (1972) Beam-induced radiation damage in polymers and its effect on the image-formation in the electron microscope, *Electron Microscopy. Proceedings of the European Congress on Electron Microscopy*, 5: 554–60.

Grünewald, Heinrich (1976) Zur Entstehungsgeschichte des Elektronenmikroskops, *Technikgeschichte* 43: 213–42; see also the rebuttal by Ernst Ruska: Gegendarstellung, ibid. 44 (1977): 86–7.

Gugerli, David (1998) Die Automatisierung des ärztlichen Blicks (Inaugural Lecture at the ETH Zürich), *Preprints zur Kulturgeschichte der Technik*, no. 4.

— (1999) Soziotechnische Evidenzen. Der 'pictorial turn' als Chance für die Geschichtswissenschaft, *Traverse* 3: 131–59 (also online).

Gugerli, D. & Barbara Orland (eds. 2002) *Ganz normale Bilder. Historische Beiträge zur visuellen Herstellung von Selbstverständlichkeit*, Zurich: Chronos.

Gunning, Tom (1999) An aesthetics of astonishment. Early film and the (in)credulous spectator, in: Leo Braudy & Marshall Cohen (eds.) *Film Theory and Criticism. Introductory Readings*, Oxford: Oxford University Press: 818–32.

Gunthert, André (2000) "La rétine du savant," *Études photographiques*, 7 (May).

Gurd, Frank R. N. (1974) The use of Corey–Pauling–Koltun space-filling models in teaching, *Biochemical Education* 2, no. 2: 27–9.

H. H. Jun[ior] (1822) Account of a steam-engine indicator, *Quarterly Journal of Science, Literature and Arts* 13: 91–6 & pl. II.

Haarmann, Harald (1991) *Universalgeschichte der Schrift*, Frankfurt: Suhrkamp.

Hachette, Jean Nicolas (1822) *Traité de géometrie descriptive*, Paris: Corby.

— (1829) Notice sur la création de l'École Polytechnique, *Journal du génie civil* 2: 251–63.

Hacking, Ian (1983) *Representing and Intervening*, University of Chicago Press.

Hadamard, Jacques (1945) *The Psychology of Invention in the Mathematical Field*, Princeton, NJ: Princeton University Press.

Hafner, Klaus (1979) August Kekulé – the architect of chemistry, *Angewandte Chemie International Edition* 18: 641–51.

Hager, Thomas (1995) *Forces of Nature. The Life of Linus Pauling*, New York: Simon & Schuster.

Hagner, Michael (2003) Bildunterschätzung – Bildüberschätzung. Ein Gespräch, in: *Bildwelten des Wissens. Kunst-historisches Jahrbuch für Bildkritik* 1, no. 1: 103–11.

Hagner, M., Claudio Pogliano & Renato Mazzolini (eds. 2009) *Biographies of Scientific Images*, special issue of *Nuncius. Journal of the History of Science* 24/2, Florence: Istituto e Museo di Storia della Scienza.

Hall, Stephen (1980) Cultural studies: two paradigms, *Media, Culture, Society* 2: 57–72.

— (1990) The emergence of cultural studies and the crisis of the humanities, *October* 53: 11–23.

Hallden, Karl William (ed. 1963) *Christopher Polhem. The Father of Swedish Technology* (Swed. orig. 1911), Engl. transl. by William A. Johnson, Hartford, CT: Trinity College.

Halley, Edmond (1686) An Historical Account of the Trade Winds, and Monsoons, Observable in the Seas between and near the Tropicks, with an Attempt to Assign the Phisical Cause of the Said Wind, *PTRSL* no. 183.

Halsband, Megan C. (2008): *Stereographs as Scholarly Resources in American Academic Libraries and Special Collections*, Master's Paper, School of Information and Library Science of the University of North Carolina at Chapel Hill.

Hamilton, Kelley (2001) Wittgenstein and the mind's eye, in: James C. Klagge (ed.) *Wittgenstein – Biography and Philosophy*, Cambridge University Press, 2001: 53–97.

Hamilton, William (1776/79) *Campi Phlegraei. Observations on the Volcanos of the Two Sicilies as They Have Been Communicated to the Royal Society of London* (Naples 1776) and *Supplement to the Campi Phlegraei being an Account of the Great Eruption of Mount Vesuvius in the Month of August 1779*, Naples 1779; both also by Reprint-Verlag, 1990.

Hammond, C. (1989) The contribution of Henry Clifton Sorby to the study of reflected light microscopy of iron and steel, *Historical Metallurgy* 23, no. 1: 1–8.

Hammond, John H. (1981) *The Camera Obscura: A Chronicle*, Bristol: Institute of Physics Publishing.

Hamou, Philippe (1995) *La Vision Perspective (1435–1740)*, Paris: Payot & Rivages, 1st ed.; 2nd exp. ed. 2007.

Hankins, Thomas (1999) Blood, dirt, and nomograms – a particular history of graphs, *Isis* 90: 50–80.

Harcourt, Glenn (1987) Andreas Vesalius and the anatomy of antique sculpture, *Representations* 17 winter: 28–61.

Hardtwig, Wolfgang & Hans-Ulrich Wehler (eds. 1996) *Kulturgeschichte heute*, Göttingen: Vandenhoeck & Ruprecht.

Harms, Wolfgang (ed. 1990) *Text und Bild, Bild und Text: DFG-Symposion 1988*, Stuttgart: Metzler.

Harrison, Stephen (1991) What do viruses look like? *The Harvey Lectures* 85: 27–152.

Harwood, John T. (1989) Rhetoric and graphics in Micrographia, in: M. Hunter & S. Schaffer (eds.) *Robert Hooke: New Studies*, Woodbridge: Boydell: 119–47.

Haüy, René-Just (1782*a*) Mémoire sur la structure des spaths calcaires, *Observations sur la Physique* 20: 33–9 & pl. I.

— (1782*b*) Mémoire sur la structure des cristaux de Granat, ibid., 366–70 & pl. II.

— (1793) Exposition de la théorie sur la structure des cristaux, *Annales de Chimie*, 17, June: 225–319 & pl. I–IV.

— (1801) *Traité de Minéralogie*, Paris: Louis (5 vols.).

— (1802) Sur de nouvelles variétés de chaux carbonaté, *Annales du Muséum National d'Histoire Naturelle* 1: 114–26.

— (1815) Mémoire sur une loi de la cristallisation, appelée loi de symétrie, *Mémoires du Muséum d'Histoire Naturelle, Paris* 1: 81–101, 206–25, 273–95, 341–52.

— (1818) Observations sur la mesure des angles des cristaux, *Annales des Mines* 3 (1818): 411–42 & pl. I.

— (1822) *Traité de Cristallographie*, Paris: Bachelier.

Hawkes, Peter (ed. 1985) Beginnings of electron microscopy, *Advances in Electronics and Electron Physics Supplement* no. 16.

Hawkesworth, John (1773) *Account of the Voyages of James Cook*. 3 vols. London: Strahan & Cadell.

Heering, Peter (2005) Jean Paul Marats öffentliche Experimente und ihre Analyse mit der Replikationsmethode, *NTM* new ser. 13: 17–32.

Heering, P., Falk Riess & Christian Sichau (ed. 2000) *Im Labor der Physikgeschichte: Zur Untersuchung historischer Experimentalphysik*, Oldenburg: BIS-Verlag.

Heidelberger, Michael & Friedrich Steinle (eds. 1998) *Experimental Essays – Versuche zum Experiment*, Baden-Baden: Nomos.

Heilbron, John L. (1993) Weighing imponderables and other quantitative science around 1800, *HSPS* 24, no. 1, suppl.

— (2009) *Galileo*, Oxford: Oxford University Press.

Hein, George E. (1966) Kekulé and the architecture of molecules, in: Benfey (ed. 1966): 1–12.

Heintz, Bettina & Jörg Huber (eds. 2001) *Mit den Augen denken. Strategien der Sichtbarmachung*, Vienna: Voldemeer.

Heller, John Lewis (1964) The early history of binomial nomenclature, (*a*) *Huntia* 1: 22–70; (*b*) reprint in: *Studies in Linnaean Method and Nomenclature*, Frankfurt, Bern & New York: Lang, 1979.

Helmholtz, Hermann von (1852) Ueber die Theorie der zusammengesetzten Farben, *Annalen der Physik* (2nd ser.) 87 = 163: 45–66.

— (1867) *Handbuch der physiologischen Optik*, Leipzig: Voss.

Helmholtz-Gesellschaft (ed. 2009): *Wunderkammer Wissenschaft. Wanderausstellung der Helmholtz-Gesellschaft*, issued in Bonn in 17 leaflets, 2009 and shown across the Federal Republic of Germany, 2009–2012.

Henderson, Kathryn (1995) The visual culture of engineers, in: Susan Leigh Star (ed.) *Cultures of Computing*, Oxford: Blackwell: 196–218.

— (1999) *On Line and on Paper: Visual Representation, Visual Culture and Computer Graphics in Design Engineering*, Cambridge, MA: MIT Press.

Henderson, Linda Dalrymple (1983): *The Fourth Dimension and Non-Euclidean Geometry in Modern Art*, Princeton, NJ: Princeton University Press.

— (1988) X-rays and the quest for invisible reality in the art of Kupka, Duchamp, and the Cubists, *Art Journal* 47: 323–40.

Hennig, Jürgen (2001) Chancen und Probleme bildgebender Verfahren für die Neurologie, *Freiburger Universitätsblätter* 3: 67–86.

— (2005) Changes in the design of scanning tunneling microscopy images from 1980 to 1990, *Techné* 8, no. 2: 1–20.

— (2011) *Bildpraxis. Visuelle Strategien in früher Nanotechnologie*, Bielefeld: Transcript.

Henri, Victor (1908) Étude cinématographique des mouvements browniens, *CRAS* 146: 1024–6.

Hentschel, Klaus (1987) Mach, Ernst, *Neue Deutsche Biographie* 15: 605–9.

— (1998*a*) *Zum Zusammenspiel von Instrument, Experiment und Theorie: Rotverschiebung im Sonnenspektrum und verwandte spektrale Verschiebungseffekte von 1880 bis 1960*, Hamburg: Kovač (2 vols.).

— (1998*b*) A breakdown of intersubjective measurement: the case of solar rotation measurements in the early 20th century, *Studies in the History and Philosophy of Modern Physics* 29B: 473–507.

— (1999*a*) Photographic mapping of the solar spectrum, 1864–1900, *JHA* 30: 93–119, 201–24.

— (1999*b*) The culture of visual representations in spectroscopic education and laboratory instruction, *Physics in Perspective* 1: 282–327.

— (2000*a*) Drawing, engraving, photographing, plotting, printing: recent historical studies of visual representations, particularly in astronomy, *Acta Historica Astronomiae* 9: 11–52.

— (2000*b*) Historiographische Anmerkungen zum Verhältnis von Experiment, Instrumentation und Theorie, in: Christoph Meinel (ed.) *Instrument–Experiment: Historische Studien*, Bassum/Stuttgart: GNT-Verlag: 13–51.

— (2002*a*) *Mapping the Spectrum. Techniques of Visual Representation in Research and Teaching*, Oxford: Oxford University Press.

— (2002*b*) Spectroscopic portraiture, *Annals of Science* 59: 57–82.

— (2002*c*) Zur Geschichte visueller Darstellungen von Spektren, *Naturwissenschaftliche Rundschau* Nov. 55: 577–87.

— (2002*d*) Spectroscopy or spectroscopies? *Nuncius* 17,2: 589–614.

— (2003*a*) Der Vergleich als Brücke zwischen Wissenschaftstheorie und -geschichte, *Journal for General Philosophy of Science* 34: 251–75.

— (2003*b*) Review of U. Klein (ed.) *Tools and Modes of Representation in the Laboratory Sciences*, *BJHS* 36: 111–2.

— (2005) Wissenschaftliche Photographie als visuelle Kultur: Die Erforschung und Dokumentation von Spektren,' *Berichte zur Wissenschaftsgeschichte* 28, no. 3: 193–214.

— (2006*a*) Zur technischen Konstituierung und historischen Analyse wissenschaftlicher Bilder, in: Martina Hessler (ed.) *Konstruierte Sichtbarkeiten. Wissenschafts- und Technikbilder seit der frühen Neuzeit*, Munich: Fink: 117–27.

— (2006*b*) Zur Rolle der Ästhetik in visuellen Wissenschaftskulturen, in: Wolfgang Krohn (ed.) *Ästhetik der Wissenschaften*, Hamburg: Meiner: 233–56 & color pl. IV.

— (2006*c*) Verengte Sichtweise: Folgen der Newtonianischen Optik für die Farbwahrnehmung bis ins 19. Jahrhundert. *Bildwelten des Wissens. Kunsthistorisches Jahrbuch für Bildkritik*, 4, no. 1: 78–89.

— (2006*d*) Das Kartieren von Spektren, *Acta Historica Leopoldina* 46: 223–46.

— (2007) *Unsichtbares Licht? Dunkle Wärme? Chemische Strahlen? Eine wissenschaftshistorische und -theoretische Analyse von Argumenten für das Klassifizieren von Strahlungssorten 1650–1925 mit Schwerpunkt auf den Jahren 1770–1850*, Diepholz, Stuttgart & Berlin: GNT-Verlag.

— (2008*a*) Visual culture in scientific practice: the case of Charles Piazzi Smyth, in: Renate Brosch (ed.) *Victorian Visual Culture*, Munich: Winter: 145–64.

— (2008*b*) Review of Daston & Galison (2007), *Centaurus* 50: 329–30.

— (2008*c*) Kultur und Technik in Engführung: Visuelle Analogien und Muster-erkennung am Beispiel der Findung der Balmerformel, in: *Themenheft Forschung* [Universität Stuttgart] 2008, No. 4: 100–9.

— (ed. 2008) *Unsichtbare Hände. Zur Rolle von Laborassistenten, Mechanikern, Zeichnern u.a. Amanuenses in der physikalischen Forschungs- und Entwicklungsarbeit*, Stuttgart, Bassum & Berlin: GNT-Verlag.

— (2009) Elektronenbahnen, Quantensprünge und Spektren, in: Bigg & Hennig (eds. 2009): 51–61.

— (2010) Die Funktion von Analogien in den Naturwissenschaften, auch in Abgrenzung zu Metaphern und Modellen, in: Klaus Hentschel (ed.) *Analogien in Naturwissenschaft und Medizin, Acta Historica Leopoldina* 56: 3–56.

— (2011*a*) Von der Werkstoffforschung zur materials science, *NTM* new ser. 19,1: 5–40.

— (2011*b*) Bildpraxis in historischer Perspektive (essay review of Burri (2008), Elkins (2008), Adelmann et al. (2009), Hessler & Mersch (eds. 2009), Rocke (2010)), *NTM* new ser. 19,4: 413–24.

— (2012) The Stuttgart Database of Scientific Illustrators 1450–1950: making the invisible hands visible, *Spontaneous Generations* 6,1, available online at http://spontaneousgenerations.library.utoronto.ca/index.php/Spontaneous Generations/article/view/17156 (last viewed April 26, 2013).

— (2014*a*) Fruitful misinterpretation: Mach and Einstein on Newton, in: Scott Mandelbrote & Helmut Pulte (eds.) *The Reception of Isaac Newton in Europe*, forthcoming.

— (2014*b*) Newton's looming shadow: color perception in the era of Newtonian optics, in Uta Hassler & Robin Rehm (eds.) *Maltechnik und Farbmittel der Semperzeit*, Zurich: ETH, forthcoming.

— (2014*c*) (Scientific) photography as a research-enabling technology – not a discipline,' in the workshop proceedings: Maren Gröning (ed.) *Photographie als Schuldisziplin/Photography as a Schooling Issue*, Vienna, forthcoming.

— (2014*d*) David Brewster, in: Thomas Hockney et al. (eds.) *Biographical Encyclopedia of Astronomers*, 2nd ed.

Hentschel, K. (ed. 1996) & Ann M. Hentschel (ed. assist. & transl.): *Physics and National Socialism: An Anthology of Primary Sources*, Basel: Birkhäuser.

— (2001) An engraver in nineteenth-century Paris: The career of Pierre Dulos,

French History 15: 64–102.

Hentschel, K. & Axel Wittmann (eds. 2000) *The Role of Visual Representations in Astronomy: History and Research Practice*, Frankfurt: Harri Deutsch (*Acta Historica Astronomiae*, vol. 9).

Hentze, Reinhard (2000) Frühe Zeugnisse zur Photographiegeschichte, *Acta Historica Leopoldina* 36: 301–28.

Herlinger, E. (1928) Review of Bragg (1928/30*a*), *Zeitschrift für angewandte Chemie* 42 (1928): 165.

Herrlinger, Robert & Putscher, Marliene (1972) *Geschichte der medizinischen Abbildung*, Munich: Moos (2 Vols).

Herschel, William (1800*a*) Investigation of the powers of the prismatic colours to heat and illuminate objects; with remarks, that prove the different refrangibility of radiant heat. To which is added an inquiry into the method of viewing the sun advantageously with large telescopes of large apertures and high magnifying powers, *PTRSL* 90: 255–83 & pl. X.

— (1800*b*) Experiments on the refrangibility of the invisible rays of the Sun, *PTRSL* 90: 284–92 & pl. XI.

— (1800*c*) Experiments on the solar, and on the terrestrial rays that occasion heat; with a comparative view of the laws to which light and heat, or rather the rays which occasion them, are subject, in order to determine whether they are the same or different, *PTRSL* 90 , 293–326 & pl. XII–XVI (part I); part II: ibid. 90 (1800*d*), 437–538 & pl. XX–XXVI.

— (1811) Astronomical observations relating to the construction of the heavens ..., *PTRSL* 101: 209–336 & pl. IV–V.

Hessler, Martina (2005) Bilder zwischen Kunst und Wissenschaft. Neue Herausforderungen für die Forschung, *Geschichte und Gesellschaft* 31: 266–92.

— (2006*a*) Von der doppelten Unsichtbarkeit digitaler Bilder, *Zeitenblicke* 5,3: 1–12.

— (2006*b*) Die Konstruktion visueller Selbstverständlichkeiten. Überlegungen zu einer Visual History der Wissenschaft und Technik, in: Gerhard Paul (ed.) *Visual History*, Göttingen: Vandenhoeck & Ruprecht: 76–95.

— (ed. 2006) *Konstruierte Sichtbarkeiten. Wissenschafts- und Technikbilder seit der Frühen Neuzeit*, Munich: Fink.

Hessler, M., Jochen Hennig & Dieter Mersch (2004) *Explorationsstudie im Rahmen der BMBF-Förderinitiative "Wissen für Entscheidungsprozesse" zum Thema Visualisierung in der Wissenskommunikation* (online).

Hessler, M. & Dieter Mersch (eds. 2009) *Logik des Bildlichen. Zur Kritik der ikonischen*

Vernunft, Bielefeld: Transcript.

Hetherington, Norris S. (1976) The British Astronomical Association and the controversy over canals of Mars, *Journal of the British Astronomical Association* 86, no. 4: 303–8.

Heumann, Ina (2013) Linus Pauling, Roger Hayward und der Wert von Sichtbarmachungen, *Berichte zur Wissenschaftsgeschichte* 36: 313–33.

Higham, Norman 1963 *A Very Scientific Gentleman: The Major Achievements of Henry Clifton Sorby*, Oxford: Pergamon Press.

Hill, K. A. (1985) Hartsoeker's homunculus – a corrective note, *Journal of the History of Behavioural Sciences* 21, no. 2: 178–9.

Hill, O. R. (1973) Medical ultrasonics: an historical review, *British Journal of Radiology* 46: 899–905.

Hille, Bertile (1984) *Ionic Channels of Excitable Membranes*, Sunderland, MA: Sinauer.

Hiller, Joh. E. (1942) Paracelsus und de Boodt als Vorläufer neuzeitlicher Mineralogie, *Die Naturwissenschaften* 30: 563–5.

— (1952) Die Mineralogie des Paracelsus, *Philosophia Naturalis* 2 (1952): 293–331, 435–78.

Hills, Richard L. (1989) *Power from Steam: A History of the Stationary Steam Engine*, Cambridge: Cambridge University Press.

— (2002) *James Watt*, Ashbourne, vol. I: *His Time in Scotland*; vol. II: *His Time in England*, 2005 and vol. III: *Triumph through Adversity*, 2006.

Hilts, V. L. (1975) A guide to Francis Galton's English Men of Science, *Transactions of the American Philosophical Society* 65, no. 5.

Hindle, Brooke (1983) *Emulation and Invention*, New York: Norton.

— (1984) Spatial thinking in the bridge era: John August Roebling versus John Adolphus Etzler, *Annals of the New York Academy of Sciences* 424: 131–47.

Hineline, Mark L. (1993) *The Visual Culture of the Earth Sciences, 1863–1970*, PhD thesis, University of California, San Diego (umi microfilm 94-20724).

Hodges, Elaine R.S. (1989) Scientific illustration: a working relationship between the scientist and artist, *Bioscience* 39: 104–11.

Hoff, Hebbel E. & Leslie A. Geddes (1959) Graphic registration before Ludwig. The antecedents of the kymograph, *Isis* 50: 5–21.

— (1960) Ballistics and the instrumentation of physiology, *Journal of the History of Medicine and Allied Sciences* 15: 133–46.

Hoffmann, Albrecht (1990) *Das Stereoskop. Geschichte der Stereoskopie*, Munich:

Deutsches Museum.

Hoffmann, Christoph (ed. 2008) *Daten sichern Schreiben und Zeichnen als Verfahren der Aufzeichnung*, Zürich: Diaphanes.

Hoffmann, C. & Peter Berz (eds. 2001) *Über Schall. Ernst Machs und Peter Salchers Geschoß-Photographien*, Göttingen: Wallstein.

Hoffmann, Roald (1990) Molecular beauty, *Journal of Aesthetics and Art Criticism* 48, no. 3: 191–204.

Hofmann, August Wilhelm (1865) On the combining power of atoms, (*a*) *Proceedings of the Royal Institution* 4: 401–30; (*b*) *Chemical News* 12: 166–9, 175–87.

Hofmann, Wilhelm (1999) Die Sichtbarkeit der Macht, in: W. Hofmann (ed.) *Die Sichtbarkeit der Macht: Theoretische und empirische Untersuchungen zur visuellen Politik*, Baden-Baden: Nomos: 7–11.

Holländer, Hans (ed. 2000) *Erkenntnis, Erfindung, Konstruktion. Studien zur Bildgeschichte von Naturwissenschaften und Technik vom 16. bis zum 19. Jahrhundert*, Berlin: Mann.

Holm, Olov Fr.: Cinematography in cerebral angiography, *Acta Radiologica* 25 (1944), 163–73.

Holmes, Frederic L. & Kathryn Olesko (1994) The images of precision. Helmholtz and the graphical method in physiology, in: Norton Wise (ed.) *The Values of Precision*, Princeton: Princeton University Press: 198–21.

Holmes, Oliver Wendell (1859) The stereoscope and the stereograph, *Atlantic Monthly* 6 (June 1859): 738–48.

— (1861) Sun-painting and sun-sculpture, *Atlantic Monthly* 8 (July 1861): 13–29.

— (1869) Holmes's handheld stereoscope, *Philadelphia Photographer* 6: 24.

Holmgren, B. S. (1945) Roentgen cinematography as a method, *Acta Radiologica* 26: 286–92.

Holthuis, Lipke Bijdeley (1996) Original watercolours donated by Cornelius Sittardus to Conrad Gesner, *Zoologische Mededelingen* 70: 169–96.

Holton, Gerald (1973) *Thematic Origins of Scientific Thought*, Cambridge, MA: Harvard University Press.

Hon, Giora & Bernard Goldstine (2008) *From Summetria to Symmetry: The Making of a Revolutionary Scientific Concept*, Berlin: Springer.

Hooke, Robert (1665) *Micrographia, or Some Physiological Descriptions of Minute Bodies Made with Magnifying Glasses, with Observations and Inquiries Thereupon*, 1st ed. London: Martyn & Allestry; reprint Lincolnwood, 1987.

Hopwood, Nick (2006) Pictures of evolution and charges of fraud. Ernst Haeckel's embryological illustrations, *Isis* 97: 260–301.

— (2007) Artist versus anatomist, models against dissection: Paul Zeiller of Munich and the revolution of 1848, *Medical History* 51: 279–308.

— (2008) Model politics, *The Lancet* 372: 1946–7.

Hoquet, Thierry (2007) *Buffon illustré: Les gravures de l'Histoire naturelle (1749–1767)*, Paris: Publications Scientifiques du Muséum national d'Histoire naturelle.

Horner, William George (1834) On the properties of the Daedaleum, a new instrument of optical illusion, *Phil. Mag.* (3) 4: 36–41.

Hoult, D. I., S. J. W. Busby, D. G. Godian, G. K. Radda, R. E. Richards & P. J. Seeley (1974) Observation of tissue metabolites using 31P nuclear magnetic resonance, *Nature* 252: 285–7.

Hounsfield, Godfrey Newbold (1980) Computed medical imaging, *Journal of Computer-Assisted Tomography* 4, no. 5: 665–74.

Hounsfield, G. N. & J. Ambrose (1973) Computerized transverse axial scanning (tomography) I & II, *British Journal of Radiology* 46: 1016–22, 1023ff.

Hoykaas, Reijer (1955) Les debuts de la théorie cristallographique de R. J. Haüy d'après des documents originaux, *Revue d'Histoire des Sciences* 8: 319–37.

Hoyt, William Graves (1976) *Lowell and Mars*, Tucson: Univ. of Arizona Press.

Huber, Daniel (1829) *Johann Heinrich Lambert: Nach seinem Leben und Wirken*, Basel: Schweighauser.

Hudson, W. (1960) Pictorial depth perception in sub-cultural groups in Africa, *Journal of Social Psychology* 52: 183–208.

Humm, Felix (1972) *J. H. Lambert in Chur 1748–1763)*, Chur: Calven.

Humphries, D.W. (1965) Sorby – the father of microscopical petrography, in: Smith (ed. 1965): 17–41.

Huygens, Christiaan (1690/1912) *Traité de la lumière*, (a) Leyden, 1690; (b) English transl. by S. P. Thompson: *Treatise on Light*, London: Macmillan, 1912.

Ihde, Don (2000) Epistemology engines, *Nature* 406: 21.

Ifrah, Georges (1991) *Universalgeschichte der Zahlen*, Frankfurt: Suhrkamp.

Inwood, Stephen (2002) *The Man Who Knew Too Much*, London: MacMillan.

Ivins, William M. (1938) *On the Rationalization of Sight*, New York; reprint 1973.

— (1973) *Prints and Visual Communication*, Cambridge, MA: MIT Press.

Jackson, Christine E. (1975) *Bird Illustrators: Some Artists in Early Lithography*, London: Littlehampton Book Services.

— (1978) *Wood Engravings of Birds*, London: H. F. & G. Witherby, 1978.

— (1989) *Bird Etchings: The Illustrators and Their Books, 1655–1855*, Ithaca: Cornell University Press.

— (1999) *Dictionary of Bird Artists of the World*, Antique Collectors' Club.

— (2011) The painting of hand-coloured zoological illustrations, *Archives of Natural History* 38: 36–52.

Jackson, Myles W. (2006) *Harmonious Triads. Physicists, Musicians, and Instrument Makers in Nineteenth-Century Germany*, Cambridge, MA: MIT Press.

James, R. F. (1935) Roentgen cinematography, *Journal of the Society of Motion Picture Engineers* 24, no. 3: 233–40.

Janker, Rudolf (1931) Zur Röntgenkinematographie, *Fortschritte auf dem Gebiete der Röntgenstrahlen* 44: 657–68.

— (1949) Das endlose röntgenkinematographische Band bei der Röntgenuntersuchung des Herzens, *Fortschritte auf dem Gebiete der Röntgenstrahlen* 71: 345–8.

Janowitz, Günther J. (1986) *Leonardo da Vinci, Brunelleschi, Dürer: Ihre Auseinandersetzung mit der Problematik der Zentralperspektive*, Einhausen: Huebner.

Janssen, Jules (1860) Sur l'absorption de la chaleur rayonnante obscure dans les milieux de l'oeil, *Annales de Physique et de Chimie* (3) 60: 71–93.

— (1873) Méthode pour obtenir photographiquement l'instant des contact avec les circonstances physiques qu'ils présentent, *CRAS* 76: 677–9.

— (1874) Présentation de quelques spécimens de photographies solaires obtenues avec un appareil construit pour la mission du Japon, *CRAS* 78: 1730–1.

— (1876) Présentation d revolver photographique et épreuves obtenus avec cet instrument, *Bulletin de la Société Française de Photographie* 22: 100–6.

— (1877*a*) Note sur la reproduction par la photographie des 'grains de riz' de la surface solaire, *CRAS* 85: 373.

— (1877*b*) Sur le réseau photosphérique solaire, *CRAS* 85: 775–6.

— (1877*c*) Sur la constitution de la surface solaire et sur la photographie envisagée comme moyen de découverte en astronomie physique, *CRAS* 85: 1249–55.

— (1878) Note sur le réseau photosphérique solaire et la photographie envisagée comme moyen de découvertes en astronomie physique, *Annuaire publié par le Bureau des Longitudes*, Paris: 689–700 & pl.

— (1882) Remarques sur la communication de M. Marey sur la photographie des diverses phases du vol des oiseaux, *CRAS* 94: 684–5 (on Marey (1882*a*)).

— (1883*a*) Les méthodes en astronomie physique, *Moniteur* 22, no. 3, Feb. 1: 23

(opening speech on Aug. 24, 1882, at the *Congrès de l'Association Française pour l'Avancement des Sciences*).

— (1883*b*) Les progrès de l'astronomie physique, *L'Astronomie. Revue mensuelle d'Astronomie populaire, de météorologie et de physique du globe* 2: 121–8.

— (1887) La photographie céleste, (*a*) *CRAS* 104: 1067–70; (*b*) in full in *Revue scientifique de la France et de l'étranger* 33, no. 2 (Jan. 14, 1888): 33–42; (*c*) as separatum Paris 1888.

— (1888) Photographies météorologiques, *Cosmos. Revue des Sciences* Année 37, no. 9: 323–6.

— (1896) Mémoire sur la photographie solaire, *Annales de l'Observatoire Physique de Paris* 1: 91–124.

— (1903) *Atlas des photographies solaires*, Paris: Observatoire de Meudon.

Japp, Francis R. (1898) Kekulé memorial lecture, *Journal of the Chemical Society* 73: 97–108.

Jardine, Lisa (2003) *On a Grander Scale: The Outstanding Career of Sir Christopher Wren*, London: Harper.

Jarrell, Richard A. (1998) Visionary or bureaucrat? T. H. Huxley, the Science and Art Department and science teaching for the working class, *Annals of Science* 55: 219–40.

Jay, Martin (1988) Scopic regimes of modernity, (*a*) in: Hal Foster (ed.) *Vision and Visuality*, Seattle: Bay Press: 3–23; (*b*) excerpted in: Mirzoeff (ed. 1998): 66–9.

— (1993) *Downcast Eyes. The Denigration of Vision in 20th Century French Thought*, Berkeley: University of California Press.

— (2002) That visual turn, *Journal of Visual Culture* 1,1: 87–92.

Jenkin, John (2011) *William and Lawrence Bragg, Father and Son: The Most Extraordinary Collaboration in Science*, Oxford: Oxford University Press.

Jenkins, Reese V. (1975) *Images and Enterprise. Technology and the American Photographic Industry, 1839 to 1925*, Baltimore: Johns Hopkins University Press.

Joblot, Louis (1718) *Descriptions et usages de plusieurs nouveaux microscopes, tant simples que composez*, Paris: Collombat.

Johns, Adrian (1998) *The Nature of the Book. Print and Knowledge in the Making*, Chicago: University of Chicago Press.

Johnson, D. & M. Cantino (1988) Artifacts of analysis in biological electron microscopy, in: Richard F. E. Crang & K. Klomparens (eds.) *Artifacts in Biological Electron Microscopy*, New York: Plenum: 219–27.

Johnson, Jeffrey Allan (2013) On molecular models, mobilization, and the paradoxes of modernizing chemistry in Nazi Germany, *Historical Studies in the Natural Sciences* 43: 391–452.

Johnson, Reuben D. & Willow Sainsbury (2012) The combined eye of surgeon and artist: evaluation of the artists who illustrated for Cushing, Dandy and Cairns, *Journal of Clinical Neuroscience* 19, no. 1: 34–8.

Johnston, Sean (2005*a*) From white elephant to Nobel Prize: Dennis Gabor's wavefront reconstruction, *HSPS* 36: 35–70.

— (2005*b*) Attributing scientific and technical progress: the case of holography, *History and Technology* 21: 367–92.

— (2005*c*) Shifting perspectives: holography and the emergence of technical communities, *Technology and Culture* 46: 77–103.

— (2006*a*) *Holographic Visions: A History of New Science*, Oxford: Oxford University Press.

— (2006*b*) Absorbing new subjects: holography as an analog of photography, *Physics in Perspective* 8: 164–88.

Johnston, Stephen (1994) *Making Mathematical Practice: Gentlemen, Practitioners and Artisans in Elisabethan England*, PhD Dissertation, Cambridge University.

Jonas, Hans (1961) Homo pictor und die Differentia des Menschen, *Zeitschrift für philosophische Forschung* 15: 161–76.

Jordanova, Ludmilla (1990) Medicine and visual culture, *Social History of Medicine* 3: 89–99.

Joyce, Kelly (2005) Appealing images: magnetic resonance imaging and the production of authoritative knowledge, *Social Studies of Science* 35, no. 3: 437–62.

Judd, John W. (1908) Henry Clifton Sorby and the birth of microscopical petrology, *The Geological Magazine*, new ser. decade V, vol. V: 193–204 & pl. VIII.

Jungnickel, Christa & Russell McCormmach (1986) *Intellectual Mastery of Nature: Theoretical Physics from Ohm to Einstein*, Chicago: University of Chicago Press.

Junker, R. (1931) Zur Röntgenkinematographie, *Fortschritte auf dem Gebiete der Röntgenstrahlen* 44: 658–68.

Jussim, Estelle (1974) *Visual Communication and the Graphic Arts: Photographic Technologies in the 19th Century*, New York: Bonker.

Jussim, E. & Gus Kayafas (eds. 2000) *Stopping Time: The Photographs of Harold Edgerton*, Boston: Abrams.

Jütte, Robert (1998) Die Entdeckung des 'inneren' Menschen 1500–1800, in: R. van Dülmen (ed.) *Erfindung des Menschen 1500–1800*, Cologne 241–58.

Kahlow, Andreas (2006) Johann August Roebling (1806–69): early projects in context, *Proceedings of the 2nd Intern. Congress on Construction History* 2: 1755–76.

Kaiser, David (1998) A Ψ is just a Ψ? Pedagogy, practice, and the reconstitution of general relativity, 1942–1975, *SHPS* part B: *Studies in History and Philosophy of Modern Physics* 29, no. 3: 321–38.

— (2000) Stick-figure realism: conventions, reification, and the persistence of Feynman diagrams, 1948–1964, *Representations*, spring 70: 49–86.

— (2005*a*) Training and the generalist's vision in the history of science, *Isis* 96, no. 2: 244–51.

— (2005*b*) *Drawing Theories Apart: The Dispersion of Feynman Diagrams in Postwar Physics*, Chicago: University of Chicago Press.

— (ed. 2005) *Pedagogy and the Practice of Science: Historical and Contemporary Perspectives*, Cambridge, MA: MIT Press.

— (2012) A tale of two textbooks: experiments in genre, *Isis* 103 (2012): 126–38.

Kalender, Willi A. (2000) *Computertomographie*, (*a*) Munich: Publicis MCD Verlag, 1st ed.; (*b*) expanded 2nd ed. 2006.

— (2005) CT: the unexpected evolution of an imaging modality, *European Radiology Suppl.* 15, suppl. 4: D21–D24.

— (2006) X-ray computed tomography, *Physics in Medicine and Biology* 51: R29–R43.

— (2011) *Computed Tomography. Fundamentals, System Technology, Image Quality, Applications*, Erlangen: Publicis, 3rd ed.

Kanefsky, John & John Robey (1980) Steam engines in 18th-century Britain: a quantitative assessment, *Technology and Culture* 21: 161–86.

Kargon, Robert (1969) Model and analogy in Victorian science: Maxwell and the French physicists, *Journal of the History of Ideas* 30: 423–36.

Kassabian, Mihran Krikor (1901) *Röntgen Rays and Electro-Therapeutics, with Chapters on Radium and Photo-Therapy*, Philadelphia: Lippincott.

Kassirer, J. P. (1992) Images of clinical medicine, *New England Journal of Medicine* 326: 829–30.

Kauffman, James (2009) *Selling Outer Space: Kennedy, the Media, and Funding for Project Apollo, 1961–1963*, Tuscaloosa: University of Alabama Press.

Kaufmann, Thomas Da Costa (1975) The perspective of shadow: the history of the theory of shadow projection, *JWCI* 38: 258–87.

Kekulé, August (1858*a*) Ueber die Constitution und die Metamorphosen der chemischen Verbindungen und über die chemische Natur des Kohlenstoffs,

Annalen der Chemie 106: 129–59.

— (1858*b*) Remarques à l'occasion d'une note de M. Couper sur une nouvelle théorie chimique, *CRAS* 47: 378–80 (commentary on Couper (1858*a*)).

— (1861/82) *Lehrbuch der organischen Chemie oder der Chemie der Kohlenstoffverbindungen*, Erlangen: Enke, vol. 1, 1861; vol. 2, 1866; vol. 3, parts 1–4, 1867–1882.

— (1865) Sur la constitution des substances aromatiques, (*a*) *Bulletin de la Société Chimique de France* (2) no. 3: 98–110; (*b*) in German: Ueber die Constitution der aromatischen Verbindungen, *Annalen der Chemie* 137 (1866): 129–96 & pl.

— (1867) Über die Constitution des Mesitylens, *Zeitschrift für Chemie* new ser. no. 3: 214–18.

— (1869) Über die Constitution der Salze und Isomormophismus, *Berichte der Deutschen Chemischen Gesellschaft* 2: 652–7.

— (1890) [Address on the jubilee of the theory of benzol], (*a*) *Berichte der Deutschen Chemischen Gesellschaft* 23: 1302–11; (*b*) partial Engl. transl. in Japp (1898), (*c*) alternative Engl. transl. and interpretation in Rothenberg (1995).

Keller, Corey (2004) The naked truth or a shadow of doubt? X-rays and the problem of transparency, *Invisible Culture* 7; available online at http://hdl.handle.net/1802/2981 (last accessed March 4, 2012).

Keller, C., Monika Faber & Maren Gröning (eds. 2009) *Fotografie und das Unsichtbare 1840–1900*, Vienna: Brandtstätter.

Keller, Evelyn Fox (1996) The biological gaze, in: George Robertson et al. (eds.) *Future Natural. Nature, Science, Culture*, London: Routledge: 107–21.

Kemner, Gerhard (ed. 1989) *Stereoskopie, Technik, Wissenschaft, Kunst und Hobby*, Berlin: Museum für Verkehr und Technik.

Kemp, Martin (1970) A drawing for the Fabrica and some thoughts on the Vesalius muscle-men, *Medical History* 14: 277–88.

— (1972) Dissection and divinity in Leonardo's late anatomies, *JWCI* 35: 200–25.

— (1977) Leonardo and the visual pyramid, *JWCI* 40: 128–49 & pl.

— (1978) Science, non-science and nonsense. The interpretation of Brunelleschi's perspective, *Art History* 1: 134–61.

— (1984) Geometrical perspective from Brunelleschi to Desargues: a pictorial means or an intellectual end? *Proceedings of the British Academy* 70: 91–132.

— (1990) *The Science of Art. Optical Themes in Western Art from Brunelleschi to Seurat*, New Haven, CT: Yale University Press.

— (1996) 'Implanted in our natures': humans, plants, and the stories of art, in:

David Philip Miller & Peter Hanns Reill (eds.) *Visions of Empire: Voyages, Botany, and Representations of Nature*, Cambridge: Cambridge University Press: 197–229.

— (2000) *Visualizations. The Nature Book of Art and Science*, Oxford: Oxford University Press; in German transl.: *Bilderwissen. Die Anschaulichkeit naturwissenschaftlicher Phänomene*, Cologne: DuMont, 2003.

— (2004) *Leonardo*, (*a*) Oxford: Oxford University Press; (*b*) in German transl.: *Leonardo*, Munich: Beck, 2005.

— (2007) *Leonardo da Vinci. Experience, Experiment and Design*, London: Victoria and Albert Museum.

Kempf, P. (1915) Oswald Lohse, *Vierteljahrsschrift der Astronomischen Gesellschaft* 50: 160–9 & pl.

Kepes, Györgi (1944) *Language of Vision*, Chicago: Paul Theobald; reprint 1995.

— (1963) *The New Landscape in Art and Science*, Chicago: Paul Theobald.

— (1965) *Structure in Art and in Science*, New York: Braziller.

Kerker, Milton (1961) Science and the steam engine, *Technology and Culture* 2: 381–90.

Kessler, Elizabeth A. (2007) Resolving the nebulae: The science and art of representing M51, *SHPS* 38: 477–91.

— (2012) *Picturing the Cosmos: Hubble Space Telescope Images and the Astronomical Sublime*, Minneapolis: University of Minnesota Press.

Ketham, Johannes de (1491) *Fasciculus Medicinae*, (*a*) Venice; (*b*) facsimile with Engl. transl. by Luke Demaitre: *The Fasciculus Medicinae*, Birmingham 1988 (Classics of Medicine Library).

Kevles, Bettyann H. (1997) *Naked to the Bone: Medical Imaging in the 20th Century*, Reading, MA: Addison-Wesley.

Kheirandish, Elaheh (1999) *The Arabic Version of Euclid's Optics, Edited and Translated with Historical Introduction and Commentary*, Berlin: Springer (2 vols.).

Kimpel, Dieter (2005) Struktur und Wandel der Mittelalterlichen Baubetriebe, in: Roberto Cassanelli (ed.) *Die Baukunst im Mittelalter*, Düsseldorf: Patmos: 11–50.

Kinkelin, Friedrich (1882) Zur Geschichte des geometrischen Zeichnens, in: *Festschrift den Mitgliedern der Deutschen Anthropologischen Gesellschaft gewidmet bei Gelegenheit der XIII. Jahresversammlung*, Frankfurt am Main: 103–16.

Kippenhahn, Rudolf (2006) *Kippenhahns Sternstunden: Unterhaltsames und Erstaunliches aus der Welt der Sterne*, ed. by Justina Engelmann, Stuttgart: Franckh-Kosmos.

Kircher, Athanasius (1646) *Ars magna lucis et umbrae*, (*a*) 1st ed. Rome: Hermann

Scheus; (*b*) 2nd ed. Amsterdam 1671.

Kitcher, Philip & Achille Varzi (2000) Some pictures are worth $2_0^\aleph$ sentences, *Philosophy* 75: 377–81.

Kittler, Friedrich (1986) *Grammophon, Film, Typewriter*, Berlin: Brinkmann & Bose.

Klein, Ursula (2001) Paper tools in experimental cultures, *SHPS* 32: 265–302.

— (ed. 2001) *Tools and Modes of Representation in the Laboratory Sciences*, Dordrecht: Kluwer.

Knight, David (1977): *Zoological Illustration: An Essay Towards a History of Printed Zoological Pictures*, London: Dawson.

— (1985) Scientific theory and visual language, in A. Ellenius (ed., 1985): 106–24.

— (1993) Pictures, diagrams and symbols: Visual language in 19th c. chemistry, in: R. Mazzolini (ed. 1993): 321–44.

Knobel, E. B. (1911) G. V. Schiaparelli, *MNRAS* 71: 283–7.

Knoll, Max & Ernst Ruska (1932) Das Elektronenmikroskop. *Zeitschrift für Physik* 78: 318–39.

Knorr, Wilbur (1991) On the principle of linear perspective in Euclid's optics, *Centaurus* 34: 193–210.

Knorr-Cetina, Karin (1991) *Die Fabrikation von Erkenntnis. Zur Anthropologie der Naturwissenschaft*, Frankfurt: Suhrkamp.

— (1999) *Epistemic Cultures. How the Sciences make Knowledge*, (*a*) Cambridge, MA: Harvard University Press; (*b*) German version: *Wissenskulturen*, Frankfurt: Suhrkamp, 2002.

— (1999) 'Viskurse' der Physik. Wie visuelle Darstellungen ein Wissenschaftsgebiet ordnen, in: Jörg Huber & Martin Heller (eds.) *Konstruktionen. Sichtbarkeiten*, Vienna: Springer: 245–63.

Koch, Robert (1876) Die Aetiologie der Milzbrand-Krankheit, begründet auf die Entwicklungsgeschichte des Bacillus Anthracis, *Cohns Beiträge zur Biologie der Pflanzen* 2,2: 277–310 & pl. XI.

— (1877) Verfahren zur Untersuchung, zum Conserviren und Photographiren der Bacterien, *Cohns Beiträge zur Biologie der Pflanzen* 2: 399–434 & pl.

— (1881) Zur Untersuchung von pathogenen Organismen, *Mittheilungen aus dem kaiserlichen Gesundheitsamte 1881* 1: 1-48 & photographic pl.

Kocka, Jürgen (1989) Probleme einer europäischen Geschichte in komparativer Absicht, in: J. Kocka: *Geschichte und Aufklärung*, Vandenhoek & Ruprecht, Göttingen: 21–8, 163–5.

Kolbe, Hermann (1865) *Das chemische Laboratorium der Universität Marburg und die seit 1859 darin ausgeführten chemischen Untersuchungen nebst Ansichten und Erfahrungen über die Methode des chemischen Unterrichts*, Braunschweig: Vieweg.

— (1877) Zeichen der Zeit, *Journal für praktische Chemie* (2) 14: 268–78; 15: 473–7.

— (1881*a*) Meine Betheiligung an der Entwickelung der theoretischen Chemie, *Journal für praktische Chemie* (2) 23: 305–23, 353–79.

— (1881*b*) Bemerkungen zu Lossen's Abhandlung: Ueber die Vertheilung der Atome in der Molekel, *Journal für praktische Chemie* (2) 23: 489–96.

Koltun, Walter L.(1965) Precision space-filling atomic models, *Biopolymers* 3: 665–79.

König, Wolfgang (1997) *Künstler und Strichezieher. Konstruktions- und Technikkulturen im deutschen, britischen, amerikanischen und französischen Maschinenbau zwischen 1850 und 1930*, Frankfurt: Suhrkamp.

Kort, Pamela et al. (2009) *Darwin, Kunst und die Suche nach den Ursprüngen*, (*a*) Cologne: Wienand; (*b*) in Engl. transl. *Darwin, Art and the Search for Origins*, ibid.

Kosslyn, Stephen M. (1990) Mental imagery, in: D. N. Osheran et al. (eds.) *Visual Cognition and Action. An Invitation to Cognitive Science*, vol. 2.

— (1994) *Image and Brain. The Resolution to the Imagery Debate*, Cambridge, MA: MIT Press.

Kragh, Helge (1990) *Dirac. A Scientific Biography*, Cambridge University Press.

— (1999) *Quantum Generations. A History of Physics in the 20th Century*, Princeton, NJ: Princeton University Press.

Kramer, Miriam, Dr. J. R. Kramer & John Benjamin (undated) *Roger Hayward: Architect, Artist, Illustrator, Inventor, Scientist*, available online at http://osulibrary.oregonstate.edu/specialcollections/coll/pauling/bond/people/hayward.html (last accessed June 15, 2009).

Kranakis, Eda (1997) *Constructing a Bridge: An Exploration of Engineering Culture, Design, and Research in Nineteenth-Century France and America*, Cambridge, MA: MIT Press.

Krieger, Murray (1984) The ambiguities of representation and illusion: An E. H. Gombrich retrospective, *Critical Inquiry* 11: 181–94.

Kroeber, Alfred L. & C. Kluckhohn (1952) *Culture: A Critical Review of Concepts and Definitions*, Cambridge, MA: Peabody Museum.

Krohn, Wolfgang (ed. 2006) *Ästhetik der Wissenschaften*, Hamburg: Meiner.

Kubbinga, Henk (2012) Crystallography from Haüy to Laue: controversies on the molecules and atomistic nature of solids, *Foundations of Crystallography* 227: 1–26.

Kuehni, Rolf G. (2002) The early development of the Munsell system, *Color Research and Application* 27: 10–27.

— (2008) *Color Ordered: A Survey of Color Systems from Antiquity to the Present*, Oxford: Oxford University Press.

Kuhn, Thomas S. & John L. Heilbron (1962): Interview with Dr. John Wheeler in Berkeley, California, March 24, 1962, American Institute of Physics, Maryland.

Kulvicki, John (2010) Knowing with images: medium and messages, *Philosophy of Science* 77: 295–313.

Kunz, George F. (1918) The life and work of Haüy, *American Mineralogist* 3: 60–89.

Kurrer, Karl-Eugen (1996) Anmerkungen zum Verhältnis von Text, Bild und Symbol in den klassischen technikwissenschaftlichen Grundlagendisziplinen von 1800 bis 1950, *Dresdner Beiträge zur Geschichte der Technikwissenschaften* 24: 74–98.

— (2008) *The History of the Theory of Structures: From Arch Analysis to Computational Mechanics*, Berlin: Ernst.

Kursunoglu, Behram N. & Eugene Paul Wigner (eds. 1990) *Paul Adrien Maurice Dirac: Reminiscences about a Great Physicist*, Cambridge University Press.

Kusukawa, Sachiko (1997) Leonhart Fuchs on the importance of pictures, *Journal of the History of Ideas* 58: 403–27.

— (2010) The sources of Gessner's pictures for Historia animalium, *Annals of Science* 67, no. 3: 303–28.

— (2012) *Picturing the Book of Nature. Image, Text and Argument in 16th Century Human Anatomy and Medical Botany*, Chicago: University of Chicago Press.

Kusukawa, S. & Maclean, Ian (eds. 2006) *Transmitting Knowledge. Words, Images, and Instruments in Early Modern Europe*, Oxford: Oxford University Press.

Kyeser, Conrad (1405) *Bellifortis*; facsimile ed. and transl. by Götz Quarg, Düsseldorf: VDI, 1967, 2 vols.

Lack, Hans Walter (2000): *Ein Garten für die Ewigkeit. Der Codex Liechtenstein*, Bern: Benteli, (*a*) 1st ed.; (*b*) 2nd rev. ed., 2003; (*c*) Engl. transl.: *Garden for Eternity. The Codex Liechtenstein*, ibid., 2001.

— (2001) *Ein Garten Eden: Meisterwerke der botanischen Illustration*, Cologne: Taschen.

— (2004) Ferdinand, Joseph und Franz Bauer: Testamente, Verlassenschaften und deren Schicksale, *Annalen des Naturhistorischen Museum in Wien* 104B: 479–551.

Lack, H. W. & Victoria Ibáñez (1997) Recording colours in late 18th century botanical drawings, *Curtis' Botanical Magazine* (6th ser.) 14: 87–100.

Lacroix, Alfred (1945) *René-Just Haüy 1743–1822)*, Paris: Masson.

Lambert, Johann Heinrich (1759) *Freye Perspective, oder Anweisung, jeden perspectivischen Aufriss von freyen Stücken und ohne Grundriss zu verfertigen*, (*a*) 1st ed. Zurich: Heidegger; (*b*) 2nd exp. ed. Zurich: Orell, Gessner & Füsslin, 1774; (*c*) as *Schriften zur Perspektive* with comm. by Max Steck and hitherto unpubl. mss. *Anlage zur Perspektive* and some of the earliest drawings by Lambert from 1752, Berlin: Lüttke Verlag 1943.

— (1760) *Photometria sive de mensura et gradibus luminis, colorum et umbrae*; in German transl. with comm. by E. Anding: *Lambert's Photometrie*, Leipzig: Engelmann, 1892 (Ostwalds Klassiker der exakten Wissenschaften, nos. 31–2).

— (1768) Mémoire sur la partie photometrique de l'art du peintre, *Histoire de l'Académie Royale des Sciences et Belles-Lettres de Berlin pour l'Année 1768*: 80–108 & pl. I, fig. 4.

— (1772*a*) Anmerkungen und Zusätze zur Entwerfung der Land- und Himmelscharten, *Beiträge zum Gebrauche der Mathematik in deren Anwendung*, part 3, section 6, Berlin: Verlag der Buchhandlung der Realschule: 105–99; (*b*) reprint ed. by Albert Wangerin, Leipzig: Engelmann, 1894 (Ostwalds Klassiker der exakten Wissenschaften, 33); (*c*) transl. by Waldo R. Tobler: *Notes and Comments on the Composition of Terrestrial and Celestial Maps*, University of Michigan Press, 1972.

— (1772*d*) *Beschreibung einer mit dem Calauischen Wachse ausgemalten Farben-Pyramide, wo die Mischung jeder Farben aus Weiß und drey Grundfarben angeordnet, dargelegt und derselben Berechnung und vielfacher Gebrauch gewiesen wird*, Berlin: Haude & Spener.

— (1786) Theorie der Parallel-Linien, *Leipziger Magazin für reine und angewandte Mathematik* I: 137–64 and 325–58.

Landa, Edward R. & Mark D. Fairchild (2005) Charting color from the eye of the beholder, *American Scientist* 93, no. 5: 436–43.

Landecker, Hannah (2005) Cellular features: microcinematography and early film theory, *Critical Inquiry* 31, no. 4: 903–37.

Lane, K. Maria D. (2006) Mapping the Mars canal mania: cartographic projection and the creation of a popular icon, *Imago Mundi* 58, no. 2: 198–211.

Lankford, John (1984) The impact of photography on astronomy, in Owen Gingerich (ed.) *Astrophysics and 20th Century Astrophysics to 1950*, Cambridge: Cambridge University Press: 16–34.

— (1997) *American Astronomy: Community, Careers, and Power, 1859–1940*, Chicago: University of Chicago Press.

Lapage, Geoffrey (1949) Draughtsmanship in zoological work, *Endeavour* 8: 70.

— (1951) Making science visible, *Medical and Biological Illustration* 10: 149–56.

— (1961) *Art and the Scientist*, Bristol: Wright & Sons.

Larder, David H. (1967) Alexander Crum Brown and his doctoral thesis of 1861, *Ambix* 14: 112–32.

Larkin, Jill J. & Herbert A. Simon (1987) Why a diagram is (sometimes) worth ten thousand words, *Cognitive Science* 11: 65–99.

Larson, Barbara & Fae Brauer (eds. 2009) *The Art of Evolution: Darwin, Darwinisms, and Visual Culture*, Hanover, NH: Dartmouth College Press.

Latour, Bruno (1986) Visualization and cognition: thinking with eyes and hands, *Knowledge and Society* 6: 1–40.

— (2002) What is iconoclash? Or is there a world beyond the image wars? in: Bruno Latour & Peter Weibel (eds.) *Iconoclash*, Karlsruhe: Center for Art and Media: 14–38.

Latour, B. & Steve Woolgar (1979) *Laboratory Life. The Construction of Scientific Facts*, Princeton, NJ: Princeton University Press.

Laudan, Larry (1977) *Progress and its Problems: Towards a Theory of Scientific Growth*, London: Routledge & Kegan Paul.

— (1989) From theories to research traditions, in: *Readings in the Philosophy of Science*, New York: Prentice Hall, pp. 368–79.

Laudan, Rachel (1987) *From Mineralogy to Geology. The Foundations of a Science 1650–1830*, Chicago: University of Chicago Press.

Laue, Max von & Richard von Mises (eds. 1926/36) *Stereoskopbilder von Kristallgittern, unter Mitarbeit von Cl. von Simson und E. Verständig*, Berlin: Springer, (*a*) 1st set 1926; (*b*) 2nd set 1936; (*c*) Engl. transl. by G. Greenwood as *Stereoscopic Drawings of Crystal Structures*, ibid.

Launey, Françoise (2001) Jules Janssen's in and out correspondence, in: C. Sterken & J. B. Hearnshaw (eds.) *100 Years of Observational Astronomy and Astrophysics*, Brussels: 159–68.

Lauterbur, Paul C. (1973) Image formation by induced local interactions: examples employing nuclear magnetic resonance, *Nature* 243: 190–1.

— (1986) Cancer detection by nuclear magnetic resonance zeugmatographic imaging, *Cancer* 57: 1899–904.

— (2003) All science is interdisciplinary – from magnetic moments to molecules to men (Nobel Lecture, Dec. 8, 2003), available online under www.nobel.org (accessed Aug. 3, 2011).

Lawrence, Snezana (2003) History of descriptive geometry in England, in: S. Huerta et al. (eds.) *Proceedings of the First International Congress on Construction History, Madrid, 20th–24th January 2003*, Madrid.

Leach, M. L. (1975) The effect of training on the pictorial depth perception of Shona children, *Journal of Cross-Cultural Psychology* 6: 457–70.

Lee, Jennifer B. (1999) *Seeing Is Believing: 700 Years of Scientific and Medical Illustration*, New York: New York Public Library.

Lees, Charles Herbert (1900) Young Thomas (1773–1829), *Dictionary of National Biography* 63.

Lefebvre, Thierry (1993) The scientia production 1911–1914. Scientific popularization through pictures, *Griffithiana* 16, May, no. 47: 137–55.

Lefèvre, Wolfgang (ed. 2004): *Picturing Machines 1400–1700*, Cambridge, MA: MIT Press:.

— (2007) Inside the camera obscura – optics and art under the spell of the projected image, *Max Planck Institut für Wissenschaftsgeschichte, Berlin, Preprints*, no. 333.

Lefèvre, W., Jürgen Renn & Urs Schoepflin (eds. 2003) *The Power of Images in Early Modern Science*, Basel: Birkhäuser.

Lemke, J. L. (1998) Multiplying meaning: visual and verbal semiotics in scientific texts, in: J. R. Martin & Robert Veel (eds.) *Reading Science. Critical and Functional Perspectives on Discourses of Science*, London: Routledge: 87–113.

Leng, Rainer (2002) *Ars belli: Deutsche taktische und kriegstechnische Bilderhandschriften und Traktate im 15. und 16. Jahrhundert*, Wiesbaden: Reichert (Imagines medii aevi, no. 12).

Lenoir, Timothy (1997) *Instituting Science. The Cultural Production of Scientific Disciplines*, Stanford University Press.

— (1998) *Inscribing Science. Scientific Texts and the Materiality of Communication*, Stanford University Press.

Lepenies, Wolf (1994) *Das Ende der Naturgeschichte*, Munich: Hanser.

Lerner, Barron H. (1992) The perils of 'x-ray vision' – how radiographic images have historically influenced perception, *Perception in Biology and Medicine* 35, no. 3: 382–97.

Levaditi, Constantin & S. Mutermilch (1913) Contractilité des fragments de coeur d'embryon de poulet *in vitro*, *Comptes Rendus hebdomadaires des Séances et Mémoires de la Société de Biologie* 65. année, vol. 1 = tome 74: 462–4.

Leverrier, U. J. J. (1850) Rapport sur l'enseignement de l'École Polytechnique, (*a*)

Paris; (*b*) in *Moniteur Universel*, Jan. 12, 1851, suppl., i–xxxiv.

Levy, Jacques R. (1973) Janssen, Pierre Jules César, *Dictionary of Scientific Biography* 7: 73–8.

Liebsch, Dimitri & Nicola Mößner (eds. 2012) *Visualisierung und Erkenntnis. Bildverstehen und Bildverwenden in Natur- und Geisteswissenschaften*, Cologne: Herbert von Halem Verlag.

Liesegang, Franz P. (1920) *Wissenschaftliche Kinematographie, einschliesslich der Reihenphotographie*, Düsseldorf: Liesegang.

Lightman, Bernard (2000) The visual theology of Victorian popularizers of science: from referent eye to chemical retina, *Isis* 91,4: 651–80.

— (ed. 2007) *Victorian Popularizers of Science. Designing Nature for New Audiences*, Chicago: University of Chicago Press.

Lima, Manuel (2014) *The Book of Trees. Visualizing Branches of Knowledge*, New York: Architectural Press.

Lima-de-Faria, José (1990) *Historical Atlas of Crystallography*, Berlin: Springer.

Lipsmeier, Antonius (1971) *Technik und Schule. die Ausformung des Berufsschulcurriculums unter dem Einfluß der Technik als Geschichte des Unterrichts im technischen Zeichnen*, Wiesbaden: Steiner.

Little, A. M. G. (1937) Perspective and scene painting, *Art Bulletin* 19: 487–95.

Lloyd, Geoffrey & Nathan Sivin (2003) *The Way and the Word. Science and Medicine in Early China and Greece*, New Haven, CT: Yale University Press.

Loewinson-Lessing, F. Y. (1954) *A Historical Survey of Petrology*, Edinburgh and London: Oliver & Boyd.

Lohne, John A. (1968) The increasing corruption of Newton's diagrams, *History of Science* 6: 69–89.

Lohse, Oswald (1883) Abbildungen von Sonnenflecken, nebst Bemerkungen über astronomische Zeichnungen und deren Vervielfältigung, *Publikationen des Astrophysikalischen Observatoriums Potsdam* 3, no. 5: 296–301 & pl. 40–2.

Lok, Corie (2011) From monsters to molecules, *Nature* 477: 359–61.

Lorenz, Dieter (1985) *Das Stereobild in Wissenschaft und Technik*, Köln & Oberpfaffenhofen: Deutsche Forschungs- und Versuchsanstalt für Luft- und Raumfahrt.

Lowell, Percival (1893) Mars, *Astronomy and Astrophysics* 13: 645–50 & pl. XXII–XXIII.

Ludwig, Heidrun (1998) *Nürnberger Naturgeschichtliche Malerei im 17. und 18. Jahrhundert*, Marburg: Basilisken-Presse.

Lüthy, Christoph & Alexis Smets (2009) Word, line, diagrams, images: towards a history of scientific imagery, *Early Science & Medicine* 14: 398–439.

Lynch, Michael (1985) Discipline and the material form of images: an analysis of scientific visibility, *Social Studies of Science* 15: 37–66.

— (1991) Science in the age of mechanical reproduction: moral and epistemic relations between diagrams and photographs, *Biology and Philosophy* 6, no. 2: 206–26.

— (1994) Representation is overrated: some critical remarks about the use of the concept of representation in science studies, *Configurations* 2,1: 140–6.

Lynch, M. & S. Y. Edgerton, Jr. (1988) Aesthetics and digital image processing: Representational craft in contemporary astronomy, in: G. Fyfe & John Law (eds.) *Picturing Power. Visual Depiction and Social Relations*, London: 184–220.

Lynch, M. & Steve Woolgar (eds. 1990) *Representation in Scientific Practice*, Cambridge, MA: MIT Press.

Lynes, John A. (1980) Brunelleschi's perspectives reconsidered, *Perception* 9: 87–99.

Lysaght, Averil (1957) Captain Cook's kangarooh, *New Scientist*, Mar. 14th: 17–20.

Maar, Christa & Hubert Burda (eds. 2004) *Iconic turn. Die Neue Macht der Bilder*, Cologne: DuMont.

Mabberley, David J. & M. P. San Pío (2012) *Haenke's Malaspina Colour-Chart: An Enigma*, Madrid: Real Jardín Botánico, CSIC.

McCabe, David P. & Alan D. Castel (2008) Seeing is believing: the effect of brain images on judgements of scientific reasoning, *Cognition* 107: 343–52.

Mach, Ernst (1886) *Beiträge zur Analyse der Empfindungen*, (*a*) Jena: Fischer; (*b*) *Die Analyse der Empfindungen und das Verhältnis der Physischen zum Psychischen*, 2nd ed., Jena: Fischer, 1900; (*c*) 9th ed. 1922.

Mach, E. & Peter Salcher (1887) Photographische Fixirung der durch Projectile in der Luft eingeleiteten Vorgänge, *Sitzungsberichte der kaiserlich-königlichen Akademie der Wissenschaften in Wien, mathem.-naturwiss. Classe* II. Abt., 95: 764–80.

MacKendrick, Archibald & Charles R. Whittaker (1927) *An X-ray Atlas of the Normal and Abnormal Structures of the Body*, Edinburgh: Livingstone, 2nd ed.

MacPherson, H. (1910) Giovanni Schiaparelli, *Popular Astronomy* 18: 467–74.

Magee, Judith (2009) *Art of Nature. Three Centuries of Natural History Art from Around the World*, London: Natural History Museum.

Maienschein, Jane (1991) From presentation to representation in E. B. Wilson's The Cell, *Biology and Philosophy* 6: 227–54.

Maitte, Bernard (2001) René-Just Haüy et la naissance de la cristallographie, *Traveaux du Comité Francaise d'Histoire de la Géologie* (3rd ser.) 15.

Malin, David & Paul Murdin (1984) *Colours of the Stars*, Cambridge Univ. Press.

Mallary, Peter & Frances Mallary (1988) *A Redoute Treasury: 468 Watercolours from Les Liliacees of Pierre-Joseph Redouté*, Seacausus: Vendome.

Manara, Alessandro (2005) Mars in the Schiaparelli–Lowell correspondence, *Memorie delle Societa Astronomia Italiana, suppl.* 6, no. 12.

Manara, A. & Franca Chlistovsky (2001) Il carteggio Lowell–Schiaparelli, *Atti del XXI Congresso Nazionale di Storia della Fisica e dell'Astronomia.*

Manetti, Antonio di Tuccio (1970) *Vita di Brunelleschi*, text-critical edition of the Italian late-15th c. manuscript with an Engl. transl. by Catherine Enggass, annot. by Howard Saalman: *The Life of Brunelleschi*, University Park & London: Pennsylvania State University Press.

Mann, Günter (1964) Medizinisch-naturwissenschaftliche Buchillustration im 18. Jahrhundert in Deutschland, *Marburger Sitzungsberichte* 86: 3–48.

Mann, H. (1987) Die Plastizität des Mondes – zu Galileo Galilei und Lodovico Cigoli, *Kunsthistorisches Jahrbuch Graz* 23: 55–9.

Mansfield, Paul & A. A. Maudsley (1977) Medical imaging by NMR, *British Journal of Radiology* 50: 188–94.

Marey, Étienne-Jules (1882*a*) Sur la reproduction, par la photographie, des diverses phases du vol des oiseaux, *CRAS* 94: 683–5; see also Janssen (1882).

— (1882*b*) Emploi de la photographie instantanée pour l'analyse des mouvement chez les animaux, *CRAS* 94: 1013–20.

— (1882*c*) La photographie du mouvement, *La Nature* July 22nd: 115–16; (*d*) in *Scientific American* 47 (Sep. 9th): 166.

— (1882*e*) Emploi de la photographie pour déterminer la trajectoire des corps en mouvements avec leur vitesse à chaque instant et leurs positions relatives, *CRAS* 95: 267–70.

— (1882*f*) Le fusil photographique, *La Nature* (Apr. 22nd): 326–30.

— (1883) Emploi des photographies partielles pour étudier la locomotion de l'homme et des animaux, *CRAS* 96: 1827–31.

— (1890) Appareil photochromographique applicable à l'analyse de toutes sortes de mouvement, *CRAS* 111: 626–9.

— (1902) The history of chronophotography, *Annual Report of the Board of Regents of the Smithsonian Institution for the Year Ending June 30, 1901*, Washington: 317–40.

Margolis, Eric & Luc Pauwels (eds. 2011) *The Sage Handbook of Visual Research Methods*, London: SAGE.

Marr, David (1982) *Vision. A Computational Investigation into the Human Representation and Processing of Visual Information*, San Francisco: W. H. Freeman.

Martinet, Alexis (ed. 1994) *Le cinema et la science*, Paris: CNRS.

Marton, Ladislaus (1943) Alice in electronland, *American Scientist* 31: 247–54.

Matschoss, Conrad (1901) *Geschichte der Dampfmaschine: ihre kulturelle Bedeutung, technische Entwicklung und ihre grossen Männer*, (*a*) Berlin: Springer; (*b*) 3rd. ed. with commentary: Hildesheim: Gerstenberg, 1983.

— (1908) *Die Entwicklung der Dampfmaschine: Eine Geschichte der ortsfesten Dampfmaschine und der Lokomobile, der Schiffsmaschine und Lokomotive*, Berlin: Springer.

Mattson, James & Merrill Simon (1996) *The Pioneers of NMR and Magnetic Resonance in Medicine. The Story of MRI*, Ramat Gan, Israel: Bar Ilan University.

Mauersberger, Klaus (1989) Leonardo da Vincis Entwürfe von Kurvenmechanismen – Studie, *Dresdner Beiträge zur Geschichte der Technikwissenschaften* 18: 53–102.

— (1994) Visuelles Denken und nichtverbales Wissen im Maschinenbau, *Beiträge zur Geschichte von Technik und technisches Bildung* 9: 3–28.

Maunder, E. Walter (1894) The canals of Mars, *Knowledge* 1: 249–52.

Maxwell, James Clerk (1855/56) On Faraday's lines of force, (*a*) *Transactions of the Cambridge Philosophical Society* 10, no. 1: 27–83; (*b*) reprint in: Maxwell (1890) vol. I: 155–229.

— (1861/62) On physical lines of force, (*a*) *Phil. Mag.* (1861) 21: 161–75, 281–91, 338–48; (1862) 22: 12–24, 85–95; (*b*) reprint in Maxwell (1890) vol. I: 451–513.

— (1870) Address to the mathematical and physical sections of the British Association, (*a*) Liverpool, Sep. 15th; (*b*) reprinted in Maxwell (1890) vol. II: 215–29.

— (1890) *Scientific Papers*, ed. by W. D. Niven, Cambridge University Press; reprint New York: Dover.

Mayer, Johann Tobias (1758) Von Messung der Farben, *Göttingische Anzeigen von gelehrten Sachen*: 1385–9.

— (1775) *Opera inedita*, ed. by Georg Christoph Lichtenberg, (*a*) Göttingen: Dieterich; (*b*) partial reprint with German transl. by Dag Nikolaus Hasse in Georg Christoph Lichtenberg: *Observationes. Die lateinischen Schriften*, Göttingen: Wallstein, 1997: 55–142.

Mayor, A. Hyatt (1946) The photographic eye, *Bulletin of the Metropolitan Museum of Art* 5: 15–26.

Mazzolini, Roberto (ed. 1993) *Non-Verbal Communication in Science Prior to 1900*, Florence: Olschki.

Meadows, Arthur Jack (1972) *Science and Controversy: A Biography of Sir Norman Lockyer*, London: Macmillan.

— (1991) The evolution of graphics in scientific articles, *Publishing Research Quarterly* 7,1: 23–32.

Meijer, E. W. (2001) Jacobus Henricus van 't Hoff und sein Einfluss auf die Stereochemie, *Angewandte Chemie* 113: 3899–905.

Meinel, Christoph (2004) Molecules and croquet balls, in: Chadarevian & Hopwood (eds. 2004): 242–75.

— (2009) Kugeln und Stäbchen, *Kultur und Technik* 2: 14–21.

Mensbrugghe, G. van der (1885) Joseph-Antoine-Ferdinand Plateau (obituary), *Annuaire de l'Académie Royale des Sciences de Bruxelles* 51: 389–486 (with biblio.).

Menzel, Randolf (2002) Schönheit in einer Bilder-Wissenschaft, *Gegenworte; Hefte für den Disput über Wissen* 9: 31–5.

Metcalfe, James (1910) Some of the uses of x rays in diagnosis and treatment, *British Medical Journal* (1) 2567: 432–4.

Metzler, J. & R. N. Shepard (1974) Transformational studies of the internal representation of three-dimensional objects, in: Robert L. Solso (ed.) *Theories of Cognitive Psychology. The Loyola Symposium.* Potomac, Maryland: Lawrence Erlbaum: 147–201.

Meyer, Victor (1890) *Ergebnisse und Ziele der stereochemischen Forschung*, Heidelberg: Winter.

Michaelis, Anthony R. (1955) *Research Films in Biology, Anthropology, Psychology and Medicine*, New York: Academic Press.

Miller, Arthur I. (1996) *Insights of Genius. Imagery and Creativity in Science and Art*, New York: Springer.

— (2001) *Einstein, Picasso: Space, Time, and the Beauty That Causes Havoc*, New York: Basic Books.

Miller, David Philip (2008) Seeing the chemical steam through the historical fog: Watt's steam engine as chemistry, *Annals of Science* 65: 47–72.

Miller, Thomas (1998) Visual persuasion – a comparison of visuals in academic texts and the popular press, *English for Special Purposes* 17, no. 1: 29–46.

Mills, Allen P. (2009) John A. Wheeler's lessons to an undergraduate student, *Physica Status Solidi* series C, 6: 2253–9.

Mirzoeff, Nicolas (ed. 1998) *The Visual Culture Reader*, London: Taylor & Francis.

— (2006) On visuality, *Journal of Visual Culture* 5,1: 53–79.

Misner, Charles (2009) In memoriam John Archibald Wheeler, *General Relativity and Gravitation* 41 (2009): 675–7.

Misner, C., Kip Thorne & John A. Wheeler (1973) *Gravitation*, San Francisco: Freeman.

Mitchell, R. F. & L. G. Cole (1935) Historical notes on x-ray cinematography, *Journal of the Society of Motion Picture Engineers* 24, no. 3: 333–45.

Mitchell, William J. T. (1984) What is an image? *New Literary History* 15,3: 503–37.

— (1992*a*) The pictorial turn, *Art Forum* 30, 89–94; (*b*) reprinted in W. J. Mitchell: *Picture Theory. Essays on Verbal and Visual Representation*, University of Chicago Press, 1994: 11–34; (*c*) German transl.: Der pictorial turn, in: Christian Kravagna (ed.) *Privileg Blick. Kritik der visuellen Kultur*, Berlin 1997: 15–40.

— (1992*d*) *The Reconfigured Eye. Visual Truth in the Post-Photographic Era*, Cambridge, MA: MIT Press.

— (1994) *Picture Theory. Essays on Verbal and Visual Representation*, Chicago: University of Chicago Press.

— (1995) Interdisciplinarity and visual culture, *Art Bulletin* 77: 540–4.

— (1996) What do pictures really want? *October* 77: 76–83.

— (2000/01) La plus-value des images, (*a*) *Études Litteraires* 33 (2000/01): 201–25; (*b*) German transl.: Der Mehrwert von Bildern, in: Stefan Andriopoulos et al. (eds.) *Die Adresse des Mediums*, Cologne: DuMont, 2001: 158–84. (*c*) The surplus value of images, *Mosaic* 35 no. 3 (2002): 1–23.

— (2002) Showing seeing: a critique of visual culture, (*a*) *Journal of Visual Culture* 1, no. 2: 165–81; (*b*) in M. A. Holly & K. Moxey (eds.) *Art History, Aesthetics, Visual Studies*, Williamstown, MA: Sterling & Francis Clark Art Institute: 231–50.

Mitman, Gregg (1999) *Reel Nature. America's Romance with Wildlife in Film*, Cambridge, MA: Harvard University Press.

Mody, Cyrus C. M. (2006) Corporations, universities, and instrumental communities. Commercial probe microscopy, 1981–1996, *Technology and Culture*: 56–80.

Mody, C. C. M. & David Kaiser (2008) Scientific training and the creation of scientific knowledge, in: Edward J. Hackett et al. (eds.) *The Handbook of Science and Technology Studies*, Cambridge, MA: MIT Press: 377–402.

Mody, C. C. M. & Michael Lynch (2010) Test objects and other epistemic things: a history of a nanoscale object, *BJHS* 43,3: 423–58.

Moiseiff, Leon S. (1933) Suspension bridges under the action of lateral forces, *American Society of Civil Engineers. Transactions* no. 1849.

Mokre, Jan (2008) *Rund um den Globus. Über Erd- und Himmelsgloben und ihre Darstellungen*, Vienna: Bibliophile Ed.

Monge, Gaspard (1798) *Géométrie descriptive*, Paris: Bachelier, (*a*) 1st ed. (an 7); (*b*) 5th exp. ed. 1827; (*c*) German transl. by Robert Haussner: *Darstellende Geometrie*, Leipzig: Engelmann, 1900.

Monmonier, Mark (1996) *How to Lie with Maps*, University of Chicago Press.

Moon, Francis C. (2007) *The Machines of Leonardo Da Vinci and Franz Reuleaux: Kinematics of Machines from the Renaissance to the 20th Century*, Berlin: Springer.

Moore, Patrick (2006) *Mars*, London: Cassell Illustrated.

Moores, Eldridge M. (ed. 1988) *The Art of Geology*, Boulder, Colorado: Geological Society of America Special Paper 225.

Mooshammer, Helge & Peter Mörtenböck (2003) *Visuelle Kultur: Körper-Räume-Medien*, Vienna: Böhlau.

Morgan, Mary S. & Margaret C. Morrison (1999) *Models as Mediators. Perspectives on Natural and Social Sciences*, Cambridge: Cambridge University Press.

Morse, Edward S. (1906) *Mars and its Mystery*, Boston: Little, Brown & Co.

Moriarty, Sandra E. (1997) A conceptual map of visual communication, *Journal of Visual Literacy* 17, no. 2: 9–24.

Morrison, Philip & Phylis Morrison (1982) *Powers of Ten. About the Relative Size of Things in the Universe*, New York: Scientific American Library.

Morrison-Low, A. D. & J. R. R. Christie (eds. 1984) *"Martyr of Science." Sir David Brewster 1781–1868*, Edinburgh: Royal Scottish Museum Studies.

Morton, Oliver (2002) *Mapping Mars. Science, Imagination, and the Birth of a World*, New York: Picador.

Morus, Iwan Rhys (2006) Seeing and believing science, *Isis* 97: 101-110.

Moser, Leonie (1939) Wie ich vor 20 Jahren das Röntgen lernte, *Blätter für Krankenpflege* [Bern] 4: 76–8; 6: 103–8.

Moxey, Keith (2001) Perspective, Panofsky, and the philosophy of history, in: K. Moxey (ed.) *The Practice of Persuasion*, Ithaca: Cornell University Press: 90–201.

Muirhead, James Patrick (1854) *The Origin and Progress of the Mechanical Inventions of James Watt: Illustrated by his Correspondence with his Friends and the Specifications of his Patents*, London: Murray, 3 vols.

Müller, Falk (2004) *Gasentladungsforschung im 19. Jahrhundert*, Stuttgart: GNT-Verlag.

Müller, Ferdinand (1872) Das atempo-Zeichnen an der Königlichen Gewerbe-Akademie zu Elberfeld, *Monatsblätter für Zeichenkunst und Zeichenunterricht* 8: 22–4.

Müller, Ingo (2007) *A History of Thermodynamics: The Doctrine of Energy and Entropy*, Berlin: Springer.

Müller, Irmgard & Heiner Fangerau (2010) Medical imaging: Pictures "as if" and the power of evidence, *Medicine Studies* 2,3: 151–60.

Müller, Kathrin (2008) *Visuelle Weltaneignung. Astronomische und kosmologische Diagramme in Handschriften des Mittelalters*, Göttingen: Vandenhoeck & Ruprecht.

— (2010) How to craft telescopic observation in a book. Johannes Hevelius's Selenography (1647) and its images, *JHA* 41, no. 3: 355–79.

Müller-Bahlke, Thomas J. (1998) *Die Wunderkammer: Die Kunst- und Naturalienkammer der Franckeschen Stiftungen zu Halle (Saale)*, Halle: Franckesche Stiftungen.

Müller-Jahncke, Wolf-Dieter (1995) Herbaria picta. Zur Tradition der illustrierten Kräuterbücher des Mittelalters, *Pharmazie in unserer Zeit* 24,4: 67–72.

Müller-Wille, Staffan (1999) *Botanik und weltweiter Handel. Zur Begründung eines natürlichen Systems der Pflanzen durch Carl von Linné (1707–78)*, Berlin: Verlag für Wissenschaft & Bildung.

Münch, Ranghild & Stephan S. Biel (1998) Experiment und Expertise im Spiegel des Nachlasses von Robert Koch, *Sudhoffs Archiv* 82: 1–29.

Murdoch, John E. (1984) *Album of Science. Antiquity and the Middle Ages*, New York: Scribner's.

Musson, Albert Edward & Eric Robinson (1969) *Science and Technology in the Industrial Revolution*, Manchester: Manchester University Press.

Mutis, José Célestino (1954ff.) *Flora de la Real Expedicion Botanica del Nuevo Reino de Granada*, Madrid: Many volumes (still in progress), 1954ff.

Muybridge, Eadweard (1878) The science of the horse's motion, *Scientific American*, Oct. 19th.

Nasim, Oman W. (2008*a*) Observations, descriptions, and drawings of nebulae: a sketch, *Max-Planck-Institut für Wissenschaftsgeschichte Berlin, Preprints*, no. 345.

— (2008*b*) Beobachtungen mit der Hand: Astronomische Nebelskizzen im 19. Jahrhundert, in: Christoph Hoffmann (ed.) *Daten sichern. Schreiben und Zeichnen als Verfahren der Aufzeichnung*, Zürich-Berlin.

— (2009) On seeing an image of a spiral nebula from Whewell to Flammarion, *Nuncius: Journal of the History of Science* 24: 393–414.

— (2010) Observation, working images and procedure: the 'great spiral' in Lord Rosse's astronomical record books and beyond, *BJHS* 43: 353–89.

— (2011) The 'landmark' and 'groundwork' of stars: John Herschel, photography and the drawing of nebulae, *SHPS* 42: 67–84.

— (2012/14) *Observing by Hand: Sketching the Nebulae in the 19th Century*, (*a*) Habilitation thesis, ETH Zürich, 2012; (*b*) University of Chicago Press, 2014.

Nasmyth, James (1862) On the structure of the luminous envelope of the Sun, *Manchester Philosophical Society Memoirs* 1: 407–11.

— (1883) *James Nasmyth Engineer: An Autobiography*, ed. and partially rewritten by Samuel Smiles, London: Murray; see also Anon. (1883).

Nasmyth, J. & James Carpenter (1874): *The Moon: Considered as a Planet, a World, and a Satellite*, London: Murray, (*a*) 1st and 2nd ed. 1874; (*b*) 3rd exp. ed. in reduced format 1885; (*c*) 4th ed. 1903.

Navier, Claude-Louis-Marie-Henri (1823) *Rapport à Monsieur Becquey Conseiller d'État, Directeur Général des Ponts et Chaussées et des Mines; et Mémoire sur les Ponts suspendus*, Paris: Imprimérie Royale, (*a*) in text & plate vols.; (*b*) 2nd exp. ed., 1830; (*c*) German transl. by J. G. Kutschera: *Bericht an Becquey ... und Abhandlung über die [Ketten-] Hängbrücken*, Lemberg, 1829.

Nedoluha, Alois (1960) *Kulturgeschichte des technischen Zeichnens*, Vienna: Springer.

Nelkin, Dorothy & M. Susan Lindee (1995) *The DNA Mystique. The Gene as a Cultural Icon*, New York: W. H. Freeman.

Nemec, Birgit (2014) Anatomical modernity in red Vienna. Julius Tandler's textbook for systematic anatomy and the politics of visual culture, *Sudhoffs Archiv*, forthcoming.

Nersessian, Nancy J. (1995) Opening the black box: cognitive science and history of science, *Osiris* 10: 194–211.

— (2008) *Creating Scientific Concepts*, Cambridge, MA: MIT Press.

Nesbit, J. & M. Bradford (2006) 2006 visualization challenge, *Science* 313: 1729.

Newbury, D. E. & D. B. Williams (2000) The electron microscope – the materials characterization tool of the millenium, *Acta Materialia* 48: 323–46.

Newhall, Beaumont (1944) Photography and the development of kinetic visualizations, *JWCI* 7: 40–5.

Newth, D. R. & Turlington, E. R. (1956) The drawings of T. H. Huxley, *Medical Biology Illustrated* 6 (1956): 71–6.

Nickelsen, Kärin (2000) *Wissenschaftliche Pflanzenzeichnungen – Spiegelbilder der Natur?* Bern: Bern Studies in the History and Philosophy of Science.

— (2004*a*) *Botanische Illustrationen des 18. und frühen 19. Jahrhunderts*, (*a*) Bern: Philoscience (thesis University of Bern); (*b*) transl.: *Draughtsmen, Botanists and Nature:*

The Construction of Eighteenth-Century Botanical Illustrations, Dordrecht: Springer 2006 (Archimedes, no. 15).

Nickelsen, K. & Gerd Graßhoff (2001+) Pflanzenzeichnungen als Ausdruck wissenschaftlicher Inhalte: penelope.unibe.ch/docuserver/compago/home_bot.html.

Nickerson, Dorothy (1976) History of the Munsell color system, company, and foundation, *Color Research and Application* 1: 7–10, 121–30.

Nickol, Thomas (2013) Zu den Beobachtungsmethoden von Achsenbildern ein- und zweiachsiger Kristalle durch Brewster, Biot und Seebeck, *Acta Historica Leopoldina* 62: 117–31.

Niedrig, Heinz (1987) Die Anfänge der Elektronenmikroskopie – eine Berliner Entwicklung, *Physik und Didaktik* 15: 152–64.

Nikolow, Sybilla & Lars Bluma (2002) Bilder zwischen Öffentlichkeit und wissenschaftlicher Praxis, (*a*) *NTM* 10: 201–8; (*b*) Engl. version: Science images between scientific fields and the public sphere. A historiographic survey, in: Bernd Hüppauf & Peter Weingart (eds.) *Science Images and Popular Images of Science*, London: Routledge, 2008: 33–51.

Nissen, Claus (1950) *Die naturwissenschaftliche Illustration. Ein geschichtlicher Überblick*, Bad Münster am Stein: Hempe.

— (1951/66) *Die botanische Buchillustration: Ihre Geschichte und Bibliographie*, Stuttgart: Hiersemann, vols. 1–2, 1951; suppl. vol. 3, 1966.

— (1969/78) *Die zoologische Buchillustration: Ihre Bibliographie und Geschichte*, Stuttgart: Hiersemann, vol. 1: *Bibliographie*, 1969; vol. 2: *Geschichte*, 1978.

Nova, Alessandro (2005) La dolce morte. Dia anatomischen Zeichnungen Leonardo da Vincis als Erkenntnismittel und reflektierte Kunstpraxis, in: Albert Schirrmeister et al. (eds.) *Zergliederungen – Anatomie und Wahrnehmung in der frühen Neuzeit*, Frankfurt: 137–63.

Nye, Mary Jo et al. (eds. 1996) *The Pauling Symposium: A Discourse on the Art of Biography. Proceedings of the Conference on the Life and Work of Linus Pauling (1901–1994)*, Corvallis: Oregon State University Libraries Special Collections.

Oestermeier, Uwe & Friedrich W. Hesse (2000) Verbal and visual causal arguments, *Cognition* 75: 65–104.

O'Gomes, Isabelle Do (1994): L'oeuvre de Jean Comandon, in: Martinet (ed. 1994): 78–86.

Ohlsson, S. (1984) Restructuring revisited, *Scandinavian Journal of Psychology* 25: 65–78, 117–29.

Olby, Robert (1974) DNA before Watson-Crick, *Nature* 248, Apr. 26th: 782–5.

— (1994) *The Path to the Double Helix*, New York: Dover.

Oldendorf, W. H. (1961) Displaying the internal structural patterns of a complex object, *Transactions of Biomedical Electronics* 8: 68–72.

Olesko, Kathryn M. (1991) *Physics as a Calling: Discipline and Practice in the Königsberg Seminar for Physics*, Ithaca: Cornell University Press.

— (2006) Science pedagogy as a category of historical analysis: Past, present, and future, *Science & Education* 15, Nov., no. 7: 863–80.

— (2009) Geopolitics and Prussian technical education in the late-eighteenth century, *Actes d'Historia de la Ciència i de la Tècnica*, new ser. 2, no. 2: 11–44.

Olson, A. & G. Goodsell (1992) Visualizing biological molecules, *Scientific American*, November: 44–51.

O'Malley, Charles D. (1964) *Andreas Vesalius of Brussels, 1514–1564*, Berkeley: University of California Press.

O'Malley, C. D. & J. B. de C. M. Saunders (1982) *Leonardo da Vinci on the Human Body. The Anatomical, Physiological and Embryological Drawings of Leonardo da Vinci*, New York: Greenwich House.

Orcel, J. (1972) Historical note, in: R. Galopin & N. R. M. Henry (eds.) *Microscopic Study of Opaque Materials*, Cambridge: Heffer: 301–5.

Oreskes, Naomi (1999) *The Rejection of Continental Drift. Theory and Method in American Earth Science*, Oxford: Oxford University Press.

Ostrow, Steven F. (1996) Cigoli's immaculate virgin and Galileo's moon, *Art Bulletin* 78: 218–35.

Ottino, Julio M. (2003) Is a picture worth 1000 words? *Nature* 421: 474–6.

Pang, Alex Soojung-Kim (1994/95) Victorian observing practices, printing technologies, and representation of the solar corona, *JHA* 25 (1994): 249–74; 26 (1995): 63–75.

— (1997*a*) Visual representation and post-constructivist history of science, *HSPS* 28, no. 1: 139–71.

— (1997*b*) Stars should henceforth register themselves: astrophotography at the early Lick Observatory, *BJHS* 30: 177–202.

Panofsky, Erwin (1927) Die Perspektive als symbolische Form, in: *Vorträge der Bibliothek Warburg 1924/25*, Leipzig & Berlin, 1927, 258–330.

— (1940) *The Codex Huygens and Leonardo da Vinci's art theory*, London: Warburg Institute.

— (1943) *The Life and Art of Albrecht Dürer*, (*a*) Princeton, NJ: Princeton University

Press; (*b*) German transl. *Das Leben und die Kunst Albrecht Dürers*, Hamburg: Rogner & Bernhard, 1995.

Paolini, Leonello (1992) Stereochemical models of benzene, 1869–1875, *Bulletin for the History of Chemistry* 12: 10–24.

Partain, C. Leon et al.(eds. 1988) *Magnetic Resonance Imaging*, Philadelphia: Saunders.

Pasachoff, Jay & Glenn Schneider & Leon Golub (2004) The black-drop effect explained, in: D. W. Kurtz (ed.) *Transits of Venus. New Views of the the Solar System and Galaxy*, Cambridge: Cambridge University Press: 242–53.

Pasch, Richard (1979) Die Geschichte der Metallographie mit besonderer Berücksichtigung der mikroskopischen Prüfverfahren, *Praktische Metallographie* 16: 26–35 etc. in several installments.

Pasveer, Bernike (1989) Knowledge of shadows: the introduction of x-ray images in medicine, *Sociology of Health and Illness* 11: 360–81.

— (2005) Representing or mediating: a history and philosophy of x-ray images in medicine, in: Luc Pauwels (ed. 2005): 41–62.

Pauling, Linus (1939) *The Nature of the Chemical Bond*, Ithaca, NY: Cornell University Press (also in several later eds., transl. etc).

— (1948) *General Chemistry*, San Francisco: Freeman & Co. (illus. by R. Hayward).

— (1957) *College Chemistry*, New York: Freeman & Co.

Pauling, L. & Robert B. Corey (1953) Structure of the nucleic acids, *Nature* 171: 346.

Pauling, L., R. B. Corey & Roger Hayward (1954) The structure of protein molecules, *Scientific American*, July: 51–9.

Pauling, L. & Roger Hayward (1964) *The Architecture of Molecules*, San Francisco : W. H. Freeman; Hermann Raaf: *Die Architektur der Moleküle*, Villingen: Neckar-Verlag, 1969.

Pauwels, Luc (ed. 2005): *Visual Cultures of Science: Rethinking Representational Practices in Knowledge Building and Science Communication*, Lebanon, NH: University Press of New England.

Peacock, George (1855) *Life of Thomas Young, M.D., F.R.S., & c.*, London: Murray.

Pedretti, Carlo (1963) Leonardo on curvilinear perspective, *Bibliothèque d'Humanisme et Renaissance* 25: 69–87.

Peiffer, Jeanne (2002) La perspective, und science mélée, *Nouvelle Revue du Seizième Ciècle* 20: 97–121.

Peitgen, Heinz-Otto, Horst Hahn & Tobias Preusser (2011) Modellbildung in der bildbasierten Medizin: Radiologie jenseits des Auges, *Nova Acta Leopoldina* new ser. 110, no. 377: 259–83.

Pellerin, Denis (2000): The origins and development of stereoscopy, in: Françoise Reynaud et al. (eds.) *Paris in 3D: From Stereoscopy to Virtual Reality 1850–2000*, Paris: Booth-Clibborn, 2000: 45–8.

Perini, Laura (2004) Convention, resemblance and isomorphism: understanding scientific visual representations, in: Grant Malcolm (ed.) *Multidisciplinary Approaches to Visual Representations and Interpretations*, Amsterdam: Elsevier (Studies in Multidisciplinarity, no. 2): 37–47.

— (2005) The truth in pictures, *Philosophy of Science* 72: 262–85.

— (2012) Depiction, detection, and the epistemic value of photography, *Journal of Aesthetics and Art Criticism* 70,1: 151–60.

— (2013) Image interpretation: bridging the gap from mechanically produced image to representation, *International Studies in the Philosophy of Science* 26,2: 153–70.

Perutz, Max (1994) Linus Pauling (1901–1994), *Nature Structural Biology* 1: 667–71.

Peters, Tom F. (1987) *Transitions in Engineering: Guillaume Henri Dufour and the Early 19th Century Cable Suspension Bridges*, Basel: Birkhäuser.

— (1998) How creative engineers think, *Civil Engineering ASCE* 68, March no. 3: 48–51.

Pfister, Arnold (ed. 1961) *De Simplici Medicina – Kräuterbuch-Handschrift aus dem letzten Viertel des 14. Jahrhunderts im Besitz der Basler Universitäts-Bibliothek*, Basel: Sandoz.

Picard, Émile (1872) *Sadi Carnot. Biographie et Manuscrit*, Paris: Gauthier-Villars.

Pickering, Andrew (ed. 1992) *Science and Practice and Culture*, Chicago: University of Chicago Press.

Pickering, William Henry (1892) Mars, *Astronomy and Astrophysics* 11: 668–75 & pl. XXX.

Pickles, John (2004) *A History of Spaces: Cartographic Reason, Mapping and the Geo-Coded World*, London: Routledge.

Picon, Antoine (1988) Navier and the introduction of suspension bridges in France, *Journal of the Construction History Society* 4: 21–34.

— (1992) *French Architects and Engineers in the Age of Enlightenment*, Cambridge: Cambridge University Press.

Piero della Francesca (1492) *De prospectiva pingendi*, (*a*) mss. before 1492; (*b*) transcr. and transl. by Constantin Winterberg, Strasbourg 1899; (*c*) critical ed. by Giusta

Nicco Fasola, Florence 1942.

Piersig, Wolfgang: (2009) *Henry Clifton Sorby – Begründer der klassischen Metallographie*, Munich: Grin.

Pinault, Madeleine (1991) *The Painter as Naturalist from Dürer to Redouté*, transl. from the French orig. by Philip Sturgess, Paris: Flammarion.

Pinder, Ulrich (1506) *Epiphanie Medicorum. Speculum videndi urinas hominum. Clavis aperiendi portas pulsuum. Berillus discernendi causas & differentias febrium*, Nürnberg: Peypus.

Pirenne, Maurice Henri (1952) The scientific basis of Leonardo da Vinci's theory of perspective, *British Journal for Philosophy of Science* 3: 169–85.

Plateau, Joseph (1829/30) *Dissertation sur quelques propriétés des impressions produites par la lumière sur l'organe de la vue*, Dissertation University of Liège: Dessain, 1829; German transl. *Annalen der Physik* 96 = (2) 20 (1830): 304–22.

— (1831) Lettre sur une illusion d'optique, *Annales de Chimie et de Physique* (2) 48: 281–90.

— (1832) Sur un nouveau genre d'illusions d'optique, *Correspondance mathematique et physique* 6: 365.

— (1833) Des illusions optiques sur lesquelles se fonde le petit appareil appelé récemment Phénakisticope, *Annales de Chimie et de Physique* (2) 53: 304–8.

Playfair, William (1786) *Commercial and Political Atlas, Representing, by Means of Stained Copper-Plate Charts, the Exports, Imports, and General Trade of England*, (*a*) London: Debrett; (*b*) reprint of the 3rd ed. 1801 with intro. by Howard Wainer & Ian Spence, Cambridge: Cambridge University Press, 2005.

Pohl, W. Gerhard (2002/03) Peter Salcher und Ernst Mach – Schlierenfotografie von Überschall-Projektilen, *Plus Lucis* 2/2002–1/2003: 22–5.

Pont, Jean Claude (ed. 2012) *Le Destin Douloureux de Walther Ritz, Physicien Théoricien de Génie*, Sion: Archives de l'État du Valais (Cahiers de Valesia 24).

Prange, Christian Friedrich (1782) *Farbenlexicon, worinn die möglichsten Farben der Natur nicht nur nach ihren Eigenschaften, Benennungen, Verhaltnissen und Zusammensetzungen sondern auch durch die wirkliche Ausmahlung enthalten sind*, Halle: Johann Christian Hendel.

Prasad, Amit (2005*a*) Scientific culture in the other theatre of modern science: an analysis of the culture of magnetic resonance imaging research in India, *Social Studies of Science* 25, no. 7: 463–89.

— (2005*b*) Making images/making bodies. Visibilizing and deciphering through magnetic resonance imaging (MRI), *Science, Technology and Human Values* 30, no.

2: 291–316.

Preissler, Johann Daniel (1721–25) *Die durch Theorie erfundene Practic: Oder Gründlich-verfasste Reguln deren man sich als einer Anleitung zu berühmter Künstler Zeichen-Wercken bestens bedienen kan*, Nürnberg: Preißler, (*a*) 1st ed., 3 vols.; (*b*) 2nd ed. 1745–50; (*c*) 3rd ed. in 4 vols. 1757–68; (*d*) 4th ed. in 4 Vols., 1778–83.

Prévost, Pierre (1806) Berechnungen und Bemerkungen über drei Reihen kyanometrischer Beobachtungen Benedikt von Saussures, *Annalen der Physik* 24: 69–84.

Probst, Jörg & Jost Philipp Keuner (eds. 2009) *Ideengeschichte der Bildwissenschaft. Siebzehn Porträts*, Frankfurt: Suhrkamp.

Proctor, Richard A. (1888) Maps and views of Mars, *Scientific American*, suppl. 26: 10659–60.

Prodger, Philip (1998) Illustration as strategy in Charles Darwin's *The Expression of the Emotions in Man and Animals*, in: Tim Lenoir (ed.) *Inscribing Science*, Stanford, CA: Stanford University Press, 140–81.

Prosser, Richard Bissell (1900) Nasmyth, James (1808–1890), *Dictionary of National Biography*, vol. 40.

Provis, W. A. (1839) Observations on the effects produced by wind on the suspension bridge over the Menai Strait, *Transactions of the Institute of Civil Engineers* 3: 357–70 & pl. XVI.

Puig-Samper, Miguel Angel (2012) Illustrators of the New World – The image in the Spanish scientific expeditions of the Enlightenment, *Culture & History Digital Journal* 1,2: 1–27.

Pulfrich, Carl (1902) Über neuere Anwendungen der Stereoskopie und über einen hierfür bestimmten Stereokomparator, *Zeitschrift für Instrumentenkunde* 22: 65–81, 133–41, 178–92, 229–46.

— (1911) *Stereoskopisches Sehen und Messen*, Jena: Gustav Fischer.

Purbrick, Louise (1998) Ideologically technical: illustration, automation and spinning cotton around the middle of the 19th century, *Journal of Design History* 11: 275–98.

Pyenson, Lewis (1985) *The Young Einstein*, London: Taylor & Francis.

— (2002*a*) Comparative history of science, *History of Science* 40: 1–33.

— (2002*b*) An end to national science, *History of Science* 40: 251–90.

Pylyshyn, Zenon W. (1973) What the mind's eye tell's the mind's brain: a critique of mental imagery, *Psychological Bulletin* 80: 1–24.

— (1981) The imagery debate: analogue media versus tacit knowledge, *Psychological Review* 87: 16–45.

Qing, Lin (1995) *Zur Frühgeschichte des Elektronenmikroskops*, Stuttgart: GNT-Verlag.

Racine, Eric, Ofek Bar-Ilan & Judy Illes (2005) fMRI in the public eye, *Nature Reviews Neuroscience* 6: 159–64.

Radelet-de Grave, Patricia & Edoardo Benvenuto (eds. 1995) *Entre Mécanique et Architecture*, Basel, Boston & Berlin: Birkhäuser.

Radon, J. H. (1917) Über die Bestimmung von Funktionen durch ihre Integralwerte längs gewisser Mannigfaltigkeiten, *Berichte über die Verhandlungen der königlich-Sächsischen Akademie der Wissenschaften, math.-physik. Klasse* 69: 262–79.

Ramberg, Peter & G. J. Somsen (2001) The young J. H. van 't Hoff: The background to the publication of his 1874 pamphlet, *Annals of Science* 58: 51–74.

Rampley, Matthew (2005) *Exploring Visual Culture: Definitions, Concepts, Contexts*, Edinburgh: Edinburgh University Press.

Ramsey, O. B. (1974) Molecules in three dimensions, *Chemistry* 47, part I, no. 1: 6–9; part II, no. 2: 6–11.

— (1975) Molecular models in the early development of stereochemistry, in: O. B. Ramsey (ed.) *Van 't Hoff LeBel Centennial*, Washington, DC: ACS: 74–96.

Rasmussen, Nicolas (1993) Facts, artifacts and mesosomes: practicing epistemology with the electron microscope, *SHPS* 24: 227–65.

— (1996) Making a machine instrumental: RCA and the wartime origins of biological electron microscopy in America, 1940–49, *SHPS* 27: 311–49.

— (1997) *Picture Control. The Electron Microscope and the Transformation of Biology in America, 1940–1960*, Stanford: Stanford University Press.

Reeves, Eileen (1997) *Painting the Heavens. Art and Science in the Age of Galileo*, Princeton, NJ: Princeton University Press.

Reichle, Ingeborg (ed. 2007) *Verwandte Bilder: Die Fragen der Bildwissenschaft*, Berlin: Kadmos.

Reichle, I., Steffen Siegel & Achim Spelten (eds. 2008) *Visuelle Modelle*, Munich: Fink.

Reimer, Ludwig (1968) *Elektronenmikroskopische Untersuchungsverfahren*, Essen:Vulkan.

Reinhardt, Carsten (2006) *Shifting and Rearranging. Physical Methods and the Transformation of Modern Chemistry*, Sagamore Beach: Science History Publications.

Reiser, Frank (2010) A lantern-slide inspired look into biology teaching's past, *American biology Teacher* 72: 557–61.

Reisner, John (1989) An early history of the electron microscope, *Advances in Electronics and Electron Physics* 73: 134–233.

Reuleaux, Franz (1875) *Theoretische Kinematik: Grundzüge einer Theorie des Maschinenwesens*, (*a*) Braunschweig: Vieweg; (*b*) transl.: *The Kinematics of Machinery: Outlines of a Theory of Machines*, London: MacMillan, 1876.

Reynaud, Françoise, Kim Timby, et al. (eds. 2000) *Paris in 3D: From Stereoscopy to Virtual Reality 1850–2000*, Paris: Booth-Clibborn.

Rheinberger, Hans-Jörg, Michael Hagner & Bettina Wahrig-Schmidt (eds. 1997) *Räume des Wissens. Repräsentation, Codierung, Spur*, Berlin: Akademie-Verlag.

Richards, Joan (1988) *Mathematical Visions: The Pursuit of Geometry in Victorian England*, Boston.

Richards, Robert J. (2008) *The Tragic Sense of Ernst Haeckel*, Chicago: University of Chicago Press.

Richardson, Ruth (2008) *The Making of Mr. Gray's Anatomy*, Oxford Univ. Press.

Rigden John S. (1990) Quantum states and precession: the two discoveries of NMR, *Reviews of Modern Physics* 58: 433–48.

Ritchin, Fred (1991) The end of photography as we have known it, in: Paul Wombell (ed.) *Photo-Video: Photography in the Age of the Computer*, London: Rivers Oram Press: 8–16.

Robertson, Frances (2006) Science and fiction: James Nasmyth's photographic images of the Moon, *Victorian Studies* 48,4: 595–623.

— (2013) David Kirkaldy (1820–1897) and his museum of destruction: the visual dilemmas of an engineer as man of science, *Endeavour* 37,3: 125–32.

Robin, Harry (1992) *The Scientific Image from Cave to Computer*, New York: Abrams.

Robinson, Andrew (2006) *The Last Man Who Knew Everything: Thomas Young*, Oxford: OneWorld.

Robinson, Eric & Douglas McKie (eds. 1970) *Partners in Science: Letters of James Watt and Joseph Black*, London: Constable.

Robinson, Henry W. (1948) Robert Hooke as a surveyor and architect, *Notes and Records of the Royal Society* 5: 46–55.

Robinson, H. W. & Walter Adams (eds. 1935) *The Diary of Robert Hooke, M.A., M.D., F.R.S., 1672–1680*, London: Taylor & Francis.

Roche, John (1993) The semantics of graphics in mathematical natural philosophy, in: Mazzolini (ed. 1993), 197–232.

Roche-Nagle, Graham, Douglas Wooster & George Oreopoulos (2010) Symptomatic thoracic aorta mural thrombus: case report, *Vascular* 18: 41–4.

Rocke, Alan J. (1981) Kekulé, Butlerov, and the historiography of the theory of chemical structure, *BJHS* 14: 27–57.

— (1985) Hypothesis and experiment in the early development of Kekulé's benzene theory, *Annals of Science* 42: 355–81.

— (1993) *The Quiet Revolution: Hermann Kolbe and the Science of Organic Chemistry*, Berkeley: University of California Press.

— (2010) *Image and Reality: Kekule, Kopp, and the Scientific Imagination*, Chicago: University of Chicago Press.

Roe, Anne (1951) A study of imagery in research scientists, *Journal of Personality* 19: 459–70.

Rohr, Moritz von (1905) Ueber perspektivische Darstellung und die Hilfsmittel zu ihrem Verständnis, *Zeitschrift für Instrumentenkunde* 25: 293–305, 329–39, 361–71.

— (1929) *Joseph Fraunhofers Leben, Leistungen und Wirksamkeit*, Leipzig: Akademische Verlagsgesellschaft.

Romé de L'Isle, Jean-Baptiste Louis de (1772) *Essai de Cristallographie*, Paris: Didot; 2nd ed. in 3 vols. with atlas under the shorter title *Cristallographie*, 1783.

— (1794) *Des caractères extérieurs des minéraux*, Paris: Didot et al.

Röntgen, Wilhelm Conrad (1895) Über eine neue Art von Strahlen, (*a*) *Sitzungsberichte der Würzburger Physikalisch-Medizinischen Gesellschaft*: 132–41; (*b*) reprint with follow-up papers in: W. C. Röntgen: *Grundlegende Abhandlungen über die X-Strahlen*, Leipzig, 1954: 5–15; (*c*) reprint in W.C. Röntgen: *Über eine neue Art von Strahlen*, Walther Gerlach & Fritz Krafft (eds.), Munich: Kindler, 1972: 27–36.

Root-Bernstein, Robert (1985) Visual thinking – the art of imagining reality, *Transactions of the American Philosophical Society* 75: 50–67.

— (1989) *Discovering: Inventing and Solving Problems at the Frontiers of Science*, Cambridge, MA: Harvard University Press.

— (2001) Van 't Hoff on imagination and genius, in: W. Hornix & S. H. W. M. Mannaerts (eds.) *Van 't Hoff and the Emergence of Chemical Thermodynamics*, Delft: Delf University Press: 295–307.

Root-Bernstein, R. & Michèle Root-Bernstein (1999) *Sparks of Genius. The 13 Thinking Tools of the World's Most Creative People*, Boston: Houghton Mifflin.

Root-Bernstein, R., Maurine Bernstein & Helen Garnier (1995) Correlations between vocations, a scientific style, work habits, and professional impact of scientists, *Creativity Research Journal* 8: 115–37.

Rösch, Herbert (1959) Christoph Scheiner, *Lebensbilder aus dem Bayerischen Schwaben* 7: 183–211.

Rose, Gillian (2001/12) *Visual Methodologies. An Introduction to Researching with Visual*

Material, London: Sage, (*a*) 1st ed. 2001, (*b*) 2nd ed. 2006; (*c*) 3rd ed. 2012.

Rosenow, Ulf F. (1995) Notes on the legacy of the Röntgen rays, *Medical Physics* 22, Nov., no. 11: 1855–67.

Rosse, Lord (1850) Observations on the nebulae, *PTRSL* 140: 499–514.

Rossell, Deac (2008) *Laterna Magica, Magic Lantern*, Stuttgart: Füsslin.

Roth, Tim Otto (2013) *Körper. Projektion. Bild. Eine Vergleichende Untersuchung zu Schattenbildern*, Diss., Kunsthochschule für Medien, Cologne, Aug. 2013.

Roth, T. O. & Robert Fosbury (2013) Colour beyond the sky: the chromatic revolution in astronomy, in: Martha Blassnigg (ed.) *Light, Image, Imagination*, Amsterdam: Amsterdam University Press; 241–68.

Roth, Wolff-Michael (2004) Emergence of graphing practices in scientific research, *Journal of Cognition and Culture* 4: 595–627.

Roth, W.-M. & G. Michael Bowen (2003) When are graphs ten thousand words worth? An expert/expert study, *Cognition and Instruction* 21: 429–73.

Roth, W.-M., M. Bowen & Michelle K. McGinn (1999) Differences in graph-related practices between high-school biology textbooks and scientific ecology journals, *Journal of Research in Science Teaching* 36: 977–1019.

Rothenberg, Albert (1995) Creative cognitive processes in Kekulé's discovery of the structure of the benzene molecule, *American Journal of Psychology* 108: 419–38.

Rothermel, Holly (1993) Images of the sun: Warren de la Rue, George Biddell Airy and celestial photography, *BJHS* 26: 137–69.

Rowland, Sidney (1896) Report on the application of the new photography to medicine and surgery, *British Medical Journal* 1, no. 1832: 361–4.

Rowlandson, T. S. (1870) *History of the Steam Hammer*, 4th ed., corr. and rev., Manchester: Heywood; 5th ed. 1875.

Rudwick, Martin J. S. (1972) *The Meaning of Fossils: Episodes in the History of Palaeontology*, (*a*) London: Macdonald, 1972; (*b*) University of Chicago Press, 1985.

— (1976) The emergence of a visual language for geology, 1760–1840, *History of Science* 14: 149–95.

— (1985) *The Great Devonian Controversy. The Shaping of Scientific Knowledge among Gentlemanly Specialists*, Chicago: University of Chicago Press.

— (1992) *Scenes from Deep Time: Early Pictorial Representations of the Prehistoric World*, Chicago: University of Chicago Press.

— (1997) *Georges Cuvier, Fossil Bones, and Geological Catastrophes: New Translations and Interpretations of the Primary Texts*, Chicago: University of Chicago Press.

— (2005) Picturing nature in the age of Enlightenment, *Proceedings of the American Philosophical Society* 149: 279–303.

— (2007) *Bursting the Limits of Time: The Reconstruction of Geohistory in the Age of Revolution*, Chicago: University of Chicago Press.

Rupke, Nicolaas (2001) Humboldtian distribution maps: the spatial ordering of scientific knowledge, in: Tory Frängsmyr (ed.) *The Structure of Knowledge. Classifications of Science and Learning since the Renaissance*, Berkeley: 93–116.

Ruska, Ernst (1979) Die frühe Entwicklung der Elektronenlinsen und der Elektronenmikroskopie, *Acta Historica Leopoldina* 12, no. 3.

— (1984) Die Entstehung des Elektronenmikroskops, *Archiv der Geschichte der Naturwissenschaften* 11/12: 525–51.

Sachs-Hombach, Klaus (ed. 2005) *Bildwissenschaft. Disziplinen, Themen, Methoden*, Frankfurt: Suhrkamp.

— (ed. 2009) *Bildtheorien: Anthropologische und kulturelle Grundlagen des Visualistic Turn*, Frankfurt: Suhrkamp.

Saeijs, W. (2004) The suitability of Haüy's crystal models for identifying minerals, in: *Dutch Pioneers of the Earth Sciences*, Amsterdam: Koninklijke Nederlandse Akademie van Wetenschappen, 59–71.

Sakarovitch, J. (1995) The teaching of stereotomy in engineering schools in France in the 18th and 19th centuries, in: Radelet-de Grave & Benvenuto (eds. 1995): 205–20.

Sargent, Pauline (1999) On the use of visualizations in the practice of science, in: *Philosophy of Science* 63 Proceedings vol.: S230–S238.

Sattelmacher, Anja (2013) Geordnete Verhältnisse. Mathematische Anschauungsmodelle im frühen 20. Jahrhundert, *Berichte zur Wissenschaftsgeschichte* 36: 294–312.

Satzinger, Helga (2006) Theodor and Marcella Boveri, *Nature Reviews* 9: 231–8.

Saunders, Gill (1995) *Picturing Plants. An Analytical History of Botanical Illustration*, London: Zemmer.

Saussure, Horace Bénédict de (1779–96) *Voyages dans les Alpes*, Geneva (4 vols.).

— (1787) *Relation abrégée d'un voyage à la Cime du Mont-Blanc: en août 1787*, Geneva: Barde & Manget; German transl. *Kurzer Bericht von einer Reise auf den Gipfel des Montblanc, im August 1787*, Strassburg: Akademische Buchhandlung, 1788.

— (1789) Schreiben an den Herausgeber, die Reise auf den Col du Géant betreffend, *Magazin für die Naturkunde Helvetiens* 4: 471–524.

— (1790*a*) Description d'un cyanomètre ou d'un appareil destiné à mesurer l'intensité de la couleur bleue du ciel, *Mémoire de l'Académie Royale des Sciences à Turin, 1788-89* 9: 409–24 & pl. X; in German: Beschreibung eines Kyanometers, oder eines Apparats zur Messung der Intensität der blauen Farbe des Himmels, *Journal der Physik* 6 (1792): 93–108.

— (1790*b*) Description d'un diaphanomètre ou d'un appareil propre à mésurer la transparence de l'air, *Mémoire de l'Académie Royale des Sciences à Turin, 1788–89* 9: 425–40.

Schaaf, Larry (1979*a*) Sir John Herschel's 1839 Royal Society paper on photography, *History of Photography* 3: 47–60.

— (1979*b*) Charles Piazzi Smyth's 1865 conquest of the great pyramids, *History of Photography* 3: 331–54.

— (1980) Herschel, Talbot and photography: Spring 1831 and Spring 1839, *History of Photography* 4: 181–204.

— (1980/81) Piazzi Smyth at Teneriffe, *History of Photography* 4 (1980): 289–307; 5 (1981): 27–50.

— (1984) Charles Piazzi Smyth's 1865 photographs of the Great Pyramid, *IMAGE: Journal of Photography and Motion Pictures* 27, no. 4, Dec., 24–32.

— (ed. 1985) *Sun Gardens. Victorian Photograms by Anna Atkins*, New York: Kraus.

— (1989) *Tracings of Light: Sir John Herschel and the Camera Lucida*, San Francisco: Friends of Photography.

— (1989) The first fifty years of British photography 1794–1844, in: Michael Pritchard (ed.) *Technology and Art: The Birth and Early Years of Photography*, Bath: Royal Photographic Society: 9–18.

— (1992) *Out of the Shadows: Herschel, Talbot and the Invention of Photography*, New Haven/London: Yale University Press.

— (ed. 1994) *Selected Correspondence of William Henry Fox Talbot 1823–1874*, London: Science Museum.

— (1996) *Records of the Dawn of Photography: Talbot's Notebooks P & Q*, Cambridge: Cambridge University Press.

Schaefer, Bradley E. (2001) The transit of Venus and the notorious black drop effect, *JHA* 32: 325–36.

Schäffer, Jacob Christian (1769) *Entwurf einer allgemeinen Farbenverein oder Versuch und Muster einer gemeinnützlichen Bestimmung und Benennung der Farben*, Regensburg: Weiß.

Schaffer, Simon (1988) Astronomers mark time: discipline and the personal equation, *Science in Context* 2: 115–45.

Scheiner, Christoph (1612*a*): *Tres epistolae de maculis Solaribus: scriptae ad Marcum Velserum; cum observationum iconismis*, Augustae Vindelicorum (Augsburg); originally under the pseudonym Apelles); for Engl. transl. with commentary see van Helden & Reeves (eds. 2010).

— (1612*b*) *De maculis solaribus et stellis circa Iovem errantibus accuratior disquisitio ad Macrum Verseum*, 2nd printing, Augsburg and 3rd printing (Rome 1613) with coppler plate summarizing Scheiner's observations from Oct.–Dec. 1611; for Engl. transl. with commentary see van Helden & Reeves (eds. 2010).

— (1619) *Oculus hoc est: Fundamentum opticum*, Innsbruck.

— (1626–30) *Rosa ursina sive Sol ex admirando Facularum & Macularum suarum Phenomeno varius ...*, Bracciani: Andreas Phaeus.

— (1631) *Pantographice, seu Ars Delineandi Res Qualibet per Parallelogrammum*, Rome: Grignani.

Scherf, Karl (1965) Kekulé und die Alchemie, *Chemie für Labor und Betrieb* 16,10: 401–8.

Schiaparelli, Giovanni Virginio (1877/78) Osservazioni astronomiche e fisiche sull' asse di rotatione planeta e sulla topographia del pianete marte fatte nella specola reale in Milano coll'equatoriale Merz durante l'opposizione de 1877, *Atti della Reale Accademia dei Lincei* Anno 275, (ser. 3) 2: 308–439 & pl. I–V.

— (1879/80) Osservazioni astronomiche e fisiche sull'asse di rotazione e sulla topografia del pianeta Marte, in: *Atti della R. Accademia del Lincei, Memoria della Classe di Scienze Fisiche. Memoria 2*, (Osservazioni dell'opposizione 1879–80), ibid. (ser. 3) 10 (1880-81): 281–388 & pl. I–VI.

— (1882) Découvertes nouvelles sur la planète Mars, (*a*) *L'Astronomie* 1: 216–21; (*b*) transl. by Garret P. Service in: *Other Worlds*, New York, 1901: 93ff.

— (1893) Il pianete Marte, *Nature ed Arte*, (*a*) 2, nos. 5–6; (*b*) 4 no. 11 (1895); (*c*) 19 no. 1 (1909); (*d*) reprint in: Pascuale Tucci, Agnese Mandrino & Antonella Testa (eds.) *Giovanni Virginio Schiaparelli: La vita sul pianeta Marte*, Milano: Interliber, 2002; (*e*) partial Engl. transl. by W. H. Pickering in *Astronomy and Astrophysics* 13 (1894): 635–40.

— (1910) Osservazioni astronomiche e fisiche sulla topographia e costituzione del pianete marte fatte nella specola reale in Milano coll'equatoriale Merz-Repsold durante l'opposizione de 1890, *Accademia Nazionale dei Lincei. Memorie della Classe di Scienza Fisiche, Mathematiche e Naturali* (ser. 5) 7: 100–56 & 12 plates.

Schickore, Jutta (2007) *The Microscope and the Eye: A History of Reflection*, Chicago: University of Chicago Press.

Schilling, Tom (2013) Uranium, geoinformatics, and the economic image of mineral exploration, *Endeavour* 37,3: 140–9.

Schinzel, Britta (2004) Digitale Bilder: Körpervisualisierungen durch bildgebende Verfahren in der Medizin, in: Wolfgang Coy (ed.) *Bilder als technisch-wissenschaftliche Medien*, workshop der Alcatel-Lucent-Stiftung 2004, online at http://mod.iig.uni-freiburg.de/fileadmin/publikationen/online-publikationen/koerpervisualisierungen.pdf (accessed 20 Mar. 2011).

Schlagintweit, Hermann (1852) Observations in the Alps on the optical phenomena of the atmosphere, *Phil. Mag.* (4) 3: 1–15 & pl.

Schlich, Thomas (2000) Linking cause and disease in the laboratory: Robert Koch's method of superimposing visual and functional representations of bacteria, *History and Philosophy of the Life Sciences* 22, no. 1: 43–58.

Schlosser, Julius, Ritter von (1908) *Die Kunst- und Wunderkammer der Spätrenaissance. Ein Beitrag zur Geschichte des Sammelwesens*, Leipzig: Klinkhardt & Biermann (*a*) 1908; (*b*) 2nd enlarged ed. 1978.

Schmidgen, Henning (2009) *Die Helmholtz Kurven. Auf der Spur der verlorenen Zeit*, Berlin: Merve.

— (2011) Passagenwerk 1850. Bild und Zahl in den physiologischen Zeitexperimenten von Hermann von Helmholtz, *Berichte zur Wissenschaftsgeschichte* 34: 139–55.

Schmidt, Georg & Robert Schenk (eds. 1960) *Kunst und Naturform*, intro. by Adolf Portmann, Basel; 3rd ed. Munich: Moos, 1967 (based on an exhibition at the Kunsthalle Basel, 20 Sep.–19 Oct. 1958).

Schnalke, Thomas (ed. 1995) *Natur im Bild. Anatomie und Botanik in der Sammlung des Nürnberger Arztes Christoph Jacob Trew; eine Ausstellung aus Anlaß seines 300. Geburtstages*, Erlangen: Universitätsbibliothek Erlangen.

Scholz, Erhard (1989) *Symmetrie, Gruppe, Dualität: Zur Beziehung zwischen theoretischer Mathematik und Anwendungen in Kristallographie und Baustatik des 19. Jahrhunderts*, Berlin: Deutscher Verlag der Wissenschaft (Science Networks, no. 1).

Schön, Erhard (1538) *Underweisung der Proporzion unnd stellung der bossen ligend und stehend, abgestolen wie man das vor augen sieht, in dem Büchlein durch Erhart Schön von Nürnberg für die Jungen gesellen und Jungen zu unterrichtung die zu kunst lieb tragen.* Nürnberg: Christoph Zell.

Schreiber, Guido (1828/33) *Kursus der darstellenden Geometrie: nebst ihren Anwendungen auf die Lehre der Schatten und der Perspektive, die Konstruktionen in Holz und Stein, das Defilement und die topographische Zeichnung*, Karlsruhe: Herder, part 1: *Reine*

Geometrie, 1828; part 2: 1833.

Schubring, Gert (ed. 1991) '*Einsamkeit und Freiheit' neu besichtigt: Universitätsreformen und Disziplinbildung in Preußen als Modell für Wissenschaftspolitik im Europa des 19. Jahrhunderts*, Stuttgart: Steiner.

Schultz, Gustav (1890) Bericht über die Feier der Deutschen Chemischen Gesellschaft zu Ehren August Kekulés, *Berichte der Deutschen Chemischen Gesellschaft* 23: 1265–312.

Schüttmann, Werner (1995) Die Aufnahme der Entdeckung Wilhelm Conrad Röntgens in Berlin, *Sudhoffs Archiv* 79: 1–21.

Schwartz, Vannessa & Jeannene M. Przyblyski (eds. 2004) *The 19th Century Visual Culture Reader*, London: Routledge.

Schwarz, Heinrich (1966) Vermeer and the camera obscura, *Pantheon* 24: 170–80.

Secord, Anne (2002) Botany on a plate: pleasure and the power of pictures in promoting early 19th c. scientific knowledge, *Isis* 93: 28–57.

Seelig, Carl (1956) *Albert Einstein und die Schweiz*, Zürich: Europa.

Sella, Andrea (2010) Classic kit: Saussure's cyanometer, *Chemistry World* 7, no. 10: 68–9.

Semczyszyn, Nola (2010) *Signal into Vision: Medical Imaging as Instrumentally Aided Perception*, PhD Dissertation, University of British Columbia.

Settle, Thomas B. (1971) Ostilio Ricci, a bridge between Alberti and Galileo, in: *XIIe Congrès International d'Histoire des Science, Actes*, Paris, vol. IIIB: 121–6.

Seymour, Charles (1964) Dark chamber and light-filled room: Vermeer and the camera obscura, *Art Bulletin* 46: 323–31.

Seymour, Frances & M. Benmosche (1941) Magnification of spermatozoa by means of the electron microscope, *Journal of the American Medical Association* 115: 2489–90.

Shapin, Steven (1994) *A Social History of Truth: Civility and Science in Seventeenth-Century England*, Chicago: University of Chicago Press.

Shaw, Geraldine A. & Stephen T. de Mers (1986/87) Relationships between imagery and creativity in high-IQ children, *Imagination, Cognition and Personality* 6, no. 3: 247–62.

Sheehan, William (1996) *The Planet Mars. A History of Observation and Discovery*, Tucson: University of Arizona Press.

— (1997) Giovanni Schiaparelli: visions of a colour blind astronomer, *Journal of the British Astronomical Association* 107: 11–15.

Shepard, Roger N. & Lynn A. Cooper (1986) *Mental Images and Their Transformation*, Cambridge, MA: MIT Press.

Sheppard, T. (1906) Prominent Yorkshire workers, I: Henry Clifton Sorby, *The Naturalist. Quarterly Journal of Natural History for the North of London* 59: 137–230.

Sherman, Paul D. (1981): *Colour Vision in the Nineteenth Century: The Young-Helmholtz-Maxwell Theory*, Bristol: Hilger.

Shields, Margaret C. (1937) The early history of graphs in physical literature, *American Physics Teacher* 5: 68–71.

— (1938) James Watt and graphs, *American Physics Teacher* 6: 162.

Shinn, Terry (1980) *L'École Polytechnique 1794–1914*, Paris: Foundation Nationale des Sciences Politiques.

Sicard, Monique (1998) Passage du Venus: Le revolver photographique de Jules Janssen, *Études photographiques* 3, no. 4: 45–63.

Siegel, Daniel M. (1991) *Innovation in Maxwell's Electromagnetic Theory. Molecular Vortices, Displacement Current, and Light*, Cambridge: Cambridge University Press.

Siegel, Steffen (2007) Einblicke. Das Innere des menschlichen Körpers als Bildproblem in der frühen Neuzeit, in: Ingeborg Reichle et al. (eds.) *Verwandte Bilder*, Berlin: Kadmos: 33–55.

Sigrist, René (ed. 2001): *H.B. de Saussure (1740–1799). Un regard sur la terre*, Geneva: Bibliothèque d'Histoire des Sciences.

Silverman, Robert J. (1993) The stereoscope and photographic depiction in the 19th century, *Technology and Culture* 34,4: 729–56.

Simon, Gérard (1988) *Le regard, l'être et l'apparence dans l'optique de l'antiquité*, Paris: Ed. Du Seuil; German transl. *Der Blick, das Sein und die Erscheinung in der antiken Optik*, Munich: Fink, 1992.

Simon, Herbert A. (1978) On the forms of mental representation, in: *Perception and Cognition*, Minneapolis: University of Minnesota: 3–18.

Simon, Josep (ed. 2012) Cross-national education and the making of science, technology and medicine, *History of Science* 50: 251–374.

— (ed. 2013) Cross-cultural and comparative history of science education, *Science and Education* 22,4: 763–866.

Siviero, Monica & Carlo Violani (2006) Drawings for an exacting author: illustrations from Giovanni Antonio Scopoli's *Deliciae florae et faunae insubricae*, *Archives of Natural History* 33: 214–31.

Smith, Cyril Stanley (1960) *A History of Metallography: The Development of Ideas on the Structure of Metals before 1890*, Chicago: Univ. of Chicago Press (reprint 1988).

— (ed. 1965) *The Sorby Centennial Symposium on the History of Metallurgy*, New York: Gordon & Breach.

— (1969) The development of ideas on the structure of metals, in: Marshall Clagett (ed.) *Critical Problems in the History of Science*, Madison, WI: University of Wisconsin Press: 467–98.

— (1977) *Metallurgy as a Human Experience*, Materials Park, OH: American Society of Metals.

Smith, Jonathan (2006) *Charles Darwin and Victorian Visual Culture*, Cambridge: Cambridge University Press.

Smith, Lawrence D. et al. (2002) Constructing knowledge: the role of graphs and tables in hard and soft psychology, *American Psychologist* 57: 749–61.

Smith, Pamela H. (2004) *The Body of the Artisan. Art & Experience in the Scientific Revolution*, Chicago: University of Chicago Press.

— (2006) Art, science and visual culture in early modern Europe, *Isis* 97: 83–100.

Smith, Robert W. & Joseph N. Tatarewicz (1985) Replacing a technology: the large space telescope and CCDs, *Proceedings of the IEEE* 75, no. 7: 1221–35.

Smyth, Charles Piazzi (1843) On astronomical drawing, (*a*) *Memoirs of the Royal Astronomical Society* 15: 71–82; (*b*) reprinted with illustrations added in Hentschel and Wittmann (ed. 2000): 66–78.

— (1858) *Teneriffe, an Astronomer's Experiment; Or, Specialities of a Residence Above the Clouds*, London: Reeves.

— (1877) Optical spectroscopy of the red end of the solar spectrum, *Astronomical Register* 15: 210–19.

— (1879) Colour in practical astronomy, spectroscopically examined, *Transactions of the Royal Society of Edinburgh* 28: 779–849 & pl. 41–3.

Smyth, William Henry (1964) *Sidereal Chromatics. Being a Re-print, with Additions from the Bedford Cycle of Celestial Objects and Its Hartwell Continuation on the Colours of Multiple Stars*, 1st ed. 1964; reprint Cambridge University Press 2010.

Snelders, H. A. M. (1975) Practical and theoretical objections to J. H. van 't Hoff's 1874-stereochemical ideas, in: O. B. Ramsey (ed.) *Van 't Hoff LeBel Centennial*, Washington, DC: ACS: 55–73.

Snyder, John P. (1987) *Map Projections – A Working Manual*, US Geological Survey Professional Paper 1395. US Government Printing Office, Washington, DC.

— (1993) *Flattening the Earth: Two Thousand Years of Map Projections*, Chicago: University of Chicago Press.

Sontag, Susan (1990) *On Photography*, New York: Anchor Books.

Sorby, Henry Clifton (1858) On the microscopical structure of crystals: indicating the origin of minerals and rocks, (*a*) *Quarterly Journal of the Geological Society of London* 14: 453–500 & pl. xvi–xix; (*b*) sep. printed in London: Taylor & Francis.

— (1865) On the application of spectrum analysis to microscopical investigations, and especially to the detection of bloodstains, (*a*) *Quarterly Journal of Science* 2: 198–214; (*b*) *Chemical News* 11 (1865): 186–8, 194–6, 232–4, 256–8.

— (1867*a*) On a definite method of qualitative analysis of animal and vegetable colouring – matters by means of the spectrum microscope, *Proceedings of the Royal Society of London* 15: 433–55.

— (1867*b*) On a new micro-spectroscope, and on a new method of printing a description of the spectra seen with the spectrum-microscope, *Chemical News* 15: 220–1.

— (1875) On new and improved microscope spectrum apparatus, and on its application to various branches of research, *Monthly Microscopical Journal* 13: 198–208.

— (1876*a*) Unencumbered research: a personal experience, in: *Essays on the Endowment of Research*, London: King: 149–75.

— (1876*b*) On the evolution of hæmoglobin, *Quarterly Journal of Microscopical Science* 16: 76–85.

— (1886) *On the Application of Very High Powers to the Study of the Microscopical Structure of Steel* ...: Ballantyne, Hanson & Co.

— (1887) On the microscopical structure of iron and steel, *Journal of the Iron and Steel Institute* new ser. 1: 255–88.

— (1897) *Fifty Years of Scientific Research: An Address Delivered before the Members of the Sheffield Literary and Philosophical Society, at Firth College, on Tuesday, February 2nd, 1897* Sheffield: Independent Press.

Spek, T. M. van der (2006) Selling a theory: the role of molecular models in J. H. van 't Hoff's stereochemistry theory, *Annals of Science* 63: 157–77.

Spillmann, Werner (2010) *Farb-Systeme 1611–2007*, Basel: Schwabe.

Spranzi, Marta (2004) Galileo and the mountains of the moon: analogical reasoning, models and metaphors in scientific discovery, *Journal of Cognition and Culture* 4, no. 3: 451–83.

Stafford, Barbara (1994) *Artful Science: Enlightenment Entertainment and the Eclipse of Visual Education*, Cambridge, MA: MIT Press.

— (1999) *Visual Analogy. Consciousness as the Art of Connecting*, Cambridge, MA: MIT Press.

Stafleu, Frans A. (1971) *Linnaeus and the Linnaeans. The Spreading of Their Ideas in Systematic Biology, 1735–1789*, Utrecht: Oosthoek's Uitgeversmaatschappij.

Stampfer, Simon Ritter von (1833) *Die stroboscopischen Scheiben oder optischen Zauberscheiben, deren Theorie und wissenschaftliche Anwendung*, Vienna & Leipzig: Trentsensky & Vieweg.

Starikova, Irina (2010) Why do mathematicians need different ways of presenting mathematical objects? The case of Cayley graphs, *Topoi* 29,1: 41–51.

Staubermann, Klaus (2007) *Astronomers at Work. A Study of the Replicability of 19th Century Astronomical Practice*, Frankfurt am Main: Harri Deutsch. (Acta Historia Astronomiae, no. 32).

Steadman, Philip (2001) *Vermeer's Camera. Uncovering the Truth behind the Masterpieces*, Oxford: Oxford University Press.

Steck, Max (1977) *Der handschriftliche Nachlass von Johann Heinrich Lambert (1728–1777)*, Basel: Öffentliche Bibliothek der Universität.

Stehr, Nico (1994) *Arbeit, Eigentum und Wissen. Zur Theorie von Wissensgesellschaften*, Frankfurt: Suhrkamp.

Stein, A. E. (1912) Ueber medizinisch-photographische und -kinematographische Aufnahmen, *Deutsche Medizinische Wochenschrift* 38: 1184.

Stein, Siegmund Theodor (1888) *Das Licht im Dienste wissenschaftlicher Forschung: Handbuch der Anwendungen des Lichtes, der Photographie und der optischen Projektionskunst in der Natur- und Heilkunde, in den graphischen Künsten und dem Baufache, ...*, Halle/Saale: Wilhelm Knapp, 2nd ed.

Steinman, David B. (1943/45) Rigidity and aerodynamic stability of suspension bridges, *Transactions of the ASCE* 110: 439–75.

— (1946) Design of bridges against wind, *Civil Engineering – ASCE* 16,1: 20–3.

— (1950) *The Builders of the Bridge: The Story of John Roebling and his Son*, New York: Harcourt & Brace.

Stenstrom, William J. (1991) Early life history and personal characteristics of medical illustrators, *Journal of Biocommunication* 18,4: 7–11.

Sternberg, Robert J. & Janet E. Davidson (eds. 1995) *The Nature of Insight*, Cambridge, MA: MIT Press.

Stichweh, Rudolf (1984) *Das System wissenschaftlicher Disziplinen. Physik in Deutschland, 1740–1890*, Frankfurt: Suhrkamp.

Stocking, Barbara & Stuart L. Morrison (1978) *The Image and the Reality*, Oxford: Oxford University Press.

Stöckmann, Hans-Jürgen (2007) Chladni meets Napoleon, *European Physics Journal*

Special Topics 145: 15–23.

Stone, E. (1753) *The Construction and Principal Uses of Mathematical Instruments*, transl. from the French by M. Bion, chief instrument maker to the French king, London: Richardson, 2nd ed.

Stone, William Henry (1879) *Elementary Lessons on Sound*, London: Macmillan.

Storey, John (ed. 1996) *What is Cultural Studies. A Reader*, London: Arnold.

Strauss, David (2001) *Percival Lowell: The Culture and Science of a Boston Brahmin*, Cambridge, MA: Harvard University Press.

Strauss, Joseph B. (1938/70): *The Golden Gate Bridge: Report of the Chief Engineer to the Board of Directors of the Golden Gate Bridge and Highway District, California*, San Francisco, CA: Golden Gate Bridge and Highway District, 1938, and *Supplement to the Final Report* by Clifford E. Paine, ibid., 1970.

Stromer, Klaus (ed. 1998): *Farbsysteme in Kunst und Wissenschaft*, Cologne: Dumont.

Stuart, Herbert (1948): Erforschung der Elektronenhüllen und der Molekülgestalt mit anderen Methoden, in: Hans Kopfermann (ed.) *Physics of the Electron Shells* (The American FIAT review of German science, 1939–1945, vol. 12), Office of Military Government for Germany Field Information Agencies, Technical: 69–91.

Sturken, Marita & Lisa Cartwright (2001) *Practices of Looking: An Introduction to Visual Culture*, Oxford: Oxford University Press.

Stüssi, Fritz (1969) Othmar H. Ammann, *Schaffhauser Biographien* 3: 33–45.

— (1974) *Othmar H. Ammann: Sein Beitrag zur Entwicklung des Brückenbaus*, Basel: Birkhäuser.

Suarez, Mauricio (2003) Scientific representation: against similarity and isomorphism, *International Studies in the Philosophy of Science* 17, no. 3: 225–44.

Süsskind, Charles (1981) The invention of computed tomography, *History of Technology* 6: 39–80.

Talbert, Richard J. A. (ed. 2008) *Cartography in Antiquity and the Middle Ages: Fresh Perspectives, New Methods*, Leiden: Brill.

— (2012) *Ancient Perspectives: Maps and Their Place in Mesopotamia, Egypt, Greece & Rome*, Chicago: University of Chicago Press.

Taton, René (1951) *L'oeuvre scientifique de Monge*, Paris: Presses Universitaires de France.

Taylor, Janelle S. (1992) The public fetus and the family car. From abortion politics to a Volvo advertisement, *Public Cultures* 4, March no. 2: 67–80.

Tempel, Wilhelm (1877) Schreiben an den Herausgeber, *Astronomische Nachrichten* 90, nos. 2138–9, cols. 27–42.

Terby, F. J. C. (1892) Physical observations of Mars, *Astronomy and Astrophysics* 11: 478–80.

Theisen, Wilfred (1982) Euclid's Optics in the medieval curriculum, *Archives Internationales d'Histoire des Sciences* 32: 159–76.

Thomas, Ann (1997) *Beauty of Another Order: Photography in Science*, New Haven, CT: Yale University Press.

Thomas, Mark Shortland (1951) Festival pattern group, *Design* 3 (1951), no. 29–30: 12–25.

Thornton, John L. & Carole Reeves (1983) *Medical Book Illustration, A Short History*, Cambridge: Oleander Press.

Tilling, Laura (1975) Early experimental graphs, *BJHS* 8: 193–213.

Timby, Kim (2000) The inventors of 3D photography in France: patents 1852–1922, in: Reynaud et al. (eds. 2000): 159–64.

— (2005) Colour photography and stereoscopy: parallel histories, *History of Photography* 24,2: 183–96.

Timoshenko, Stephen P. (1983) *History of Strength of Materials: With a Brief Account of the History of Theory of Elasticity and Theory of Structures*, New York: Dover.

Tkaczyk, Viktoria (2014) The making of acoustics around 1800: Ernst F. F. Chladni, in: Mary-Helen Dupree & Sean B. Franzel (eds.) *Performing Knowledge in the Long Eighteenth Century*, Berlin and New York: De Gruyter.

Tobin, Richard (1990) Ancient perspective and Euclid's optics, *JWCI* 53: 14–42.

Touret, Lydie (2004) Crystal models: milestone in the birth of crystallography and mineralogy as sciences, in: *Dutch Pioneers of the Earth Sciences*, Amsterdam: Koninklijke Nederlandse Akademie van Wetenschappen: 43–58.

Truesdell, Clifford (1980) *The Tragicomical History of Thermodynamics 1822–1854*, New York: Springer.

Trumpler, Maria (1997) Converging images: techniques of intervention and focus of representation of sodium-channel proteins in neve-cell membranes, *Journal of the History of Biology* 30: 55–89.

Tsivian, Yuri (1996): Media fantasies and penetrating visions: some links between x-rays, the microscope, and film, in: *Laboratory of Dreams. The Russian Avant-Garde and Cultural Experiment*, Stanford, CA: Stanford University Press.

Tsvetkov, M. & K. et al. (1999) Lohse's historic plate archive, *Astronomische Nachrichten* 320, cols. 63–70.

Tufte, Edward R. (1983) *The Visual Display of Quantitative Information*, Cheshire, CT: Graphics Press.

— (1997) *Visual and Statistical Thinking. Displays of Evidence for Making Decisions*, Cheshire, CT: Graphics Press.

Turner, Dorothy M. (1953) Thomas Young on the eye and vision, in: *Science, Medicine and History*, vol. 2, Oxford: Oxford University Press,: 243–55.

Turner, Norman (1992) Some questions about E. H. Gombrich on perspective, *The Journal of Aesthetics and Art Criticism* 50: 139–50.

Ullmann, Dieter (1996) *Chladni und die Entwicklung der Akustik 1750–1860*, Basel: Birkhäuser (Science Networks no. 19).

Ullrich, Wolfgang (2003*a*) Schwarze Legenden, Wucherungen, visuelle Schocks. Der Kunsthistoriker Horst Bredekamp im Gespräch mit Wolfgang Ullrich, *Neue Rundschau* 114: 9–25.

— (2003*b*) Das Vorzimmer Gottes und andere Sensationsbilder. Über die Rolle der Geisteswissenschaften im Zeitalter bildseliger Naturwissenschaften, *Neue Rundschau* 114, no. 3: 75–87.

Ulrich, Franz (1958) Die technische Zeichnung. Eine Studie über den Wandel ihrer Funktion und ihrer Bedeutung, *VDI-Nachrichten*, no. 12: 9–10.

Urban, Knut (2007) Der späte Nobelpreis [on Ernst Ruska], *Physik-Journal* 6, no. 2: 37–41.

Utzt, Susanne (2004) *Astronomie und Anschaulichkeit. Die Bilder der populären Astronomie des 19. Jahrhunderts*, Frankfurt: Harri Deutsch.

Van Der Zee, John (1986) *The Gate: The True Story of the Design and Construction of the Golden Gate Bridge*, New York: Simon & Schuster.

van Dijck, José (1998) *Imagenation. Popular Images of Genetics*, New York: New York University Press.

Van Diver, Bradford B. (1988) *Imprints of Time: The Art of Geology*, Missoula, MT: Mountain Press.

van Helden, Albert (1974*a*) Saturn and his anses, *JHA* 5: 105–21.

— (1974*b*) 'Annulo cingitur': the solution of the problem of Saturn, ibid., 155–74.

— (1977) The invention of the telescope, *Transactions of the American Philosophical Society* 67, no. 4: 1–67.

van Helden, A. & Eileen Reeves (eds. 2010) Galileo Galilei & Christoph Scheiner: *On Sunspots*, Chicago: University of Chicago Press.

van Helden, A. & Sven Dupré et al. (eds. 2010) *The Origins of the Telescope*, Amsterdam: Koninklijke Nederlandse Akademie van Wetenschappen.

Vaucouleurs, Gerard de (1961) *Astronomical Photography, From the Daguerreotype to the Electron Camera*, New York: Macmillan.

VEB (1989) *Technische Illustrationen des 16. bis 18. Jahrhunderts. Aus den Beständen des Deutschen Buch- und Schriftmuseums [Leipzig]*, Leipzig: Fachbuchverlag.

Veltman, Kim (1979) Military surveying and topography: the practical dimension of Renaissance linear perspective, *Revista da Universidade de Coimbra* 27: 329–68.

— (1980*a*) Ptolemy and the origins of linear perspective, in: Maria Dalai Emiliani (ed.) *La prospettiva rinascimentale. Codificazione e trasgressione*, vol. 1: 403–7.

— (1980*b*) Panofsky's perspective: a half-century later, ibid.: 565–84.

— (1986) *Studies on Leonardo da Vinci. I. Linear Perspective and the Visual Dimensions of Science and Art*, Munich: Deutscher Kunstverlag.

Villiard, Ray & Zoltan Levay (2002) Creating Hubble's technicolor universe, *Sky & Telescope*, September: 28–34.

Vincenti, Walter G. (1993) *What Engineers Know and How They Know It: Analytical Studies from Aeronautical History*, Baltimore: Johns Hopkins University Press.

Visual Culture Questionnaire (1996) *October* 77: 25–70.

Vögtli, Alexander & Beat Ernst (2007) *Wissenschaftliche Bilder. Eine kritische Betrachtung*, Basel: Schwabe.

Volkmer, Ottomar (1894) *Die photographische Aufnahme von Unsichtbarem*, Halle: Knapp.

Voss, Julia (2007/10) *Darwins Bilder*, Frankfurt: Fischer, 2007; transl. as *Darwin's Pictures: Views of Evolutionary Theory, 1837–1874*, New Haven, CT: Yale University Press, 2010.

Wade, Nicholas J. (1983) *Sir Brewster and Wheatstone on Vision*, New York: Academic Press.

Waldby, Catherine (2000) Virtual anatomy: from the body in the text to the body on the screen, *Journal of Medical Humanities* 21, no. 2: 85–107.

Walker, John (1923) Obituary notice for Alexander Crum Brown, (*a*) *Proceedings of the Royal Society of Edinburg* 43: 268–74; (*b*) *Journal of the Chemical Society*: 3422–31.

Wallace, William (1836) An account of the inventor of the pantograph, *Transactions of the Royal Society of Edinburgh* 13: 418–39 & pl. XIV.

Warner, Brian (1983) *Charles Piazzi Smyth: Astronomer–Artist: His Cape Years 1835–1845*, Cape Town: Balkema.

Warwick, Andrew (2003) *Masters of Theory: Cambridge and the Rise of Mathematical*

Physics, Chicago: University of Chicago Press.

Watson, James D. (1964) *The Double Helix: A Personal Account of the Discovery of the Structure of DNA*, (*a*) New York; (*b*) *A Critical Edition*, Norton, 1980.

Wautier, Kristell, Alexander Jonckheere & Danny Segers (2012): The life and work of Joseph Plateau, father of film and discoverer of surface tension, *Physics in Perspective* 14: 258–76.

Weber, Ernst Heinrich & Wilhelm Weber (1836) *Die Mechanik der menschlichen Gehwerkzeuge Eine anatomisch-physiologische Untersuchung*, Göttingen: Dieterich; also reprinted in *Wilhelm Webers Werke*, edited by Königliche Gesellschaft der Wissenschaften zu Göttingen, Berlin: Springer, 1894: 1–305.

Weber, Leonhard (1922) René Just Haüy, *Zeitschrift für Kristallographie* 57: 129–44.

Weber-Unger, Simon (ed. 2009) *Der naturwissenschaftliche Blick: Fotografie, Zeichnung und Modell im 19. Jahrhundert.* Vienna: Wissenschaftliches Kabinett.

Wechsler, Judith (ed. 1988) *On Aesthetics in Science*, Basel: Birkhäuser.

Weiner, Charles & Gloria Lubkin (1967) Interview with Dr. John A. Wheeler at Princeton University, April 5, 1967, American Institute of Physics, Maryland.

Weiser, Martin (1919) *Medizinische Kinematographie*, Leipzig & Dresden: Steinkopff.

Wellmann, Janina (2011) Science and cinema, *Science in Context* 24,3: 311–28.

Wells, Jonathan (2000) *Icons of Evolution. Science or Myth? Why Much of What We Teach about Evolution Is Wrong*, Lanham: Regnery.

Wengenroth, Ulrich (1993) *Enterprise and Technology: The German and British Steel Industries, 1897–1914*, Cambridge: Cambridge University Press.

Werner, Abraham Gottlob (1774/1814) *Von den äusserlichen Kennzeichen der Fossilien*, (*a*) Leipzig, 1774; (*b*) transl. by Patrick Syme (flower-painter): *Werner's Nomenclature of Colours: With Additions, arranged so as to render it highly useful to the arts and sciences. Annexed to which are examples selected from well-known objects in the animal, vegetable, and mineral kingdoms*, Edinburgh: Ballantyne & Co, 1814.

— (1778) Von den verschiedenerley Mineraliensammlungen, aus denen ein vollstaendiges Mineralienkabinett bestehen soll, *Sammlungen zur Physik und Naturgeschichte* 1,4: 387–420.

Werner, Petra (2006) Himmelsblau. Bemerkungen zum Thema Farben in Humboldts Alterswerk 'Kosmos,' *International Review for Humboldtian Studies* 7, 12.

Wertheimer, Max (1945) *Productive Thinking*, Chicago: University of Chicago Press, 1st ed.; reprinted 1982.

Wertheimer, Michael (1982) Gestalt theory, holistic psychologies and Max Wertheimer, *Zeitschrift für Psychologie* 190: 125–40.

— (1985): A Gestalt perspective on computer simulations of cognitive processes, *Computers in Human Behavior* 1: 19–33.

West, Keith R. (1988) *How to Draw Plants: Techniques of Botanical Illustration*, London: Herbert.

Wheaton, Bruce (1983) *The Tiger and the Shark. Empirical Roots of Wave-Particle-Duality*, Berkeley: University of California Press.

Wheatstone, Charles (1838/52): Contributions to the physiology of vision: on some remarkable, and hitherto unobserved, phenomena of binocular vision, *PTRSL* 142 (1852): 1–17.

Wheeler, John Archibald & Kenneth Ford (1998) *Geons, Black Holes, and Quantum Foam: A Life in Physics*, New York: Norton.

Wheeler, Tom & Tim Gleason (1995) Photography or photofiction? *Visual Communication Quarterly* 2, no. 1: 8–12.

Wheelock, Arthur K. (1977*a*) *Perspective, Optics, and Delft Artists around 1650*, New York: Garland.

— (1977*b*) Constantijn Huygens and early attitudes towards the camera obscura, *History of Photography* 1: 93–101.

White, John (1949/51) Developments in Renaissance perspective, *JWCI* 12 (1949): 58–79; 14 (1951): 42–69.

— (1956) *Perspective and Ancient Drawing and Painting*, London: Society for the Promotion of Hellenic Studies.

Whitehead, P. J. P. & P. L. Edwards (1974) *Chinese Natural History Drawings from the Reeves Collection*, London: Natural History Museum.

Wiederkehr, Karl Heinrich (1977) Von frühen Ideen über eine regelmäßige Gestalt kleinster Materieteilchen bis zu Delisles und Bergmanns Vorarbeiten für Haüys Kristallstrukturtheorie, *Centaurus* 21: 27–43.

— (1977/78) René-Just Haüys Strukturtheorie der Kristalle, Centaurus 21: 278–99; René-Just Haüys Konzeption vom individuellen integrierenden Molekül, ihre Widerlegung und seine Ansichten über kristallbildende Kräfte, *Centaurus* 22: 131–56.

— (1978) Das Weiterwirken der Haüyschen Idee von der Polyedergestalt der Moleküle in der Chemie, die Umgestaltung der Haüyschen Strukturtheorie durch Seeber und Delafosse, und Bravais' Entdeckung der Gittertypen, *Centaurus* 22: 177–86.

Williams, L. Pearce (1978) *Album of Science. The Nineteenth Century*, New York: Scribner's.

Wilson, Alexander (1774) Observations on solar spots, *PTRSL* 64, part I.

Wilson, Catherine (1995) *The Invisible World – Early Modern Philosophy and the Invention of the Microscope*, Princeton, NJ: Princeton University Press.

Wilson, Wendell E. (1994) The history of mineral collecting 1530–1799, *The Mineralogical Record* 25,6: 1–243.

Wimsatt, William (1990) Taming the dimension – visualization in science, *PSA 1990*: 111–35.

Winkler, Mary & Albert Van Helden (1992) Representing the heavens. Galileo and visual astronomy, *Isis* 83: 195–217.

Wise, Norton (2006) Making visible, *Isis* 97: 75–82.

Wittfoht, Hans (1983) *100 Jahre Brooklyn-Brücke. Leben und Werk des Ingenieurs John A. Roebling*, Düsseldorf: VDI.

Wittgenstein, Ludwig (1953) *Philosophical Investigations*, first published posthumously in Engl. translation by G. E. M. Anscombe 1953; the German original publ. at Oxford: Blackwell, 2001, and Frankfurt: Suhrkamp, 1972.

Wittkower, R. (1953) Brunelleschi & proportion in perspective, *JWCI* 16: 275–91.

Wizinger-Aust, Robert (1966) August Kekulé, Leben und Werk, in: *Kekulé und seine Benzolformel*, Weinheim: VCH: 7–32.

Wohlfeil, Rainer (1991) Methodische Reflexionen zur Historischen Bildkunde, in: Brigitte Tolkemitt & Rainer Wohlfeil (eds.) *Historische Bildkunde. Probleme – Wege – Beispiele*, (*Zeitschrift für historische Forschung* suppl. no. 12) Berlin: 17–35.

Wolf, Charles & Charles André (1874ff.) Recherches sur les apparences singulières qui ont souvent accompagné l'observation de Mercure et de Vénus avec le bord du Soleil. *Recueil de mémoires, rapports et documents relatifs à l'observation du passage de Vénus sur le soleil*, Paris: Didot, 1874ff., vol. I, 2nd part, suppl. (1876/77): 115–72 & pl. I.

Wolf, Jörn Henning & Franz Härle (eds. 1994) *Krankheiten des Gesichts in künstlerischen Illustrationen des 19. Jahrhunderts*, Neumünster: Wachholtz-Verlag.

Wollons, Roberta L. (ed. 2000) *Kindergartens and Cultures: The Global Diffusion of an Idea*, New Haven, CT: Yale University Press.

Woodward, David (1975) *Five Centuries of Map Printing*, Chicago Univ. Press.

— (2007) Techniques of map engraving, printing and colouring in the European Renaissance, in: D. Woodward (ed.) *The History of Cartography*, vol. 3: *Cartography in the European Renaissance*, Chicago: University of Chicago Press, 591–610.

Wotiz, John H. & Susanna Rudofsky (1984) Kekulé's dreams: fact or fiction? *Chemistry in Britain* 20: 720–3.

Wright, Aaron Sidney (2011) Visual Representation and the 'Renaissance' of General Relativity, 1955–1975, unpubl. talk at the 6th European Spring School on History of Science and Popularization, Menorca, Spain, May 21, 2011.

— (2013) The origins of Penrose diagrams in physics, art, and the psychology of perception, 1958–62, *Endeavour* 37,3: 133–9.

Wright, Sewall (1977) *Evolution and the Genetics of Populations: Genetics and Biometric Foundations*, vol. 3: *Experimental Results and Evolutionary Deductions*, Chicago: University Chicago Press; rev. 2nd ed. 1984.

Young, Thomas (1793) Observations on vision, *PTRSL* 83: 169–81; reprinted in Young (1807*a*) vol. II: 525–9 & pl. I.

— (1801) On the mechanism of the eye, *PTRSL* 91: 23–50; reprinted in Young (1807*a*) vol. II: 579–606 & pl.

— (1807) *A Course of Lectures on Natural Philosophy and the Mechanical Arts*, (*a*) London: Johnson (2 vols.); (*b*) new ed. with references and notes by P. Kelland, Taylor & Walton, 1845 (2 vols.).

Yoxen, Edward (1987) Seeing with sound: a study of the development of medical images, in: W. E. Bijker, Thomas Hughes & Trevor Pinch (eds.) *The Social Construction of Technological Systems: New Directions in the Sociology and History of Technology*, Cambridge, MA: MIT Press: 281–303.

Zeki, Semir (1999) *Inner Vision. An Exploration of Art and the Brain*, Oxford: Oxford University Press.

Zilsel, Edgar (2003) *The Social Origins of Modern Science*, ed. by D. Raven, W. Krohn & R.S. Cohen, Dordrecht, Netherlands: Kluwer.

Zoller, Paul (2000) The steam engine indicator, *Bulletin of the Scientific Instrument Society* 67: 9–22.

— (2002) Instruments and methods for the evaluation of indicator diagrams, *Bulletin of the Scientific Instrument Society* 75: 11–17.

Select webpages on central issues pertinent for this book

For the sake of brevity, the prefix http:// and the dates of last access have been omitted. Unless noted otherwise, all cited webpages were last checked on February 14, 2014 before completion of the book manuscript.

www.uni-stuttgart.de/hi/gnt/dsi for the Stuttgart Database of Scientific Illustrators 1540–1950 approximately 8,000 entries (July 2014) and 20 search fields

www.uni-stuttgart.de/hi/gnt/dsi/websites/websites.php, on natural history images, illustrators & photographers

www.eikones.ch/, and further links given there

en.wikipedia.org/wiki/Iconic_turn , .../Visual_culture, and further links given there

staff.science.uva.nl/ leo/lego/linkages.html and web.mat.bham.ac.uk/C.J.Sangwin/ howroundcom/contents.html, on visual thinking in technology

www.psychology.tcd.ie/other/Ruth_Byrne/mental_models/theory.html

home.tiscalinet.ch/biografien/biologen/vesalius.htm

vesalius.northwestern.edu, on Vesalius's *De humani corporis fabrica* 1543

images.umdl.umich.edu/w/wantz/vesd1.htm, with plates from his oeuvre 1543

special.lib.gla.ac.uk/exhibns/month/oct2002.html, on Fuchs's *Great Herbal* 1542

www.epaves.corsaires.culture.fr/mobile/en/uc/02_04, on early-modern shipwrightery

www.kuttaka.org/ JHL/Main.html, for Lambert's collected works

www.colorsystem.com/?lang=en and irtel.uni-mannheim.de/colsys/, on historical color systems and color encoding schemes

special.lib.gla.ac.uk/exhibns/month/nov2009.html, Darwin and others on the expression of emotions

www.wcume.org/wp-content/uploads/2011/05/Tsung.pdf, on ultrasound

www.zum.de/stueber/haeckel/kunstformen/natur.html

www.bnv-bamberg.de/home/ba2282/main/faecher/biologie/haeckel.htm

styleskilling.com/2006/12/31/design-and-organic-forms/

legacy.mblwhoilibrary.org/leuckart/, on zoological wall charts

www.artic.edu/aic/collections/exhibitions/American/resource/1241 and geologyinart. blogspot.de/, on geology and art

www.minrec.org/, mineralogical record of collections, collectors, samples

www.geh.org/fm/stm/htmlsrc4/teneriffe_sum00001.html, on Charles Piazzi Smyth

www.londonstereo.com/modern_stereos_moons.html

en.wikipedia.org/wiki/Anaglyph_3D, on anaglyph 3D images

courses.ncssm.edu/gallery/collections/toys/opticaltoys.htm, on optical gadgets

libweb5.princeton.edu/visual_materials/maps/websites/thematic-maps/contents.html, historic maps

www.sorby.org.uk/hcsorby.shtml, on Henry C. Sorby

vlp.mpiwg-berlin.mpg.de/experiments, on the kymograph in physiology

www.micrographicarts.com and www.denniskunkel.com, on electron microscopy

scholar.lib.vt.edu/ejournals/SPT/v8n2/hennig.html, on scanning tunneling microscopy 1980–90

authors.library.caltech.edu/5456/1/hrst.mit.edu/hrs/materials/public/Binnig& Rohrer.htm, an interview with Binnig and Rohrer and various papers

www.nsf.gov/news/special_reports/scivis/, International Visualization Challenge

www.fonar.com/fonar_timeline.htm, the history of MRI in a timeline

www.refindia.net/rlinks/reviewedlinks/mri.htm

nobelprize.org/nobel_prizes/medicine/laureates/2003/index.html

www.nature.com/physics/looking-back/lauterbur/index.html, on Lauterbur 1973

en.wikipedia.org/wiki/Magnetic_resonance_imaging

www.meb.uni-bonn.de/epileptologie/cms/upload/homepage/lehnertz/CT1.pdf, with a lot of information about CT scans and their history

airto.hosted.ats.ucla.edu/BMCweb/SharedCode/MRArtifacts/MRArtifacts.html

www.nlm.nih.gov/archive/20120612/research/visible/vhp_conf/le/haole.htm and

www.mathworks.de/products/demos/image/ipexblind/ipexblind.html, on deblurring-algorithms

www.lefevre.darkhorizons.org/articles/proctutorial2.htm, on CCD imaging

starizona.com/acb/ccd/ccd.aspx, also on CCD imaging

messier.obspm.fr/xtra/Bios/rosse.html and www.maa.clell.de/Messier/E/m051.html

www.mikeoates.org/lassell/lassell_by_a_chapman.htm

www.archive.org/details/monographofcentr00holdrich

heritage.stsci.edu/, the Hubble Heritage Page

hubblesite.org/gallery/behind_the_pictures/meaning_of_color

solarsystem.dlr.de/Missions/express/second/09.02.2005.shtml, 3D video of the Martian surface

www.mpa-garching.mpg.de/galform/data_vis/index.shtml, millenium simulation

www.bl.uk/whatson/exhibitions/beautiful-science/

NAME INDEX

This index lists the names of all persons mentioned or quoted in the narrative. Composite terms incorporating proper names (e.g., Max Planck Society) are excluded. Italicized emphasis indicates that biographical information (living dates) is provided on the relevant page. An f. following a page number signifies that and the following page; ff. includes the subsequent page. For ranges exceeding three successive pages, the first and last pages are indicated. Initials occur only where a duplicate surname is separately cited.

ABOUT THE AUTHOR

Klaus Hentschel studied physics and philosophy at the University of Hamburg, earning his *Magister* in philosophy and his *Diplom* in theoretical high-energy physics. His doctoral thesis examines philosophical interpretations of Einstein's theory of relativity (Birkhäuser, 1990) and his habilitation thesis the interplay between instrumentation, experimentation and theory in the discovery of redshift in the solar spectrum (Hamburg 1998). Since 2006, he has directed the Section for History of Science and Technology at the History Department of the University of Stuttgart. Pertinent publications of relevance include *Mapping the Spectrum: Techniques of Visual Representation in Research and Teaching* (Oxford, 2002), and the introduction to the proceedings volume he edited together with Axel Wittmann on *The Role of Visual Representations in Astronomy: History and Practice* (Frankfurt, 2000).

A complete publications list of his 8 books, 16 anthologies or editions and numerous scholarly articles in leading international journals is accessible online at: www.uni-stuttgart.de/hi/gnt/hentschel.

SUMMARY

This book attempts a synthesis. It delves into a rich reservoir of case studies on visual representations in scientific, medical and technological practice that have been accumulated over the past couple of decades by historians, sociologists, and philosophers of science and technology. It adopts an integrative view of recurring general features of visual cultures in science and technology. By systematic comparison of numerous case studies, the purview broadens away from myopic microanalysis in search of overriding patterns. The many different disciplines and research areas involved encompass mathematics, technology, natural history, medicine, the geosciences, astronomy, chemistry and physics. The chosen examples span the period from the Renaissance to the late 20th century. Some pioneers of new visual cultures are portrayed, along with modes of skill transfer and development. A broad range of visual representations in scientific practice is treated, as well as schooling in pattern recognition, design and implementation of visual devices, with a narrowing in on the special role of illustrators and image specialists. A mixed audience from the various natural sciences and humanities is envisioned. At the same time, the book should be perfectly readable for beginners seeking basic orientation in the maze of more recent pertinent analyses. The extensive bibliography references selected primary and secondary literature to pave a thoroughfare through these clusters of historical documentation and analysis.